Physics and Chemistry in Space   Vol. 21

Space and Solar Physics

Series Editors: M. C. E. Huber · L. J. Lanzerotti · D. Stöffler

R. Schwenn   E. Marsch (Eds.)

Physics of the Inner Heliosphere II

Particles, Waves and Turbulence

With 152 Figures

Springer-Verlag
Berlin Heidelberg New York
London Paris Tokyo
Hong Kong Barcelona
Budapest

Dr. Rainer Schwenn
Dr. Eckart Marsch
Max-Planck-Institut für Aeronomie, Postfach 20,
W-3411 Katlenburg-Lindau, Fed. Rep. of Germany

Series Editors:

Professor Dr. M. C. E. Huber
European Space Research and Technology Centre,
Keplerlaan 1, NL-2200 AG Noordwijk, The Netherlands

Dr. L. J. Lanzerotti
AT&T Bell Laboratories, 600 Mountain Avenue,
Murray Hill, NJ 07974-2070, USA

Professor Dr. D. Stoeffler
Institut für Planetologie, Universität Münster,
Wilhelm-Klemm-Str. 10, W-4400 Münster, Fed. Rep. of Germany

ISBN-13: 978-3-642-75366-4   e-ISBN-13: 978-3-642-75364-0
DOI: 10.1007/978-3-642-75364-0

Typesetting: Data conversion by Springer-Verlag

56/3140 – 5 4 3 2 1 0 – Printed on acid-free paper

Preface

The idea of producing a book on the inner heliosphere dates back to 1982. At that time, shortly after the activity maximum of solar cycle 21, the scientific evaluation of the data returned by the *Helios* solar-probe mission had also reached peak intensity. It was a common feeling of the scientists involved that this was an appropriate time to summarize in a tutorial fashion the results from *Helios*. The basic idea was put into action by the designated editors F.M. Neubauer and L.J. Lanzerotti, and it has not changed much since then, except in one crucial respect: the mission went on, surprisingly enough, for another four years, and the data evaluation has also continued to produce new results at a fast pace. A substantial expansion of the book project in terms of both page allocation and delivery deadlines was therefore inevitable.

After F.M. Neubauer had withdrawn as an editor, L.J. Lanzerotti appointed us as new co-editors of the book. We were fortunate in that we found Springer a very cooperative and patient partner in the project. They agreed with our proposal to spread the material over two volumes, and they accepted an extension of the time schedule. The authors turned out to be highly motivated and productive as well, though fairly slow in some cases. However, they all made proper use of the time and put a lot of effort into maturing their work.

All articles were scrutinized by well-known experts in their fields: B. Bavassano, W.C. Feldman, S.P. Gary, J.T. Gosling, J.T. Hoeksema, R.B. McKibben, M. Neugebauer, D.A. Roberts, T.R. Sanderson, S.J. Schwartz, N.R. Sheeley, Jr., S.T. Suess, B.T. Tsurutani, R. Woo, and H.A. Zook. It is a particular pleasure for us to thank here these referees for their immense efforts, which were highly appreciated and carefully taken into consideration by the authors and were, we think, to the benefit of the readers.

It is certainly appropriate to express here our gratitude to the many individuals, organizations, and companies who have rendered possible such ambitious space missions as the *Helios* solar probes. Thus, they have laid the basis for the many new scientific achievements that finally led to the compilation of this book. As representatives for all them, let us just mention the names of the *Helios* project scientists H. Porsche and J. Trainor. Further thanks are due to various persons who promoted the final production of the book: F.M. Neubauer and L.J. Lanzerotti for inaugurating it, W. Engel and H.U. Daniel at Springer, M.K. Bird and P.W. Daly for linguistic advice, and Mrs. G. Bierwirth and Mrs. U. Spilker for their secretarial services.

H. Porsche deserves special credit, since we were particularly inspired and motivated in editing this book by the impressive booklet he had edited at the tenth anniversary of the *Helios 1* launch. We realize that his work has gained the considerable attention of the public and has served as a major source of attraction to our field of research for many students. We hope that the present book will, in turn, continue to inspire students and scientists to further expand our views on the physics of the inner heliosphere.

Katlenburg-Lindau   *Eckart Marsch*
April 1991   *Rainer Schwenn*

Contents

Index of Contributors

Burlaga, Leonard F. E.
Code 692, NASA/GSFC, Greenbelt, MD 20771, USA

Green, Günter
Institut für Kernphysik, Universität Kiel, Otto-Hahn-Platz 1,
W-2300 Kiel, Fed. Rep. of Germany

Gurnett, Donald A.
Department of Physics and Astronomy, University of Iowa,
Iowa City, IA 52242, USA

Kallenrode, May-Britt
Institut für Kernphysik, Universität Kiel, Otto-Hahn-Platz 1,
W-2300 Kiel, Fed. Rep. of Germany

Kunow, Horst
Institut für Kernphysik, Universität Kiel, Otto-Hahn-Platz 1,
W-2300 Kiel, Fed. Rep. of Germany

Marsch, Eckart
Max-Planck-Institut für Aeronomie, Postfach 20,
W-3411 Katlenburg-Lindau, Fed. Rep. of Germany

Müller-Mellin, Reinhold
Institut für Kernphysik, Universität Kiel, Otto-Hahn-Platz 1,
W-2300 Kiel, Fed. Rep. of Germany

Richter, Arne K.
Max-Planck-Institut für Aeronomie, Postfach 20,
W-3411 Katlenburg-Lindau, Fed. Rep. of Germany

Wibberenz, Gerd
Institut für Kernphysik, Universität Kiel, Otto-Hahn-Platz 1,
W-2300 Kiel, Fed. Rep. of Germany

6. Magnetic Clouds

Leonard F.E. Burlaga

6.1 Introduction

The existence of "plasma" clouds propagating from the sun to the earth was proposed as a cause of geomagnetic storms even before the existence of the solar wind was contemplated [6.49, 15]. Such plasma clouds (now usually called ejecta or ejections) are routinely observed in the solar wind. As a consequence of the frozen-in property of magnetic fields in a fully ionized plasma, a beam of plasma from the sun carries along the solar magnetic field [6.2]. The idea of a plasma cloud and that of frozen-in magnetic fields were combined in the concept of a "magnetized plasma cloud" [6.54, 55, 26, 17]. The magnetic field in a magnetized cloud might be either turbulent [6.54] or in the form of a smooth loop or tongue [6.17, 26]. There is a question as to whether a Forbush decrease in cosmic ray intensity is caused by scattering from turbulent magnetic fields [6.54] or drifting in strong ordered magnetic fields [6.17]. In the original model of a tongue the magnetic field lines connect to the sun. It was later suggested that the magnetic field lines could disconnect from the sun by the process of magnetic reconnection [6.58]. Thus, thirty years ago, the concept of a magnetized plasma cloud was being considered, and the possibilities of turbulent clouds versus ordered clouds and connected clouds versus disconnected clouds were being debated. Figure 6.1 summarizes the early views of magnetic clouds and ejecta.

6.2 Definition and Properties of Magnetic Clouds

The term "magnetic cloud" was introduced [6.10] to describe a particular type of interplanetary ejection with the following properties: (1) the magnetic field direction rotates smoothly through a large angle during an interval of the order of one day; (2) the magnetic field strength is higher than average; and (3) the temperature is lower than average. All three of these criteria must be satisfied if an event is to be identified as a magnetic cloud. Any one of the above criteria can be observed in the absence of a magnetic cloud. For example, strong magnetic fields and low temperatures occur after some shocks even though no magnetic cloud is present [6.24, 9, 6]. Low temperatures often occur at sector boundaries in the absence of strong magnetic fields [6.24]. North–south fields [6.59] and rotations in the magnetic field direction [6.63, 64, 50] can occur without strong

Physics and Chemistry in Space - Space and Solar Physics, Vol. 21

Physics of the Inner Heliosphere II Editors: R. Schwenn · E. Marsch

© Springer-Verlag Berlin Heidelberg 1991

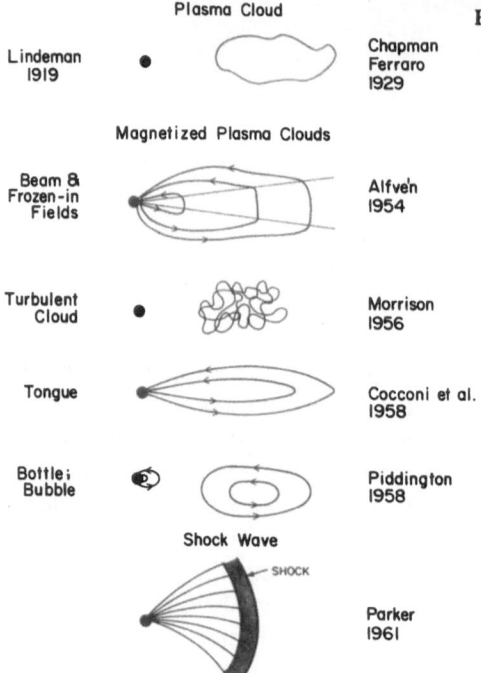

Plasma Cloud

Lindeman 1919

Chapman Ferraro 1929

Magnetized Plasma Clouds

Beam & Frozen-in Fields

Alfvén 1954

Turbulent Cloud

Morrison 1956

Tongue

Cocconi et al. 1958

Bottle; Bubble

Piddington 1958

Shock Wave

SHOCK

Parker 1961

Fig. 6.1. Early views of clouds

magnetic fields and low temperatures. Magnetic clouds are just a subset of the transient interplanetary ejecta [6.28, 66], but they are important subset, as this chapter aims to demonstrate.

An example of a magnetic cloud, observed by *Helios 1* on 20 June 1980 [6.11] is shown in Fig. 6.2. The monotonic variations in the latitude angle δ and the longitude angle λ indicate a rotation in the magnetic field direction (δ and λ are measured in heliographic coordinates). The early papers on magnetic clouds discussed only events with a large monotonic variation in the delta angle in order to avoid confusion with rotations that might occur at sector boundaries. However, the magnetic field direction in a magnetic cloud can rotate in any plane, including the ecliptic plane. The enhanced magnetic field strength is evident in the magnetic cloud shown in Fig. 6.2, and the proton temperature falls to a low value within the magnetic cloud. In that case the magnetic cloud was moving faster than the solar wind ahead of it – sufficiently fast to drive a shock wave, which was observed on 19 June.

In the absence of dynamical interactions, the magnetic field strength inside a magnetic cloud near 1 AU is higher than that outside; thus the magnetic pressure $B^2/8\pi$ in a magnetic cloud is higher than the ambient pressure. Unless there is an additional force, either within the magnetic cloud or outside it, a magnetic cloud would expand in response to the force associated with the gradient in the magnetic pressure. Figure 6.3 shows the enhanced pressure P_T (= proton thermal pressure + magnetic pressure) of the magnetic cloud presented in Fig. 6.2. Figure 6.3 also shows that β, the ratio of the proton thermal pressure to the magnetic

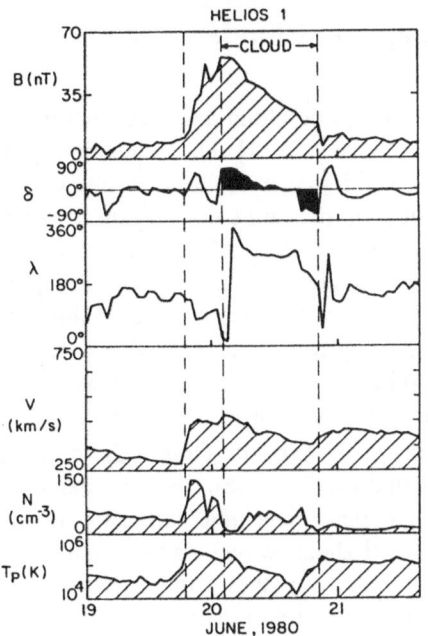

HELIOS 1

Fig. 6.2. A magnetic cloud. The magnetic field direction rotates from large positive angles (northern directions) to large negative angles (southern directions) in the magnetic cloud. This magnetic cloud was moving faster than the ambient flow, and it was driving a shock

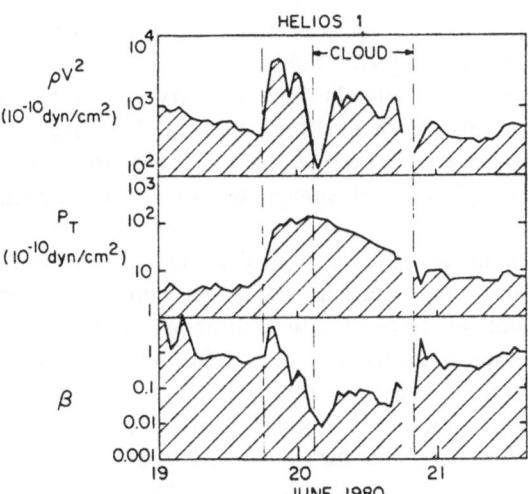

HELIOS 1

Fig. 6.3. The momentum flux, magnetic pressure plus proton thermal pressure, and the ratio of the proton thermal pressure to the magnetic pressure for the magnetic cloud shown in Fig. 6.2

pressure, is relatively low in that magnetic cloud. β is characteristically low in magnetic clouds, because by definition the magnetic field strength is high and the temperature reaches a low value in a magnetic cloud, and because the density is typically not enhanced in a magnetic cloud. In fact, it is better to use low β rather than low temperature as a defining property of a magnetic cloud. Since the plasma pressure is relatively small, the structure is appropriately called a "magnetic cloud", rather than a "magnetized plasma cloud".

3

6.3 Origin and Composition of Magnetic Clouds

Direct evidence of an association between a magnetic cloud and a coronal mass ejection was obtained for the event shown in Fig. 6.2 [6.11]. The magnetic cloud was observed on 20 June 1980, by *Helios 1* when it was over the west limb of the sun, relative to the earth. A coronal mass ejection moving from the sun towards *Helios 1* was observed on 18 June 1980 by the spacecraft *P78-1*, which was in orbit around the earth. The speed of the magnetic cloud at *Helios 1* is consistent with the observed time delay between the passage of the magnetic cloud through the corona and its arrival at *Helios 1*. The magnetic cloud was also detected remotely by the zodiacal light photometer on *Helios 1* [6.41], and it was found that the ejection retained its basic structure and speed out to 0.2–0.4 AU.

Statistical evidence for an association between magnetic clouds and coronal mass ejection transients has also been reported [6.77, 78]. Six of nine magnetic clouds preceded by shocks were preceded an appropriate time earlier by meter-wave type II radio bursts indicative of coronal shock waves and coronal mass ejections. The solar data are consistent with the hypothesis that all magnetic clouds are manifestations of solar coronal mass ejection transients.

The mapping from the coronal mass ejections observed near the sun to the clouds in the heliosphere is not understood. It is not known how to determine which coronal mass ejections will be seen in the solar wind as magnetic clouds. There are no direct measurements of the magnetic field configuration in coronal mass ejections which might be compared with the interplanetary observations. The indirect measurements of the density profile in coronal mass ejections suggest a three-part structure – a bright loop, a dark region, and a bright core [6.36, 42]. A corresponding pattern in the density profile of a magnetic cloud has not been reported.

Coronal mass ejections are associated with both solar flares and disappearing filaments [6.35]. Likewise, some magnetic clouds are associated with solar flares [6.10, 13] and other magnetic clouds are associated with disappearing filaments [6.11, 78]. In a study which identified five magnetic clouds from 1977 through 1979 [6.70] three of the magnetic clouds were associated with flares and two were associated with prominence eruptions. Studies combining solar data, coronagraph data for CMEs near the sun [6.35], photometer data for ejecta farther from the sun [6.76] and *in situ* observations are needed to fully understand the relations between solar events and magnetic clouds in the solar wind.

The composition of the plasma in magnetic clouds and the surrounding flows has not been investigated extensively, although it should provide significant information on the sources of magnetic clouds. One magnetic cloud associated with a quiescent eruptive filament was found to have an anomalously large He^{++}/H^+ ratio, 0.1 [6.13]. Other cases of magnetic clouds with and without He^{++}/H^+ enhancements have been reported, but we will not attempt to survey all of the measurements. Composition anomalies can occur either ahead of magnetic clouds or behind them [6.23]. It would be particularly interesting to look for a possible

4

relation between enhancements of He$^+$ [6.31, 65] and magnetic clouds. Obviously, there is a need for more detailed measurements of the composition of the plasma in magnetic clouds and in the flows preceding and following them.

6.4 Model of the Local Structure of Magnetic Clouds

It was suggested that magnetic clouds are force-free magnetic field configurations [6.27]. A force-free magnetic field is defined as one in which the Lorentz force $J\kappa B$ vanishes, which implies that the current J is parallel to the magnetic field B, i.e. $J = \alpha B$, where α is in general a scalar function of the coordinates. The magnetic field lines in a force-free configuration form a family of helices with a flux-rope geometry illustrated in Fig. 6.4. The magnetic field on the symmetry axis at the center of the rope is a straight line, and the pitch angle of the other field lines increases with increasing distance of the field line from the axis, reaching the asymptotic form of circles on the outer boundary of the magnetic cloud (see Fig. 6.5). An observer who passes through the axis of the flux rope will see the magnetic field to rotate in a plane, as illustrated in the top of Fig. 6.4. If the axis of the flux-rope is in the ecliptic and at right angles to the earth–sun line, then the normal to the plane of rotation is in the radial direction, as is frequently observed. When a magnetic cloud moves past an observer, the magnetic field vector rotates smoothly through a large angle, by definition [6.8, 10, 43]. In fact, this rotation was the motivation for the force-free flux-rope model of magnetic clouds [6.27].

Other models of magnetic clouds have been proposed. Some observations of magnetic clouds are consistent with the passage of a planar loop or tightly wound helix [6.43], but these data are also consistent with the passage of a flux rope. The model of a closed planar loop implies the existence of a zero point in the magnetic field at the center of the loop, whereas the magnetic field strength

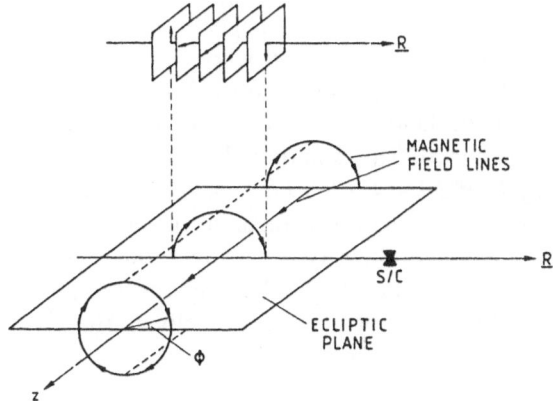

Fig. 6.4. The magnetic field lines tend to rotate parallel to a plane when a magnetic cloud moves past a spacecraft, as shown at the top. Such a pattern can be produced by a magnetic flux rope in which the field lines form a family of helices with the limiting forms shown at the bottom

5

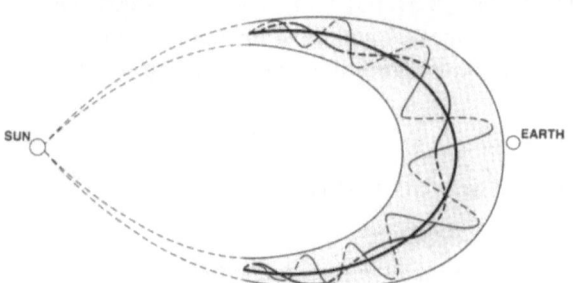

Fig. 6.5. The magnetic field lines in a force-free magnetic field configuration form a familiy of helices. The global configuration near 1 AU is similar to that shown here. The dashed lines near the sun indicate that connectivity to the sun is unknown

is generally maximum near the center of a magnetic cloud. Thus, qualitatively the flux rope model of a magnetic cloud is more consistent with the observations than either the model of a closed planar loop or the model of a tightly wound helix. It has been suggested that magnetic clouds are simply distortions of the interplanetary magnetic field produced by shocks [6.19], but this model does not explain the highly ordered pattern of magnetic field observed in magnetic clouds in general.

The flux rope associated with a magnetic cloud probably curves as shown in Fig. 6.5. However, for the purpose of describing the local observations of a magnetic cloud obtained by a single spacecraft, it is convenient to model the magnetic cloud as a cylindrically symmetric structure to first approximation. Solutions of the cylindrically symmetric force-free field equation were obtained for the case in which the pitch angle of the magnetic field increases as the square of the distance from the axis of symmetry [6.51], and the magnetic field profiles in two magnetic clouds were described in this way.

It has been suggested that magnetic clouds are approximately force-free fields with constant α [6.7]. Force-free fields with α = constant have a special place in the literature on force-free configurations. The cylindrically symmetric constant-α solution gives the types of magnetic field profile observed during the passage of magnetic clouds [6.7]. Since this solution describes a cloud with zero plasma pressure, it is appropriate to refer to the cloud as a magnetic cloud, rather than a magnetized plasma cloud. The observed variations of the magnetic field direction depend on the orientation and position of the magnetic cloud relative to the spacecraft. The left panel of Fig. 6.6 shows the magnetic field profile of the constant-α force-free solution for a magnetic cloud whose axis is inclined 20° with respect to the ecliptic and whose projection in the ecliptic is 60° from the radial direction [6.7]. The magnetic field points northward initially, rotates into the ecliptic and then turns southward. This solution closely resembles the magnetic field profile observed by *IMP 8* and *ISEE 3* on 19 and 20 March 1980 as shown in the panel on the right of Fig. 6.6. The theoretical magnetic field strength profile is symmetric, the maximum intensity occurring in the middle of the magnetic cloud, whereas the observed magnetic field strength is asymmetric with maximum field strength toward the front of the magnetic cloud. The asymmetry in the magnetic field strength profile may be due to a dynamical effect

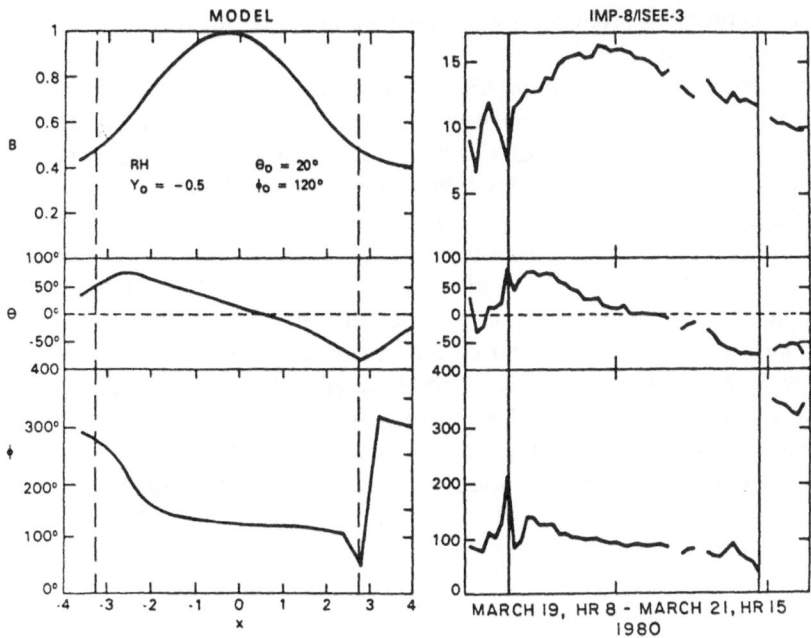

Fig. 6.6. A model of a force-free magnetic cloud whose symmetry axis is close to the ecliptic plane (*left panel*) and observations of a magnetic cloud which is similar to that of the model (*right panel*)

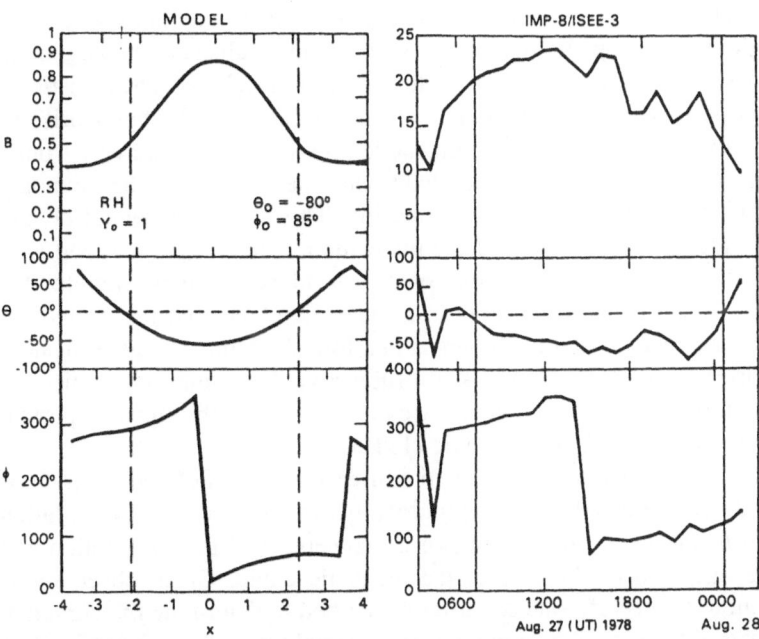

Fig. 6.7. A model of a force-free magnetic cloud whose symmetry axis is nearly normal to the ecliptic (*left panel*) and observations of a magnetic cloud which is similar to that of the model (*right panel*)

7

such as a compression produced by the interaction of the magnetic cloud with slower moving plasma ahead of it.

The predicted magnetic field profile for a constant-α cylindrically symmetric magnetic cloud whose axis of symmetry is inclined 85° with respect to the ecliptic [6.7] is shown in the left panel of Fig. 6.7. In this case one does not expect to observe the pattern of northern fields followed by southern fields (or vice versa) that was sought in the early studies of magnetic clouds, even though the magnetic cloud geometry relative to the axis of the cloud is identical to that in Fig. 6.6. The right panel of Fig. 6.7 shows observations of a magnetic structure whose profile is very similar to the theoretical profile in the left panel of that figure. Thus, the observations represent the detection of a magnetic cloud whose axis is highly inclined with respect to the ecliptic. Recently another highly inclined magnetic cloud was identified in the *ISEE-3* data [6.18]. The early searches for magnetic clouds deliberately excluded highly inclined magnetic clouds. Now that a model for magnetic clouds is available, future searches should include highly inclined magnetic clouds.

6.5 Large-Scale Structure of Magnetic Clouds

In order to determine the size of a magnetic cloud accurately one must have an objective criterion for identifying its boundaries. Unfortunately, such a criterion has not been established. One can unambiguously identify the points where the elevation angle of the magnetic field reaches its extreme values, at the beginning and end of the rotation of the magnetic field vector which is characteristic of a magnetic cloud [6.10], but these points are not necessarily the boundaries of the magnetic cloud. Theoretically, the magnetic pressure decreases away from a maximum on the axis of the magnetic cloud, and one might take the boundary as the point where the pressure stops decreasing. This condition cannot always be applied in practice, e.g. because of the interaction of the magnetic cloud with another flow. Another possible signature of the boundary of a magnetic cloud is a gradient in β, since β is always low in a magnetic cloud, but this cannot be observed if the magnetic cloud is immersed in a low-β region. The identification of the boundaries of magnetic clouds is still subjective. Determining an objective criterion for the boundary of a magnetic cloud is one of the most important problems for future research on magnetic clouds.

Since magnetic clouds appear to be force-free magnetic field configurations which can be described locally by the cylindrically symmetric constant-α solution [6.7], one can fit the local observations of a magnetic field to this solution and determine the direction of the axis of the magnetic cloud. Given observations from two or more spacecraft, one can use this method to determine the curvature of the outer part of the magnetic cloud. Using a special fitting procedure [6.48] on the data for the January 1978 magnetic cloud [6.10] from each of the spacecraft

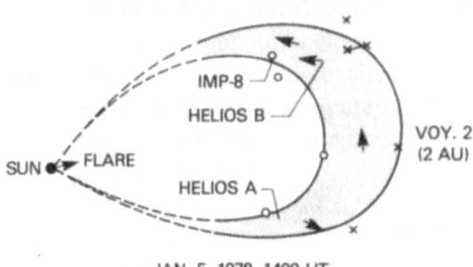

MAGNETIC CLOUD

IMP-8

HELIOS B

VOY. 2
(2 AU)

SUN ☀ FLARE

HELIOS A

JAN. 5, 1978, 1400 UT

Fig. 6.8. The large-scale geometry of a magnetic cloud determined by multispacecraft measurements. The orientation of the axis of the magnetic cloud, shown by the arrows, was obtained from fits of the data to the solution for a cylindrically symmetric constant-α force-free magnetic field

IMP, Helios 1, Helios 2, and *Voyager 1* gave the results shown in Fig. 6.8 from [6.14]. For the chosen boundaries the minor radius of the outer part of the flux tube corresponding to the magnetic cloud is 0.15 AU at 1 AU. The radius of curvature of the axis of the flux rope is approximately $\frac{1}{3}$ AU, which is consistent with the boundaries that were selected.

The multispacecraft method is not sufficient to determine the global topology of a magnetic cloud. One possible topology for a magnetic cloud is that of a flux tube which "ends" at a pair of compact regions on the surface of the sun. Of course, the magnetic field lines in the tube do not end on the solar surface, but it is still appropriate to think of one region as the source of the *interplanetary* magnetic field lines in the flux tube and the other region as a sink of the magnetic field lines. One can think of the integral of the magnetic flux over the region as a magnetic charge or a topological period. The magnetic flux which leaves one region is equal to that which enters the other region, and the flux is invariant throughout the flux tube. A second possible topology for a magnetic cloud is that of a flux tube which bifurcates into several smaller tubes, each of which is rooted in the sun. A third possibility is that the flux tube is closed to form a torus. Presumably a closed tube would have to be produced by magnetic reconnection, but the formation of a torus in this way has not been investigated.

Another approach to determining the curvature of the axis of a magnetic cloud is to assume that the magnetic cloud has a toroidal geometry and fit the data to the constant-α solution for a torus [6.7, 39]. A good fit to the data from two spacecraft was obtained in this way for one magnetic cloud [6.39]. In principle this method can determine the curvature of the axis of the magnetic cloud using observations from a single spacecraft. It is unlikely that magnetic clouds have the topology of a torus, since that would imply that two passages through the magnetic cloud would be observed when the axis of the magnetic cloud is near the ecliptic, whereas no such double-passages have been reported. Nevertheless, it might be appropriate to approximate the local geometry of a magnetic cloud as that of a torus when the radius of curvature of the axis is comparable to the radius of the flux-tube.

It is difficult to identify magnetic clouds smaller than approximately 0.1 AU, because of the presence of Alfvénic fluctuations (see Chap. 4, Vol. 1) and other small-scale features that may be associated with large variations in the magnetic

9

field direction (see Chap. 10). The first statistical study of magnetic clouds [6.43] selected magnetic clouds which passed the spacecraft during an interval greater than or approximately equal to one day, corresponding to a radial extent of approximately 0.25 AU at 1 AU. Such magnetic clouds occur at a rate of the order of 0.5 to 1 magnetic cloud per month. A study of events with bidirectional proton anisotropies, which included some magnetic clouds [6.50] determined that the mean scale size of the events with shocks was 0.20 AU and the mean scale size of events without shocks was 0.14 AU; the rate of occurrence of the events was 0.6/month. The differences between the results of the two studies can be attributed to the different selection criteria and the differences in the completeness of the data sets.

A shock wave may occur ahead of a magnetic cloud. One expects a shock to be driven by a magnetic cloud when the momentum flux of the magnetic cloud is high. However, one may also observe a shock ahead of a magnetic cloud in which there is no increase in the momentum flux [6.10]. In principle, one can observe a shock propagating freely ahead of a magnetic cloud at some distance from the sun if a magnetic cloud driving a shock near the sun decelerates in transit.

A forward–reverse shock pair can occur in association with a magnetic cloud at 1 AU [6.34, 37, 47]. The forward shock propagates ahead of the magnetic cloud, and the reverse shock propagates into the magnetic cloud, producing a very strong magnetic field in the magnetic cloud. Figure 6.9 shows a shock pair and a magnetic cloud observed at 1 AU [6.47] drawn to scale on the sun–earth line. The dots in the figure represent energetic particles accelerated by the shock, which do not penetrate into the magnetic cloud. A forward–reverse shock pair associated with a magnetic cloud was observed near 11 AU [6.12], but the reverse shock was probably produced by a corotating stream that was being overtaken by the magnetic cloud. A forward–reverse shock pair associated with a magnetic cloud was observed by *ISEE 3* at 1 AU [6.34]; the magnetic cloud was interacting

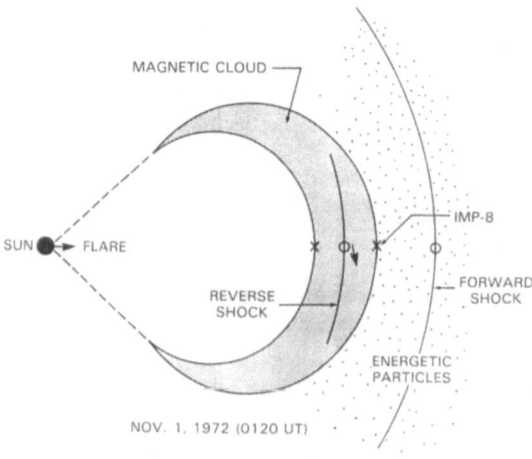

Fig. 6.9. A shock pair associated with a magnetic cloud. The positions of the various structures are drawn to scale on the earth–sun line, but the structure away from that line is an extrapolation based on other results such as that in Fig. 6.8

with a corotating stream [6.13], so that we cannot exclude the possibility that the reverse shock was associated with the interaction rather than produced by the magnetic cloud itself. A forward–reverse shock pair driven by transient ejecta is rarely observed at 1 AU, because it requires a strong solar event near the central meridian [6.67].

A shock wave does not necessarily precede a magnetic cloud. Only thirty percent of the magnetic clouds identified in the interval from 1967 to 1978 are associated with shocks [6.43]. On the other hand, eighty percent of the magnetic clouds identified in one study [6.82] of the data from 1978 to 1982 and seventy percent of the magnetic clouds in another study [6.50] of the data in the same period are associated with shocks.

6.6 Motion and Stability of Magnetic Clouds

A magnetic cloud expands as it moves away from the sun. This is obvious, since the size (diameter) of a magnetic cloud at 1 AU is approximately 0.25 AU, which is more than 50 times the radius of the sun. Expansion at a speed of about one half of the Alfvén speed can account for the observed size of a magnetic cloud at 1 AU [6.43]. There are models for the motion and expansion of nonequilibrium hydromagnetic gas clouds, which might be applicable near the sun [6.57, 40] and for the motion of a magnetic loop near the sun [6.16], but there are no dynamical models for the motion of magnetic cloud in the solar wind. Since the Alfvén speed is largest close to the sun, it is possible that most of the expansion takes place there. The expansion of magnetic clouds apparently continues beyond 1 AU, since the magnetic clouds observed there are larger than those observed at 1 AU [6.8]. The largest known magnetic cloud, with a diameter of approximately 1 AU, was observed at 11.5 AU; again the size is consistent with expansion at approximately half the Alfvén speed. Another observation that suggests the expansion of magnetic clouds at about $V_A/2$ is the linear decrease in bulk speed which is often seen during the passage of a magnetic cloud [6.43].

If a magnetic cloud expands relative to the ambient solar wind as it moves away from the sun, then the density in a magnetic cloud decreases faster than the density of the ambient material ahead of the magnetic cloud. Thus, the momentum flux of the magnetic cloud decreases with increasing distance from the sun, which can lead to a deceleration of the magnetic cloud as it moves away from the sun if it does not encounter any obstacles or fast flows. Some magnetic clouds have speeds characteristic of the "slow solar wind", but they may have originated with much higher speeds. It is customary to assume that slow solar wind at 1 AU originates as slow solar wind at the sun, but clearly this assumption is not always valid. If a magnetic cloud which was driving a shock near the sun decelerates appreciably, then there will be a point where the shock precedes the magnetic cloud but is no longer driven by the magnetic cloud. Such a shock may have been observed [6.10] at 2 AU.

The existence of magnetic clouds implies that they are in some sense stable. One magnetic cloud was identified at 11 AU; if the cloud moved from the sun to 11 AU at an average speed of 400 km/s, then it was stable for 35 days. The stability of magnetic clouds presents many interesting physical questions. It has been suggested that a force-free field with constant α represents a minimum energy configuration, which should therefore be stable [6.80]. Thus the model of magnetic clouds as constant-α force-free fields may be consistent with their apparent stability for a long period of time.

Magnetic clouds in the heliosphere appear to be near a state of equilibrium, yet they are expanding. Equilibrium of a magnetic cloud is consistent with expansion as it moves away from the sun only if the relaxation time is much shorter than the expansion time of about 0.25 AU/day [6.69, 7, 81]. As a magnetic cloud expands, the pressure-gradient force and the magnetic curvature force diminish, and there may come a point at which the cloud is no longer in a quasi-equilibrium state under the action of these forces, but is rather a passive remnant of such an equilibrium.

Under some conditions including zero resistivity, a force-free field with constant-α represents a minimum energy configuration, which should therefore be stable [6.80]. However, it is not clear that a force-free configuration with arbitrarily small resistivity is stable [6.22]. A force-free field can be unstable with respect to the kink mode [6.75], but the growth time in the solar wind is so long that this effect is negligible. There is often a large gradient in β at the boundary of a magnetic cloud (see Fig. 6.3), and it is known that a gradient in β can produce a ballooning instability [6.68], so the boundary of a magnetic cloud might be unstable with respect to a balloon instability. This possibility has not been investigated.

A turbulent plasma generated by a pinch discharge might relax to a constant-α force-free configuration [6.69, 71, 72]. Support for this relaxation hypothesis is found in laboratory experiments, and this idea has been applied to astrophysical jets [6.44]. One can imagine that a magnetic cloud observed in the solar wind is the result of the relaxation of a turbulent magnetic field produced by a pinch discharge at the sun. On the other hand, a magnetic cloud might be the result of the ejection of a magnetic flux rope from the sun. The ejection could be the result of an instability which does not disrupt the flux rope. Magnetic flux ropes can be accelerated to supersonic speeds in this way by the Lorentz force [6.16].

6.7 Nonthermal Particles and Magnetic Clouds

The large-scale geometry and the topology of a magnetic cloud cannot be determined directly from measurements of the magnetic field made by one spacecraft, and there are few opportunities for multispacecraft observations. Thus, it is important to consider indirect means of obtaining such information.

It was suggested that a low temperature in a fast flow in the solar wind might indicate a disconnected (closed) magnetic field configuration [6.30, 33, 53, 29, 24, 25]. Of course, even if it could be shown that a low-temperature–high-speed region is a necessary condition for a disconnected closed magnetic field configuration, it would not follow that it is a sufficient condition for such a configuration. The class of structures with low temperatures might include disconnected magnetic field configurations, but it might include other magnetic field configurations as well.

The observations of bidirectional anisotropies in the energetic protons and suprathermal electrons [6.5, 32, 33, 46, 50, 59, 60, 61, 62, 73] provide a more promising means of obtaining information about the topology of a magnetic cloud. The suprathermal (heat-flux) electrons are almost always present, whereas more energetic protons and electrons are less frequently seen. The basic idea is that if nonthermal particles are somehow injected into magnetic field lines which are either anchored at both ends in the sun or closed in the interplanetary medium, then they should exhibit a bidirectional anisotropy. A bidirectional anisotropy is not a sufficient condition for the identification of a closed magnetic structure. For example, the reflection of energetic particles by a shock wave or by a diffusing region [6.52] produces a bidirectional anisotropy of energetic particles moving along an open magnetic field line. Moreover, the detection of a closed magnetic field structure does not necessarily imply the presence of a magnetic cloud, because there may exist closed magnetic field structures which are not magnetic clouds (e.g. loops formed by reconnection at sector boundaries).

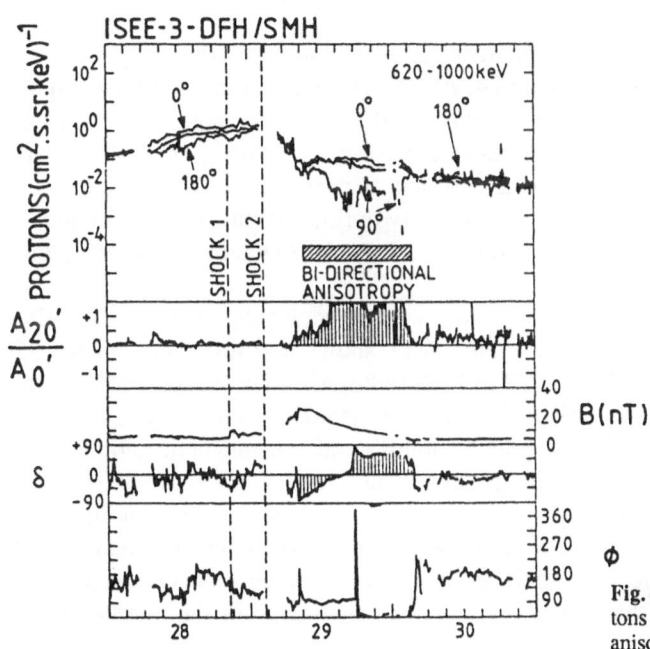

Fig. 6.10. Fluxes of energetic protons from three directions and the anisotropy of the fluxes observed in a magnetic cloud

A bidirectional anisotropy in the flux of 620–100 MeV protons was observed in a magnetic cloud that moved past *ISEE 3* on 29 September 1978 [6.50]; Figure 6.10 shows the data for this event. The magnetic cloud followed two shocks, and the intensity of energetic protons was maximum at the shocks. The proton intensity dropped by more than an order of magnitude inside the magnetic cloud, suggesting that the magnetic field lines carrying energetic protons away from the shock did not penetrate through the boundary of the magnetic cloud. A bidirectional anisotropy was observed throughout the magnetic cloud. Its magnitude varied inside the magnetic cloud, suggesting possible filamentary structure. In this study, forty-one percent of the sixty-six bidirectional anisotropy events were associated with magnetic clouds [6.50].

A bidirectional anisotropy in the flux of > 80 eV (heat-flux) electrons was observed in the magnetic cloud that moved past *ISEE 3* on 29 September 1987 [6.33]. Bidirectional electron heat-flux anisotropies were observed more frequently than bidirectional anisotropies in energetic protons during the same interval as the proton study, and there is no simple relation between the two types of anisotropy [6.33]. The lack of a one-to-one relationship between the low-energy electrons and the more energetic protons is possibly related to the different sources and acceleration mechanisms of the two populations, rather than to the characteristics of magnetic clouds or other interplanetary structures.

6.8 Effects of Magnetic Clouds on Geomagnetic Activity

There is a correlation between geomagnetic activity and a southward component of the interplanetary magnetic field [6.4, 20, 21]. During the passage of a magnetic cloud the magnetic field is usually southward during the passage of at least one part of the magnetic cloud and northward during the passage of another part, depending on the inclination of the axis of the magnetic cloud. Thus one expects to observe an increase in geomagnetic activity during the passage of a magnetic cloud. Such an increase was observed in association with the first magnetic cloud that was reported [6.10].

Strong geomagnetic activity associated with magnetic clouds occurred during the period 1973 to 1978 [6.79]. Figure 6.11 shows a similar relation for 19 magnetic clouds observed from 1978 to 1982 [6.82], based on a superposed epoch analysis. This figure distinguished between the magnetic clouds with southward fields arriving first (negative magnetic clouds) and the magnetic clouds with northward fields arriving first (positive magnetic clouds). As expected, the Dst index decreases when the magnetic field turns southward. Thus for negative magnetic clouds the geomagnetic activity increases when the magnetic cloud arrives at the earth, whereas for positive magnetic clouds the geomagnetic activity does not increase until the magnetic field turns southward in the middle of the magnetic cloud.

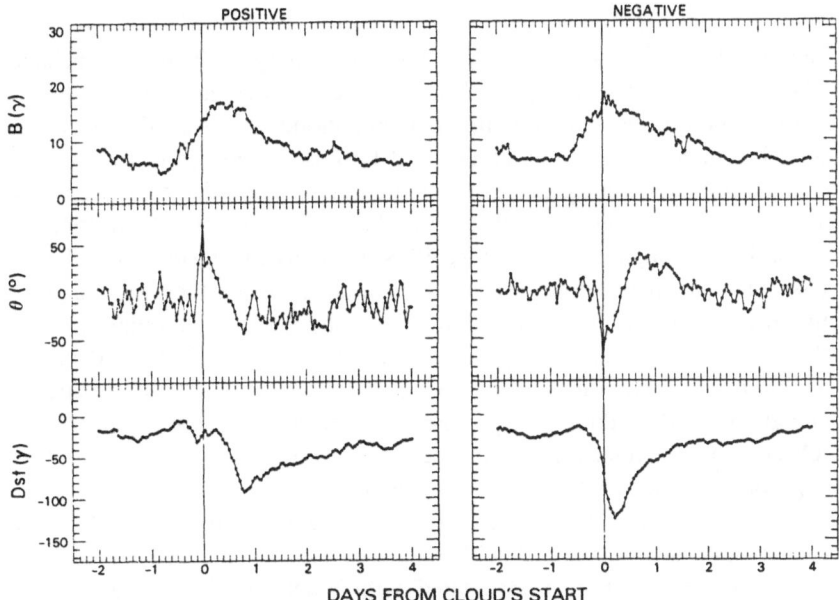

Fig. 6.11. Superposed epoch plots of positive and negative magnetic clouds and the Dst index. The Dst index decreases when that part of a magnetic field in which the field is southward passes the earth

Several processes, such as the compression of ambient magnetic field fluctuations by shocks, can produce strong southward magnetic fields in the solar wind [6.74] and hence geomagnetic storms. Magnetic clouds can also produce intense geomagnetic storms, because the southward fields in magnetic clouds are relatively large in magnitude and persist for several hours. Fifty-nine percent of the largest magnetic storms (storms with $Ap^* > 90$) for which interplanetary data are available in the interval from 1972 to 1982 were associated with magnetic clouds [6.13]. Half of the 10 largest magnetic storms during the 500-day period from 16 August 1978 to 28 December 1979 [6.74] were associated with magnetic clouds.

The effect of magnetic clouds on auroral, ionospheric, and magnetospheric processes has not been studied extensively. An unusual auroral feature was observed during the passage of a magnetic structure which appears to be a magnetic cloud [6.1], although the authors did not identify it as such.

15

6.9 Effects of Magnetic Clouds on Galactic Cosmic Rays

The magnetic cloud observed by *Helios 1* on 5 January 1978 was associated with a Forbush decrease [6.10] (see Chap. 11), but the decrease in cosmic ray intensity began before the arrival of the magnetic cloud. In a statistical study of the relationship between magnetic clouds and Forbush decreases it is necessary to distinguish between a magnetic cloud which is preceded by a shock and a magnetic cloud which is not preceded by a shock. A magnetic cloud contains strong, smoothly varying magnetic fields with a relatively low level of turbulence. In the case of a magnetic cloud preceded by a shock wave there is usually a turbulent sheath between the shock and the magnetic cloud. If turbulence is the cause of Forbush decreases, then one should observe a relatively large decrease in the cosmic ray intensity during the passage of the sheath. If drifting in strong, smooth magnetic fields is the primary cause of Forbush decrease [6.17], then one should observe only a relatively small decrease in cosmic ray intensity during the passage of a magnetic cloud without a shock.

For the magnetic clouds identified from 1967 to 1978 [6.43], the decrease in cosmic ray intensity associated with magnetic clouds preceded by a shock is greater on average than that associated with magnetic clouds that are not preceded by a shock [6.3]. The Forbush decrease starts at the time of arrival of a shock at the earth and continues during the passage of the sheath. Similar results were obtained for the period 1978 to 1982 [6.82]. Figure 6.12 shows superposed epoch plots of the magnetic field intensity, the variation of the magnetic field components and the cosmic ray intensity measured by the neutron monitor at Deep River for fifteen magnetic clouds that were preceded by a shock and for four magnetic clouds that were not preceded by a shock [6.82]. The cosmic ray intensity decrease for the set of magnetic clouds with shocks is significantly larger than that for the magnetic clouds without shocks. The greater effect of schock-associated magnetic clouds on the cosmic ray intensity is probably not primarily the effect of the magnetic field strength, since the magnetic field strength profiles are similar in the two cases. On the other hand, the fluctuations in the magnetic field direction are large at the front of the shock-associated magnetic clouds, whereas smaller enhancements in the fluctuations occur ahead of the magnetic clouds which do not follow shocks. Thus, the results do not support the hypothesis that most Forbush decreases are primarily due to the drifting of cosmic rays in strong, smooth magnetic fields. The results are consistent with the hypothesis that most Forbush decreases are due to the scattering of particles by a region of enhanced magnetic turbulence.

The turbulence occurs ahead of the magnetized cloud, rather than inside the cloud as originally proposed [6.55]. In other words, a correct picture of the cause of a Forbush decrease combines features of both the original models [6.17, 55]. A magnetized plasma cloud with ordered magnetic fields can produce a large Forbush decrease if it is moving fast enough to drive a shock which produces enhanced turbulence at the front of the cloud.

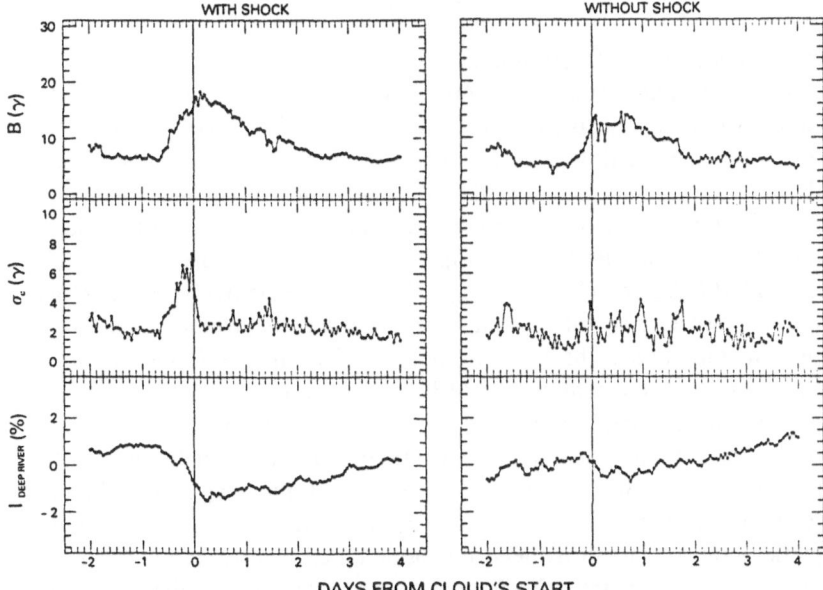

Fig. 6.12. Superposed epoch plots of the magnetic field strength and the variance in the magnetic field components for magnetic clouds with a shock and magnetic clouds without a shock, together with plots of the neutron-monitor relative counting rate observed at Deep River. A magnetic cloud preceded by a shock produces a Forbush decrease, presumably caused by scattering of the cosmic rays from turbulence in the sheath between the shock and the magnetic cloud. A magnetic cloud without a shock produces only a small decrease in the cosmic ray intensity

A magnetic cloud is not necessary for a Forbush decrease. Any ejection moving fast enough to produce a shock might produce a large Forbush decrease. Perhaps the essential feature responsible for a large Forbush descrease is an increase in the level of turbulence such as that which is usually caused by the compression of the ambient magnetic field fluctuations during the passage of a shock. It is not clear what the effect of a shock will be when it moves through undisturbed magnetic fields. One example of a shock that did not cause a large Forbush decrease has been identified [6.61], but the level of turbulence behind the shock was not determined.

Sometimes *increases* in the cosmic ray intensity at the earth occur in association with geomagnetic activity [6.45]. Similarly, short-lived cosmic ray intensity increases occur occasionally during the passage of magnetic clouds [6.38]. This effect is probably associated with a change in the cutoff rigidity. When a magnetic cloud or any structure with prolonged intense southward magnetic fields moves past the earth, it produces an increase in the ring current, which causes a decrease in the magnetospheric field strength, thereby allowing entry to particles with a lower rigidity and consequently a greater cosmic ray intensity to reach the ground.

6.10 Magnetic Clouds and Compound Streams

We have assumed up to this point that magnetic clouds are isolated objects moving through an unstructured medium. Actually, the solar wind within 1 AU is very nonuniform and consists of fast flows of various types (see Chap. 3, Vol. 1 for a discussion of such flows). A fast flow can overtake a magnetic cloud; in one study [6:43] approximately a third of the magnetic clouds were in this class. Similarly, a fast magnetic cloud can overtake a slower flow. A schematic example of the latter phenomenon is shown in Fig. 6.13 from [6.13]. The magnetic cloud observed by *IMP/ISEE* and *Helios 1* and its shock (S2) were overtaking a transient flow (T1) that moved past *IMP 8* and *ISEE 3* earlier, but which did not extend far enough in longitude to be seen at *Helios B*. The magnetic cloud was also being overtaken by a transient stream (T3) and shock (S3).

When a magnetic cloud collides with another stream, and when a shock or stream overtakes and interacts with a magnetic cloud, the magnetic field strength and the magnitude of the southward component of the magnetic field in the magnetic cloud increase. Such a compound system is more effective in producing geomagnetic activity than an isolated magnetic cloud. Thirty-five percent of the largest storms for which interplanetary data are available for the period from 1972 to 1982 were caused by compound streams which included a magnetic cloud [6.13].

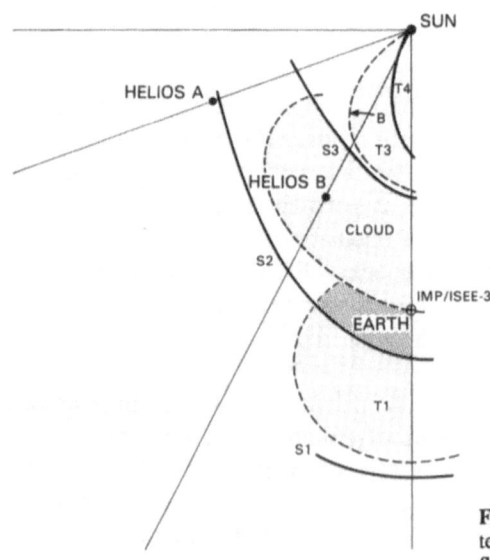

APRIL 3, 2200 UT, 1979

Fig. 6.13. A compound flow formed by the interaction of a magnetic cloud with one transient flow T1 and the interaction of another transient flow T3/T4 with the magnetic cloud

6.11 Questions Concerning Magnetic Clouds

The existence of magnetic clouds as distinct and important physical entities with important effects on the earth and on cosmic rays is now widely accepted. There is increasing support for the idea that magnetic clouds are flux ropes which are related to force-free field configurations. Although there have been many studies of magnetic clouds, many questions remain to be answered. How does one identify the boundary of a magnetic cloud? Is the boundary stable? What are the sources of magnetic clouds? How are magnetic clouds related to the broader class of coronal mass ejections? What is the composition of magnetic clouds, and how does it relate to their sources? How are magnetic clouds accelerated? Do they represent the extension of a solar flux rope or are they the result of the relaxation of a turbulent configuration produced by a pinch discharge? How do magnetic clouds move through the heliosphere, and how do they interact with other solar wind flows? What is the fine structure of magnetic clouds and their boundaries? What is the global topology of magnetic clouds? Are they connected to the sun or disconnected? What is the ultimate fate of magnetic clouds? What is the large-scale effect of a magnetic cloud on the distribution of cosmic rays in space? What are the magnetospheric, ionospheric, and auroral effects of magnetic clouds?

Acknowledgements. The author wishes to thank J. Gosling for many valuable comments on a draft of this manuscript. One of the editors, R. Schwenn, also provided important suggestions for improving the manuscript.

References

6.1   Akasofu, S.-I., B. Tsurutani, Unusual auroral features observed on January 10–11, 1983, and their possible relationships to the interplanetary magnetic field, Geophys. Res. Lett., **11**, 1086, 1984.
6.2   Alfvén, H., Tellus, **6**, 232, 1954.
6.3   Badruddin, R., S. Yadov, N.R. Yadov, S.P. Agarawal, Influence of magnetic clouds on cosmic ray intensity variations, *Proc. 19th ICRC*, SH, 5.1–12, 1985.
6.4   Baker, D.N., S.-I. Akasofu, W. Baumjohann, J.W. Bieber, D.H. Fairfield, E.W. Hones, Jr., B. Mauk, R.L. McPherron, T.E. Moore, Substorms in the magnetosphere, in *Solar Terrestrial Physics: Present and Future*, ed. by D.M. Butler and K. Papadopoulos, NASA Reference Publication 1120, 1984.
6.5   Bame, S.J., J.R. Asbridge, W.C. Feldman, J.T. Gosling, R.D. Zwickl, Bidirectional streaming of solar wind electrons > 80 eV: ISEE evidence for a closed field structure within the driver gas of an interplanetary shock, Geophys. Res. Lett., **8**, 173, 1981.
6.6   Borrini, G., J.T. Gosling, S.J. Bame, W.C. Feldman, Analysis of shock wave disturbances observed at 1 AU from 1971 through 1978, J. Geophys. Res., **87**, 4365, 1982.
6.7   Burlaga, L.F., Magnetic clouds: constant alpha force-free configurations, J. Geophys. Res., **93**, 7217, 1988.
6.8   Burlaga, L.F., K.W. Behannon, Magnetic clouds: Voyager observations between 2 and 4 AU, Solar Phys., **81**, 181, 1982.

6.9   Burlaga, L.F., J.H. King, Intense interplanetary magnetic fields observed by geocentric space-craft during 1963–1975, J. Geophys. Res., **84**, 6684, 1981.

6.10   Burlaga, L.F., E. Sittler, F. Mariani, R. Schwenn, Magnetic loop behind an interplanetary shock: Voyager, Helios and IMP 8 observations, J. Geophys. Res., **86**, 6673, 1981.

6.11   Burlaga, L.F., L.W. Klein, N.R. Sheeley, Jr., D.J. Michels, R.A. Howard, M.J. Koomen, R. Schwenn, H. Rosenbauer, A magnetic cloud and a coronal mass ejection, Geophys. Res. Lett., **9**, 1317, 1982.

6.12   Burlaga, L.F., F.B. McDonald, M.L. Goldstein, A.J. Lazarus, Cosmic ray modulation and turbulent interaction regions near 11 AU, J. Geophys. Res., **90**, 12, 127, 1985.

6.13   Burlaga, L.F., K.W. Behannon, L.W. Klein, Compound streams, magnetic clouds and major magnetic storms, J. Geophys. Res., **92**, 5727, 1987.

6.14   Burlaga, L.F., R.P. Lepping, J.A. Jones, Global configuration of a magnetic cloud, in *Physics of Magnetic Flux Ropes*, ed. by E.R. Priest, L.C. Lee and C.T. Russel, AGU Geophysical Monograph 58, 1990.

6.15   Chapman, S., V.C.A. Ferraro, Solar streams of corpuscles, their geometry, absorption of light and penetration, Monthly Notices Roy. Astron. Soc., **89**, 470, 1929.

6.16   Chen, J., Effects of toroidal forces in current loops embedded in a background plasma, Ap. J., **453**, 1989.

6.17   Cocconi, G., T. Gold, K. Greisen, S. Hayakawa, J.P. Morrison, The cosmic ray flare effect, Nuovo Cimento, **8**, 161, 1958.

6.18   Crooker, N.U., J.T. Gosling, E.J. Smith, C.T. Russell, A coronal mass ejection flux rope in the solar wind?, in *Physics of Magnetic Flux Ropes*, ed. by E.R. Priest, L.C. Lee and C.T. Russell, to appear, 1989.

6.19   Dryer, M.S., T. Wu, G. Gislason, Some simulated characteristics of magnetic clouds, in *Proceedings of the Sixth International Solar Wind Conference*, ed. by V.J. Pizzo, T.E. Holzer and D.G. Sime, NCAR/TN-306+Proc, 735, 1987.

6.20   Dungey, J.R., Interplanetary magnetic field and auroral zones, Phys. Rev. Lett., **6**, 47, 1961.

6.21   Fairfield, D.H., L.J. Cahill, Jr., Transition region magnetic field and polar magnetic disturbances, J. Geophys. Res., **71**, 155, 1966.

6.22   Field, G., Magnetic helicity in astrophysics, in *Magnetospheric Phenomena in Astrophysics*, ed. by R.I. Epstein and W.C. Feldman, American Institute of Physics, New York, 1986.

6.23   Galvin, A.B., F.M. Ipavich, G. Gloeckler, D. Hovestadt, S.J. Bame, B. Klecker, M. Scholer, B. Tsurutani, Solar wind iron charge states preceding a driver plasma, J. Geophys. Res., **92**, 12069, 1987.

6.24   Geranios, A., Magnetically closed regions in the solar wind, Astrophys. Space Sci., **81**, 103, 1982.

6.25   Geranios, A., Statistical analysis of magnetically closed structures, Planetary Space Science, **35**, 722, 1987.

6.26   Gold, T., Magnetic storms, Space Science Rev., **1**, 100, 1962.

6.27   Goldstein, H., On the field configuration in magnetic clouds, in *Solar Wind Five*, NASA Conference Publ. 2280, p. 731, 1983.

6.28   Gosling, J.T., The magnetic topology of coronal mass ejections in interplanetary space, in *Physics of Magnetic Flux Ropes*, ed. by E.R. Priest, L.C. Lee and C.T. Russell, AGU Geophysical Monograph 58, 1990.

6.29   Gosling, J.T., E.C. Roelof, A comment on the detection of closed magnetic structures in the solar wind, Sol. Phys., **39**, 405, 1974.

6.30   Gosling, J.T., V. Pizzo, S.J. Bame, Anomalously low proton temperatures in the solar wind following interplanetary shock waves – Evidence for magnetic bottles?, J. Geophys. Res., **78**, 2001, 1973.

6.31   Gosling, J.T., J.R. Asbridge, S.J. Bame, W.C. Feldman, R.D. Zwickl, Observations of large fluxes of He^+ in the solar wind following an interplanetary shock, J. Geophys. Res., **85**, 3431, 1980.

6.32   Gosling, J.T., D.N. Baker, S.J. Bame, R.D. Zwickl, Bidirectional solar wind electron heat flux and hemispherically symmetric polar rain, J. Geophys. Res., **91**, 352, 1986.

6.33   Gosling, J.T., D.N. Baker, S.J. Bame, W.C. Feldman, R. Zwickl, E.J. Smith, Bidirectional solar wind electron heat flux events, J. Geophys. Res., **92**, 8519, 1988.

6.34   Gosling, J.T., S.J. Bame, E.J. Smith, M.E. Burton, Forward–reverse shock pairs associated with transient disturbances in the solar wind at 1 AU, J. Geophys. Res., **93**, 8741, 1988.

20

6.35   Hundhausen, A.J., The origin and propagation of coronal mass ejections, in *Proc. of the Sixth International Solar Wind Conference*, ed. by V.J. Pizzo, T.E. Holzer and D.G. Sime, NCAR Technical Note, NCAR/TN 306+Proc, 181, 1988.

6.36   Illing, R.M.E., A.J. Hundhausen, Observations of a coronal transient from 1.2 to 6 solar radii, J. Geophys. Res., **90**, 275, 1985.

6.37   Ipavich, F.M., R.P. Lepping, Analysis of the October 31, 1972 interplanetary shock wave and associated unusual phenomena, in *Proceedings of the 14th International Cosmic Ray Conference*, Munich, Federal Republik of Germany, August 15–19, 1568, 1975.

6.38   Iucci, N., M. Parisi, C. Signorini, M. Storini, G. Villorese, Anomalous short-term increases in the galactic cosmic ray intensity: Are they related to interplanetary magnetic cloud-like structures?, in *Proc. of the 19th ICRC*, SH 5.1–5.3, 1985.

6.39   Ivanov, K.G., A.F. Harshiladze, E.G. Eroshenko, V.A. Styazhkin, Configuration, structure and dynamics of magnetic clouds from solar flares in light of measurements on board VEGA 1 and VEGA 2 in January–Feburary, 1986, Solar Physics, **120**, 407, 1989.

6.40   Ivanov, K.G., A.F. Harshiladze, Dynamics of hydromagnetic clouds from powerful solar flares, Solar Physics, **92**, 351, 1984.

6.41   Jackson, B.V., R.A. Howard, N.R. Sheeley, Jr., D.J. Michels, M.J. Kooman, R.M. Illing, Helios spacecraft and earth perspective observations of three looplike solar mass ejection transients, J. Geophys. Res., **90**, 5075, 1985.

6.42   Kahler, S., Observations of coronal mass ejections near the sun, in *Proc. of the Sixth International Solar Wind Conference*, ed. by V.J. Pizzo, T.E. Holzer and D.G. Sime, NCAR Technical Note, NCAR/TN 306+Proc, 215, 1988.

6.43   Klein, L.W., L.F. Burlaga, Interplanetary magnetic clouds at 1 AU, J. Geophys. Res., **87**, 613, 1982.

6.44   Konigl, A., A. Choudhur, Force-free equilibria of magnetized jets, Astrophys. J., **289**, 173, 1985.

6.45   Kudo, S., M. Wada, T. Tanskannen, Transient cosmic ray increases associated with a geomagnetic storm, in *Proc. 19th ICRC*, SH-5.1–8, 1985.

6.46   Kutchko, F.J., P.R. Briggs, T.P. Armstrong, The bidirectional particle event of October 12, 1977, possibly associated with a magnetic loop, J. Geophys. Res., **87**, 1419, 1982.

6.47   Lepping, R.P., F.M. Ipavich, L.F. Burlaga, A flare-related shock pair at 1 AU and related magnetic cloud, submitted to J. Geophys. Res., 1991.

6.48   Lepping, R.P., J.A. Jones, L.F. Burlaga, Magnetic Field structure of interplanetary magnetic clouds at IAU, J. Geophys. Res., **95**, 11957, 1990.

6.49   Lindeman, F.A., Phil. Mag., **38**, 669, 1919.

6.50   Marsden, R.G., T.R. Sanderson, C. Tranquille, K.-P. Wenzel, ISEE-3 observations of low energy proton bidirectional events and their relation to interplanetary magnetic structures, J. Geophys. Res., **92**, 11, 009, 1987.

6.51   Marubashi, K., Structure of the interplanetary magnetic clouds and their solar origins, Adv. Space Res., **6**, 335, 1986.

6.52   Meyer, P., E.N. Parker, J.A. Simpson, Solar cosmic rays of February 1956 and their propagation through interplanetary space, Phys. Rev., **104**, 768, 1956.

6.53   Montgomery, M.D., J.R. Asbridge, S.J. Bame, W.C. Feldman, Solar wind electron temperature depressions following some interplanetary shock waves: Evidence for magnetic merging?, J. Geophys. Res., **79**, 3103, 1974.

6.54   Morrison, P., Solar-connection variations of the cosmic rays, Phys. Rev., **95**, 646, 1954.

6.55   Morrison, P., Solar origin of cosmic ray time variations, Phys. Rev., **101**, 1397, 1956.

6.56   Palmer, I.D., F.R. Allum, S. Singer, Bidirectional anisotropies in solar cosmic ray events: Evidence for magnetic bottles, J. Geophys. Res., **83**, 75, 1978.

6.57   Parker, E.N., The gross dynamics of a hydromagnetic gas cloud, Ap. J. Supplement, **25**, 51, 1957.

6.58   Piddington, J.H., Phys. Rev., **112**, 589, 1958.

6.59   Pudovkin, M.I., S.A. Zaitseva, E.E. Benevolenska, The structure and parameters of flare streams, J. Geophys. Res., **83**, 6649, 1979.

6.60   Sanderson, T.R., R.G. Marsden, R. Reinhard, K.-P. Wenzel, E.J. Smith, Correlated particle and magnetic field observations of a large-scale magnetic loop structure behind an interplanetary shock, Geophys. Res. Lett., **10**, 916, 1983.

6.61   Sanderson, T.R., J. Beeck, R.G. Marsden, K.-P. Wenzel, R. B. McKibben, E.J. Smith, Energetic ion and cosmic ray characteristics of magnetic clouds, in *Physics of Magnetic Flux Ropes*, ed. by E.R. Priest, L.C. Lee and C.T. Russel, AGU Geophysical Monograph 58, 1990.

6.62   Sarris, E.T., S.M. Krimigis, Evidence for solar magnetic loops beyond 1 AU, Geophys. Res. Lett., **90**, 19, 1985.

6.63   Schatten, K.H., Evidence for coronal magnetic bottles at 10 solar radii, Solar Phys., **12**, 484, 1970.

6.64   Schatten, K.H., J.E. Schatten, Magnetic field structure in flare-associated solar wind disturbances, J. Geophys. Res., **77**, 4858, 1972.

6.65   Schwenn, R., H. Rosenbauer, K.-H. Mühlhäuser, Singly-ionized helium in the driver gas of an interplanetary shock wave, Geophys. Res. Lett., **7**, 201–204, 1980.

6.66   Schwenn, R., Direct correlation between coronal transients and interplanetary disturbances, Space Sci. Rev., **34**, 85, 1983.

6.67   Steinolfson, R.S., M. Dryer, Y. Nakagawa, Interplanetary shock pair disturbances: Comparison of theory with space probe data, J. Geophys. Res., **80**, 1989, 1975.

6.68   Strauss, H.R., The effect of ballooning modes on thermal transport and magnetic field diffusion in the solar corona, Geophys. Res. Lett., **16**, 219, 1989.

6.69   Suess, S.T., Magnetic clouds and the pinch effect, J. Geophys. Res., **93**, 5437, 1988.

6.70   Tang, F., B.T. Tsurutani, W.D. Gonzalez, S.-I. Akasofu, E.J. Smith, Solar sources of 10 southward B_z events responsible for geomagnetic storms (1978–1979), J. Geophys. Res., **94**, 3535, 1989.

6.71   Taylor, J.B., Relaxation of toroidal plasma and generation of reverse magnetic fields, Phys. Rev. Lett., **33**, 1974.

6.72   Taylor, J.B., Relaxation and magnetic reconnection in plasmas, Rev. Modern Physics, **58**, No. 3, 741, 1986.

6.73   Tranquille, C., T.R. Sanderson, R.G. Marsden, K.-P. Wenzel, E.J. Smith, Properties of a large-scale interplanetary loop structure as deduced from low-energy proton anisotropy and magnetic field measurements, J. Geophys. Res., **92**, 6, 1987.

6.74   Tsurutani, B.T., W.D. Gonzalez, F. Tang, S.-I. Akasofu, E.J. Smith, Origin of interplanetary southward magnetic fields responsible for major magnetic storms near solar maximum (1978–1979), J. Geophys. Res., **93**, 8519, 1988.

6.75   Voslambe, D., D.K. Callebaut, Stability of force-free magnetic fields, Phys. Rev., **128**, 2016, 1962.

6.76   Webb, D.F., B.V. Jackson, Detection of CMEs in the interplanetary medium from 1976–1979 using Helios-2 Photometer data, in *Proceedings of the Sixth International Solar Wind Conference*, ed. by V.J. Pizzo, T.E. Holzer and D.G. Sime, NCAR/TN-306+Proc, 267, 1987.

6.77   Wilson, R.M., E. Hildner, Are interplanetary magnetic clouds manifestations of coronal transients at 1 AU?, Solar Physics, **91**, 168, 1984.

6.78   Wilson, R.M., E. Hildner, On the association of magnetic clouds with disappearing filaments, J. Geophys. Res., **91**, 5867, 1986.

6.79   Wilson, R.M., Geomagnetic response to magnetic clouds, Planet. Space Sci., **35**, 329, 1987.

6.80   Woltjer, L., A theorem on force-free magnetic fields, Proc. National Academy of Sciences, **44**, No. 6, 389, 1958.

6.81   Yang, W.-H., Expansion of solar-terrestrial low-β plasmoid, Ap. J., **344**, 966, 1989.

6.82   Zhang, G., L.F. Burlaga, Magnetic clouds, geomagnetic disturbances, and cosmic ray decreases, J. Geophys. Res., **93**, 2511, 1988.

7. Interplanetary Slow Shocks

Arne K. Richter

7.1 Introduction

Ever since the identification of the first interplanetary shock wave from the *Mariner 2* plasma and magnetic field measurements [7.38], shock research has received great attention in solar system plasma physics, and this has resulted in an outstanding collaboration between laboratory and space experimentalists, theorists, and specialists in numerical simulation. This type of research, however, has been devoted, at least from an observational point of view, more or less exclusively to the investigation of fast-mode shock waves, whereas the corresponding study of slow-mode shocks has remained in a fairly primitive state. To some extent this is rather surprising, as MHD *per se* does not predict any major problems in generating slow-mode waves or shocks, in particular not in the solar wind, where, as we will show, ample "perturbations" are present for their generation. Yet, although long-standing records of solar wind measurements ranging from 0.3 AU out to several tens of AU are at our disposal, the number of observations of interplanetary slow-mode MHD turbulence or slow-mode shocks published over the last few years is remarkably small. In fact, there exists only one published example of low-frequency slow-mode MHD turbulence identified in the solar wind [7.31], and the number of published interplanetary slow shocks, regardless of the level of sophistication with respect to their unique identification, can still be counted on the fingers of one hand [7.9, 28].

Over the last few years this situation has changed perceptibly, in particular with respect to investigations regarding the constraints on the development, evolution, and propagation of interplanetary slow shocks in combination with actual spacecraft observations, mainly in the inner part of the heliosphere. The aim of this review is to summarize these recent efforts from the perspective of previous studies and in connection with observational results. Three separate topics will be discussed. The first is a collection of those source perturbations, occurring close to the sun and in the solar wind, which are specifically suited to generating slow shocks. The second summarizes the various constraints on the development and evolution of solar wind slow shocks, and it therefore deals with those arguments in support of a deficiency of slow shocks, at least over that part of the interplanetary medium at present accessible to spacecraft observations. And finally, the third discusses the observational difficulties and ambiguities concerning the unique identification of slow shocks within the framework of MHD, thereby

Physics and Chemistry in Space - Space and Solar Physics, Vol. 21
Physics of the Inner Heliosphere II Editors: R. Schwenn · E. Marsch
© Springer-Verlag Berlin Heidelberg 1991

supporting further the fact that the number of "acceptable" interplanetary slow shocks is and will remain small, unless a new generation of plasma experiments is placed much closer to the sun.

We hope that this review might also stimulate theorists and specialists in numerical simulation in, for example, considering possible ways to generate slow shocks other than those previously investigated [7.39, 46], and/or also simulating a certain class of tangential discontinuity and comparing it with the slow shock solution.

7.2 Generation of Interplanetary Slow Shocks

In this section we summarize the theoretical and observational studies concerned with the generation of slow shocks near the sun and in the interplanetary medium.

7.2.1 Steepening of Slow-Mode Waves

In view of MHD the most straightforward way to generate slow shocks is by the process of steepening of large-amplitude, low-frequency slow-mode MHD waves (see, e.g. [7.20]). Nonlinear waves steepen because the local wave speed depends on the local wave amplitude. The fast parts of the nonlinear pressure pulse overtake the slower parts, and this process continues until dissipation and/or dispersion prevents further steepening. There are, however, certain constraints on the accomplishment of this process. First, Landau damping of the slow-mode wave has to be small, and, second, the linear damping time must exceed the actual steepening time. These constraints imply that the wave has to propagate into a plasma regime where (i) the ratio of the ion temperature to the electron temperature has to be small, i.e., $T_i/T_e \leq 1$ [7.2], and (ii) where the ratio of the total plasma kinetic pressure to the magnetic field pressure has to be small, i.e., $\beta < 1$ [7.3, 15], respectively. In Sect. 7.3 we show that, in general, it is rather difficult for the solar wind to accomplish these two conditions simultaneously.

By chance such a steepening process has been verified in the solar wind at 0.31 AU from the *Helios* plasma and magnetic field measurements [7.28, 31]: in Fig. 7.1 the time profiles of the solar wind proton velocity, density, and temperature and of the magnetic field intensity are depicted. The occurrence of a slow forward shock is indicated by a dashed line. It has been shown that the medium upstream of this event does fullfill the conditions $T_i/T_e < 1$ *and* $\beta < 1$ [7.28]. Employing power and coherence spectral analysis to the fluctuations observed over the time interval τ_2 downstream of this event, it has been shown that these fluctuations are highly compressive and slow-mode in nature [7.31]. For frequencies $f \leq 5 \times 10^{-3}$ Hz the coherence values between magnetic field and density (between velocity and density) fluctuations were less than -0.6 (larger than 0.6). Moreover, Fig. 7.1 shows that these fluctuations are actually superimposed on a large-amplitude, low-frequency slow-mode wave train, indicated by heavy lines

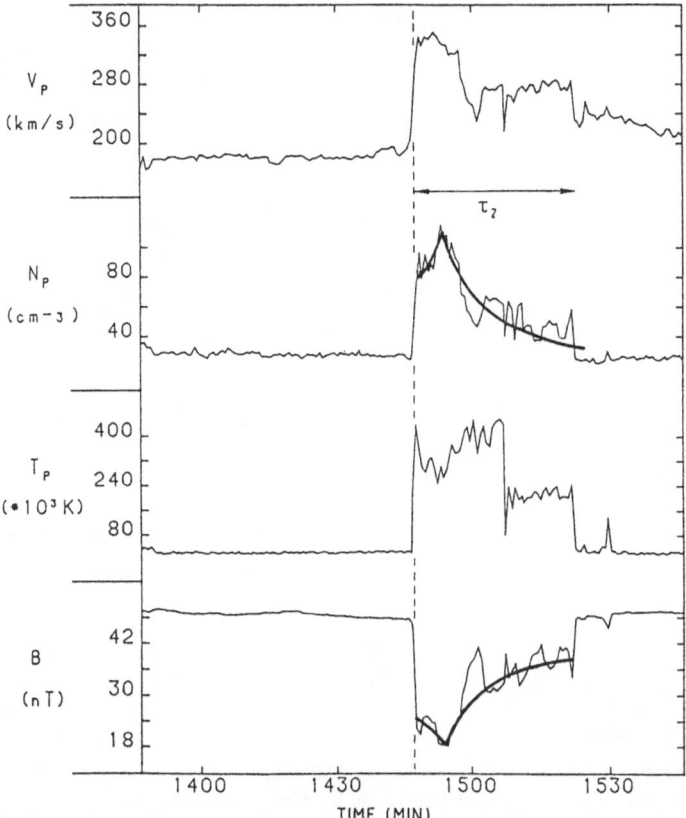

Fig. 7.1. Time histograms of the solar wind proton bulk speed, number density, temperature, and magnetic field intensity across a forward slow shock observed by *Helios 1* at 0.31 AU [7.28]. The downstream fluctuations over the time period τ_2 are compressional and slow-mode in nature for frequencies $\leq 5 \times 10^{-3}$ Hz [7.31]

in the N_p and B plots. It may have been this wave that actually generated the slow shock by the process of steepening. However, at the time of observation this wave had already fallen back behind the shock and, most probably, its amplitude had already decreased at the same time.

There are a number of interaction processes occurring in the solar wind near the sun and throughout interplanetary space which, according to MHD, should be able to generate the full spectrum of MHD turbulence and therefore also slow-mode waves, which, in principle, could result in slow shocks by the process of wave steepening. In the following, we want to summarize just a few candidates.

1. Interaction of Sub-Alfvénic with Super-Alfvénic Streams. Owing to the variability in the expansion of the solar corona, sub-Alfvénic solar wind streams, i.e., plasma portions in which the local Alfvén speed exceeds the solar wind bulk speed, will exist at least in regions close to the sun. It has been shown that

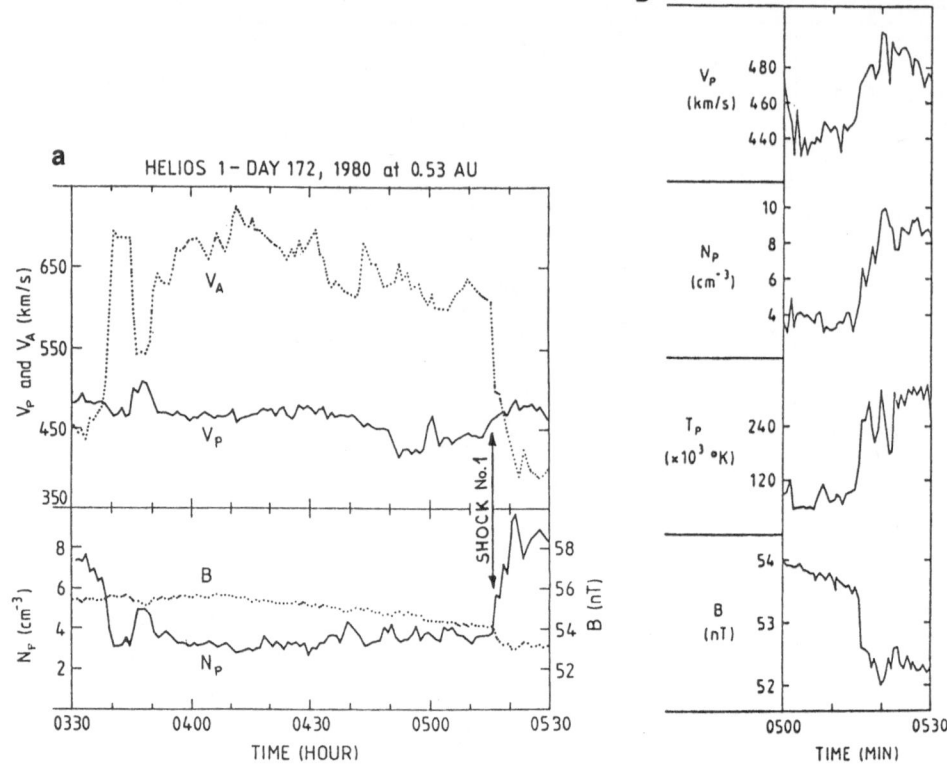

Fig. 7.2. (a) Example of a sub-Alfvénic solar wind region (SSWR) observed by *Helios 1* at 0.53 AU. Across this region the Alfvén speed V_A is larger than the solar wind bulk speed V. The time histograms of the solar wind number density N and magnetic field strength B are shown in the *lower panel*. (b) Expanded time histograms of the solar wind proton bulk speed, number density, temperature, and magnetic field strength across the right-handed (a) interaction region of the sub- with the super-Alfvénic solar wind stream. Notice that this transition resembles the signature of a forward slow-shock-type discontinuity

the interaction of such sub-Alfvénic with the (normally) super-Alfvénic solar wind streams could generate standing slow shocks, the so-called "coronal slow shocks" [7.43]. In part, this process has been verified from *Helios* observations at 0.53 AU, which are depicted in Figs. 7.2a and b. In Fig. 7.2a the time profiles of the Alfvén speed, V_A, and the solar wind bulk speed, V, are shown in the top panel, and those of the number density, N, and the magnetic field strength, B, in the bottom panel. Notice that across the sub-Alfvénic region the Alfvén speed exceeds the solar wind bulk speed by almost 200 km s^{-1}. The situation on the right-hand edge of the sub-Alfvénic region where V starts to exceed V_A is shown in an expanded plot in Fig. 7.2b. This figure clearly indicates the occurrence of a slow-shock-type discontinuity, as the velocity, density, and temperature are increasing while the magnetic field strength is decreasing discontinuously and at the same time across this event. We have labeled this event "1", as it will be further discussed in the Sect. 7.3 of this review.

2. Coronal Transients/Coronal Mass Ejecta. Coronal mass ejections (CMEs) observed with white-light coronographs are often associated with large interplanetary fast shocks at 1 AU [7.14]. There exists an almost 1:1 correlation between CMEs and transient interplanetary fast shocks, provided the CMEs and the spacecraft detecting the shocks are in the same longitudinal and latitudinal range and provided the expansion speed of the CMEs exceed about 400 km s^{-1} (see [7.35] and references therein). Based on these results, it is reasonable to assume that CMEs with expansions speeds less than the local Alfvén speed could be associated with slow shocks. A first indicative example for this type of association has been provided from the SMM coronograph observations [7.19]. A complete correlation of slowly expanding CMEs with interplanetary slow shocks, however, is still awaiting its observational verification.

3. Leading Edges of High-Speed Streams. The interaction of high-speed and low-speed solar wind streams in the interplanetary medium leads to a compression and pile-up of the plasma and magnetic field lines in front of the leading edges of the fast streams and to an enhancement in the longitudinal gradients in all plasma parameters in front of and at the leading edges. The importance of both effects increases with increasing heliocentric distances from the sun [7.30]. It is the compression and/or the enhanced gradients which will generate low-frequency MHD waves and therefore corotating fast shocks [7.36]. Depending on the actual value of the local Alfvén speed, it is reasonable to assume that slow shocks also could be generated in these regions. In this respect it is important to notice that the three forward slow shocks identified some years ago [7.7,9] did occur in regions of increasing longitudinal velocity gradients. This situation is depicted in Fig. 7.3 [7.6]. This association is also true for the two slow forward shocks to be discussed in Sect. 7.3.

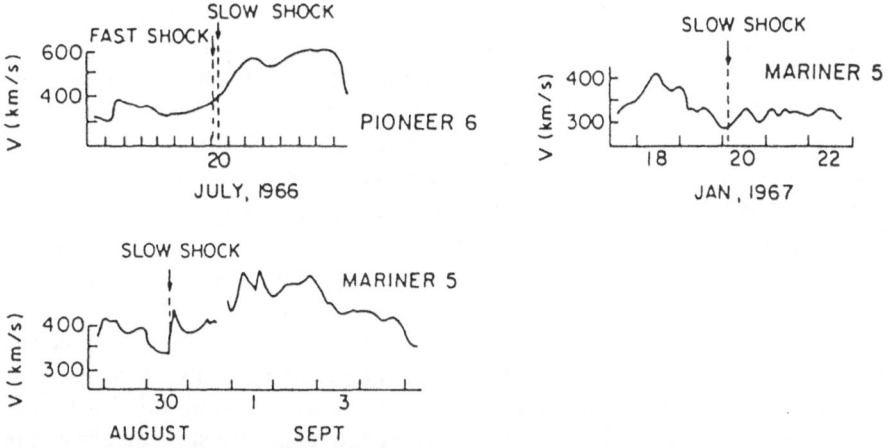

Fig. 7.3. Positions of slow shocks identified earlier with respect to the transition from low- to high-speed solar wind portions [7.6,7,9]. The slow shocks I and II listed in Table 7.1 and depicted in Fig. 7.9a also occurred in such transient regions

27

4. Miscellaneous. Within MHD there exists the possibility that slow shocks can be generated behind fast shocks with non-zero normal magnetic field components (the so-called Riemann problem) [7.20, 34]. It is likely that this generation mechanism has been verified in the solar wind [7.7] (see also [7.6] and Fig. 7.3). Moreover, it is worth noticing that, according to the *Helios* observations, transient fast shocks are more frequently followed within a few hours by slow-shock-type discontinuities than by fast reverse shocks, as expected by theoretical considerations (e.g., [7.18]).

In situ observations have indicated that the interaction of the solar wind with magnetized bodies, such as the earth or Jupiter, and with unmagnetized bodies, such as Venus, Comet Giacobini-Zinner or Comet Halley, can provide a full spectrum of MHD turbulence. In this respect it is worth mentioning that, presumably, slow-mode MHD turbulence may have been detected by the *Giotto* spacecraft upstream of the bow shock of Comet Halley [7.21].

7.2.2 Magnetic Field Line Merging

Since Petschek's work [7.25] it has been well known that the merging of portions of magnetic field of opposite polarity may generate slow-mode shocks. This merging process does occur at various places in the solar photosphere and atmosphere and in the magnetospheres of planets and small bodies, and it may occur at interplanetary current sheets and near sector boundaries. To date, however, the existence of interplanetary slow shocks in association with magnetic field line merging could not be demonstrated. In the tail of the earth's magnetosphere, however, this possibility of generating planetary slow shocks has been verified (see [7.13], and references therein).

7.2.3 Interaction of Interplanetary Discontinuities

According to MHD the interaction of, e.g., two fast forward shocks will lead to two sets of shock pairs separated by a tangential discontinuity: a forward- and a reverse-propagating pair each consisting of one fast and one slow shock separated by an Alfvénic discontinuity [7.20]. In this way slow forward and slow reverse shocks could also be generated in the solar wind. In a similar way the interaction of transient and corotating fast shocks with interplanetary tangential discontinuities could include the generation of both slow forward and slow reverse shocks [7.24].

These possibilities of generating interplanetary slow shocks deserve special attention for future investigations. First, tangential discontinuities and also, to a lesser extent, fast shocks do occur quite frequently in the solar wind; second, their interaction and therefore the possibility of generating slow shocks may, in principle, take place throughout interplanetary space; and third, this is the only explicit mechanism according to which *reverse* interplanetary slow shocks can be produced. Unfortunately, none of these processes has been verified in the solar wind so far.

7.2.4 Summary

In summary we may conclude that there exists an ample number of "perturbations" and interaction processes occurring in the vicinity of the sun and throughout the interplanetary medium which could generate slow-mode MHD turbulence and/or slow shocks. Thus, the deficiency of interplanetary slow shocks is certainly not the cause of a deficiency of appropriate processes concerning their generation.

7.3 Solar Wind Constraints on the Development and Evolution of Interplanetary Slow Shocks

This section is devoted to more recent studies, elaborating the various, rather severe constraints the solar wind itself places on the development and, once they are generated, on the evolution and propagation of interplanetary slow shocks in the region beyond about 0.3 AU. It should be kept in mind, however, that these constraints will only hold "on average", and that under certain, more local conditions the situation may rather favor the existence of slow shocks. Nonetheless, the combination of the discussions that follow with actual spacecraft observations will enable us (i) actually to understand the deficiency of interplanetary slow shocks, (ii) to regard slow shocks as a local rather than a global phenomenon (as, e.g., interplanetary fast shocks), and (iii) to realize that slow shocks should occur more frequently, by far, in the innermost part of the heliosphere, presently inaccessible to spacecraft observations, and in the winds of stars other than the sun.

7.3.1 Constraints on the Development

In the Sect. 7.2 a large number of "perturbations" were discussed, which, according to MHD, should produce low-frequency, slow-mode MHD waves. Now, we want to show that it is the large-scale behavior of the average, stream-structure solar wind over sizeable parts of the inner heliosphere that places rather severe constraints on the evolutions of slow waves in the interplanetary medium and on their probability to steepen into slow shocks.

The evolution of the slow (SL) and the fast magnetoacoustic (F) wave in a magnetized plasma is represented by the Friedrich-II diagrams, i.e., the polar diagrams of the group velocities of these two compressional waves with respect to the direction of the background magnetic field. These diagrams are shown in the top part of Fig. 7.4 (taken from [7.5]). From these diagrams we find that the area of propagation for slow waves is very much reduced in size and that the angular width under which slow waves are allowed to propagate relative to the direction of the background magnetic field is rather limited, too. Moreover, the actual size of this area crucially depends on the values of the local sound

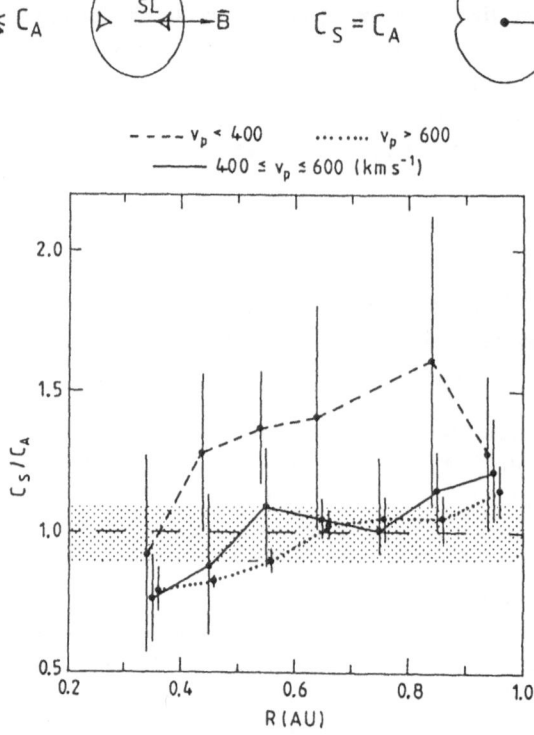

Fig. 7.4. Polar diagrams of the group velocity for fast (F) and slow (SL) magnetoacoustic waves with respect to the magnetic field direction and showing their dependence on the value of the plasma sound speed (C_S) relative to the value of the Alfvén speed (C_A) (*upper panel*). Also shown is the radial dependence of the ratio C_S/C_A as determined from *Helios* observations (*lower panel*). Data are binned according to the solar wind speed indicated at the top, and averages over radial distance bins of 0.1 AU in width are shown with their respective standard deviations [7.32]

speed C_S and Alfvén speed C_A, respectively. If these two velocities become comparable to each other, this area actually degenerates to a point. Notice that the corresponding area for fast waves remains large and more or less unaffected. Taking the *Helios* solar wind plasma (protons, electrons, and alpha particles) and magnetic field observations into account, we have calculated the two speeds C_S and C_A between 0.3 and 1 AU. Thereby, we have averaged their values in bins of 0.1 AU and, by taking the stream structure of the solar wind into account, we have classified them into three categories corresponding to low speed (< 400 km s^{-1}), intermediate-speed (400–600 km s^{-1}), and high-speed (> 600 km s^{-1}) flows, respectively. The corresponding radial dependences of the ratio C_S/C_A are shown in the lower part of Fig. 7.4 [7.32]. Allowing for a 10% uncertainty (gray shaded area), we find that $C_S \approx C_A$ holds below 0.4 AU in the solar wind, but even between about 0.45 to 0.8 AU and 0.55 to 0.95 AU in the intermediate- and in the high-speed wind, respectively. Thus, slow-mode waves are substantially hindered in freely evolving over extended regions in the inner heliosphere.

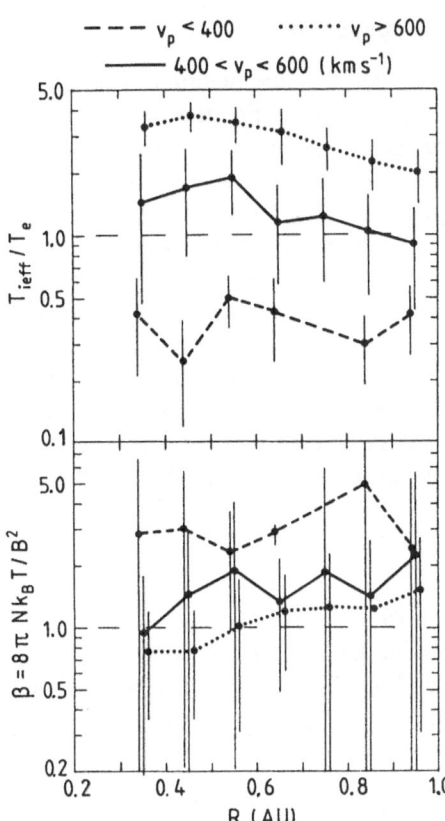

Fig. 7.5. Radial dependence in AU of the ratio of the effective ion temperature $T_{i,\text{eff}}$ (taking protons and alpha particles into account) and the electron temperature T_e (*top*) and of the ratio β of the total plasma kinetic pressure (considering protons, alpha particles, and electrons) and the magnetic field pressure (*bottom*). For further information see Fig. 7.4 [7.32]

For low-frequency slow waves to overcome ion Landau damping and to steepen into slow shocks the background solar wind plasma has to fullfill the conditions $T_i/T_e < 1$ and $\beta < 1$ simultaneously [7.2, 3, 15]. As in Fig. 7.4 we have plotted the radial profiles of T_i/T_e and of the total plasma β in Fig. 7.5 [7.32]. To account for the alpha particles, we used the effective ion temperature $T_{i,\text{eff}}$. From Fig. 7.5 it immediately follows that the above two conditions counteract each other across the different portions of the stream-structured solar wind. On average, steepening of slow waves into slow shocks is likely to occur only at and within 0.4 AU and only in the intermediate wind, i.e., in those regions of the interplanetary medium that are mainly characterized by the stream-stream or the corotating interaction regimes. This result has been verified by a direct investigation of the leading-edge regions of high-speed solar wind streams between 0.3 and 1 AU [7.32]. Taking into account that almost all slow shocks identified so far in the solar wind did occur in association with these regions, some agreement between theory (combined with *Helios* measurements) and observations has been established.

7.3.2 Constraints on the Evolution and Propagation

Once slow shocks have been generated in the interplanetary medium by processes such as steepening of slow waves, magnetic field line merging, or interaction of interplanetary discontinuities, these structures will only be able to propagate relative to the expanding solar wind, if the so-called "evolutionary conditions" for slow shocks are fulfilled [7.22]. We now want to show that because of these "evolutionary condition'" severe constraints are imposed on the evolution of slow shocks in interplanetary space.

According to the "evolutionary conditions" the upstream shock-normal speed in the shock frame of reference, $V_n^* = (V - V_S) \cdot \hat{n}$, where V is the solar wind bulk velocity, V_S the shock velocity, and $\hat{n}$ the unit vector along the shock-normal direction, must be larger than the local slow-magnetoacoustic speed, C_{SL}, but less than the normal Alfvén speed $C_A \cos\theta$, where θ is the angle between the shock normal and the direction of the upstream magnetic field. Thus the conditions

$$C_{SL} < V_n^* \leq C_A \cos\theta \tag{7.1}$$

must hold simultaneously in front of the shock.

By multiplying (7.1) with V_n^*, by applying the appropriate inequalities, and by taking the explicit expressions of C_{SL} and the Alfvén speed C_A into account, we can cast (7.1) into the following form:

$$\frac{\left[1 + \alpha - \{(1+\alpha)^2 - 4\alpha\cos^2\theta\}^{1/2}\right]}{2\cos^2\theta} < M_A^2 \leq 1 . \tag{7.2}$$

Here $\alpha = (C_S/C_A)^2 = \gamma\beta/2$, γ the specific heat ratio, and $M_A = V_n^*/C_A\cos\theta$ the Alfvénic Mach number. By employing β as a free parameter and by increasing its value in steps from 0.01 to 10, we can determine the corresponding area in the (M_A, θ) parameter space for which, in accordance with (7.2), "evolutionary" slow-shock solutions may exist [7.45]. (For a similar treatment see also [7.11].) The result is summarized in Fig. 7.6. As the value of β increases, the size of the area of accepted slow-shock solutions decreases rapidly, while M_A and θ are shifted towards their upper limits $M_A = 1$ and $\theta \lesssim 90°$, respectively. The most dramatic reduction and changes occur when β increases from 0.1 to 1.

Traditionally, the strength of a shock is characterized by the jump in density or in kinetic pressure across the shock. Considering the ratio of the downstream to the upstream density ϱ_2/ϱ_1, one can roughly classify slow shocks into weak $(\varrho_2/\varrho_1 < 1.5)$, intermediate $(\varrho_2/\varrho_1 = 1.5-2.5)$, and strong $(\varrho_2/\varrho_1 > 2.5)$ shocks. The ranges of each of these three classes in the (M_A, θ) parameter space and as a function of β are also depicted in Fig. 7.6 [7.45]. As the value of β increases, accepted slow shocks become, generally speaking, weaker. For $\beta \geq 1$ only weak slow shocks exist.

The severe negative implication of these results concerning the evolution of interplanetary slow shocks immediately emerges from Fig. 7.5. Reasonably small

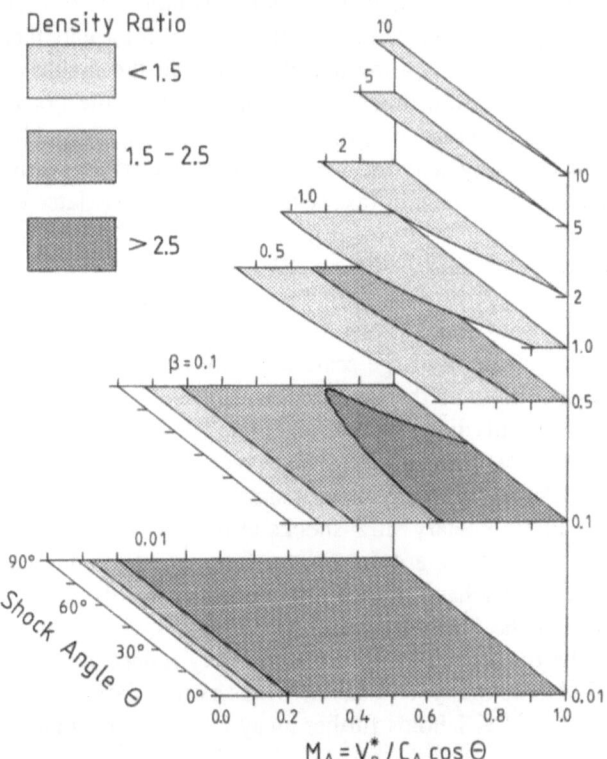

Density Ratio

	<1.5
	1.5 – 2.5
	>2.5

$M_A = V_n^* / C_A \cos \Theta$

Fig. 7.6. Results of a parametric study of the upstream "evolutionary conditions" and of the strength (defined by the downstream to the upstream density ratio) of slow shocks showing their dependence on β, the ratio of the total plasma (kinetic and magnetic field) pressure (see Fig. 7.5 for comparison). While β increases, slow shocks become weaker and the allowed area in the (M_A, θ) parameter space becomes smaller. M_A is the upstream Alfvénic Mach number ($M_A = V_n^*/C_A \cos \theta$) and θ is the angle between the shock-normal vector and the direction of the upstream magnetic field. At $\theta = 90°$ slow shocks become nonpropagating discontinuities, and for $M_A \geq 1$ slow shocks may turn into fast shocks [7.45, 44]

values of β are only achieved below about 0.4 AU in the intermediate and fast winds and only much closer to the sun in the low-speed wind. Thus, the chance to observe slow shocks over a wider range with respect to their parameters M_A and θ and/or with regard to a well documented increase in their plasma parameters across the shock will be extremely small at distances larger than about 0.4 AU from the sun!

The increase in β or, correspondingly, the shift in the Alfvénic Mach number M_A towards its upper limit $M_A = 1$ is mainly due to the rapid decrease in the Alfvén speed C_A for increasing radial distances from the sun. The velocity C_A can be as large as 800–1000 km s^{-1} in the corona, but it will decrease rather fast to about 100 km s^{-1} at around 0.3 AU and more slowly to about 40 km s^{-1} at 1 AU. Taking into account that only more oblique shock solutions are expected when β increases, the "evolutionary condition" $V_n^* \leq C_A \cos \theta$ will soon be

33

violated in the sense that the shock normal velocity will become larger than the projected Alfvén speed. This, of course, may also occur in interplanetary space owing to the longitudinal gradients in the density and/or in the magnetic field strength in the vicinity of corotating interaction regions. Once the above condition starts to be violated and provided $\beta \leq 2/\gamma$, slow shocks will gradually turn into fast shocks [7.44]. In this way, shocks, which may have started as slow shocks near the sun, will be transformed into fast shocks before being detected by a spacecraft near 0.3 AU or beyond.

7.3.3 Summary

In summary, it is fair to state that it is the large-scale behavior of the solar wind plasma parameters over sizeable parts of the inner heliosphere near 0.3 AU and beyond that places severe constraints both on the development of slow shocks by the steepening of low-frequency, slow-mode MHD waves and on the evolution and propagation of slow shocks in interplanetary space. In this respect, the deficiency of observations of interplanetary slow shocks at distances larger than 0.3 AU from the sun can be regarded as a quite natural fact, which is inherent to the solar wind itself. On the other hand, the results presented in this and the previous section clearly support the idea that closer to the sun, i.e., in regions below about 40 solar radii, the occurrence of slow shocks may indeed be fairly frequent. Certainly, this will also be the case in stellar winds even at larger radial distances, provided the condition $\beta \lesssim 1$ holds further away from the star than in case of the sun.

7.4 Constraints on the Unique Identification of Interplanetary Slow Shocks from Space Observations

This section will be devoted to the discussion regarding the difficulties in uniquely identifying interplanetary discontinuities as MHD slow shocks from high-time resolution solar wind measurements.

7.4.1 The Classical Way

Computer-aided programs for the identification of shocks from satellite measurements have been developed more or less exclusively for the investigation of fast-mode shock waves. Irrespective of their level of sophistication with respect to the set of Rankine–Hugoniot or shock jump conditions employed and the mathematical procedure used to optimize the (arbitrarily selected) upstream and downstream time averages of the various plasma parameters within their bounds of uncertainty to these conditions, they agree in view of the following main objectives: to determine the shock-normal direction and the shock velocity

34

as accurately as possible. In order to simplify the procedure, two rather severe restrictions have been imposed. First, to employ (besides the magnetic field observations) the solar wind proton measurements only, which have been the easiest to obtain in the past, and, second, to keep the set of shock jump conditions to be optimized as limited as possible. In the simplest approach the shock-normal direction is determined directly from one of the equations expressing coplanarity [7.10, 1] and the shock velocity from the equation expressing the conservation of the mass flux (see [7.40, 42] and references therein). Besides, a more sophisticated program has been developed [7.8, 23], where the determination of these two shock parameters is actually embedded into a closed loop routine within which the jump conditions expressing continuity of the normal mass flux, the tangential momentum flux, the tangential electric field, and the normal magnetic field across the discontinuity are optimized simultaneously. Recently, this type of procedure was further generalized by applying a faster and more unique optimization procedure for the underlying equations (see e.g. [7.41]), or by taking the time variations in the various plasma parameters generally associated with shock waves into account by incorporating a flexible, self-optimizing selection of their appropriate upstream and downstream time averages [7.17, 29].

In the more elaborate procedures the "evolutionary conditions" for shock waves [7.22] are also checked, once a solution to the shock model has been found. However, to do so, the corresponding magnetoacoustic velocity and thus the plasma sound speed, which crucially depends on the actual value of the electron temperature, must be determined. It should be stressed that in many cases measurements of the electron temperature were not available! This was, for example, the case when the first forward and reverse interplanetary slow shocks were identified in the *Mariner 5* and *Pioneer 6* solar wind data [7.9, 7].

7.4.2 Additional Conditions

As was indicated earlier [7.8], an optimum fit of the observed plasma parameters to the Rankine–Hugoniot conditions plus a positive test of the "evolutionary conditions" actually may not be *sufficient* for a unique identification of a slow shock in view of other MHD discontinuities, such as the tangential discontinuity, the reason being that across a tangential discontinuity the velocity, density, and temperature, and the magnetic field strength, may change (independent of one another) in an arbitrary way. This therefore also includes the changes characterizing slow shocks.

In order to cope with this ambiguity, a number of additional conditions have been proposed [7.8] which have to be fulfilled simultaneously, e.g. to assist in the identification of slow shocks. The most decisive conditions are (i) that the entropy S and therefore the plasma kinetic pressure p has to increase across a shock ($\Delta S > 0$ or $\Delta p > 0$), (ii) that the shock must be compressive with $0 < \Delta \varrho / \varrho_1 \leq 3$, (iii) that for the angular change of the magnetic field direction the condition $\sphericalangle (B_1, B_2) \leq 90°$ must hold, and (iv) that the magnetic field vector must be refracted towards the shock normal direction across a slow shock, or more precisely that $|\Delta B_T|B_1 \leq \sin \Theta$ must hold, where B_T is the magnitude of the tangential component of the magnetic field.

With the aid of some of these additional conditions, slow shocks were identified in the distant geomagnetic tail [7.12, 37], whereas others (e.g. [7.7, 8, 9]) also included a check of the Rankine–Hugoniot conditions and the "evolutionary conditions" in their identification of interplanetary slow shocks. For the slow shocks identified from the *Helios* observations within 1 AU of the sun, the jump conditions, the "evolutionary conditions", and the *complete* set of the above additional conditions have been verified [7.27]. And still, even such a grand and rather complicated procedure is by no means sufficient, and in view of the results to follow it may even be considered inferior.

7.4.3 Implications of Checking Against Tangential Discontinuities

Inherent to the fact that the plasma and magnetic field parameters may change in an arbitrary way across a tangential discontinuity (TD), really none of the above additional conditions can definitely exclude the possibility that a slow shock-type discontinuity may instead be a TD. Such a possibility can only be excluded if we explicitly show that the two and only conditions for TDs are violated. These two conditions are (i) that the total plasma (kinetic (p) plus magnetic field ($B^2/8\pi$)) pressure is conserved across a TD, i.e.,

$$\Delta(p_{\text{tot}}) := \Delta(p + B^2/8\pi) = 0 , \qquad (7.3)$$

and (ii) that there exists no normal component of the magnetic field across a TD, i.e.,

$$\boldsymbol{B}_1 \cdot \hat{\boldsymbol{n}} = 0 = \boldsymbol{B}_2 \cdot \hat{\boldsymbol{n}} \qquad (7.4)$$

(see, e.g., [7.4]).

In particular, the first condition places severe constraints with respect to the unique identification of slow shocks as well as on the completeness and accuracy of the plasma parameters that have to be available for both the upstream and downstream regions, the reason being that across a slow shock (in contrast to a fast shock) the plasma density and temperature, and thus the plasma kinetic pressure p, on the one hand, and the magnetic field magnitude B, on the other, change in opposite directions. This could imply that the total pressure p_{tot} might actually be conserved across such an event within the bounds of uncertainty in the individual data points. In turn, this would mean that such an event would be a slow shock as well as a TD. Thus, for the sake of uniqueness one must show that

$$\Delta(p_{\text{tot}}) \neq 0 . \qquad (7.5)$$

In view of the conservation equation of the normal momentum flux across a slow shock one must even show that

$$\Delta(p_{\text{tot}}) > 0 . \qquad (7.5')$$

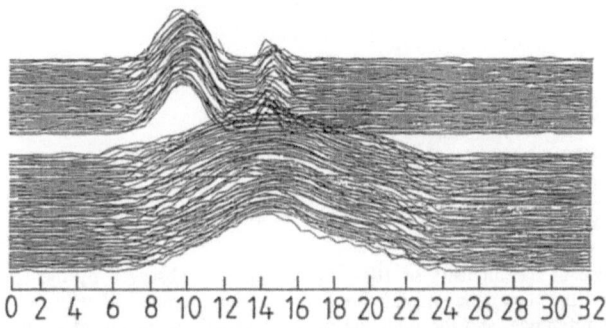

0 2 4 6 8 10 12 14 16 18 20 22 24 26 28 30 32

Fig. 7.7. Typical time histogram of ion distribution functions (intensity vs. energy-over-charge) as measured by the *Helios* ion electrostatic analyzer in front of (*upper part*) and behind (*lower part*) an interplanetary shock wave. In the cold, upstream solar wind the distributions for protons (*left*) and alpha particles (*right*) are well separated, which is no longer the case in the hot, downstream plasma. The channel numbers in energy-over-charge of the experiment are shown on the x-axis

However, in order to do so one has to realize that in the solar wind the total plasma kinetic pressure is

$$p = k\{N_{\mathrm{p}}(T_{\mathrm{p}} + T_{\mathrm{e}}) + N_{\alpha}(T_{\alpha} + 2T_{\mathrm{e}})\} , \tag{7.6}$$

where k is the Boltzman constant, N_{p} and N_{α} the proton and alpha-particle number densities, and T_{p}, T_{α}, and T_{e} the proton, alpha-particle, and electron temperatures, respectively. Thus, in view of (7.6) definite and realiable measurements of $(N_{\mathrm{p}}, T_{\mathrm{p}})$, (N_{α}, T_{α}), and T_{e} are mandatory both before and after the discontinuity. Normally, $(N_{\mathrm{p}}, T_{\mathrm{p}})$ and (N_{α}, T_{α}) are determined as the moment of the corresponding velocity distribution function. These, however, are obtained in parallel from one common ion energy-over-charge (E/Q) measurement (see Fig. 7.7). In the "cold" solar wind upstream of a shock (upper part) the velocity distributions for the protons (left) and for the alpha particles (right) are normally well separated, and therefore $(N_{\mathrm{p}}, T_{\mathrm{p}})$ and (N_{α}, T_{α}) can be determined quite accurately. However, downstream of a shock (lower part) where the plasma is hot the two distributions may overlap each other in such a way that the determination of $(N_{\mathrm{p}}, T_{\mathrm{p}})$ may imply severe ambiguities. For the majority of fast and slow shocks identified from the *Helios* observations a determination of the downstream values of (N_{α}, T_{α}) from the E/Q measurements was not possible. In these cases one therefore has to rely on other, more indirect arguments (e.g. [7.28]).

In case the set of upstream and downstream plasma parameters is not complete, at least two other possibilities exist to check against TDs involving (7.4) directly or indirectly. In the first case one assumes that the slow shock-type discontinuity is a TD and determines the direction of its normal by $\hat{n} := (B_1 \times B_2)/|B_1 \times B_2|$ (see [7.10]). If the proton bulk velocity V_{p} has been measured either upstream or downstream, one then calculates $(V_{\mathrm{p}} \cdot \hat{n})_1$ or $(V_{\mathrm{p}} \cdot \hat{n})_2$. If one (or both) of these values deviate significantly from zero, then the TD is excluded, as the normal mass flux across a TD for any plasma species must be zero. Such a scheme was applied earlier for fast shocks [7.16].

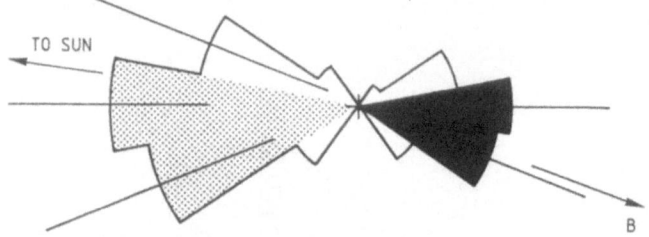

Fig. 7.8. Directional dependence of the intensity of 105–117 keV ions as observed by the *Helios* ion spectrometer directly behind the slow shock depicted in Fig. 7.1. Measurements have been transformed in the shock frame of reference, and the direction of the magnetic field projected into the plane of the ecliptic has been inserted. Notice the bidirectional streaming of these ions towards and away from the sun, thereby crossing the slow shock [7.33]

Sometimes, however, the bulk velocity cannot be determined very accurately or only the bulk speed is available. In these cases one can apply an indirect method by investigating the directional dependence of the electron distribution function and its suprathermal tail or of the flux of low energy ions directly in front of and behind the discontinuity: if a net flux of these particles is observed to cross the discontinuity, the TD is excluded in view of (7.4). In this way, TDs were excluded for the slow shocks observed in the geomagnetic tail from electron observations [7.13] and from low-energy ion observations in association with the slow shock depicted in Fig. 7.1 [7.33]. In Fig. 7.8 the angular distribution in the plane of the ecliptic and the shock frame of reference of the flux of 105–117 keV ions is shown, as measured by the *Helios* ion spectrometer directly behind the slow shock. A well-defined, bidirectional streaming of these ions along the magnetic field and both towards and away from the sun (thereby crossing the discontinuity) is observed.

7.4.4 Examples

From the set of interplanetary slow shock-type discontinuities observed by *Helios* we have selected five examples, in order to elucidate the points of discussion presented in the last three sections. These examples are numbered I and II, and 1, 2, and 3, and they are depicted in Figs. 7.9a and 7.9b, respectively. Obviously, all five discontinuities resemble the signature for forward slow shocks, as the proton bulk speed V_p, number density N_p, and temperature T_p increase while the magnetic field strength B decreases discontinuously across these events. Moreover, each of these events did pass the grand procedure, i.e., the optimum fitting

---

Fig. 7.9a,b. Time histograms of the solar wind proton bulk speed, density, temperature, and of the magnetic field strength across five forward slow shock-type discontinuities observed by *Helios 1* and 2 at different heliocentric radial distances (see Table 7.1). According to Table 7.1, the events labeled I and II are slow shocks, event 1 is a tangential discontinuity, and events 2 and 3 are non-MHD-type discontinuities associated with a *decrease* in the total plasma (kinetic plus magnetic field) pressure

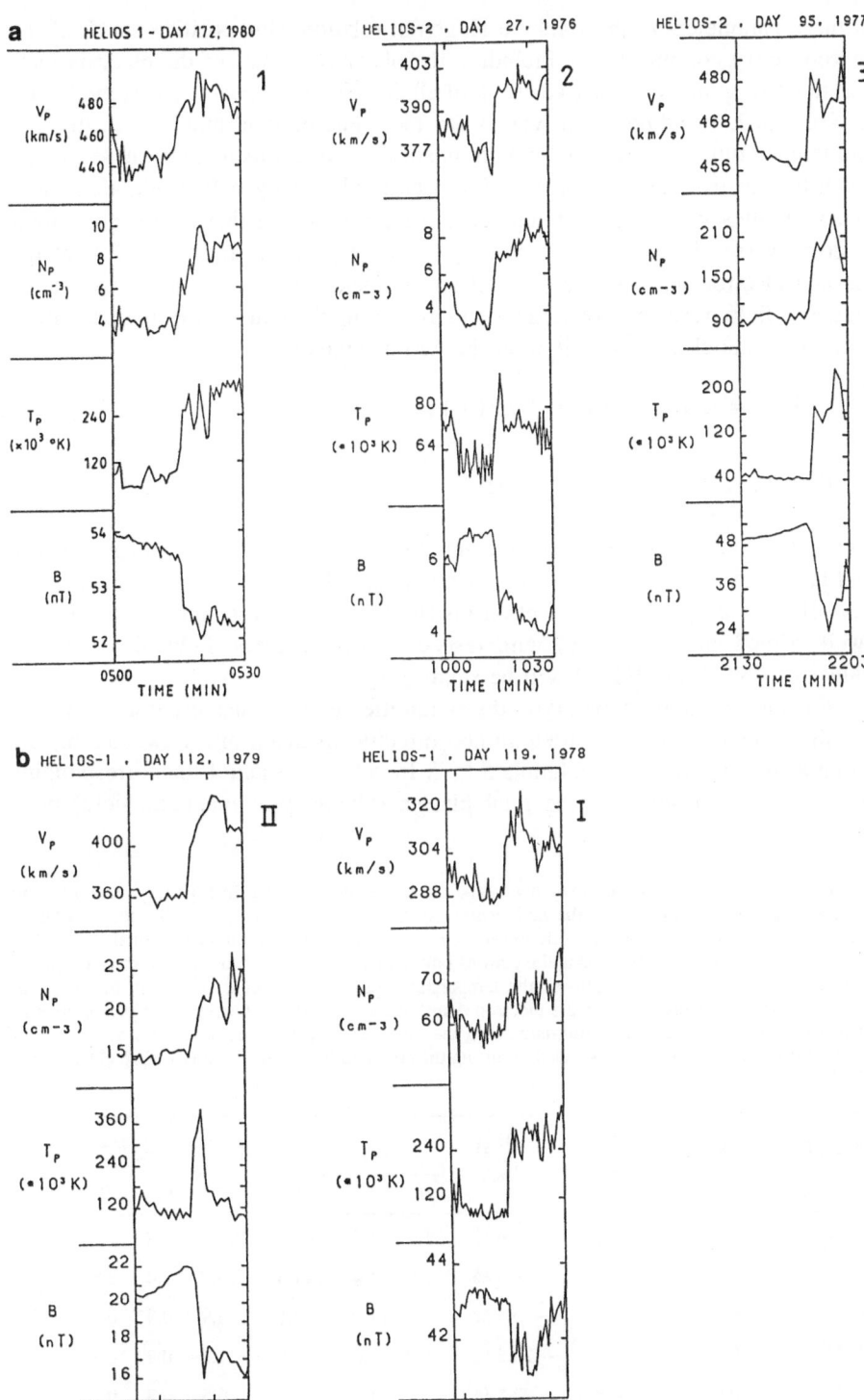

of these parameters to the Rankine-Hugoniot relations, the positive check of the "evolutionary conditions" by including the observed values of the electron temperatures $T_{e_{1,2}}$, and the positive check of all the additional conditions summarized in Sect. 7.4.2. In additon, the values for the alpha-particle number density N_α and temperature T_α are available both upstream and downstream of these events. From the electron distribution functions observed directly before and after these discontinuities various parameters have been determined that characterize their anisotropy (see [7.26]): first, regarding the core part, the ratio $T_{e\parallel}/T_{e\perp}$ of the thermal electron temperature parallel and perpendicular to the direction of the magnetic field, respectively; and second, regarding the halo part or the suprathermal tail of the electron distribution the two parameters

$$\text{EST 1} := \left[F(v_\parallel) - F(-v_\parallel)\right]/F(v_\parallel) , \tag{7.7}$$

$$\text{EST 2} := \int \left[(F - F_0)^2/F_0\right]d\Omega , \tag{7.8}$$

where F is the electron distribution function, F_0 the bi-Maxwellian for the core and halo, and $v_\parallel$ the velocity of the electrons parallel to the magnetic field direction. If the halo part of the electron distribution function is small and isotropic (well-defined and strahl-like) with respect to the magnetic field direction, the values of EST 1 and EST 2 will be small (large).

For the five slow shock-type discontinuities under consideration the values for these three parameters (determined directly upstream and downstream) are summarized in Table 7.1, columns 5 to 7. In columns 3 and 4 the corresponding values for the changes in the total plasma (kinetic plus magnetic field) pres-

Table 7.1. For the five forward slow-shock-type discontinuities depicted in Figs. 7.9a and 7.9b this table summarizes the following: the heliocentric distance of their observation (R); the changes in the total plasma (kinetic plus magnetic field) pressure across the discontinuities, normalized to their upstream values including protons and electrons only ($\Delta p_{tot}/p_{tot\,1}$) and including protons, electrons, and alphas ($\Delta p'_{tot}/p'_{tot\,1}$); the ratio of the temperatures parallel and perpendicular to the magnetic field direction of the core part of the electron distribution ($T_{e\parallel}/T_{e\perp}$); the skewness of the electron distribution in the antisolar direction and along the magnetic field (EST 1); and the deviation of the electron distribution from a bi-Maxwellian fitting the core and halo parts directly before (1) and after (2) the discontinuity, respectively.

Shock/Type	R	$\dfrac{\Delta p_{tot}}{p_{tot\,1}}$	$\dfrac{\Delta p'_{tot}}{p'_{tot\,1}}$	$T_{e\parallel}/T_{e\perp}$		EST 1		EST 2	
	[AU]			1	2	1	2	1	2
I (SF)	0.31	16.8 %	27.6 %	1.2	1.3	104.3	196.5	7.8	7.5
II (SF)	0.54	7.6 %	39.4 %	1.4	1.4	121.4	73.0	7.4	3.8
1 (TD)	0.53	− 0.5 %	− 0.4 %	1.0	1.0	3.8	3.9	0.3	0.3
2 (? ?)	0.97	5.4 %	− 30.3 %	1.0	1.0	1.9	1.2	0.3	0.1
3 (? ?)	0.47	− 25.4 %	− 22.8 %	1.11	1.1	0.0	0.5	0.3	0.7

sure normalized to its upstream value are listed by taking only the proton and electron parameters into account (column 3) and by considering the proton, electron *and* alpha-particle parameters (column 4), respectively. The corresponding heliocentric radial distances are provided in column 2.

Discontinuity no. 3 proves the existence of slow shock-type discontinuities across which the total plasma kinetic plus magnetic field pressure can *decrease*, i.e., across which $\Delta(B^2/8\pi)$ is larger than Δp. Thus, it really is mandatory to show that $\Delta(p_{tot}) > 0$ and not just that $\Delta(p_{tot}) \neq 0$ for a unique identification of slow shocks!

Discontinuity no. 2 indicates that it is also essential to include the measurements of (N_α, T_α) in the determination of p_{tot}. Without alpha particles we find that $\Delta p_{tot} > 0$; however, by including the alpha particles we obtain $\Delta p_{tot} < 0$ (the reason being that $\Delta N_\alpha < 0$). For slow shocks I and, in particular, II we find that an increase in p_{tot} is further emphasized if alpha-particle measurements are included.

Discontinuity no. 1 really supports the necessity for the discussions presented in this section. Recalling Subsect. 1 Sect. 7.2.1, this discontinuity occurs at the interaction boundary of a sub-Alfvénic with a super-Alfvénic solar wind stream (Figs. 7.2a,b). According to theory, it would therefore be a prime candidate for a slow shock. Applying the grand procedure outlined in Sects. 7.4.1 and 7.4.2, we find that all of its conditions are satisfied, and that the shock velocity is even as high as 241 km s^{-1}. And yet the check on the total plasma (kinetic plus magnetic field) pressure discloses that this pressure is actually conserved wether or not the alpha-particle measurements are included. Thus, this event is a TD and not a slow shock! Moreover, this example clearly proves that the entire procedure described in Sects. 7.4.1 and 7.4.2 is of little, if not of no, use at all with respect to a unique identification of interplanetary slow shocks.

Finally, by comparing the various parameters describing the anisotropy of the electron distribution function we find that their values are rather large and well preserved from the upstream to the downstream regions for the slow shocks (I and II), and much smaller and even close to unity for the nonslow-shocks (1, 2, and 3), respectively. Thus, these measurements provide a distinct indication of whether or not thermal and/or suprathermal electrons are able to propagate freely across a discontinuity or, equivalently, of whether or not the magnetic field has a nonzero normal component intersecting the plane of the discontinuity (in violation of (7.4)).

7.4.5 Summary

With respect to the characteristic changes in the plasma and magnetic field parameters, tangential discontinuities and slow shocks are in many respects alike. This ambiguity will not be removed even if one can show that the procedures (originally developed for the investigation of interplanetary fast shocks) to optimize the high-resolution plasma data within their bounds of uncertainty to the appropriate Rankine-Hugoniot conditions for slow shocks are applied success-

41

fully, and that the check of the "evolutionary conditions" for slow shocks and of other additional conditions summarized in Sect. 7.4.2 turn out to be positive. This also includes those cases in which large shock speeds are determined. For an unambiguous identification of slow shocks it is mandatory to show explicitly that the total plasma (kinetic plus magnetic field) pressure increases across the slow shock-type discontinuity. This, however, implies that reliable and accurate measurements of the number densities and temperatures of the protons *and* alpha particles, the electrons, and the magnetic field strengths both upstream and downstream must be available. The inclusion of alpha-particle measurements is absolutely necessary. Though in the undisturbed solar wind the kinetic pressure of the alphas may only amount to 10–20% of the kinetic pressure of the protons, this percentage could be much larger in the downstream regions of slow shocks. The necessity of providing alpha-particle densities and temperatures also downstream in a hot plasma environment places severe constraints on the reliability of the determination of the ion-plasma parameters and on the experimental techniques to separate the proton and alpha-particle velocity distribution functions.

The identification of slow shocks can be further assisted by demonstrating that thermal and/or suprathermal electrons and/or low-energy ions are able to propagate across the discontinuity in the shock frame of reference. Such cross-streaming is excluded in the case of a tangential discontinuity, as the magnetic field directions upstream and downstream are parallel to the plane of the discontinuity. Only slow (and fast) shocks have a nonzero normal magnetic field component which connects the upstream and the downstream regions.

7.5 Summary

The deficiency of slow shocks identified so far in the interplanetary medium has been one of the open problems in solar wind studies. The aim of this review was to investigate the reasons for this deficiency from the perspective of past and recent theoretical and observational studies. In Sect. 7.1 various source perturbations and interaction processes which are able to generate slow-mode turbulence and/or shocks near the sun and throughout interplanetary space were summarized. The richness of this collection indicates that the deficiency of interplanetary slow shocks is certainly not the cause of a deficiency of appropriate processes concerning their generation. In the second chapter various dynamical constraints were discussed regarding the evolution and steepening of slow-mode waves and the propagation of slow shocks in the solar wind. In view of the *Helios* observations, these constraints turned out to be extremely severe and obstructive and to extend over sizable parts of the heliosphere beyond 0.3 AU. Thus, we argued that the deficiency of slow shocks at distances larger than 0.3 AU can be regarded as being natural and inherent to the large-scale behavior of certain solar wind parameters. Within 0.4 AU, however, slow shocks should occur quite frequently. In Sect. 7.4 the observational constraints concerning the unique identification of slow shocks with respect to other MHD discontinuities were discussed. The necessity of providing the complete set of upstream and downstream solar wind plasma parameters (including alpha-particle densities and temperatures) further supported the fact that the number of "genuine" interplanetary slow shocks will remain very small, unless a new generation of plasma experiments is placed closer to the sun.

References

7.1 Abraham-Shrauner, B., Determination of magnetohydrodynamic shock normals, J. Geophys. Res., **77**, 736, 1972.
7.2 Barnes, A., Collisionless damping of hydromagnetic waves, Phys. Fluids, **9**, 1483, 1966.
7.3 Barnes, A., J.V. Hollweg, Large-amplitude hydromagnetic waves, J. Geophys. Res., **79**, 2302, 1974.
7.4 Boyd, T.J.M., J.J. Sanderson, *Plasma Dynamics*, Barnes and Noble, New York, 1969.
7.5 Burlaga, L.F., Hydromagnetic waves and discontinuities in the solar wind, Space Sci. Rev., **12**, 600, 1971.
7.6 Burlaga, L.F., Interplanetary stream interfaces, J. Geophys. Res.,**79**, 3717, 1974.
7.7 Burlaga, L.F., J.-K. Chao, Reverse and forward slow shocks in the solar wind, J. Geophys. Res., **76**, 7516, 1971.
7.8 Chao, J.-K., Interplanetary collisonless shock waves, Rep. CSR TR-70-3, Mass. Inst. of Technol. Cent. for Space Res., Cambridge, Mass., 1970.
7.9 Chao, J.-K., S. Olbert, Observation of slow shocks in interplanetary space, J. Geophys. Res., **75**, 6394, 1970.
7.10 Colburn, D., C.P. Sonett, Discontinuities in the solar wind, Space Sci. Rev., **5**, 439, 1966.
7.11 Edmiston, J.P., C.F. Kennel, A parametric study of slow shock: Rankine-Hugoniot solutions and critical Mach numbers, J. Geophys. Res., **91**, 1361, 1986.
7.12 Feldman, W.C., S.J. Schwartz, S.J. Bame, D.N. Baker, J. Birn, J.T. Gosling, E.W. Hones Jr., D.J. McComas, J.A. Slavin, E.J. Smith, R.D. Zwickl, Evidence for slow-mode shocks in the deep geomagnetic tail, Geophys. Res. Lett., **11**, 599, 1984.
7.13 Feldman, W.C., R.L. Tokar, J. Birn, E.W. Hones, Jr., S.J. Bame, C.T. Russell, Structure of a slow-mode shock observed in the plasma sheet boundary layer, J. Geophys. Res., **92**, 83, 1987.
7.14 Gosling, J.T., E. Hildner, R.M. McQueen, R.H. Munro, A.I. Poland, C.L. Ross, Direct observations of a flare related coronal and solar wind disturbance, Solar Phys., **40**, 439, 1975.
7.15 Hada, T., C.F. Kennel, Nonlinear evolution of slow waves in the solar wind, J. Geophys. Res., **90**, 531, 1985.
7.16 Hsieh, K.C., A.K. Richter, Generalized technique to estimate shock parameters from single-spacecraft observations, Rep. MPAE-W-79-84-44, Max-Planck-Institut für Aeronomie, Katlenburg-Lindau, Federal Republic of Germany, 1984.
7.17 Hsieh, K.C., A.K. Richter, The importance of being earnest about shock fitting, J. Geophys. Res., **91**, 4157, 1986.
7.18 Hundhausen, A.J., *Coronal Expansion and Solar Wind*, Springer-Verlag, New York, Berlin, Heidelberg, 1972.
7.19 Hundhausen, A.J., T.E. Holzer, B.C. Low, Do slow shocks precede some coronal mass ejections?, J. Geophys. Res., **92**, 11173, 1987.
7.20 Jeffrey, A., T. Tanuiti, *Non-linear Wave Propagation*, Academic Press, New York, 1964.
7.21 Johnstone, A., K. Glassmeier, M. Acuña, H. Borg, D. Bryant, A. Coates, V. Formisano, J. Heath, F. Mariani, G. Musmann, F. Neubauer, M. Thomsen, B. Wilken, J. Winningham, Waves in the magnetic field and solar wind flow outside the bow shock of Comet P/Halley, Astron. Astrophys., **187**, 47, 1987.
7.22 Kantrowitz, A., H.E. Petschek, MHD characteristics and shock waves, in *Plasma Physics in Theory and Application*, ed. by W.B. Kunkel, McGraw-Hill, New York, 1966.

7.23   Lepping, R.P., P.D. Argentiero, Single-spacecraft method of estimating shock normals, J. Geophys. Res., **76**, 4349, 1971.

7.24   Neubauer, F.M., Nonlinear interaction of discontinuities in the solar wind and the origin of slow shocks, J. Geophys. Res., **81**, 2248, 1976.

7.25   Petschek, H.E., Magnetic field annihilation, AAS-NASA Symposium on the Physics of Solar Flares, NASA Spec. Publ., SP-50, 425, 1964.

7.26   Pillip, W.C., H. Miegenrieder, M.D. Montgomery, K.-H. Mühlhäuser, H. Rosenbauer, R. Schwenn, Characteristics of electron velocity distribution functions in the solar wind derived from the Helios plasma experiment, J. Geophys. Res., **92**, 1075, 1987.

7.27   Richter A.K., K.C. Hsieh, A.H. Luttrell, E. Marsch, R. Schwenn, Review of interplanetary shock phenomena near and within 1 AU, paper presented at Chapman Conference on Collisionless Shock Waves in the Heliosphere, AGU, Napa Valley, Calif., 1984.

7.28   Richter, A.K., H. Rosenbauer, F.M. Neubauer, N.G. Ptitsyna, Solar wind observations associated with a slow forward shock wave at 0.31 AU, J. Geophys. Res., **90**, 7581, 1985.

7.29   Richter, A.K., K.C. Hsieh, H. Rosenbauer, F.M. Neubauer, Parallel fast-forward shock waves within 1 AU: Helios-1 and -2 observations, Annales Geophysicae, **4**, 3, 1986.

7.30   Richter, A.K., A.H. Luttrell, Superposed epoch analysis of CIR's at 0.3 and 1.0 AU: A comparative study, J. Geophys. Res., **91**, 5873, 1986.

7.31   Richter A.K., A.H. Luttrell, Evidence of slow-mode MHD turbulence in the solar wind: Post slow shock observations at 0.31 AU, J. Geophys. Res., **92**, 13653, 1987.

7.32   Richter A.K., E. Marsch, Helios observational constraints on the development of interplanetary slow shocks, Annales Geophysicae, **6**, 319, 1988.

7.33   Richter, A.K., K.C. Hsieh, Evidence for gradient and possibly curvature drifts at a slow shock: Helios observations at 0.31 AU, Rep. MPAE-W-79-88-36, Max-Planck-Institut für Aeronomie, Katlenburg-Lindau, Federal Republic of Germany, 1984.

7.34   Rosenau, P., S.T. Suess, Slow shocks in interplanetary medium, J. Geophys. Res., **82**, 3643, 1977.

7.35   Schwenn R., Relationship of coronal transients to interplanetary shocks: 3D aspects, Space Sci. Rev., **44**, 139, 1986.

7.36   Smith, E.J., J.H. Wolfe, Observations of interaction region and corotating shocks between one and five AU: Pioneer 10 and 11, Geophys. Res. Lett., **3**, 137, 1976.

7.37   Smith, E.J., J.A. Slavin, B.T. Tsurutani, W.C. Feldman, S.J. Bame, Slow-mode shocks in the earth's magnetotail: ISEE-3, Geophys. Res. Lett., **11**, 1054, 1984.

7.38   Sonett, C.D., D.S. Colburn, L. Davis, Jr., E.J. Smith, P.J. Coleman, Jr., Evidence for a collision-free magnetohydrodynamic shock in interplanetary space, Phys. Rev. Lett., **13**, 153, 1964.

7.39   Swift, D.W., On the structure of the magnetic slow switch-off shock, J. Geophys. Res., **88**, 5685, 1983.

7.40   Tsurutani, B.T., R.P. Lin, Acceleration of > 47-keV ions and > 2-keV electrons by interplanetary shocks at 1 AU, J. Geophys. Res., **90**, 1, 1985.

7.41   Viñas, A.F., J.D. Scudder, Fast and optimal solution to the "Rankine-Hugoniot problem", J. Geophys. Res., **91**, 39, 1986.

7.42   Volkmer, P.M., F.M. Neubauer, Statistical properties of fast magnetoacoustic shock waves in the solar wind between 0.3 and 1 AU: Helios-1 and -2 observations, Annales Geophysicae, **3**, 1, 1985.

7.43   Whang, Y.C., Slow shocks around the sun, Geophys. Res. Lett., **9**, 1081, 1982.

7.44   Whang, Y.C., Slow shocks and their transition to fast shocks in the inner solar wind, J. Geophys. Res., **92**, 4349, 1987.

7.45   Whang, Y.C., Evolution of CME associated slow shocks, presented at the Sixth International Solar Wind Conference, Estes Park, Colorado, USA, 23-28 August, 1987.

7.46   Winske, D., E.K. Stover, S.P. Gary, The structure and evolution of slow-mode shocks, Geophys. Res. Lett., **12**, 295, 1985.

8. Kinetic Physics of the Solar Wind Plasma

In partial fulfilment of the requirements for the venia legendi in astronomy and astrophysics at Göttingen University

Eckart Marsch

8.1 Introduction

8.1.1 General Remarks

The interplanetary medium has been continuously explored for more than two decades and substantial progress has been made, both with regard to *in situ* measurements and a theoretical understanding of the solar wind. Concurrently, measurement techniques for remote sensing of its source regions in the solar corona by means of photons covering the full electromagnetic spectrum from γ-rays to radar have also become mature. The results obtained have greatly improved and corroborated our knowledge about the solar wind from the very coronal base to several solar radii and further out into interplanetary space (see the reviews [8.15, 66–68, 76, 109, 119, 120, 126, 131, 134, 215, 245] and [8.22, 23, 26, 286, 288]). However, even after completion of the *Helios* mission there remains an important but poorly explored region in the inner heliosphere below 0.3 AU to be investigated and, of course, the wide space out of the ecliptic plane and above the solar poles. Whereas a solar probe, the feasibility of which has already been demonstrated [8.239], may possibly be realized in the far future, the out-of-ecliptic mission [8.199] is soon to become reality and will certainly help to comprehend better the three-dimensional structure of the heliosphere.

In spite of all these advances, the basic mechanisms for accelerating fast and slow solar wind streams are as yet not fully understood. Certainly, a hot corona ($T \approx 10^6$K) surrounding the sun necessarily leads to supersonic solar wind expansion, as was demonstrated early on within polytropic and other fluid models by the pioneering work of Parker [8.226–229] and also by exospheric kinetic models [8.144, 163]. However, neither the microphysical processes (wave–particle interactions and Coulomb collisions) and the associated transport, governing the actual radial evolution of the internal energy of the multispecies solar wind plasma and the underlying velocity distributions, are well understood; nor do we know the detailed state of the coronal plasma and its flow in association with magnetic field structures on various spatial scales. High-speed streams are now known to originate from coronal holes [8.35, 133, 218, 258, 262]. Low-speed streams appear to come from coronal streamers [8.34, 107, 213] and closed-field structures, but how they form the interplanetary slow solar wind remains obscure. Other main questions concern the total energy input into the solar wind, the various energy fluxes in the corona, and the mechanism by which all the energy ultimately resides in the bulk motion of the supersonic wind. The purpose of this article is to review more recent work related to the fluid aspects and kinetic physics of the solar wind plasma, to emphasize firm results obtained so far, to summarize pertinent open questions, and to indicate unsolved problems.

8.1.2 The Importance of Kinetic Physics

The extensive reviews in [8.23, 38, 176, 257] deal with the large-scale structure of the inner heliosphere, in particular the solar wind stream structure in association with the coronal magnetic field and the resultant interplanetary sectors, with the global invariants of the flow during the solar cycle, and the average angular and radial profiles of solar wind parameters. This fluid-like description of the solar wind plasma has been surprisingly successful in view of the fact that at several solar radii collisions between particles become so rare that a simple fluid description is expected to fail [8.15, 67, 84, 108, 131, 151, 173, 249, 267]. For example, it was predicted and then verified observationally that, as a result of low collision rates in the solar wind, pressure or temperature anisotropies evolve (though not as predicted), or that particle velocity distributions become skewed and develop tails and heat fluxes along the local magnetic field direction. In fact, in order to account fully for these observed microscopic details of the solar wind plasma a kinetic description is certainly necessary.

Nonthermal characteristics and velocity-space anisotropies are permanently present. However, they do not grow indefinitely, but tend to regulate themselves since they contain sufficient free energy to drive the plasma until it becomes microscopically unstable, thus supplying various sources for the excitation of high-frequency electromagnetic waves and electrostatic noise. These self-excited waves and microturbulences are believed to provide the main dissipation mechanism required to maintain a fluid-like behavior of the plasma in spite of the fact that collisional free paths become comparable with the large-scale dimensions of the inner heliosphere. Studying kinetic processes in the solar wind and the related binary collisions and plasma microinstabilities is therefore motivated by several things. As outlined above, to understand the "fluid" solar wind is an obvious incentive.

Another more direct motivation stems from the pleasing fact that we are able to measure the full and detailed features required to describe the plasma on the kinetic level; namely, the entire three-dimensional velocity distributions of electrons, protons, and helium ions as the major species and the associated electromagnetic fluctuations on all relevant kinetic scales. The solar wind as an anisotropic, multispecies, micro-turbulent, weakly collisional medium provides a unique testing ground for many theoretical concepts of basic plasma physics. For example, measuring distribution functions with unprecedented time resolution, sufficient even to trace the evolution of nongyrotropic features and the response of the particles to large amplitude MHD turbulence existing in the solar wind, is one of the many possibilities for doing basic plasma research in the interplanetary medium.

The adequate description of the solar wind plasma also demands the development of a new nonclassical transport paradigm which may then perhaps be applied and transferred to the coronal and turbulent envelopes and associated winds of other sun-type stars. This is another major reason for analyzing solar wind plasma properties in depth, besides the obvious good reason that many interesting plasma phenomena deserve to be studied in their own right with all the physical tools available.

8.1.3 Outline of the Article and Previous Reviews

The philosophy and organization of the following presentation of kinetic and fluid physics of the solar wind is as follows. Throughout the whole article the discussion of new experimental results from *Helios* and of the phenomenology of velocity distributions and derived moments will be in the foreground. The intention is to place emphasis on the basic concepts and to verbalize the relevant physics without relying on extensive mathematical formulations and theory. For more detailed elaborations of the many topics covered here we must refer the reader to the literature. Several reviews on kinetic aspects of solar wind physics are available which provide valuable overviews of the relevant plasma parameters and their variations (as known from 1 AU). These papers also contain tables of the typical spatial and temporal kinetic scales and textbook-like outlines of the fluid or moment and of the Vlasov or kinetic equations needed for the description of the solar wind plasma on different levels of theoretical sophistication. Without intending completeness we quote here some of these general papers or books, which together are to a certain extent prerequisite to following the subsequent theoretical issues and empirical material.

The microscopic structure of the solar wind was first reviewed comprehensively by Scarf [8.244]. This review has already addressed some of the main problems and subjects of debate and research which have been with us ever since. It is a collection of theoretical plasma-physics concepts, a report on the experimental data known at that time, and a discussion of interplanetary wave–particle interactions. Several later reviews by Feldman et al. [8.66, 67, 70, 76] contain extensive discussions of the observed ion velocity distributions, electron characteristics, transport features and mechanisms for electron heat-flux regulation, and helpful parameter tables compiled by means of *in situ* plasma observations made in the earth's orbit. Another detailed account of observations and ideas concerning the role of kinetic plasma processes in the solar wind is given in [8.50], reflecting the knowledge before the *Helios* epoch. Our paper may be partly considered a continuation of all these works, as it reviews critically the observational evidence that exists after more than ten years of *Helios* measurements, on those plasma characteristics that surely need to be explained within the framework of kinetic theory.

Several reviews [8.15, 32, 119] on hydromagnetic waves and turbulence exist which provide excellent introductions and overviews of solar wind magnetohydromagnetic turbulence and kinetic waves and fluctuations, and which assess their relevance for the non-Maxwellian features of the particle velocity distributions.

The theoretical review written by Schwartz [8.250] on plasma microinstabilities gives a comprehensive and still very much up-to-date compilation of instabilities in the context of linear plasma theory as well as of nonlinear phenomena, which are believed to be relevant to our understanding of solar wind kinetic physics. For a more general theoretical account of instabilities in space plasma we recommend Melrose's recent book [8.203] and the basic references therein. The relevant kinetic and MHD-wave observations made on the *Helios* spacecraft are reviewed and can be found in [8.110, 182].

The present work starts with a phenomenological survey of particle velocity distributions and then addresses the role of Coulomb scattering and nonclassical collisional transport. Subsequently, wave–particle interactions and the associated anomalous transport are discussed, and then the nonequilibrium thermodynamics resulting from these processes is examined in the light of the observed radial gradients of thermal solar wind parameters. The dissipation of MHD turbulence and the associated interplanetary heating is briefly examined. Finally, the energy input into the solar wind and the constants of motion associated with the fluxes of mass, energy, and momentum are discussed. The paper finishes with a short summary of the conclusions and an outline of future research perspectives.

8.2 Particle Velocity Distributions

8.2.1 Introduction

After more than twenty years of *in situ* observations of ion and electron velocity distributions a fairly complete, but also rather complex, picture of the interplanetary plasma state has emerged. The data acquired by many missions have revealed that the solar wind is highly structured in space and time and correspondingly variable on multiple kinetic scales [8.32, 109, 157, 245, 279]. The underlying velocity distributions, as a rule, deviate strongly from Maxwellians [8.70, 72, 135, 185, 186, 232–234, 241] as the particles must react to the average global forces and the small-scale fluctuating interplanetary electromagnetic fields. Coulomb collisions are usually not sufficient to establish thermodynamic equilibrium [8.214, 215]. Nevertheless, they play a role, and collisional properties vary considerably in association with the stream structure [8.171, 191, 214].

The spatial inhomogeneity of the solar wind, the variability of its coronal boundaries and plasma sources, and the complex topology of the solar and interplanetary magnetic field manifest themselves in a variety of nonthermal features of the measured velocity distributions, which are mediated and shaped by many dissipative processes and wave–particle interactions. As a result of this complexity, transport coefficients and the role of plasma microinstabilities are notoriously difficult to evaluate [8.50, 55, 177, 178, 222, 244, 250]; consequently the radial evolution of solar wind internal energy remains one of the major issues of solar wind physics that is still not fully comprehended.

This section presents a phenomenological survey of proton, alpha-particle and electron velocity distributions as measured under various plasma conditions and in different streams [8.185, 186, 232–234, 241]. Emphasis is placed on the inner heliosphere since the data, mostly pertaining to 1 AU and beyond, of other, previous missions are well documented in the ample scientific literature. It will become apparent that the *Helios* mission has considerably enlarged our knowledge and understanding of the theoretically very challenging and puzzling microscopic properties of the solar wind plasma.

8.2.2 Ion Velocity Distributions

The morphology of proton distributions is illustrated in Fig. 8.1 by iso-density contours in velocity space [8.186]. Among the most typical features are: a marked and persistent skewness, related to nonthermal tails or even secondary peaks along the magnetic field direction (dashed lines), and the pronounced total temperature anisotropies with $T_\perp > T_\parallel$ and anisotropic cores in high-speed velocity distributions (right) and with $T_\parallel > T_\perp$ and isotropic cores (left column) in low-speed velocity distributions. On the average, the proton temperature increases on approaching the sun (see Sect. 8.5 and [8.186, 254]). The drift speed V_D of the secondary proton peak varies typically between 1 and 1.5 times the local Alfvén speed V_A. At intermediate solar wind flow speeds, ranging between 400 and 600 km s^{-1}, an anisotropy with $T_\parallel > T_\perp$ is observed, very frequently in close connection with a time-variable proton heat-flux tail [8.70, 135, 246]. The relevant proton parameters of the distributions shown in Fig. 8.1 are put together in the Table 8.1. The double-peak phenomenon also appears in alpha-particle distributions [8.7, 185, 276], often with drifts again of the order of V_A as shown in Fig. 8.2.

All these nonthermal features of ion velocity distributions appear to be correlated with the large-scale solar wind streams and the related interplanetary magnetic field structure and with small-scale inhomogeneities of the bulk flow. Clearly resolved double-peaked proton distributions are often found in the rarefaction regions of high-speed streams [8.68, 70]. Most frequently they occur, however, throughout the body of fast streams [8.186], where the largest values of $T_\perp/T_\parallel$ are also found. The temperature signature $T_\perp > T_\parallel$ was first observed at 1 AU [8.8] and interpreted as evidence for local heating. In particular, at the *Helios* perihelion near 0.3 AU values of 3–4 for $T_\perp/T_\parallel$ are quite common and long lasting. This observation is one of the strongest pieces of direct evidence for the occurrence of wave–particle interactions in the interplanetary medium.

In addition to these morphological features, a persistent differential streaming between the two ionic species is often observed [8.185, 220, 276], in particular in fast solar wind streams. Figure 8.3 illustrates this behavior, showing from top to bottom: the proton speed (dots indicate alpha-particle speeds); the ion differential speed and (indicated by dots) the Alfvén speed; the scalar product between magnetic field B and $\Delta V_{\alpha p}$; and the azimuthal and elevation angles of the magnetic field vector and of the proton and alpha-particle velocities, re-

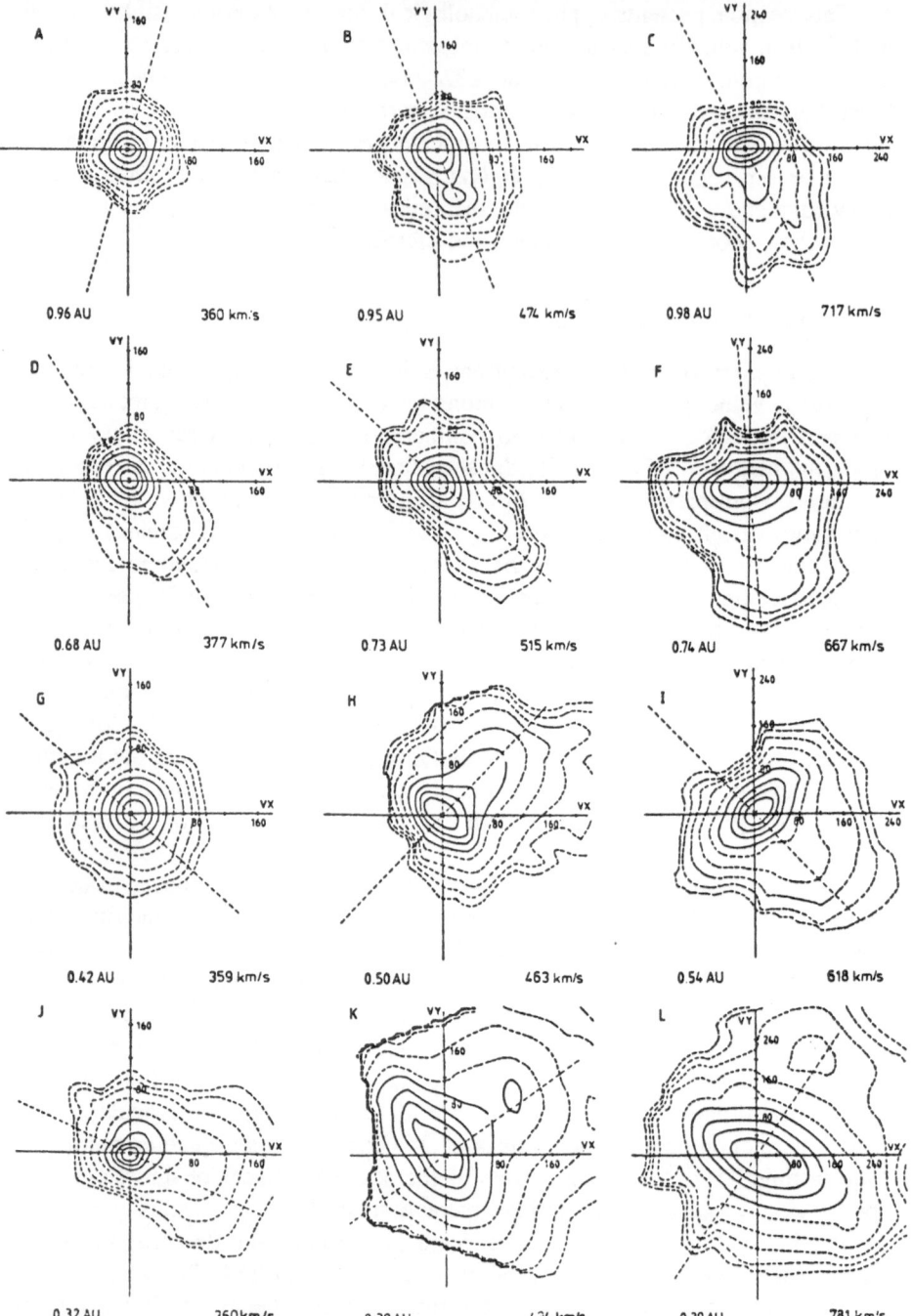

Fig. 8.1. Survey of proton velocity distribution functions as measured by the *Helios* plasma experiment for different solar wind speeds (increasing from *left-* to *right-hand side*) and heliocentric distances (decreasing from *top* to *bottom*). The cuts through the distributions are in the plane defined by the flow velocity (*VX*-axis) and magnetic field vector (*dashed line*, axis of gyrotropy). Isodensity contours correspond to fractions of 0.8, 0.6, 0.4, 0.2, and of 0.1, 0.03, 0.01, 0.003, 0.001 (*dashed contours*), respectively. Scales are in km s^{-1} and the origin of velocity space relates to the maximum phase-space density [8.186]

Table 8.1. Solar Wind Parameters for Fig. 8.1 [8.186]

Day 1976	Time UT	R AU	V_p km s^{-1}	n_p cm^{-3}	$\lvert Q_p \rvert$ 10^{-4} erg cm^{-2}s^{-1}	V_D km s^{-1}	$T_{p\parallel}$ 10^{5}K	$T_{p\perp}$ 10^{5}K	B 10^{-5}G	α_B deg	ε_B deg	$f(\upsilon_M)$ 10^{-20}cm^{-6}s^3	Letter
30	1634:21	0.964	360	17.3	0.20	—	0.48	0.39	8.2	71.7	7.2	18.93	A
35	1021:43	0.949	479	6.7	1.83	67.7	1.81	1.12	6.8	-67.6	21.1	1.33	B
23	1150:54	0.978	717	2.1	5.17	—	3.43	2.62	4.7	134.3	45.3	0.34	C
73	2134:54	0.681	377	20.6	5.08	—	0.94	0.36	12.7	124.3	6.8	22.80	D
71	1952:25	0.703	515	4.4	4.15	—	2.84	0.81	9.4	-39.9	-2.8	1.87	E
67	2232:33	0.742	667	5.6	9.35	—	3.13	3.30	9.4	-80.5	44.2	0.30	F
122	0157:09	0.421	359	139.6	2.08	—	0.75	0.80	17.6	145.8	34.4	59.43	G
88	0351:25	0.504	463	21.2	45.15	—	4.15	1.92	17.8	36.3	17.2	2.75	H
85	1132.18	0.546	618	10.4	15.00	—	2.67	2.88	17.2	-16.5	38.8	0.96	I
102	0933:49	0.319	360	129.0	111.90	—	2.23	0.88	33.5	156.1	-15.9	70.92	J
119	2114:15	0.391	494	26.2	81.15	122	5.23	4.10	28.0	-166.5	27.6	0.97	K
107	0750:54	0.291	781	28.3	118.60	242	5.66	9.67	44.9	-141.5	-21.0	0.43	L

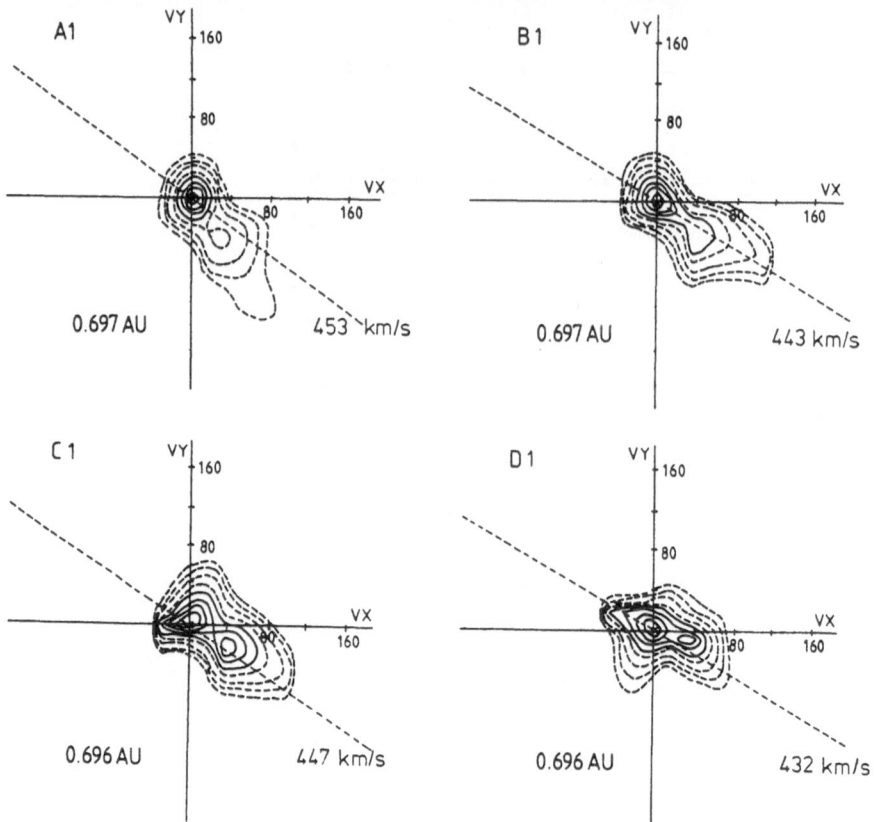

Fig. 8.2. Helium-ion double-peak distributions as measured by *Helios 2* in slow solar wind at speeds of 430–450 km s^{-1} near 0.7 AU. The resolved second component is drifting along the magnetic field direction (*dashed line*) with about the Alfvén speed of 60 km s^{-1}. Associated with this He^{2+} beam is a marked temperature anisotropy of $T_{\|\alpha} \approx 2\, T_{\perp\alpha}$ [8.185]

spectively. Note the fact that protons participate strongly in the Alfvénic wave motion, which is clearly discernible by a close correlation of the angular fluctuations. It appears as if protons carry the waves whereas alpha particles tend to "surf" or ride the waves. Their velocity reveals much smaller angular variations. Alpha particles seem to stream radially away from the sun while disregarding the ambient Alfvénic turbulence (also indicated by the constancy of their speed profile in the top panel). The last panel documents the field-alignment of the differential motion. In contrast to these observations, the low-speed regions are usually characterized by alpha particles lagging slightly behind the protons or by very small speed differences [8.215]. The whole subject of ion differential streaming is at present still not well understood [8.138]. Some of the existing theoretical explanations are discussed in the subsequent sections.

Striking differences also appear between the ion temperature anisotropies. Figure 8.4 shows twelve days with the data (20-min averages) from the first *Helios 2* perihelion passage at 0.3 AU in a broad high-speed stream [8.186]. The

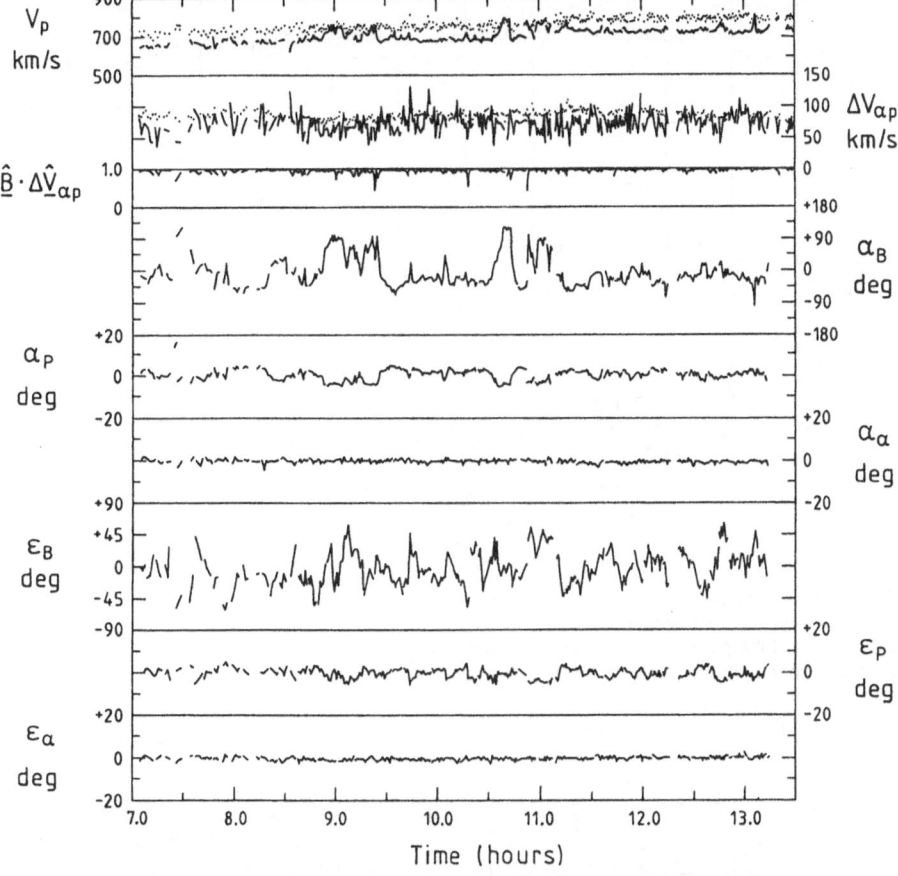

Fig. 8.3. Ion differential streaming, Alfvén wave activity, and alpha-particle surfing. Six hours of *Helios 2* data measured at day 66 in 1976 at 0.75 AU. Reading from above, the proton (*dots*: the alpha-particle) mean speed, the ion differential speed $| \Delta V_{\alpha p} |$ (*dots*: the Alfvén speed), the dot product between B and $\Delta V_{\alpha p}$, and the azimuthal and elevation angles of B, and of the mean velocities of protons and alpha particles with respect to the solar-ecliptic coordinate system are shown [8.185]

overall temperature anisotropies are indicated by continuous traces and the core anisotropy by isolated points. The sizable proton velocity fluctuations observed are mainly due to Alfvén waves but might also be addressed partly to various neighboring flux tubes with different expansion speeds [8.272]. Large fluctuations occur in $T_{\|p}/T_{\perp p}$ in close association with the highly time-variable proton heat flux [8.69, 70, 73, 135, 186]. However, the core (above the 10% level) anisotropy is fairly stable ranging at values from 0.25 to 0.3. In contrast, the alpha-particle total-temperature anisotropy fluctuates about 1, while their core anisotropy is larger than 1 on the average. These results suggest that the processes shaping ion distributions act selectively, e.g. they are dependent on the charge-per-mass ratio, and therefore affect the various species in different ways. Tentative explanations for the temperature regulation mechanism are offered in Sect. 8.4.

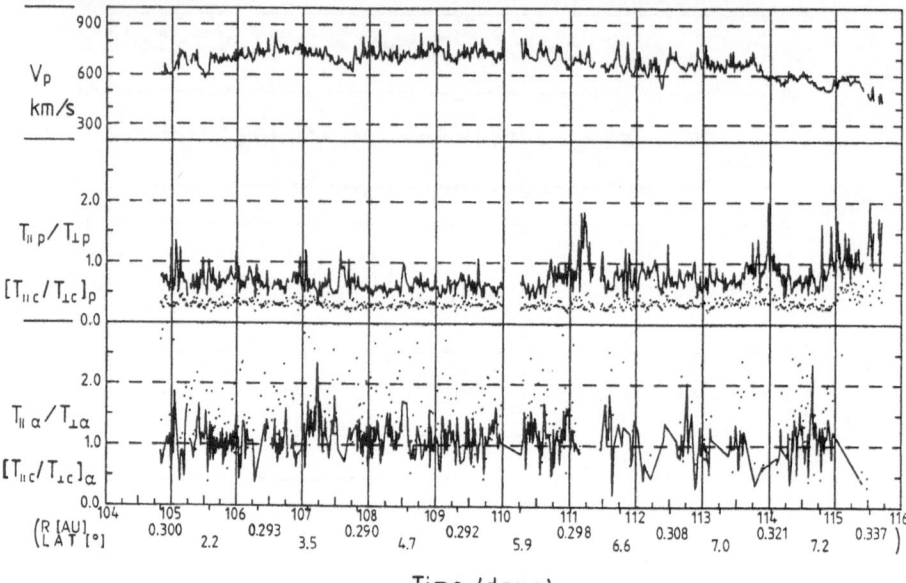

Fig. 8.4. Half hour averages of *Helios 2* ion parameters as observed near perihelion (0.29 AU) in a very fast solar wind stream. Proton bulk speed and total (*dots*: core) temperature anisotropy of the protons and alpha particles are shown for twelve successive days. Note the opposite values of the ion core temperature anisotropies [8.186]

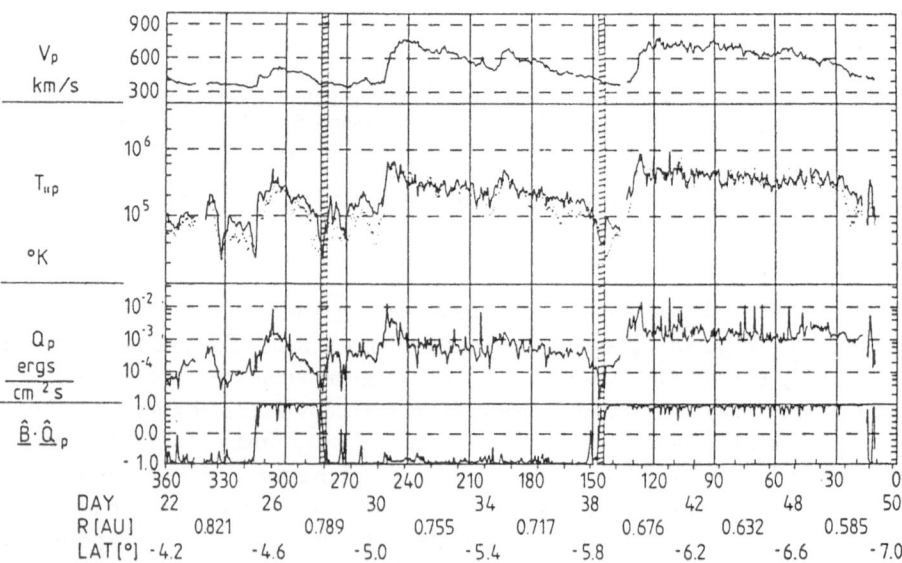

Fig. 8.5. One-hour averages of the proton bulk speed, of the parallel (*dots*: perpendicular) temperature and heat flux, and of the dot product between B and Q_p. A full Carrington rotation corresponding to radial distances between 0.82 and 0.58 is displayed. The two *shaded stripes* correspond to crossings of the heliospheric current sheet [8.186]

A survey of proton bulk speed, temperature, and heat flux is provided in Fig. 8.5, which shows data of a complete solar rotation plotted versus Carrington longitude. The bottom trace proves that the proton heat flux is aligned with and parallel or antiparallel to the magnetic field depending on the sector polarity. The largest values of the temperatures and heat fluxes are observed in the recurrent high-speed streams. The sizable dips occurring in the temperature profiles correspond to the very cold plasma that is usually measured right at a magnetic sector boundary. The isotropic distributions A and G in Fig. 8.1 refer to such regions. The low-speed environment, in which sector boundaries are normally embedded, is characterized by marked temperature anisotropies, with $T_\parallel > T_\perp$, arising from double-peaks or extended tails in the proton distributions (letters B, E, H, J in Fig. 8.1). Finally, one may see the increase of the proton temperature and heat flux while the spacecraft approaches the sun from 0.8 AU to about 0.55 AU. These temperature gradients are discussed in more detail below.

8.2.3 Electron Velocity Distributions

We now turn our attention to the electron distribution functions. The relevance of electrons for heliospheric plasma physics arises for several reasons. Firstly, the electron heat flux is assumed to be the dominant energy flux in the corona for a thermally driven solar wind [8.126, 131, 222, 261]. Secondly, the more energetic electrons may serve well as test particles exploring the interplanetary magnetic field topology [8.231, 234] and probing the solar wind stream structure [8.74, 78, 132, 136] and associated nonsteady phenomena. The electron temperature in the tail of the distribution as measured *in situ* may still be reminiscent of the coronal temperature in the plasma source regions [8.233, 260]. And finally, as will be discussed in Sect. 8.4, the detailed electron distribution is indispensible in analyzing plasma microinstabilities and transport [8.250].

Basically, the electron distribution can be considered to consist of a core part and a hotter halo, which experimentally can be fitted approximately by two slightly displaced Maxwellians of different temperatures and densities [8.72, 208, 232]. The salient features of these two components [8.72] are the following. Within experimental errors one finds the zero current condition, $n_C V_C + n_H V_H = 0$ in the solar wind frame, to hold for the core and halo drift speeds along the magnetic field and for their densities. The density ratio is variable with an average of $n_H/n_C \approx 0.05$. The core temperature T_C and the halo temperature T_H are correlated with an average ratio of $T_H/T_C \approx 6$ at 1 AU. Finally, the core drift speed V_C seems to be closely correlated with the local Alfvén speed [8.74] and to be enhanced in high-speed solar wind regions. However, here the core–halo model tends to become a poorer characterization of the true distribution.

The reason this model breaks down is that in addition a narrow magnetic-field-aligned "strahl" is frequently observed [8.230, 233, 241]. The resulting skewness of the distribution is highly variable and can be very distinct. The phase-space density at energies typically beyond 100 eV is often by orders of magnitude

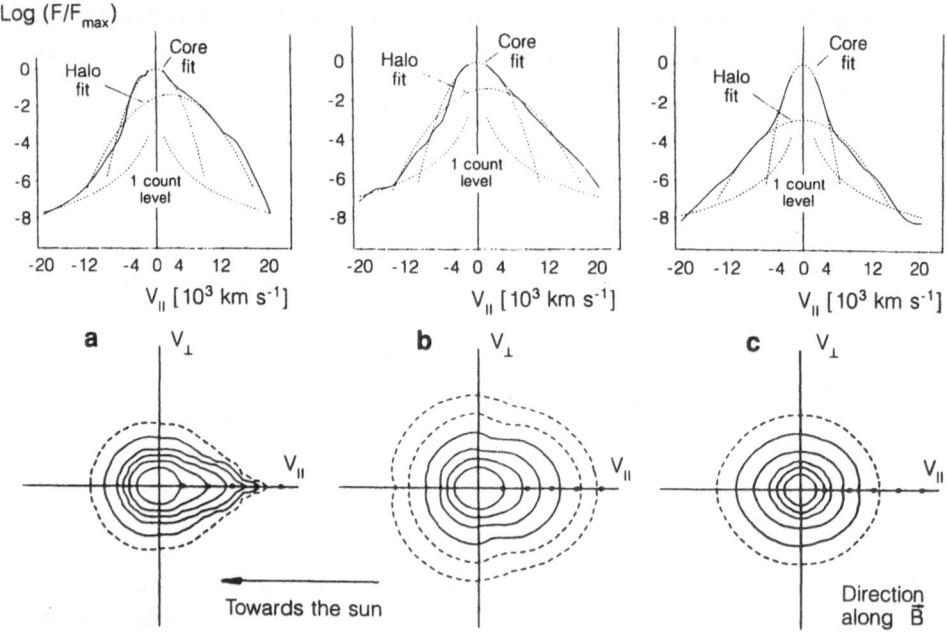

Fig. 8.6a–c. Electron distribution functions with high angular resolution constructed from data of several subsequent measurement cycles. Contours correspond to fractions of 10^{-1}, 10^{-2},..., 10^{-6} of the maximum phase-space density corresponding to the origin of velocity space with scales as indicated. Three typical examples are shown: with a narrow strahl (a), a broad strahl (b), and an isotropic distribution (c) [8.233]

smaller in the sunward direction than in the direction away from the sun. The strahl may at times narrow down to only several degrees in width at energies above 600 eV [8.232]. For such strongly anisotropic electron distributions the major part of the heat flux is carried directly by the strahl. For less skewed distributions the remaining small heat flux then arises mostly from a weak drift between core and halo. The location of maximum phase-space density (core peak) is to a varying degree shifted relative to the solar wind bulk velocity (i.e. the ion center-of-mass frame) in qualitative agreement with predictions of exospheric theory [8.118, 144, 163]. However, the *Helios* observational values of this core drift (and of the related halo drift) are definitely smaller than these theoretical predictions, which indicates the existence of collisional coupling and/or other kinds of friction between ions and electrons [8.72, 232–234].

In Fig. 8.6 one-dimensional cuts parallel to the magnetic field direction (coincident with the strahl direction or axis of gyrotropy) through differently skewed electron distributions are shown. The related isodensity contour plots are also shown as a function of the corresponding velocity component. The core–halo structure and the strahl are readily apparent. Dashed parabolas represent bi-Maxwellian fits to the core and halo, respectively. The significant break in the slope of the symmetric distribution (c) of Fig. 8.6 at $V_B \approx 6000\,\mathrm{km\,s^{-1}}$ ($E_B \approx 100$ eV) has been associated theoretically with the value of the interplanetary

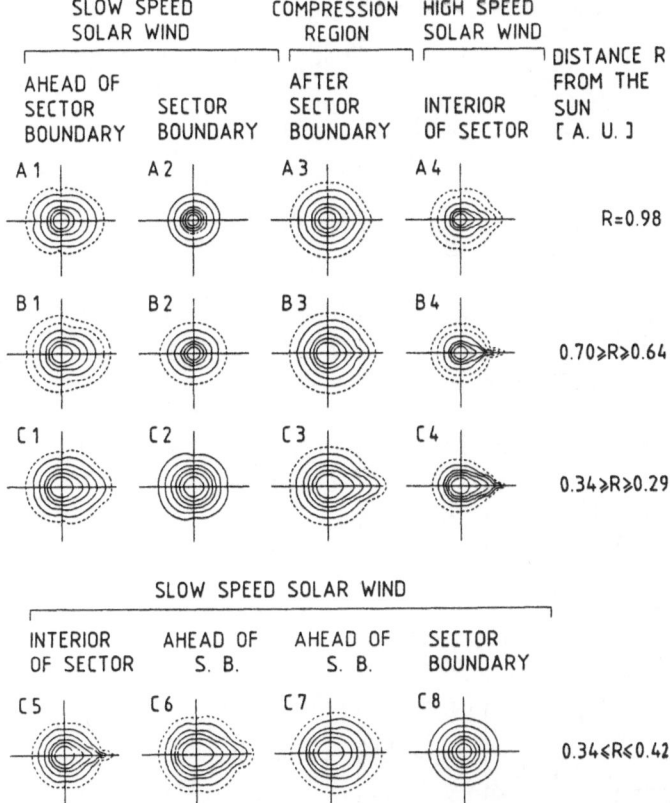

Fig. 8.7. Contour plots of *Helios* 2 electron velocity distribution functions as observed at different heliocentric distances and in various large-scale solar wind structures. Contours are spaced by order-of-magnitude fractions. Notice the variety of shapes as a function of solar wind speed and magnetic topology [8.232]

potential that is related to the electric field induced by gradients of the electron pressure. According to exospheric theory the value of E_B should roughly correspond to the escape energy out of this electrostatic potential well. Detailed analyses of many electron spectra [8.72, 232, 233], however, have not brought about any final conclusive evidence whether the core–halo division and the break in the slope of the distributions is in fact caused by the potential and/or mainly by various scattering processes.

Comprehensive correlative studies of the morphology of electron distributions in association with the magnetic sector and plasma stream structure have been carried out [8.74, 78, 132, 136, 231–234]. Some of the *Helios* results are condensed in Fig. 8.7, which shows data obtained at four heliocentric distance intervals. Electron velocity distributions are plotted as isodensity contours spaced by order-of-magnitude fractions, respectively. Within the interior of magnetic sectors the distributions are extremely anisotropic and skewed along the field direction at higher energies. This narrow strahl points away from the sun and

57

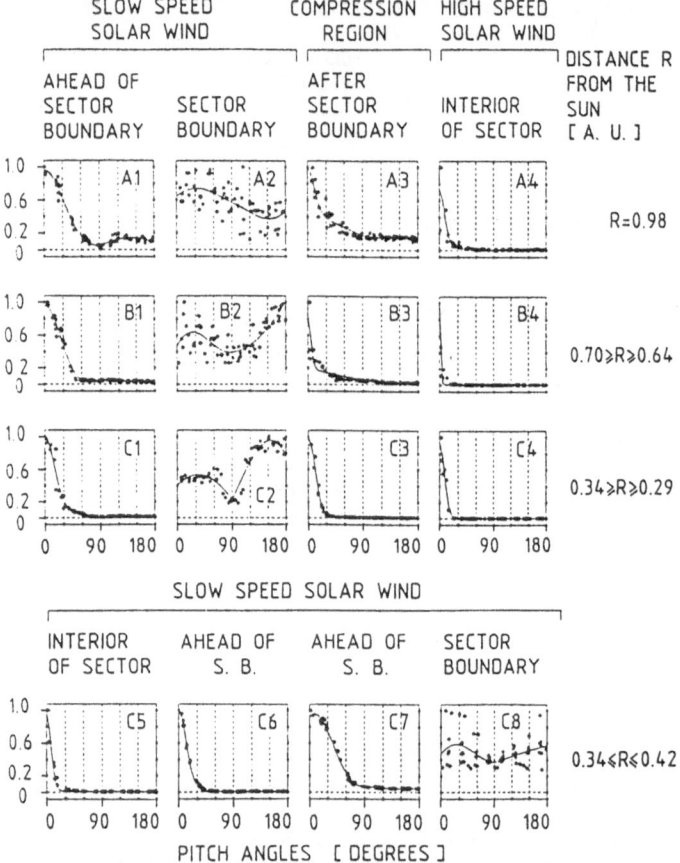

SLOW SPEED SOLAR WIND

INTERIOR OF SECTOR	AHEAD OF S. B.	AHEAD OF S. B.	SECTOR BOUNDARY	
C5	C6	C7	C8	0.34≤R≤0.42

PITCH ANGLES [DEGREES]

Fig. 8.8. Pitch-angle distributions for the same electron velocity distributions as shown in Fig. 8.7. The format of the presentation is also similar. The electron energy ranges between 304 and 312 eV. Notice the strong correlation of the electron strahl (pitch angle about zero) with the magnetic sector structure [8.232]

closely traces the local field direction. Towards sector boundaries the distributions become more isotropic and cooler (columns 1 and 3). Right at a sector boundary, the electron distributions usually appear isotropic, but sometimes also show a slight bidirectional anisotropy [8.234]. This may be interpreted as evidence for closed field loops at and near to sector boundaries and possibly for reconnection, occurring in the associated current sheet.

Even more sensitive indicators of the interplanetary magnetic field topology, as well as for scattering processes [8.166], are the individual pitch-angle distributions at a fixed electron energy. Examples are shown in Fig. 8.8 for the same distribution functions as given in Fig. 8.7. Here the phase-space density in the energy range 304–312 eV is plotted versus pitch angle with respect to the magnetic field. A pitch angle of zero degrees corresponds to particle propagation away from the sun along the field direction. At these energies the electrons are not trapped by the interplanetary electrostatic potential barrier but can freely es-

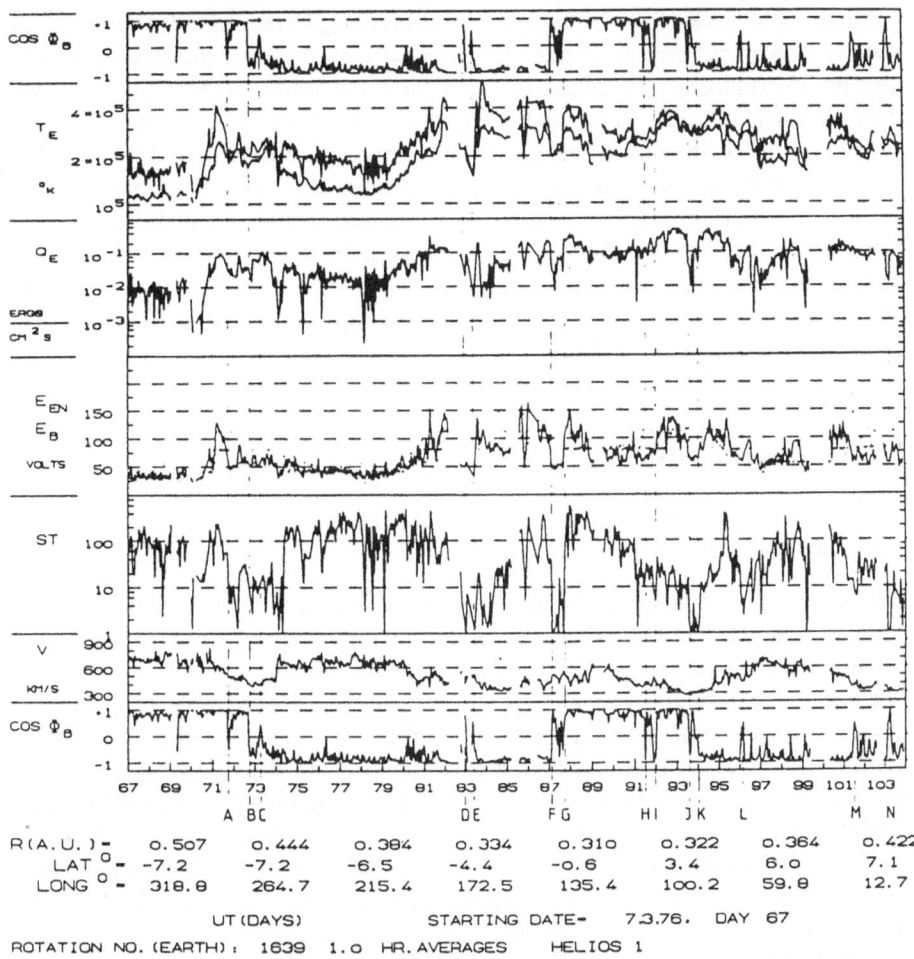

Fig. 8.9. Electron plasma parameters as measured by *Helios 1* between days 67 and 104 in 1976 during perihelion (0.31 AU) passage. In order from *top* to *bottom*, the parameters are: the cosine of the azimuthal angle of the magnetic field, the parallel (*top trace*) and perpendicular temperatures, the heat flux, the escape energy (interplanetary potential calculated from the energy equation) E_{EN} and the breakpoint energy E_B indicated by *dots*, the "strahl strength" ST, the proton speed, and finally again $\cos \phi_B$. Notice the close association between electron thermal parameter variations and the solar wind streams and magnetic field structure [8.230]

cape to the outer heliosphere. The width of the pitch angle distribution is highly correlated with the magnetic sector structure. The distributions broaden in the regions neighboring the current sheet and then spread out over all angles at the sector boundaries. In contrast, in open field configurations related to fast streams emanating from coronal holes, the strahl becomes most clearly visible by a very narrow pitch angle distribution.

A survey of electron parameters in association with the large-scale solar wind structures as observed by *Helios* is shown in Fig. 8.9 from [8.230] which presents, as a function of time from the top to the bottom: the angle of the magnetic field with respect to the outward-pointing Parker-spiral direction; the parallel (top trace) and perpendicular electron temperatures; the heat flux density; the interplanetary potential E_{EN} and the break-point energy E_B (dots); the electron strahl strength (ST); and the proton bulk speed for reference. The parameter ST is defined as minus one plus the maximum ratio of the phase-space densities for electrons moving along the magnetic field away from and towards the sun. By definition ST vanishes for a symmetric velocity distribution. The most clear-cut sector boundaries in Fig. 8.9 are indicated by vertical dashed lines and can be identified by jumps in the cosine traces. The close correlation of all parameters with the solar wind stream structure is obvious. A further, more detailed discussion can be found in [8.232–234].

8.3 Coulomb Collisions and Nonclassical Transport

8.3.1 General Remarks

The preceding section provides an overview of ion and electron velocity distribution functions. The inhomogeneity of the interplanetary plasma, a complicated magnetic field geometry, and the radial gradients of fluid parameters all result in forces which should, in a collisionless expansion, lead to very strong deviations from a Maxwellian shape in the distributions. Actually, these anticipated nonthermal features do prevail in the data, yet they are less distinctly developed than would be expected for an entirely collisionless medium. Consequently, some dissipative processes must exist, which at least partly isotropize the distributions, leading to a redistribution of internal energy between the various species and kinetic degrees of freedom, and thus ensuring the observed fluid-like behavior of the solar wind plasma as a whole.

Coulomb collisions are a theoretically well-understood means of thermalizing particle distributions, see e.g. [8.28, 116, 242]. However, in the solar wind the Knudsen number $Kn = \lambda/L$, i.e. the thermal free path λ over a typical scale length L of the fluid, like the density scale length, is usually found to be of the order of one or larger (with the exception of the heliospheric current sheet), and therefore classical transport theory is certainly not applicable [8.131, 151, 221]. This does not mean, however, that collisions can be neglected entirely. Even 2 to 3 collisions per AU suffice to reduce dramatically the temperature anisotropy compared with the extreme values obtained by assuming that the particles' magnetic moments are strictly conserved [8.108]. On the other hand, collisional relaxation times are energy dependent, and therefore "runaway" must be expected to occur above a certain threshold energy [8.116]. Experimentally, solid evidence has been provided for Coulomb collisional shaping of particle distributions [8.191], although other dissipative processes can be more important and often dominate the transport [8.55, 250].

8.3.2 Observations on the Role of Coulomb Collisions

Ample observational evidence exists for the regulative effects of Coulomb collisions on ion differential temperatures and speeds [8.71, 149, 185, 214, 215]. Statistical analyses on T_α/T_p and $\Delta V_{\alpha p}$ as a function of the ratio of the solar wind expansion time (τ_{exp}) and the collisional energy exchange and slowing-down time (τ_c) yield the empirical result that $T_\alpha/T_p \lesssim 2$ and $\Delta V_{\alpha p} \approx 0$ or a few km s^{-1} if $\tau_{exp}/\tau_c \geq 1$. Whereas Coulomb collisions seem to play a minor role in high-speed solar wind, owing to high-temperature and low-density conditions, it appears that in the slow plasma, with speeds below about 400 km s^{-1}, collisional friction couples the different ionic species with each other more strongly and is almost able to enforce equal temperatures and speeds. Whenever the number of ion collisions ($N = \tau_{exp}/\tau_c$) is larger than one, the corresponding mean free path λ is considerably shorter than 1 AU. An overview of the variation of λ and N with the solar wind stream structure can be found in [8.171, 191], in which the Coulomb collisional domains have been delimited.

How the number of collisions and the particles' free path vary with the ion speed in the bulk frame is shown in Fig. 8.10 from [8.191], which also displays a one-dimensional cut through and contours of the distribution indicated by letter C in Fig. 8.1. Notice the isotropic core and the Maxwellian shape within the collision-dominated regime, defined as the gray shaded region bound by the lines $N(v) = 1$. At such high density ($N_p = 140$ cm^{-3}) and low temperature ($T_p = 7.5$–8×10^4 K) the number of collisions, in particular in the core above 10% of the maximum ($f_M = 6 \times 10^{-19}$ cm^{-6}s^3) of the ion distribution, apparently sufficed to enforce a thermal equilibrium distribution. A detailed analysis of many other distributions (similar to the ones shown in Fig. 8.1) reveals that all kinds of shapes and intermediate forms between isotropic Maxwellians and extremely elongated (cigar-shaped) almost collision-free distributions are observed.

The collisional effects in slow solar wind on the He^{2+} to H$^+$ temperature ratio $T = T_\alpha/T_p$ were recognized early on and analyzed [8.71] and have recently been investigated in much detail [8.114, 115]. On the basis of a simplified three-fluid model including energy exchange and heat-flux degradation the ion temperature ratio T has been calculated and integrated as a function of heliocentric distance. The resulting dynamic equilibrium value of T in the case of a steady energy and momentum exchange between differentially streaming protons and alpha particles could be shown to be mainly a function of the relative abundance $n = n_\alpha/n_p$ and the locally determined number of collisions N.

As the slow solar wind is inhomogenous and weakly collisional ($N \approx 1$), forces related to differential pressure gradients and the rest-frame electric field will always tend to produce a finite ion differential speed and in connection with that an ion temperature ratio $T \neq 1$. Coulomb collisions may partly counterbalance these forces to the extent that T ranges observationally between 1.5 and 2.5, but the thermodynamic equilibrium value of $T = 1$ is very seldom reached. Figure 8.11 illustrates this behavior. For various values of N the observed temperature ratios fall nicely on the theoretically predicted equilibrium curves for $T(n)$ as

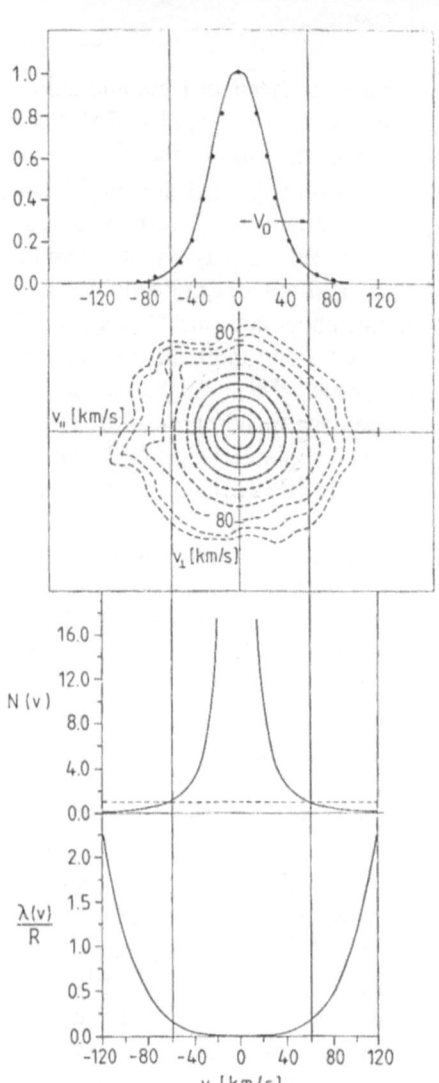

Fig. 8.10. Proton velocity distribution function shown as a one-dimensional spectrum (*top*) and as a contour plot (*second panel*). The mean speed is $360\,\mathrm{km\,s^{-1}}$, the density $140\,\mathrm{cm^{-3}}$, and the temperatures $T_{\perp\mathrm{p}} = 8 \times 10^4$ and $T_{\|\mathrm{p}} = 7.5 \times 10^4\,\mathrm{K}$. A Maxwellian fit based on $T_{\perp\mathrm{p}}$ is shown in the *top panel*. Isodensity contours are spaced as in Fig. 8.1, and the maximum value is $f_M = 5.9 \times 10^{-19}\,\mathrm{cm^{-6}s^3}$. The collision-dominated area is *shaded gray* and bordered by the lines at $V_D = 46\,\mathrm{km\,s^{-1}}$, corresponding to the number of collisions N being unity. This number N is plotted as a function of the speed in the *third panel*. The *fourth panel* displays the proton free path λ in units of solar distance R versus speed. Note the steep increase of $N(V)$ at low speed (below the thermal speed of $36\,\mathrm{km\,s^{-1}}$), indicating that collisions predominantly shape the core part of the distribution [8.191]

determined in [8.115]. In a stationary state, increasing N would correspond to growing solar distance.

Equally well, the N-sequence may be interpreted as a sequence in time since N is nothing but the transit time through a solar wind density scale height in units of the collision time. Having these definitions in mind, one may in particular view the upper-left box of Fig. 8.11 as showing the collisional relaxation of T seen by an observer comoving with the solar wind. The bottom line is that the dynamic equilibrium value $T(n)$ is not equal to unity and is not sensitive to the spatial inhomogeneity. Furthermore, owing to a prexisting speed difference $\Delta V_{\alpha\mathrm{p}}$, the possibility arises of preferentially heating the minor alpha

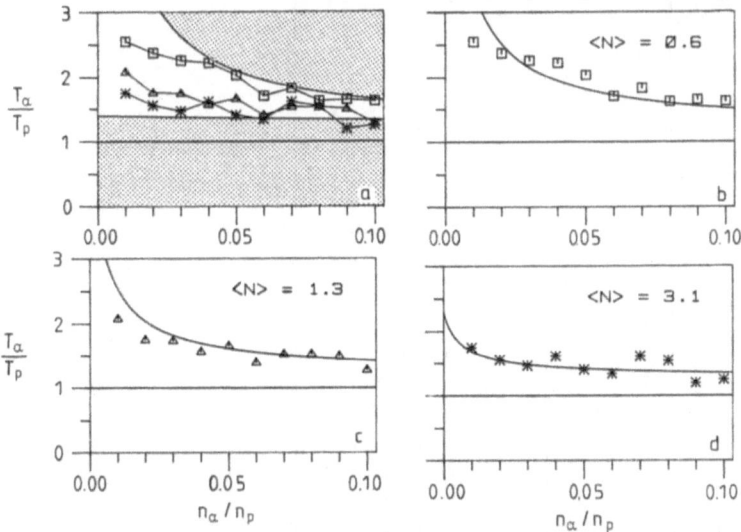

Fig. 8.11a–d. The He^{2+} to H$^+$ temperature ratio T versus the density ratio n for various mean numbers of ion collisions as indicated. Box (**a**) presents a summary of boxes (**b**), (**c**), and (**d**), and gives in addition the inferred "initial" and predicted "final" profiles for $T(n)$, if we assume that the three curves correspond to a sequence in time. The "forbidden" area, *shaded gray*, in the T–n-plane is bordered by those profiles. Note that the fewer the alpha particles the higher their temperature is with respect to the dominant protons. The N-values give the means for the intervals $0.3 \leq N \leq 1$, $1 \leq N \leq 2$, $2 \leq N \leq 5$, respectively. The total number of spectra used is about 1300, with 800, 350, and 150 for boxes (**b**), (**c**) and (**d**) [8.115]

particles (with lower mass density) by Coulomb collisions, given that the required differential speed can be steadily maintained against collisional friction by, for example, the interplanetary electric field.

8.3.3 Kinetic Modeling of the Effects of Coulomb Collisions

It has been proven by kinetic modeling of velocity distributions that under various collisional conditions even a few collisions per AU suffice to prevent the formation of very extended tails and extreme temperature anisotropies [8.173]. These kinetic studies (supported by observational evidence) also indicate that a few collisions are sufficient to maintain an isotropic thermal core of varying size in velocity space. However, at higher than thermal energies energetic ion runaway still occurs and leads naturally to a formation of double beams that cannot be slowed down by Coulomb friction alone. Some results of these numerical simulations are presented in Fig. 8.12, showing the evolution of a tail and, subsequently, of a second peak in the proton distribution while the particles are traveling from 0.05 to 1 AU.

Apparently, the combined action of the mirror force of the large-scale interplanetary magnetic field and of collisional scattering produces variously skewed proton velocity distributions, which can undergo considerable reshaping while expanding from the outer corona to 1 AU, if the number of collisions is as small

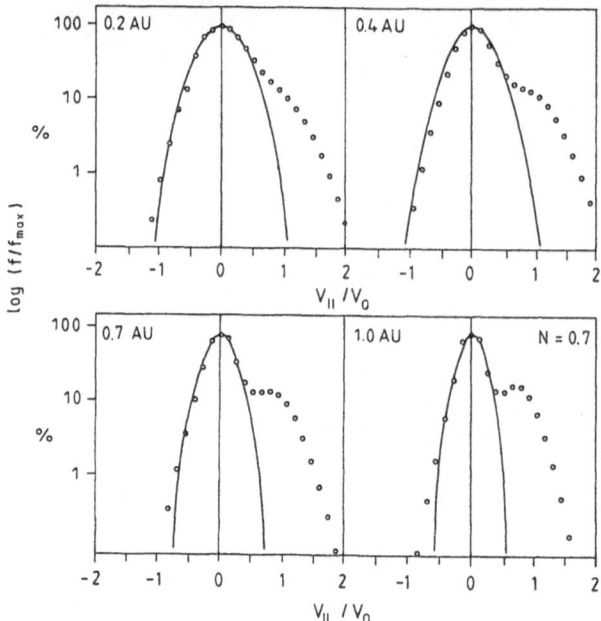

Fig. 8.12. Generation of solar wind proton double beams by Coulomb collisions in the diverging interplanetary magnetic field. One-dimensional cuts through the distributions are shown for four solar distances. At 1 AU the number of collisions is $N = 0.7$, corresponding to $n_p = 13\,\mathrm{cm}^{-3}$, $V_p = 320\,\mathrm{km\,s}^{-1}$, and $T_p = 4 \times 10^4\,\mathrm{K}$. The pronounced tail at 0.2 AU evolves into a resolved second peak of 20 % in density at 1 AU [8.173]

as $N \approx 1$. In Fig. 8.12 we see that at 0.4 AU a marked heat-flux-carrying tail has developed, and at 1 AU even a resolved secondary peak in the one-dimensional cut through the distribution function has occurred at a 20% level of the maximum phase-space density. Detailed modeling under different coronal or 1 AU constraints yields satisfying quantitative agreement between observations (see Fig. 8.1 again) and simulations for the Coulomb collisional domains of the solar wind.

A similar phenomenon is observed in the case of the electrons. Basically, the strahl owes its existence to an almost collision-free escape of suprathermal electrons from the solar corona [8.144, 163]. Exospheric theory predicts the principal features of the electron distribution functions, some remnants of which, like the strahl, are indeed observed. However, by inspection of Fig. 8.7 one becomes convinced that the quantitative agreement is quite poor, because the velocity distributions are less anisotropic and skewed than expected for a collisionless expansion. To explain these observations Coulomb-collision-mediated transport has been invoked extensively in kinetic models of solar wind electrons [8.79, 260, 261]. However, observations do not show all the features predicted by these theories [8.81, 166, 233]. Some researchers have achieved numerical solutions of the Boltzmann equation with a Krook collision operator, which are capable of reproducing a suprathermal and strahl population. These theoretical

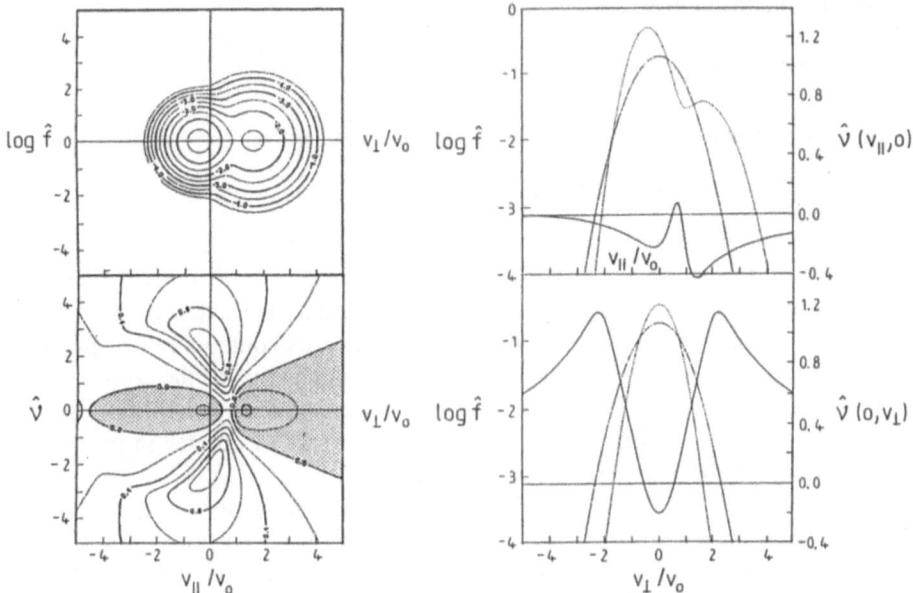

Fig. 8.13. Instantaneous Coulomb relaxation rate $\hat{\nu}$ for a proton double-beam function. Isodensity contours (*upper left*) and one-dimensional cuts through the distribution function parallel (*upper right*) and perpendicular (*lower right*) to the magnetic field are shown together with the corresponding isodensity contours of $\hat{\nu}$ (*lower left*) and dependence of $\hat{\nu}$ on $V_{\parallel}$ and $V_{\perp}$ indicated by the *continuous lines* on the right-hand side. *Shaded areas* indicate negative $\hat{\nu}$ (phase-space density needs to be reduced). The *dot-dashed* lines on the right give the equivalent Maxwellian towards which the double-beam distribution will ultimately relax [8.195]

studies lead to the conclusion that suprathermal electrons might play an important role in the dynamics of a thermally driven wind [8.221].

A possibly serious shortcoming of the existing models is, however, the collision operator used. The Landau collision integral or the equivalent Fokker–Planck operator, if employed instead of a simple relaxation-time approach, yields dramatically different collision frequencies compared with Spitzer's rates [8.194]. A thorough evaluation of model velocity distributions, adopted from those shown in Figs. 8.1 and 8.6, reveals the high sensitivity of the effective collision rate to their detailed shape [8.195]. This is illustrated by Fig. 8.13 for a proton double-beam model distribution. Cuts parallel and perpendicular to the magnetic field (right) and isodensity contours of the distribution and of the collision rate (left) are shown. The parabolas relate to an equivalent Maxwellian with equal particle density and overall temperature. The effective collisional relaxation rate $\hat{\nu}$ is measured in units of the basic Coulomb time scale τ.

We quote explicitly the formula for $\hat{\nu}$ as it is derived in [8.170, 172, 194, 195] for the rate of relaxation by self-collisions of any velocity distribution function f. The Fokker–Planck operator may be written as

$$\left.\frac{df(v)}{dt}\right|_c = \frac{1}{\tau}\hat{\nu}(v)f(v) \ , \tag{8.1}$$

65

with the dimensionless rate

$$\hat{\nu} = 2\pi \hat{f}(\boldsymbol{y}) + \frac{1}{4}\frac{\partial^2}{\partial \boldsymbol{y}\,\partial \boldsymbol{y}} : \left(\int d^3 y' |\boldsymbol{y} - \boldsymbol{y}'| \hat{f}(\boldsymbol{y}') \right) \frac{1}{\hat{f}} \frac{\partial^2 \hat{f}}{\partial \boldsymbol{y}'\,\partial \boldsymbol{y}'} \qquad (8.2)$$

where the velocity has been normalized, $\boldsymbol{y} = \boldsymbol{v}/v_0$, as well as f itself,

$$\hat{f}(\boldsymbol{y}) = v_0^3/n\; f(\boldsymbol{v}/v_0)\,. \qquad (8.3)$$

The mean thermal speed is here denoted by v_0, and n is as usual the particle number density. In Fig. 8.13 the effective rate $\hat{\nu}$ for protons turns out to be positive or negative according as the actual phase-space density is smaller than or greater than the Maxwellian value. Note, that the trace of $\hat{\nu}$ and its contours reflect precisely the structural details of the distribution. Model calculations of the collisional reshaping and decay of distributions towards thermodynamic equilibrium can be found in [8.170]. Similar studies have also been carried out for more realistic electron model distributions [8.195].

However, even the validity of the Fokker–Planck equation itself or the Landau collision integral for dilute space plasmas has been put into question by a recent thorough evaluation of the underlying Boltzmann collision integral [8.266] if it is expanded to higher order in the postcollision values of the particle velocities. The main result of this reevaluation is that for velocity distributions far from equilibrium the large-angle scattering plays a more important role in the collisional relexation than suggested by the Fokker–Planck diffusion-like approach. A significant number of particles distributed in a nonequilibrium fashion in velocity space (as in Fig. 8.1 and Fig. 8.6) favor large-angle-scattering events. This should have a particular effect on the collisional properties of the electron strahl and the ion beams. Future modeling of these effects is highly desirable. The development of an adequate transport theory is still in its infancy.

8.3.4 Towards a Nonclassical Collisional Transport Theory

Even if the Fokker–Planck equation is employed, because of its widely appreciated relatively simple mathematical form, one certainly has to abandon the assumption of local thermodynamic equilibrium (LTE) in modeling electron or ion velocity distributions in the plasma source layers of the solar atmosphere as well as in the interplanetary medium. For example, even in the solar transition zone the angle-averaged electron distribution function does not remain Maxwellian, but instead forms an anisotropic high-velocity tail [8.265]. This bearing has multiple effects, for example, on collisional EUV-line excitation, ionization rates [8.224] and the thermal and electrical conductivities of the plasma.

Similarly, the electron strahl in the solar wind requires a non-LTE description. Presenting a significant correction to a Maxwellian, the kappa distribution has been adopted [8.44] as a possible starting point for nonlocal formulations of the solar wind transport problem. Even in these models the directed tails (and the skewness) of the distribution function still need to be described adequately. As shown in Fig. 8.7, and many other figures contained in [8.232–234], the electron pitch-angle distribution at several 100 eV can vary by orders of magnitude between the solar and antisolar direction. The strahl may become as small as 10° in width in the interior of a magnetic sector. Therefore, it is hard to conceive how an expansion about an isotropic, "locally" determined quasi-equilibrium distribution should ever work [8.55].

In a paper devoted to heat conduction in a spherically symmetric expansion of the solar or of stellar winds [8.44] it has been proposed to remedy the failure of classical transport theory by incorporating even higher moments of the velocity distributions, accompanied by a set of higher-order fluid equations. In this theory a new parameter is introduced (a fifth velocity moment) describing the excess or deficiency of particles compared to their number in exponential Maxwellian tails. The rationale is to carry out higher-order expansions in terms of Legendre polynomials [8.45]. In order to obtain a convergent series these higher-order normalized moments should become decreasingly smaller than one. As demonstrated with the help of measured particle velocity distributions [8.57], the appropriate Legendre-type expansions are often at best semi-convergent and require up to more than ten polynomials; even then they can badly fail to represent acceptable fits to the observed ion beams and strahl electrons. Therefore these improvements of classical theory are only relevant for very moderately nonthermal distributions and do not go essentially beyond the standard transport theory for multispecies anisotropic magnetized plasmas as reviewed extensively, e.g. in [8.9]. They are even less general as far as the assumed magnetic field configuration is concerned.

Although considerable progress has been made in better understanding the nonclassical Coulomb transport in the inhomogeneous solar wind plasma, a coherent picture and mathematical recipes for explicitly calculating transport coefficients are still lacking. Some general ideas on how one might possibly proceed towards a combined global–local concept for electron heat flow have been formulated. Nevertheless, it is also rather difficult to establish convincing evidence from the observations that the conceived specific dissipative processes are really at work. Concerning energetic halo electrons, for example, any collision process that rapidly weakens with increasing energy (like Coulomb collisions or ion sound-wave scattering) yields collision rates too low to explain the isotropization of the halo and broadening of the strahl [8.166, 232]. However, these judgements are based merely on test-particle collision rates. Better assessments relying on the Fokker–Planck or, better, the Boltzmann collision operator are urgently needed.

8.4 Wave–Particle Interactions and Anomalous Transport

8.4.1 General Remarks

The importance of anomalous transport and wave–particle interactions in the solar wind was increasingly recognized during the *Helios* mission. Dedicated plasma-wave experiments have detected copious kinetic waves with a frequency range extending from the ion gyrofrequency up to the electron plasma frequency. The low-frequency MHD waves and turbulence are reviewed in [8.182], and the high-frequency turbulence, being more local in nature and of shorter wavelength, is extensively reviewed in [8.109, 110]. This collective small-scale turbulence (with wave numbers obeying $k\lambda_D < 1$, where λ_D is the Debye length) of varying intensity and character is an important means of shaping and mediating particle velocity distributions in addition to the unavoidable, but mostly weak, collisions, which may be regarded as being due to the electrostatic field fluctuations (with $k\lambda_D > 1$) associated with the particles' discreteness.

Besides the fact that the solar wind provides a nearly uniform, infinite plasma ideally suited to the study of small-scale microinstabilities, the main motivation for studying wave–particle interaction arises from the need to identify the relevant dissipative processes capable of regulating and limiting the anisotropies in velocity space [8.50, 250] discussed in Sect. 8.2 and thus to understand better transport phenomena and the success of simple fluid concepts in solar wind modeling. Figures 8.1 and 8.6 illustrate the main features of velocity distributions that need to be explained: the ion temperature anisotropy; the ion heat flux and differential streaming (double beams); temperature and velocity differences between alpha particles and protons; the electron heat flux and its regulation; the isotropization of the electron core population and the broadening of the electron strahl; and the anomalous coupling between ions and electrons ($T_e \neq T_p$) and the related energy and momentum transfer.

The first question always arising is whether the observed skewed and anisotropic distributions are stable. Secondly, if instabilities are found it is to be asked which instability mode prevails and whether it provides an effective relaxation mechanism for the dissipation of free energy by wave excitation. The observed wave activity of course helps in identifying the most relevant modes [8.17, 18, 260, 261], but owing to sizable Doppler shifts it often does not allow a definite identification. Finally, the transfer rates for anomalous transport need to be evaluated or estimated [8.55, 90]. Frankly speaking, present-day understanding is far from enabling us to give definite answers to all of these questions or to allow final conclusions. However, some progress has been made, and some results achieved are encouraging.

8.4.2 Electrostatic Waves and Plasma Instability

The most thoroughly studied electrostatic mode is the ion acoustic mode [8.110, 112], which leads to predominantly elastic scattering for the core electrons. A stability analysis for ion sound waves which used the measured proton and electron distributions directly has been carried out [8.57, 58]. The results obtained were then compared with simultaneously observed wave activity. The most interesting time periods for a stability analysis were selected on the basis of statistical correlations between plasma parameters and wave activity. Namely, wave intensities were found to be closely correlated with the electron–proton temperature ratio T_e/T_p, the electron heat flux Q_e, and the occurrence of ion double peaks, thereby already indicating the two main possible sources of free energy for wave excitation [8.59, 88]. Theoretically, these are associated with the sunward electron core shift (inversely related to the heat flux) and the proton beam drift. In the numerical stability analysis the observed electrostatic fluctuations of short wavelength could, for large T_e/T_p, be identified definitely as ion acoustic waves. Growth and dispersion characteristics were found to depend very strongly on the shape of the actual distributions. The resulting instabilities appeared to be weak and the growth to be only marginal with a growth rate $\gamma_k \approx 0$, probably indicating, by relaxation oscillations, a steady energy and momentum exchange between ions and (mostly) core electrons [8.58, 59].

Without going into mathematical details we quote a few expressions pertinent to the linear stability analysis of electrostatic waves. For further information we refer the reader to monographs and the original literature as quoted in [8.15, 203, 250]. For our present purpose we closely follow the work in [8.57, 58, 88, 89]. The unmagnetized electrostatic dispersion relation is obtained by equating the dielectric constant with zero. If this equation is resolved with respect to the wave vector k, the following result obtains:

$$k^2 + \sum_j k_j^2 \, \varepsilon_j(k,\omega) = 0 \, , \tag{8.4}$$

where the dielectric susceptibility for particle species j depends only on the phase speed and propagation direction of the wave and is given by

$$\varepsilon_j(k,\omega) = v_j^2 \int_{-\infty}^{\infty} d^3v \frac{1}{\omega/k - \hat{k} \cdot v} \, \hat{k} \cdot \frac{\partial}{\partial v} f_j(v) \tag{8.5}$$

as a functional of the particle distribution function $f_j(v)$. Here $v_j = \sqrt{k_B T_j/m_j}$ is the mean thermal speed, and $k_j = \omega_j/v_j$, the Debye wave number with $\omega_j = (4\pi e_j^2 n_j/m_j)^{1/2}$, the plasma frequency of species j. As usual, ω denotes the frequency of the wave, which is assumed to be complex, $\omega = \omega_r + i\gamma$, and γ is the growth (damping if negative) rate. For resonant modes and small γ we can recast (8.4) after its decomposition into real and imaginary parts as follows:

$$(k/k_p)^2 = - \left(\mathrm{Re}\, \{\varepsilon_p(k,\omega_r)\} + (T_p/T_e)\, \mathrm{Re}\, \{\varepsilon_e(k,\omega_r)\} \right) \, , \tag{8.6a}$$

$$\gamma/\omega_r = - \left(\mathrm{Im}\, \varepsilon_p + (T_p/T_e)\, \mathrm{Im}\, \varepsilon_e \right) \Big/ \omega_r \frac{\partial}{\partial \omega_r} \left(\mathrm{Re}\, \varepsilon_p + (T_p/T_e)\, \mathrm{Re}\, \varepsilon_e \right) \, , \tag{8.6b}$$

which applies to a proton–electron plasma.

Note that according to (8.5) the dielectric constants $\varepsilon_j(k,\omega)$ are only functions of the phase velocity ω/k and the wave-propagation unit vector $\hat{k}$, i.e. of the propagation direction. By evaluation of ε_j for theoretical or measured particle velocity distributions the wave dispersion characteristics and growth rates can be calculated. In the case of a Maxwellian, (8.5) is given by the plasma dispersion function [8.203]. For any gyrotropic distribution a Legendre expansion

$$f(v) = \sum_{l=0}^{\infty} P_l \left(\hat{k} \cdot \hat{B} \right) f_l(v) \tag{8.7}$$

may be employed [8.57], where P_l is the Legendre polynomial as a function of the wave pitch angle $\cos \theta = \hat{\boldsymbol{k}} \cdot \hat{\boldsymbol{B}}$ with respect to the magnetic field, and $f_l(v)$ denotes the speed-dependent expansion coefficient for any order l. In many observational cases (8.7) provides a rather convenient way to represent the measurements without imposing biased model assumptions.

Most of the stability analyses that have been carried out deal with idealized model distributions and use observational data to provide some constraints on wave–particle interactions rather than intending specific agreement between theory and observation. Therefore, we will concentrate on some unique developments to assess the relevance of electrostatic wave–particle processes on the basis of measured velocity distributions and wave turbulence [8.57–59]. Detailed results of a particular numerical plasma stability analysis for ion acoustic waves are shown in Fig. 8.14. The parameters are: $\omega_p = 3.32\,\mathrm{kHz}$, $\lambda_D = 4.9\,\mathrm{m}$, $v_{p\parallel} = 20.5\,\mathrm{km\,s^{-1}}$, and $T_p/T_e = 0.2$. The top part (a) gives isodensity contours of a cold, measured double-beam proton distribution ($v_B = 40\,\mathrm{km\,s^{-1}}$) along with two circles corresponding to a wave phase velocity of $\omega/k = 27\,\mathrm{km\,s^{-1}}$ (b) and $\omega/k = 48\,\mathrm{km\,s^{-1}}$ (c). The corresponding frames (b) and (c) below show the normalized growth rate (right-hand scale) together with its constituents, the imaginary parts of the ion and electron dielectric constants, which determine the Landau damping, the wave vector times the Debye length squared, and the real parts of the electron and proton dielectric constants [8.58]. These quantities are plotted versus wave propagation angle with respect to the magnetic field, corresponding to the circles in frame (a).

As can be seen in Fig. 8.14, for these particular velocity distributions the electron core shift (c) as well as the ion double-beam (b) instability are operating simultaneously. The resonant protons at the positive slope of the beam drive a strong instability with γ at several percent of ω_p, corresponding to growth times in the ms range in close accord with simultaneous observations of bursts of wave activity [8.110–112]. Also, the resonant core electrons drive ion sound waves, which grow by about a factor of ten less rapidly. During the time of these plasma measurements intense electrostatic noise was indeed observed [8.59]. These results strongly support the validity of quasi-linear theory for ion acoustic waves and indicate that the solar wind plasma remains mostly close to marginal stability. This conclusion also seems to apply to other instabilities as discussed later.

The above analysis suggests that it is very promising to spend more future effort in determining wave growth from measured distributions rather than from stability calculations with idealized model functions. As far as ion-sound-moderated transport is concerned, however, the main drawback of this mode is that it occurs relatively rarely. Owing to the stringent contraint of $T_e/T_p > 1$ its niche of existence is spatially limited to the current-sheet regions where the rather cold ions (see Fig. 8.1 and [8.85, 112, 171]) contribute only weakly to Landau damping. In conclusion, although much more effective than Coulomb collisions, particle scattering by ion sound waves occurs too seldom to dominate solar wind transport.

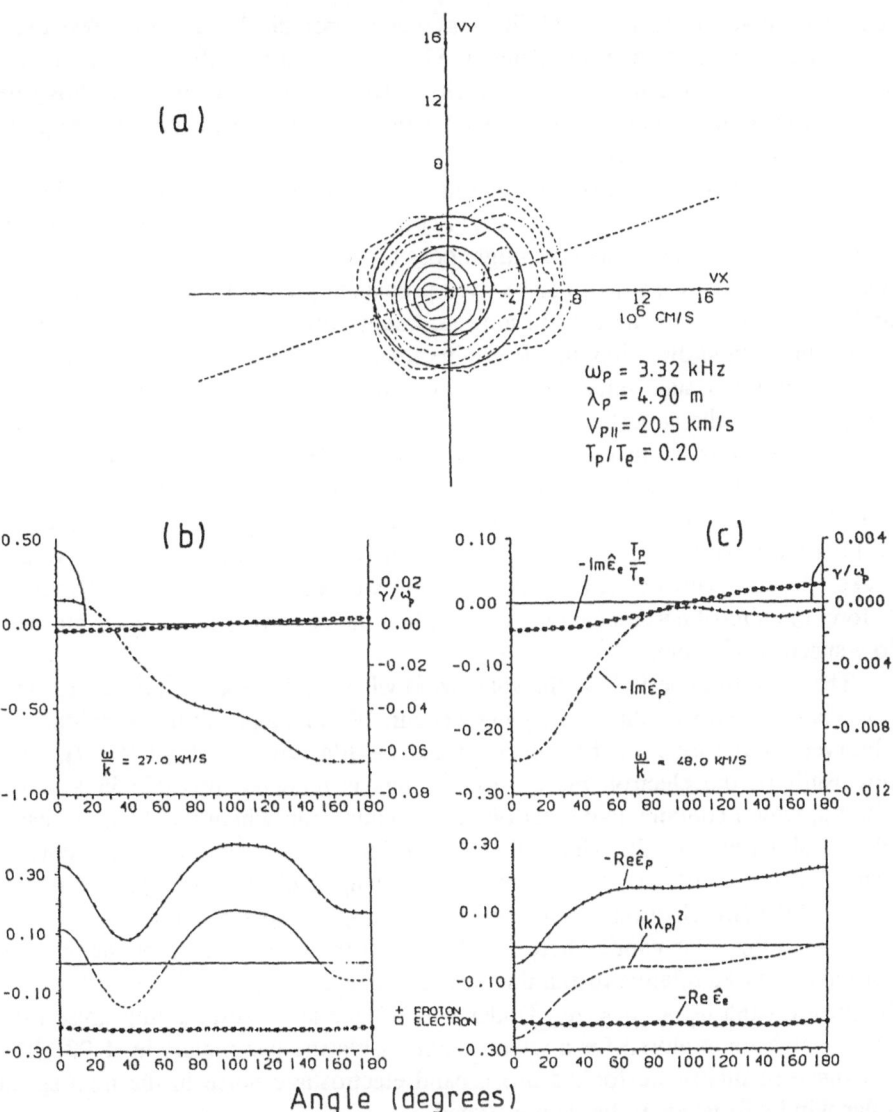

Fig. 8.14a–c. Electron heat flux and proton-double-beam-driven ion acoustic instability. *Top:* Isodensity contours like in Fig. 8.1 of a double-peaked proton distribution function measured at 0.84 AU. The two *circles* correspond to wave phase speeds of 27 and 48 km s⁻¹, for which the detailed dependence of the dispersion characteristics on the wave propagation direction is shown at the bottom. Note the positive growth rates (γ/ω_p) in units of the proton plasma frequency (ω_p = 3.32 kHz) for waves propagating in the sunward (180°) as well as in the anti-sunward (0°) direction [8.59]

In particular in high-speed streams, one finds $T_e/T_p < 1$ [8.73, 77], implying strong Landau damping there. Generally, ion sound turbulence cannot provide a direct scattering mechanism for the strahl electrons. Only under favorable plasma conditions might ion sound activity yield effective scattering of core electrons [8.55] and thereby indirectly influence the strahl electrons as well. Indeed the ion-beam-driven acoustic instability may occur sporadically and manifest itself in short (ms) electrostatic noise bursts. However, high overall ion temperatures will prevent its operation in high-velocity solar wind streams and will thus rule out the ion sound turbulence there as a continuous mechanism for the regulation of ion double-beam drifts (see Fig. 8.1).

Some doubts have also been raised as to whether all of the noise in the frequency range $\omega_p < \omega < \omega_e$ [8.110, 111] is due to ion acoustic noise, or if part of these electrostatic fluctuations might possibly be highly Doppler-shifted lower-hybrid noise [8.187, 189]. As LH-waves propagate predominantly perpendicular to the magnetic field, whereas ion acoustic waves propagate mainly in cones about the magnetic field direction, the determination of the full wave vector k is crucial in deciding this issue. Since the unfavorable temperature ratios with $T_e < T_p$ do not even allow the acoustic waves to exist as normal modes in the plasma, there remains a serious controversy about the nature of this high-frequency acoustic-like noise in the fast solar wind. Therefore, a future reexamination of this noise would be highly desirable in those regions where $T_e/T_p < 1$. As shown in [8.112] the high correlation of the most intensive noise with large T_e/T_p-values makes the ion acoustic wave, driven by the two possible instabilities discussed before, the almost certain candidate to explain the observations in moderate- and low-speed solar wind.

There are time periods in the solar wind when fairly broad-band electrostatic noise below the local plasma frequency occurs [8.111], a phenomenon frequently observed in the electron foreshock of the earth's bow shock [8.75, 87]. This downshift of the electron plasma oscillations is rendered possible in a warm plasma, which contains two electron components or in which an electron-beam-plasma situation prevails. Under these conditions the so-called electron acoustic wave [8.91, 92, 179] can become a weakly damped electrostatic normal mode, provided the two electron components have strongly disparate temperatures, as is the case in the solar wind, where $T_H/T_C \approx 6$. If these electron components have sufficiently large relative drifts, the electron acoustic instability may arise in the frequency band $\omega_p < \omega < \omega_e$. Under which general conditions this noise may occur has been explored for a wide regime of plasma parameters in [8.92]. The relevance of this mode for the broad-band electrostatic noise in the high-speed solar wind still needs to be demonstrated.

8.4.3 Electromagnetic Waves and Plasma Instability

Having discussed the ion acoustic instability at some length, we now turn our attention to electromagnetic waves. It is far beyond the scope of this article to cover all relevant references on this topic. For the enormous amount of papers on microinstabilities we must refer to the specific literature and some general reviews [8.15, 50, 119, 157, 250, 279]. The most recent and still relevant one is the theoretical review in [8.250]. The existing literature relies predominantly on model distributions adopted from *in situ* observations. Parametric studies of this type are undoubtedly of value for a qualitative understanding of the mode under consideration. However, they are not sufficient to establish conclusively whether an instability is actually important or of purely academic interest. Namely, it is equally valid for electrostatic and electromagnetic waves, as for example ion-cyclotron waves or whistlers, that the true growth rates depend sensitively upon the precise shape of the particles' distribution functions in the resonance region of velocity space [8.57]. This is the principal conclusion to be drawn from the wave growth determinations that have been carried out for the measured velocity distribution functions to be discussed below.

Before we go into the details of the numerical analyses we compile some relevant mathematical expressions pertinent to electromagnetic waves propagating along the magnetic field. In this context [8.15, 50, 57, 89, 93, 197, 209, 210, 250] are particularly helpful. The dispersion relation for right- and left-hand polarized waves reads

$$\left(k_{\|} V_{\mathrm{A}}\right)^2 = \sum_j \hat{\varrho}_j^2 \Omega_j^2 \varepsilon_j^{\pm}(k_{\|}, \omega) + \omega^2 \left(V_{\mathrm{A}}/c\right)^2 , \tag{8.8}$$

where the dielectric susceptibility may be expressed by

$$\varepsilon_j^{\pm}(k_{\|}, \omega) = -\int_{-\infty}^{\infty} d^3 w \frac{1}{\omega' - k_{\|} w_{\|} \pm \Omega_j} \left(\omega' + \frac{k_{\|} w_{\perp}}{2} \frac{\partial}{\partial \theta}\right) f_j(w_{\|}, w_{\perp}) \tag{8.9}$$

with the frequency $\omega' = \omega - k_{\|} u_j$ Doppler-shifted into the rest frame of species j. This is assumed to move with speed u_j along the magnetic field with respect to the center-of-mass frame given by

$$\sum_j \hat{\varrho}_j u_j = 0 , \tag{8.10}$$

where $\hat{\varrho}_j = \varrho_j/\varrho$ is the fractional mass density. As usual, Ω_j denotes the cyclotron frequency, $\Omega_j = e_j B/m_j c$. In (8.9) we assume a gyrotropic distribution and use a short-hand notation for the pitch angle gradient

$$\frac{\partial}{\partial \theta} = w_{\|} \frac{\partial}{\partial w_{\perp}} - w_{\perp} \frac{\partial}{\partial w_{\|}} \tag{8.11}$$

in the species' rest frame. By imposing the zero-current and charge-neutrality condition

$$\sum_j \hat{\varrho}_j u_j \Omega_j = 0 , \quad \text{and} \quad \sum_j \hat{\varrho}_j \Omega_j = 0 \tag{8.12}$$

on (8.8) we can derive the MHD limit, i.e. the low-frequency and small-wave-number limit, for cold and isotropic particle distributions:

$$V_{\mathrm{A}}^2 = \sum_j \hat{\varrho}_j \left(\omega/k_{\|} - u_j\right)^2 / \left(1 \pm \left(\omega - k_{\|} u_j\right)/\Omega_j\right) , \tag{8.13}$$

corresponding to the cold plasma dispersion relation. The situation of marginal stability for small $|\gamma|$ is characterized by the expansion of (8.8) about the real axis. This yields for a warm plasma the result

$$\frac{\gamma}{\omega_{\mathrm r}} = -\left(\sum_j \hat{\varrho}_j \Omega_j^2 \operatorname{Im} \varepsilon_j^{\pm}\right) \bigg/ \left(\sum_j \hat{\varrho}_j \Omega_j^2 \omega_{\mathrm r} \frac{\partial}{\partial \omega_{\mathrm r}} \operatorname{Re} \varepsilon_j^{\pm}\right) \qquad (8.14)$$

for the normalized growth (damping if negative) rate, where $\operatorname{Im} \varepsilon_j^{\pm}$ and $\operatorname{Re} \varepsilon_j^{\pm}$ are obtained by applying the Dirac identity [8.203] to the resonant denominator of (8.9) in the limit $\gamma \to 0$ and evaluated at the real roots of the real part of (8.8). If strongly growing or damped modes are considered, one ought to solve (8.8) fully in the complex plane. If the distribution function $f_j(\boldsymbol{w})$ is a bi-Maxwellian, then $\varepsilon_j^{\pm}$ can be entirely expressed by means of the plasma dispersion function and the temperature ratio $T_{j\parallel}/T_{j\perp}$. Even in the most general situation [8.57] only the two reduced distributions

$$F_{j\parallel}(w_{\parallel}) = 2\pi \int_0^{\infty} dw_{\perp} w_{\perp} f_j(w_{\parallel}, w_{\perp}) \ , \qquad (8.15)$$

$$F_{j\perp}(w_{\parallel}) = 2\pi \int_0^{\infty} dw_{\perp} w_{\perp} \left(w_{\perp}^2/2v_{j\parallel}^2\right) f_j(w_{\parallel}, w_{\perp}) \qquad (8.16)$$

are required ($v_{j\parallel} = \sqrt{k_{\mathrm B} T_{j\parallel}/m_j}$). These functions can be directly obtained from the measurement by fitting appropriate interpolation functions to the particles' velocity distributions after an integration over the perpendicular velocity component. In stability calculations drifting bi-Maxwellians have also been used very often. In this case the following relation holds:

$$F_{j\perp} = \left(T_{j\perp}/T_{j\parallel}\right) F_{j\parallel} \ , \qquad (8.17)$$

where $F_{j\parallel}$ itself is a Gaussian of the parallel velocity component. In most of the work to be discussed below these model assumptions have been incorporated. As far as the *Helios* data analysis is concerned both representations (8.7) and (8.15, 16) have been used.

8.4.4 Electron Heat Flux Regulation

Ample literature exists that is concerned with the regulation of the electron heat flux [8.1, 51, 57, 74, 75, 88, 89, 93, 94, 250]. However, no conclusive picture has yet emerged. Very detailed studies have been carried out on the whistler mode and on its electrostatic relative the lower-hybrid mode, which, in the solar wind, has a significant magnetic component [8.40, 146, 147, 187, 188]. The whistler waves occur ubiquitously in high-speed streams [8.18], in which the electron strahls also are most pronounced. This instability is driven by temperature anisotropy, $T_{e\perp} > T_{e\parallel}$, or within the framework of the core–halo model, by the core drift. Detailed parametric investigations lead to the conclusion that one cannot unambiguously decide if the whistler mode regulates the electron heat flux.

In Fig. 8.15 various instability threshold curves are composed as derived from a measurement correlation [8.74], left-hand side (a), and from instability calcu-

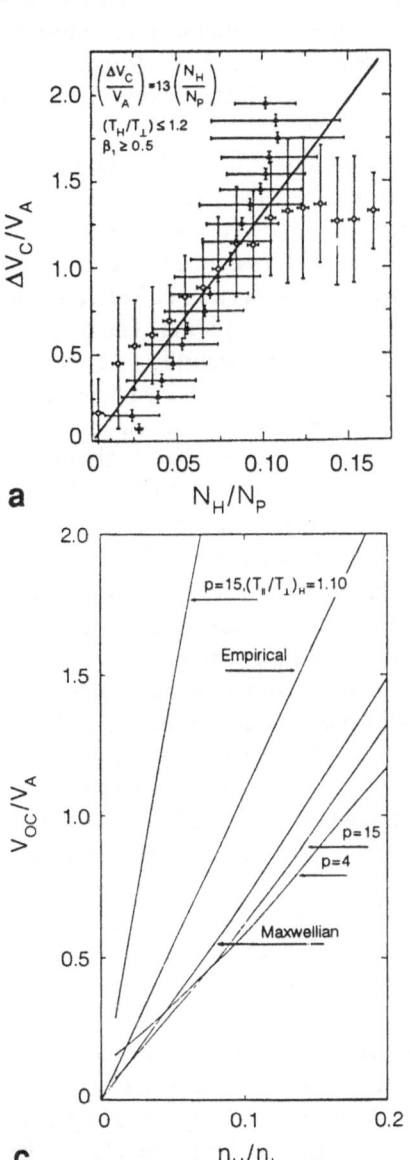

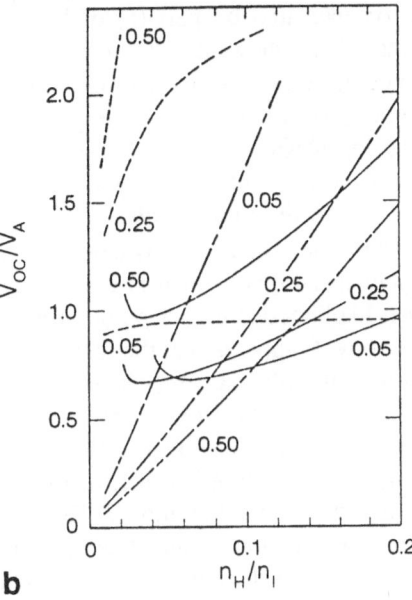

a

b

c

Fig. 8.15a–c. Electron heat flux instability. (a) Correlation between the core drift speed and density. The data were subject to the constraint of weak halo-temperature anisotropies and $\beta_i \geq 0.5$. The straight line is given by $\Delta V_C = 13 V_A n_H/n_p$ [8.74]. (b) Threshold core drift speed versus halo density in units of the ion density with β_i variations indicated at each curve. *Solid lines* refer to the magnetosonic, *uniformly dashed lines* to the Alfvén, and *long-dash–short-dash lines* to the whistler instabilities. Calculations are based on the model of drifting bi-Maxwellians [8.94]. (c) Whistler instability threshold curves for the normalized core drift as a function of core density. The parameter p refers to the different shapes of bi-Lorentzians used to model the electron distribution. The ion plasma beta is $\beta_i = 0.5$. The curve labeled "empirical" refers to the straight line of (a) [8.1]

lations employing the core–halo model with bi-Maxwellians [8.94] (b) or with drifting bi-Lorentzians [8.1] (c), used to model two possible, different shapes of the distributions. As is to be expected for a resonant instability, the required critical drift speed in units of the Alfvén speed depends sensitively upon the halo–core density ratio. Similarly, but not shown here, the halo parallel to the perpendicular temperature ratio has a decisive influence on the stability threshold. As indicated by part (c) of Fig. 8.15 the bi-Lorentzian shape parameter p characterizing the power-law tails also plays a critical role, whereby the large-p case comes closest to the straight empirical line taken from the left-hand observational plot (a). In the middle panel (b) the additional parameter β_i, referring to the ion plasma beta, is also given. The three curves pertain to Alfvén-mode (dashed), the magnetosonic-mode (continuous), and the whistler-mode (long and short dashed line) thresholds.

As the beta rarely drops below 0.25 the Alfvén mode is usually stable. The whistler instability, which is essentially independent of the ions, occurs with a higher likelihood at higher β_i, but it is extremely sensitive to the halo temperature anisotropy. It appears that the fast magnetosonic instability may commence if the halo becomes almost isotropic and low in density. An inspection of Figs. 8.6 and 8.7 shows that these conditions do not prevail in high-speed wind. The analysis of Fig. 8.15(c) is concerned exclusively with the whistler instability [8.1], which has straight lines (see Fig. 8.15(a)) for the threshold in the $V_C - n_H$-plane in the case of a bi-Maxwellian model. The results in (c) clearly show that the growth is a sensitive function of the shape parameter p of the bi-Lorentzian. It should be emphasized that the empirical linear relation in (a) is only adequate for a limited data set with $(T_\parallel / T_\perp)_H \lesssim 1.2$ and $\beta_i \gtrsim 0.5$, which does not represent well the majority of electron observations, in particular in fast solar wind.

All these calculations have demonstrated that instability threshold values are sensitive to the precise shape of the electron distribution function at the resonant velocity. By fitting various models to the observations one can at best conclude that the whistler instability is marginal, but yet it remains a possible candidate for electron scattering [8.93, 94]. Similar conclusions were drawn from a stability analysis based on measured distributions [8.57]. Aside from fitting problems, a principal difficulty lies in the fact that in the core–halo model the resonating core electrons mainly travel sunward and are linked only indirectly to the mainly earthward-propagating halo electrons that actually carry most of the heat flux, apart from the strahl electrons.

It would therefore be better to search for a wave mechanism that directly affects the halo and strahl electrons. Lower-hybrid waves are such candidates. A detailed model analysis [8.187, 188] reveals that the solar wind can easily support these waves as weakly damped eigenmodes, although an experimental identification is difficult because single-point wave measurements do not allow a determination of the wave vector that is decisive on this issue. In principle, resonant halo and strahl electrons can drive the instability which is damped in turn by the core electrons and the ions in the thermal tails perpendicular to the magnetic field. Lower-hybrid waves possess the attractive feature that they can provide effective momentum and energy transfer between ions and electrons, and further that they can affect all of the suprathermal electrons because of the broad band in allowed resonance speeds (phase velocities). However, their very existence in the undisturbed solar wind has yet to be confirmed [8.146, 147]. A detailed stability analysis, like the one outlined for ion acoustic waves in Sect. 8.4.2, is desirable and urgently needed. In conclusion, we must admit that the whole issue of electron heat-flux regulation and energetic-electron scattering remains under debate and requires further work to be done.

Instead of deriving instability growth rates from velocity distributions one may use the observed properties of fluctuating field spectra to estimate possible wave–electron momentum scattering rates. This approach yields some valuable numbers for the rates that can be compared with the Coulomb collision rates. It was found in [8.252] that waves with frequencies near the ion gyrofrequencies and wave vectors comparable with inverse ion Larmor radii can indeed provide strong electron–wave coupling. It was further concluded that, given the observed interplanetary density and magnetic fluctuation spectra, the interactions between a small subset of the broad-band waves and the thermal electrons are competitive with Coulomb scattering. This statement is valid regardless of whether the observed spectrum of essentially magnetosonic waves is assumed to be strictly electrostatic or electromagnetic. The momentum transfer rates discussed are based on expressions derived in a second-order theory of field-aligned electromagnetic instabilities [8.95]. Use was also made of equations that are considered relevant for wave–particle transport from electrostatic instabilities. A corresponding theory has been worked out by Gary and coworkers and is described in a general review [8.90], which also has some relevance to the solar wind plasma.

8.4.5 Regulation of the Ion Temperature Anisotropy and Heat Flux

We now concentrate on plasma instabilities associated with nonthermal ion velocity distributions. The two prevailing characteristics found in proton distributions in high-speed streams (Fig. 8.1) are the core temperature anisotropy with variable temperature ratios, $T_\perp \approx 1\text{--}4\, T_\|$, and the heat flux or double beam at a typical speed of 1–1.5 V_A [8.8, 21, 68–70, 104, 135, 177, 186]. In addition, the observed differential speeds between alpha particles and protons ($\Delta V_{\alpha p} \approx 0.5\text{--}1.0\ V_A$) and the differences in their temperatures ($T_\alpha \approx 2\text{--}4\ T_p$) are other salient features that still need to be explained [8.184–186] from first principles by kinetic plasma physics. Existing explanations range from heating by electrostatic turbulence excited by colliding plasma streams [8.255] to preferential heating by electromagnetic waves. The wave modes, which are probably most important for establishing and regulating these nonthermal features, are the high-frequency extensions of the Alfvén and magnetosonic mode. These are the electromagnetic ion-cyclotron mode in the frequency range between Ω_α and Ω_p and the fast magnetosonic mode in the same frequency regime. At higher frequencies this wave mode links to the whistler branch. The dispersion characteristics of the two modes have been studied extensively [8.3, 15, 41, 43, 51, 56, 57, 60, 63, 139, 154–156, 165, 167–169, 188, 209, 210, 250, 251]. Detailed computer simulations on these cyclotron-resonant instabilities in a multicomponent plasma have also been carried out [8.41].

Some results of parametric studies [8.209, 210] on the electromagnetic instabilities driven by unequal proton beams are compiled in Fig. 8.16. It shows the normalized beam drift versus the plasma beta of the main proton component (a)

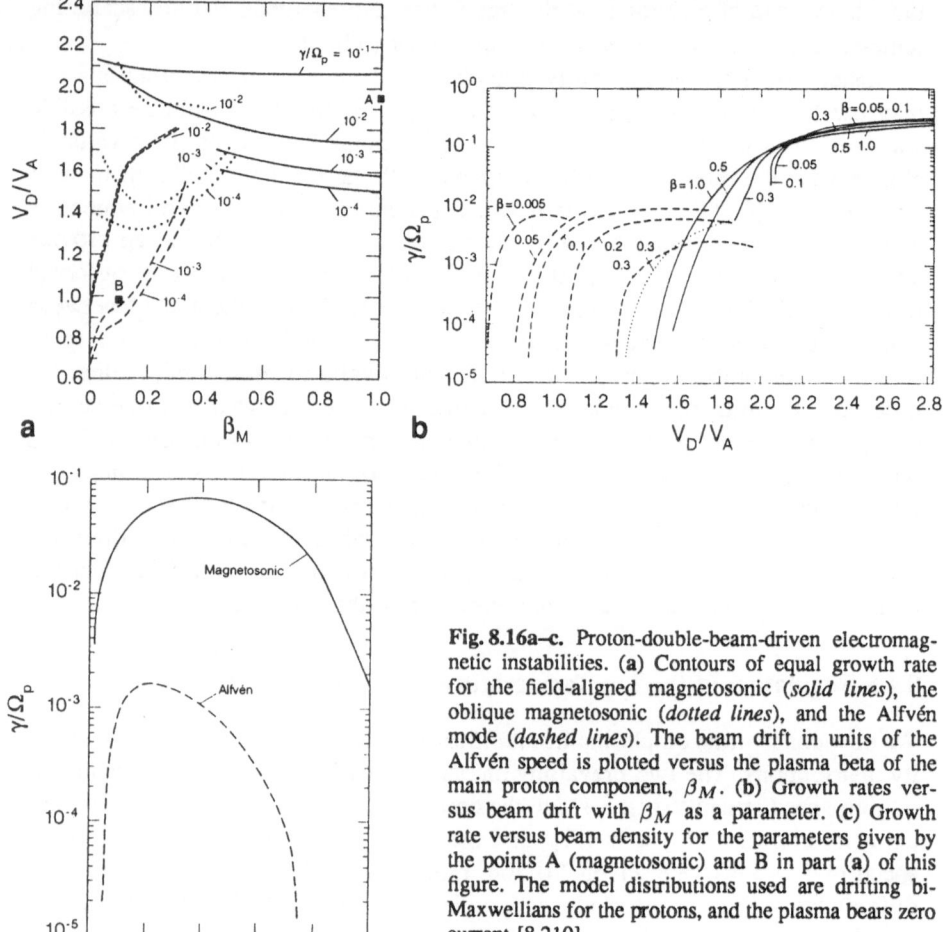

Fig. 8.16a–c. Proton-double-beam-driven electromagnetic instabilities. (a) Contours of equal growth rate for the field-aligned magnetosonic (*solid lines*), the oblique magnetosonic (*dotted lines*), and the Alfvén mode (*dashed lines*). The beam drift in units of the Alfvén speed is plotted versus the plasma beta of the main proton component, β_M. (b) Growth rates versus beam drift with β_M as a parameter. (c) Growth rate versus beam density for the parameters given by the points A (magnetosonic) and B in part (a) of this figure. The model distributions used are drifting bi-Maxwellians for the protons, and the plasma bears zero current [8.210]

and below the normalized growth rate as a function of the beam drift (b) and relative density (c) for the field-aligned magnetosonic wave and Alfvén wave. Dotted lines refer to the oblique magnetosonic mode that turns out to be of minor importance. The parameter in (b) is the plasma beta $\beta_M = 8\pi n_M T_{M\parallel}/B^2$, based on the parallel temperature of the main peak, which is evaluated by assuming drifting bi-Maxwellian velocity distributions. Typical observed parameters for the beam density and drift can be found in [8.186] and for the total plasma beta in [8.193].

Several general conclusions can be drawn from Fig. 8.16(a). Three instabilities are found when the beam drift speed V_D approaches the Alfvén speed V_A. At relatively high beta ($\beta_M \approx 1$) the pure field-aligned magnetosonic mode has the highest growth rates. The oblique (15° − 30° to the magnetic field) magnetosonic mode grows fastest only if β_M is near 0.3 in a narrow β_M-range, and it proves to be less important than the other two modes under normal solar wind conditions.

An oblique Alfvén mode (left-hand polarized) dominates at low betas smaller than 0.3 and at small drifts (see Fig. 8.16(a) and (b)). Linear growth rates are sensitive to the beam density and also vary with the temperature anisotropy of the main and secondary protons. In addition, the electron parameters play a role and the magnetosonic growth responds sensitively to changes in the electron heat flux and core–halo temperature anisotropies. Although they are helpful for a general survey, these parametric studies indicate clearly that the results are not necessarily applicable to a stability analysis of measured particle distributions, since these exhibit more complicated structures not well represented by the models.

Therefore, instead of doing parametric surveys one may better fit the model distributions directly to the measurements and thus uniquely determine the relevant plasma parameters. Such an analysis comes closer to the real situation and considerably restricts the freedom in parameter space. Based on twelve *Helios* ion distributions such a study has been carried out [8.169] for both types of wave discussed above, for which field-aligned propagation was found to be most important.

Owing to the persistent core-proton temperature anisotropy the left-hand ion-cyclotron instability is typically found in high-speed streams and occurs independently of a secondary beam (Fig. 8.17(b)). In contrast the low- and intermediate-speed distributions were found to be only marginally unstable with $\gamma/\Omega_p < 10^{-5}$. The investigation of double-peaked distributions revealed a distinct correlation between the instability onset of the right-hand mode and the beam drift speed and relative density. An example is given in Fig. 8.17(a), which shows normalized real and imaginary parts of ω versus k in units of the inverse Larmor radius $v_{p\perp}/\Omega_p$, where $v_{p\perp}$ is the perpendicular thermal speed of the main proton peak. It was found that a change of 10% in fractional beam density could make all the difference between a marginal and a strongly unstable situation. This conclusion from the analysis of a few individual distributions is further supported by the results that follow in Sect. 8.4.6, in which a larger statistical sample is considered.

In conclusion, solid observational evidence for the marginal stability of proton double beams has been provided in several studies. The early investigations [8.168, 209, 210, 227, 246, 247] were concerned with parametric analyses based on typical solar wind ion parameters used to define model distributions in terms of multiple drifting bi-Maxwellians. Subsequently, the detailed measured distributions were analyzed [8.57, 181] in the regime of marginal stability. These results were obtained without any model assumptions about the velocity distributions. Figure 8.18 shows the proton-beam-driven magnetosonic instability and Fig. 8.19 the temperature-anisotropy-induced ion-cyclotron instability. In both cases the distributions were plotted as isodensity contours in a twelve-term Legendre expansion adapted to the measurements. This expansion in terms of the particles' pitch angle is particularly well suited to retaining the detailed structure of the distributions in the $V_\parallel$–$V_\perp$-plane and to reproduce meticulously the pitch

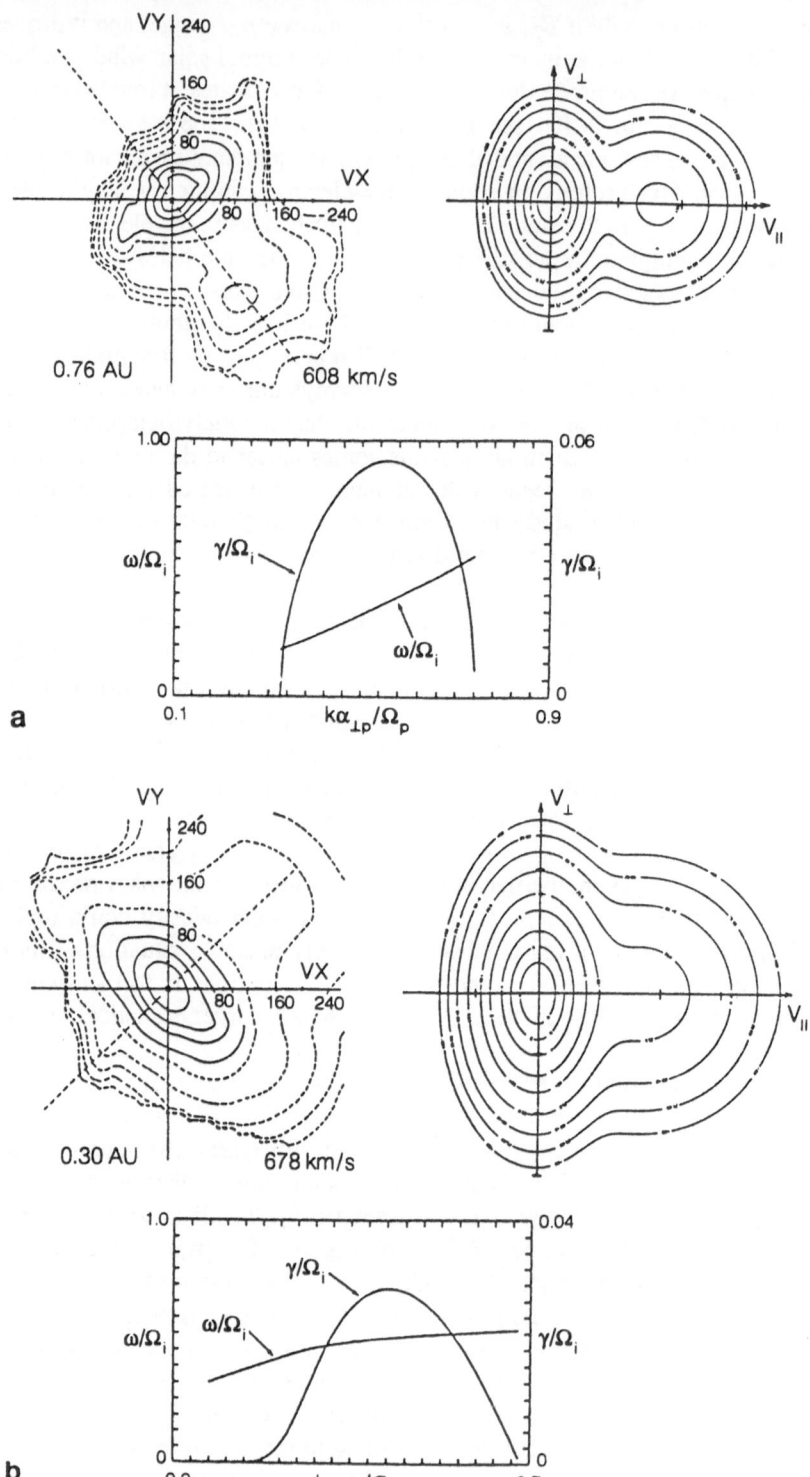

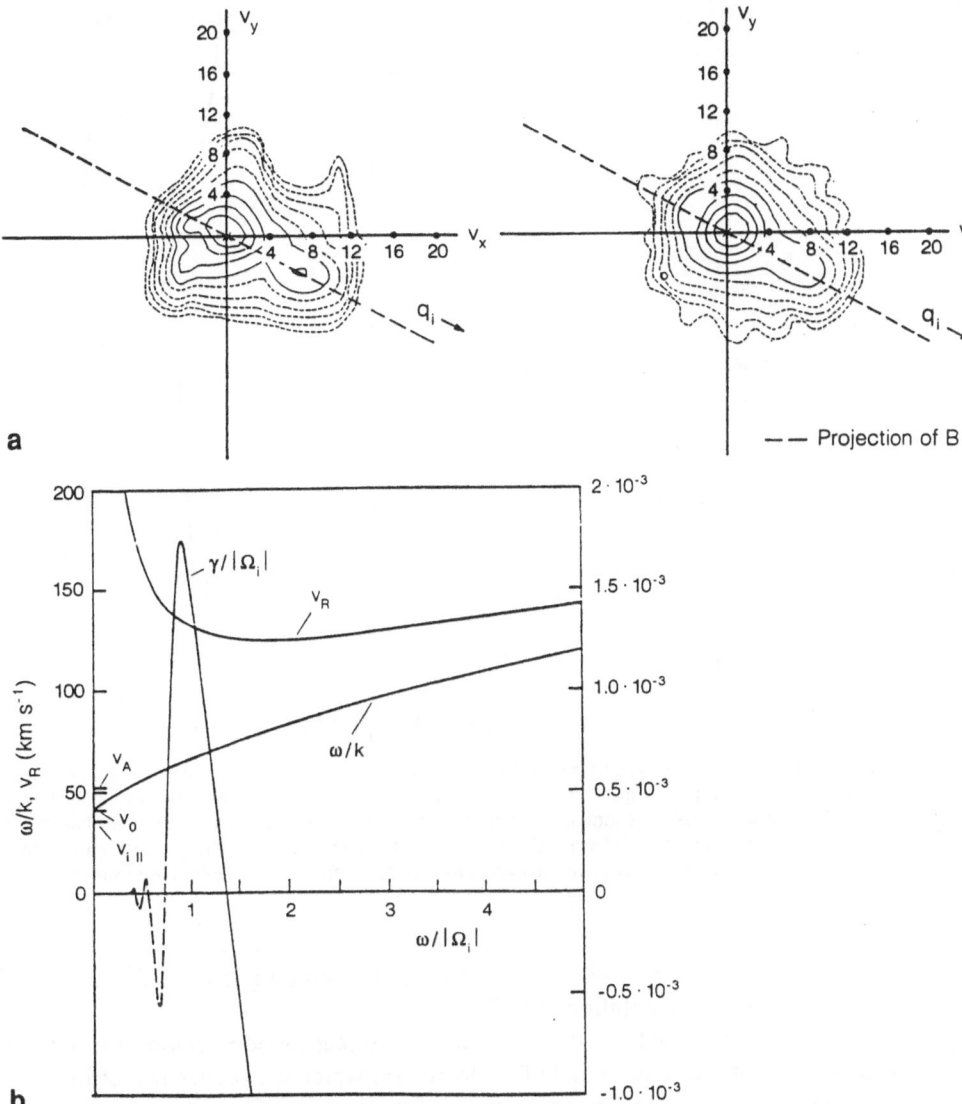

Fig. 8.18a,b. Detailed magnetosonic stability analysis: (a) Contours of a heat-flux-carrying, anisotropic solar wind proton distribution and the corresponding twelve-term Legendre expansion (*right*). Velocity is given in units of $10\,\mathrm{km\,s^{-1}}$. (b) Phase velocity, normalized growth rate, and ion resonance speed (V_R) for right-hand circularly polarized waves propagating along the field or heat-flux (q_i) direction. Parameters are $V_A = 52\,\mathrm{km\,s^{-1}}$, $V_{i\parallel} = 35\,\mathrm{km\,s^{-1}}$ and the phase speed corrected for the fire-hose factor $V_0 = V_A(1 - 4\pi(p_\parallel - p_\perp)/B^2)^{1/2} = 41\,\mathrm{km\,s^{-1}}$. The *dashed part* of the γ-curve is not reliable owing to very large ion resonance speeds [8.57]

Fig. 8.17. (a) Electromagnetic beam instability, and (b) ion-cyclotron instability driven by the proton core-temperature anisotropy. Isodensity contours of individual proton distributions are shown together with their representation by two drifting bi-Maxwellians. In addition the real (ω) and imaginary (γ) parts of the frequency in units of the proton-cyclotron frequency are shown versus the wave vector in units of the inverse Larmor radius $\Omega_p/\alpha_{\perp p}$, with $\alpha_{\perp p}$ being the perpendicular proton thermal speed [8.169]

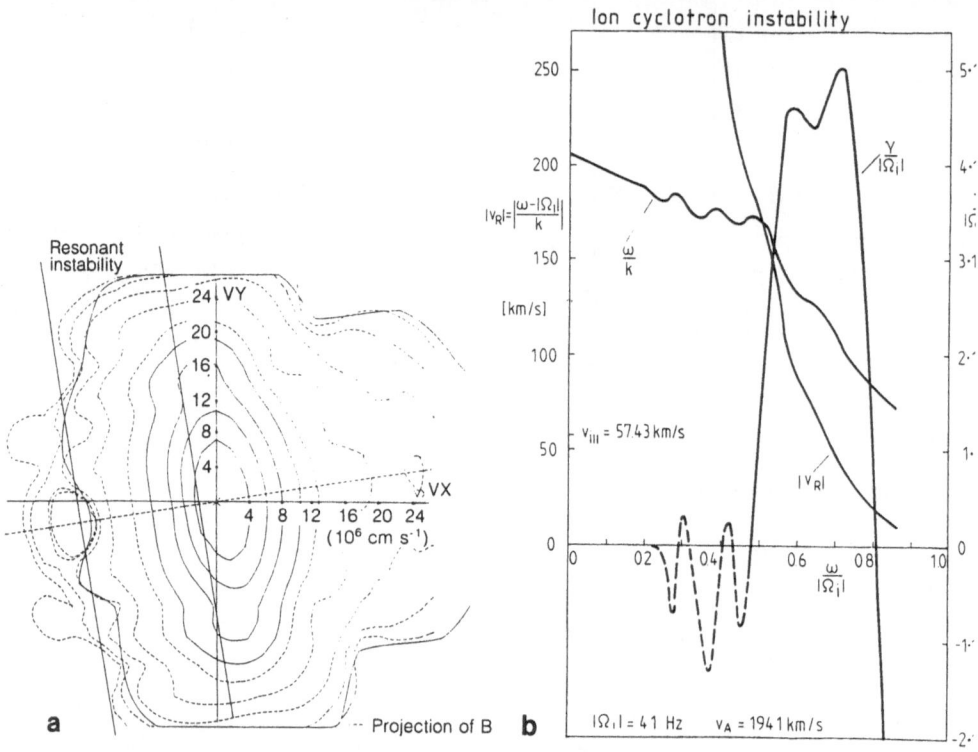

Fig. 8.19a,b. Detailed ion-cyclotron (Alfvén wave) instability: (a) Twelve-term Legendre expansion of a very anisotropic proton distribution function measured near 0.3 AU in a high-speed solar wind stream. The instrument one-count level is indicated by the outermost *continuous line*. Notice the broad stripe of negative parallel velocities ($20 \lesssim |V_R| \lesssim 180\,\mathrm{km\,s^{-1}}$) where positive growth rates are obtained. (b) Phase velocity, normalized growth rate, and ion resonance speed are displayed versus wave frequency [8.56]

angle gradients which determine the growth and damping rates at the margin of cyclotron-resonant instabilities [8.57].

These stability analyses demonstrate clearly that the ion-cyclotron instability is a function of the solar wind bulk velocity, on which the occurrence of the free-energy source, i.e. the thermal anisotropy in the core (see Fig. 8.1 again), depends. The right-hand polarized ion-beam-driven mode is independent of the solar wind speed, but its occurrence varies strongly with tiny changes in the beam drift or density. The growth rates for the left-hand circularly polarized ion-cyclotron wave may reach a few percent of the local proton-cyclotron frequency. Therefore, this instability provides a fast regulating mechanism for the proton core-temperature anisotropy. In the earth's magnetospheric plasma environment [8.41] this mode has also been demonstrated to be of prime importance, especially by means of comparison with nonlinear computer experiments, which also provide helpful insights into the solar wind situation.

To explain the variations of the observed proton double beams (heat flux) and core-temperature anisotropy is of major importance for various aspects of solar wind physics. For an understanding of ionic transport in a microturbulent plasma the question as to what creates these nonthermal features in the first place must be answered. Furthermore, how do the particle velocity distributions respond to and maintain self-consistently the simultaneous large-amplitude Alfvénic turbulence (see e.g. [8.15, 182]) characterizing in particular the high-speed solar wind flows? Quasi linear as well as nonlinear plasma theory is expected to provide the ultimate answers to these problems. So far only a few papers have been concerned with analyzing the linear stability of actually measured ion distributions. Even fewer papers have addressed the question of how a nonlinear dynamic wave–particle equilibrium might be established [8.2]. In this context one must answer the question of how the concurrent MHD turbulence is finally dissipated, and how the measured average shapes of observed power spectra can be explained.

8.4.6 Regulation of the Electromagnetic Ion-Beam Instability

The ion beam instability seems to be the key regulator for the drift speed of proton double beams and of ion differential streaming. A statistical analysis carried out with thousands of ion spectra [8.197] showed that the free energy of proton double streams often sufficed to drive the field-aligned magnetosonic wave (see Fig. 8.18 again) unstable. The results are shown in Fig. 8.20 in the parameter plane of the beam drift speed and fractional mass density. The two panels at the top refer to a collisional solar wind with $N > 0.3$ (left) and to a collisionless plasma, $N < 0.3$, where the number of collisions per expansion time is defined by $N = \tau_{exp}/\tau_c$ (for the details see [8.197]). The right-hand blue line gives the fire-hose limit in the $\Delta V/V_A - \varrho_B/\varrho$-plane, and the horizontal line refers to the instability threshold as calculated in [8.96, 97] for the electromagnetic resonant proton beam instability. The white line represents the mean normalized beam drift speed, if the data are binned in relative mass densities of the beam. The bottom panels present the results of the numerical stability analysis of a proton–alpha-particle-beam configuration (triple-peak ion distribution, bottom-right corner) and of an alpha-particle beam on a single-peak proton distribution (left).

The model distribution used in [8.197] to analyze the data is a superposition of drifting Maxwellians with parameters adjusted to the measured velocity distributions. Assuming that the denominator of (8.14) for right-handed waves can simply be replaced by its cold plasma value, the following normalized growth rate is obtained:

$$\frac{\gamma}{\omega_r} = -\left(\frac{\pi}{8}\right)^{1/2} \sum_j \hat{\varrho}_j \left(\frac{\Omega_j}{k_{\|}V_A}\right)^2 \left(\frac{\omega'}{k_{\|}v_{j\|}}\right)$$

$$\times \exp\left(-\frac{1}{2}\left(\frac{\omega' \pm \Omega_j}{k_{\|}v_{j\|}}\right)^2\right), \tag{8.18}$$

where $\omega' = \omega_r - k_{\|}u_j$, and the same definitions as in the previous sections have been used. In Fig. 8.20 the red points indicate positive growth rates and thus

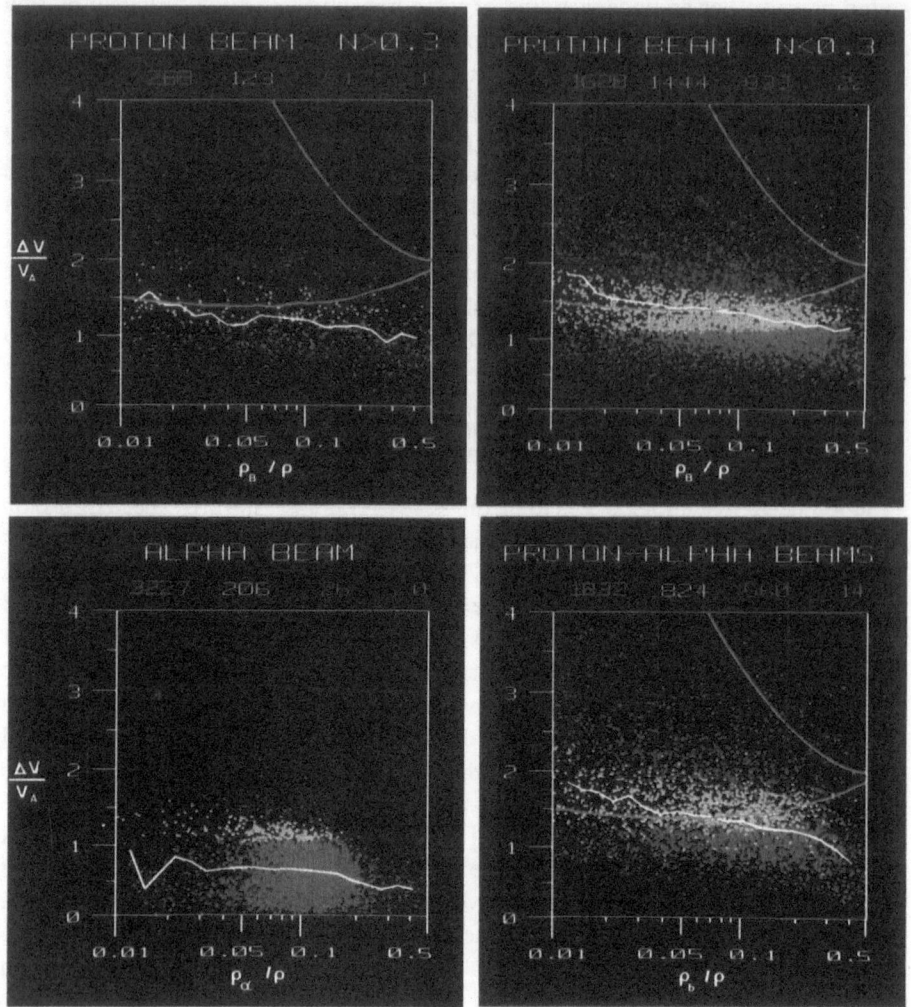

Fig. 8.20. Marginal stability of solar wind ion beams. The ion beam drift speed ΔV in units of the Alfvén speed V_A is displayed versus the fractional mass density of the beam component. The *lower left panel* gives $\Delta V = V_{\alpha p}$, the ion differential speed. The *upper blue curves* indicate the stability threshold for the fire-hose, and the *lower blue curves* indicate the boundary lines for the resonant electromagnetic beam instability. The color coding is as follows: *red* means positive and *green* negative γ, while *yellow* means that γ/ω ranges between 0 and 5×10^{-3}. *Violet dots* refer to spectra found to be fire-hose unstable. The *heavy white line* represents the average drift as a function of relative beam density. The total number of points in each color class is given above each panel. In the *upper two boxes* alpha particles have been neglected whereas in the *lower-right box* they have been included in the stability calculations [8.197]

instability, yellow means $0 < \gamma/\omega_r < 5 \times 10^{-3}$, and green points indicate a negative γ, i.e. stability of the beam configuration.

The results contained in Fig. 8.20 can be briefly summarized as follows. It appears that in low-speed solar wind (with $N > 0.3$) resolved proton double beams are at the margin of stability whereas in fast streams ($N < 0.3$) more strongly unstable distributions are encountered (the cloud of yellow points surrounds the instability-threshold line). In this sense collisions are found to influence or possibly to limit the amount of free energy available for the electromagnetic beam instability. In high-speed wind a large number of unstable proton velocity distributions are found, but the mean value of $\Delta V/V_A$ remains close to the stability boundary located in the marginal yellow region. Still, the persistence of the proton beams seems to contradict the predictions of quasi-linear theory, according to which the beam should slow down or pitch angle broaden and thus become rapidly marginally stable or even dissolve in a plateau or monotonic tail. The solution to this problem may require a nonlinear analysis. Alpha particles by themselves were found to be unable to excite the magnetosonic waves; however, they turned out to be of crucial importance in that they tend to stabilize an existing proton double-stream configuration by enhancing the cyclotron damping of the proton core population [8.197, 244].

8.4.7 Some Nonlinear Aspects of Wave–Particle Interactions

Electromagnetic ion beam instabilities were found in the past few years to be of broad relevance for a variety of physical systems, in particular for field-aligned ion beams at the earth's bow shock, where the beams often originate as an additional feature of supercritical shocks [8.96, 283]. Beam instabilities also occur in the cometary environment associated with heavy-ion pick-up [8.284] and the release of their free energy, owing to streaming relative to the solar wind, in violent bursts of large-amplitude MHD-wave turbulence. Numerical simulations [8.97, 285] of this situation have provided valuable physical insight into the nonlinear evolution of the beam instability which may be partly transferable to the solar wind context.

The nonlinear evolution of solar wind ion beams and their generation in the collisionless high-speed streams stands as an entirely unresolved problem. In numerical simulations [8.96, 97, 283–285] some predictions of quasi-linear theory proved to hold surprisingly well. For the solar wind several authors have proposed that ion temperature anisotropies and streaming originate as the result of preferential heating and acceleration (in particular of minor ions) by cyclotron resonance with preexisting electromagnetic waves about and slightly below the proton (alpha-particle) cyclotron frequency [8.60, 139–141, 188]. In a critical review of the various proposed scenarios, related to physical processes in the outer turbulent solar envelope, and their merits Isenberg [8.138] came to the conclusion that the left-hand cyclotron-resonant interaction is incapable of producing the observed ion differential speeds.

However, the subtleties of the dispersion characteristics of ion-cyclotron waves in the warm H^+–He^{++} solar wind plasma need further investigation, as is suggested by similar studies related to the earth's magnetosphere [8.103]. Furthermore, computer simulations [8.46] of the nonlinear behavior of this plasma showed clearly that ion-cyclotron waves do propagate through the cold plasma stopband [8.139] under various conditions. Yet strong damping occurs about

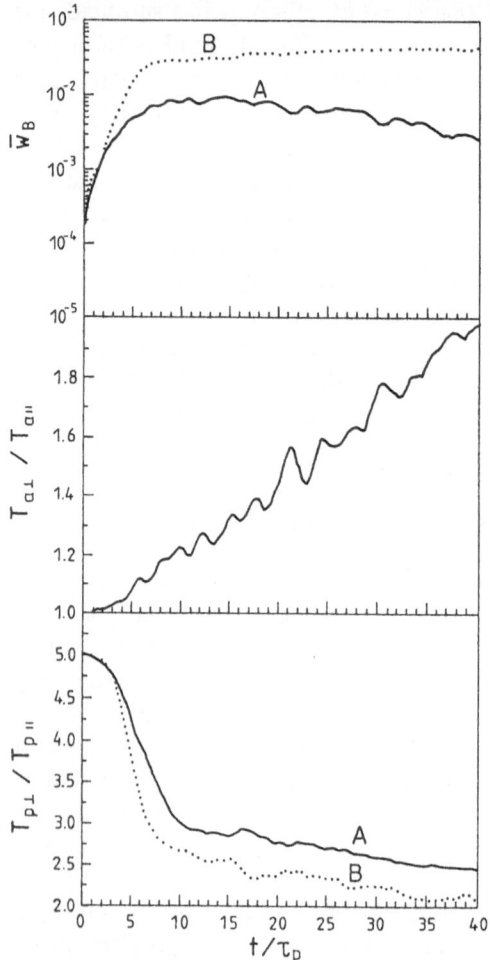

Fig. 8.21. Time evolution of the normalized ion-cyclotron wave energy (*top*), alpha-particle temperature ratio (*middle*), and proton temperature ratio (*bottom*) for a case (*B*) without He^{2+} ions and a case (*A*) with 10 % alpha particles. The curves shown are the result of a numerical simulation of the relaxation of a strongly anisotropic proton velocity distribution. Notice that the nonlinear wave saturation beyond 5–10 proton cyclotron periods coexist with a final stabilization of $T_{p\perp}/T_{p\parallel}$ at about 2.5 in accord with the observation shown in Fig. 8.4. Concurrently, alpha particles are being heated perpendicularly [8.46]

Ω_α to the effect that alpha particles are considerably heated up to a ratio of $T_{\alpha\perp}/T_{\alpha\parallel} \approx 2$, where their thermal anisotropy itself may in turn give rise to a cyclotron instability. At the same time the relaxation of the proton temperature anisotropy stabilizes nonlinearly at about $T_{p\perp}/T_{p\parallel} \approx 2$–2.5, values in close accord with observations [8.186]. The time evolution of the wave energy and the ion temperature ratios is illustrated in Fig. 8.21 taken from [8.46] for two situations, with (*A*), and without (*B*), a ten-percent abundance of alpha particles.

Right-hand polarized magnetosonic waves at sufficient power (energy densities at a reasonable fraction of the required beam kinetic-energy density in the solar wind frame) may be able to produce fast protons or alpha particles streaming at the local Alfvén speed relative to the core protons [8.188]. A serious drawback in the numerical calculations published so far is the assumption of drifting bi-Maxwellian model distributions and the resulting insufficient flexibility in the shape of these velocity distributions. As shown in the previous sections, wave damping and excitation and the resulting transfer of energy and momentum from the preexisting waves to the particles depends rather sensitively on the actual shape of the velocity distributions (for electrostatic waves see for example [8.55]). Therefore hybrid kinetic models based on model distributions described by some moments for which fluid equations are then applied must be judged with caution. On the other hand, no attempts have yet been made to embark on the admittedly titanic task of modeling the inhomogeneous solar wind on the basis of kinetic equations with a complete quasi-linear collision operator that incorporates all the relevant wave–particle interactions.

8.5 Nonequilibrium Thermodynamics

8.5.1 Introduction

While developing early solar wind theory, researchers were inclined toward one of two opposite theoretical positions. Parker [8.226–229] advanced the basic understanding of the supersonic solar wind expansion in terms of single-fluid equations. Subsequently, magnetohydrodynamic theory [8.280] and classic Coulomb transport theory [8.28,42] were widely but uncritically applied. The alternative point of view was to consider the solar wind as a collision-free solar exosphere [8.29, 64, 117, 144, 158, 159, 163, 164]. Within the fluid picture this exosphere was first modeled by the original double-adiabatic equations [8.39] and subsequently by including appropriate modifications [8.65, 129, 196, 281]. However, the detailed *in situ* observations over the last decades have shown clearly that neither of these simple pictures holds, but that reality demands a more complex description incorporating the characteristics of the plasma as a fluid and a kinetic entity.

As far as modeling the transport of the total mass, linear and angular momentum, and energy of the multicomponent, anisotropic solar wind is concerned, one wishes to rely on simple fluid equations with appropriate transport coefficients. It is not surprising that the overall description of the solar wind expansion is accessible via the single-fluid equations since they express merely the conservation of the mass, momentum, and energy associated with the matter lost from the solar atmosphere by the expanding wind. If so, however, a solution of the real problems rests indeed with the proper description of the internal energy state of the plasma and of the binary and collective interactions among its constituents. What needs to be properly understood is the nonequilibrium thermodynamics and relevant processes, which we finally hope to describe adequately, perhaps by a manageable internal energy equation and/or transport coefficients that can be readily handled in numerical calculations. There is still a long way to go to achieve these goals, and much more information has to be extracted out of existing or future data to develop the desired solar wind transport theory.

Considerable progress in this direction has been made through the *Helios in-situ* measurements of the average profiles of the particles' temperatures, differential speeds, and heat fluxes and other parameters characterizing the internal or thermal energy state of the plasma (see Sect. 8.2). Similarly, the radial and spectral evolution of magnetohydrodynamic waves and turbulence [8.15, 109, 110, 182] in the inner heliosphere has been thoroughly investigated. The dissipation of this turbulence is intimately linked with the internal energy budget of the particles. Waves represent the major energy input into the ions beyond the critical points, besides the minor indirect thermal contact that is still maintained with their coronal sources via the electron heat flux and its possible coupling into the ions through the damping of plasma microinstabilities. In the following we concentrate on describing the observations of radial gradients of parameters related to the internal energy of the entire plasma or to properties of the individual species (with respect to their rest frames), such as heat fluxes and differential speeds. Concurrently, theoretical questions related to these phenomena will be addressed, and modeling attempts, provided they exist, will be discussed.

8.5.2 Differences Between Ion Parameters

Velocity and temperature differences between helium and hydrogen ions are among the most prominent and extensively investigated nonthermal features of the solar wind plasma. A comprehensive review of the observations has been composed in [8.215], and the physical implications of these observations have been reviewed in several papers [8.121, 130, 138], where further references can be found. Here we concentrate on selected *Helios* observations with emphasis on the perihelion passages near 0.3 AU. The observed ion differential speeds are closely correlated with the solar wind stream structure. With the exception of the heliospheric current sheet plasma [8.27], He^{2+} tends to move faster than H^+ and reaches the highest relative speeds of the order of the local Alfvén speed in coronal-hole-associated high-speed flows. The observed close correlation of the differential speed with the Alfvén speed [8.183–185] suggests that waves play an important role in regulating the speed of helium and other minor ions, such as oxygen and iron ions. Practically all the minor ions which have been reliably measured were found to move at the same speed as helium in high-speed solar wind streams [8.137, 183, 215, 247, 248]. Figure 8.22 shows the average dependence of the differential speed $\Delta V_{\alpha p}$ upon heliocentric radial distance for various solar wind speed classes as indicated. Whereas for slow wind no gradient can be discerned, a distinct increase appears in $\Delta V_{\alpha p}$ while approaching the sun within fast streams. To explain these radial trends and the obvious preferential acceleration of minor ions in the first place remains one of the theoretical puzzles to be solved.

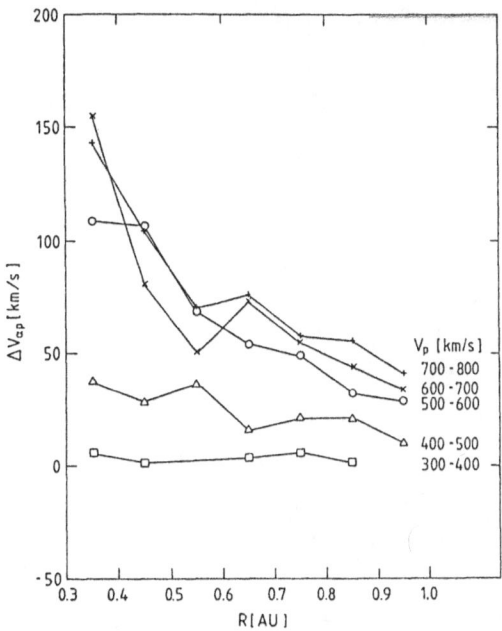

Fig. 8.22. Alpha-particle–proton differential speed versus heliocentric solar distance. The data pertain to various solar wind speed regimes as indicated at the respective curves. The *points* represent averages calculated for radial distance bins of 0.1 AU in width. The points have been connected by *straight lines* to guide the eye. Differential speeds of up to 150 km s^{-1} are observed near the *Helios* perihelion at 0.3 AU [8.185]

Several explanations have been published and critically reviewed [8.60, 141, 188, 201, 202]. It seems clear that the desired models must incorporate the microphysics of resonant wave–particle interactions and also pick the relevant wave modes (presumably left- and right-hand-polarized electromagnetic waves, as already discussed in Sect. 8.4) If the theories are not fully kinetic, then they should at least adopt model velocity distributions with a sufficient number of internal kinetic degrees of freedom to mimic the expected response of the particles to the turbulent wave spectrum. Furthermore, a mechanism is to be found which replenishes and maintains the required waves at a level sufficient to pitch-angle-scatter and accelerate the ions and thus to produce the observed distributions. The energy cascade to be discussed in Sect. 8.6 on the dissipation of Alfvénic turbulence may provide such a process. Wave propagation and dispersion characteristics crucially depend on the nonthermal features of the velocity distributions [8.139]. Therefore, it appears mandatory to employ the warm-plasma dispersion relation and to account for the observed differential properties in the proton–alpha-particle system. Furthermore, nonlinear phenomena must also be taken care of. Their importance is suggested by the detailed results from numerical simulations [8.41].

Very similar statements apply to the phenomenon of preferential ion heating as observed in the interplanetary medium [8.215]. Kinetic temperatures

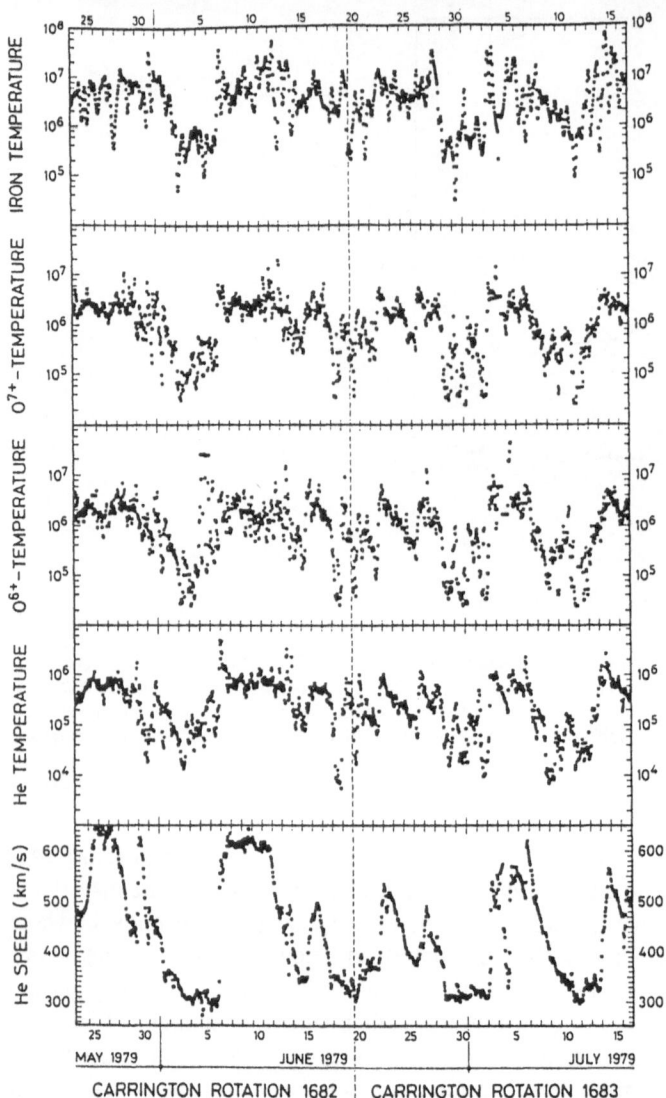

Fig. 8.23. Kinetic temperatures of solar wind minor ions as determined at 1 AU through two consecutive solar rotations in 1979. The traces give from *top* to *bottom* the temperatures of Fe, O^{7+}, O^{6+}, He^{2+}, and the Helium ion bulk speed. The period shown is characterized by two pronounced high-speed streams in which the ion temperatures are elevated distinctly. Often, Fe has a temperature of more than 3×10^6K, which is well above mean coronal values [8.24]

of heavy species tend to follow the rule $T_i \approx T_p m_i/m_p$, i.e. all ions have approximately the same mean thermal spread of their velocity distributions [8.24, 82, 185, 217, 220, 247]. Exceptions to this rule are found in the cool and dense solar wind where at times collisions tend nearly to equalize the ion temperatures [8.27, 171]. Detailed investigations of the ion temperature ratio $T = T_\alpha/T_p$, within a three-fluid model with energy exchange by Coulomb collisions and heat flux degradation, have shown that for various relative ion densities the *Helios* measurements are in good agreement with a purely collisional regulation of T regardless of any wave input [8.115]. Ratios of T between 1 and 2 are still in complete agreement with collisional energy and momentum exchange, albeit the inhomogeneity of the solar wind plays a major role in driving the plasma towards a dynamic equilibrium configuration. The temperature ratios resulting from balancing collisional cooling with heating, related to temperature and heat-flux gradients, depend sensitively on the relative ion abundance $n = n_\alpha/n_p$, which largely determines the collision rates.

The situation is quite different in high-speed solar wind streams. No convincing explanation for the observed T-values exists yet in the literature. It is hard to conceive of a mechanism other than one associated with waves. Whether it is nonlinear, like nonresonant heating due to the sloshing of the particles trapped in the large-amplitude Alfvén waves, or quasi-linear, like heating caused by cyclotron-resonant damping of higher frequency waves at moderate amplitudes, remains to be shown and demonstrated theoretically. Also, the location of the preferential heating source has to be found. It is unclear whether this process takes place in the corona itself, in its outer turbulent envelope, where Alfvén wave amplitudes may saturate nonlinearly [8.15, 120, 126], or if the heating is a weak but continuous interplanetary process. This problem is further elaborated in Sect. 8.6.

We conclude this topic with Fig. 8.23, which shows the helium speed and the kinetic temperatures of He^{2+}, O^{6+}, O^{7+}, and Fe as determined for two consecutive solar rotation periods on the *ISEE* spacecraft [8.24]. Similar Carrington-rotation plots with alpha-particle parameters can be found in [8.185, 215]. Figures 8.22 and 8.23 signify the need to develop new theoretical concepts to explain the observations of minor ions in the solar wind. Other topics related to the present issue are the relative abundances of helium and minor ions, their coronal sources, and variations over the solar cycle and wether these depend on solar activity. This is a broad field of research on its own and is beyond the scope of this article. The interested reader may consult [8.25, 31, 99, 102, 130], where other detailed references can be found, in particular on the abundance measurements as a diagnostic tool to probe the physical state of the inner solar corona [8.6, 7, 100, 101, 131].

8.5.3 The Equation of State and Nonequilibrium Thermodynamics

To tackle the problem of macroscopically modeling the coronal and solar wind plasma flows requires a detailed knowledge of the internal energy state of the plasma. Unfortunately, as pointed out before, there exists nothing like a simple equation of state for the multicomponent, anisotropic solar wind plasma. In the early days of solar wind modeling often a simple polytropic equation of state was assumed. This alleviates the problem considerably since the energy equation can then be discarded. On the other hand in such an approach the microphysics has disappeared and is rather roughly accounted for in a single parameter γ, the polytropic index. In view of the complexity of the solar wind plasma, as described in the previous sections, such a treatment is not acceptable.

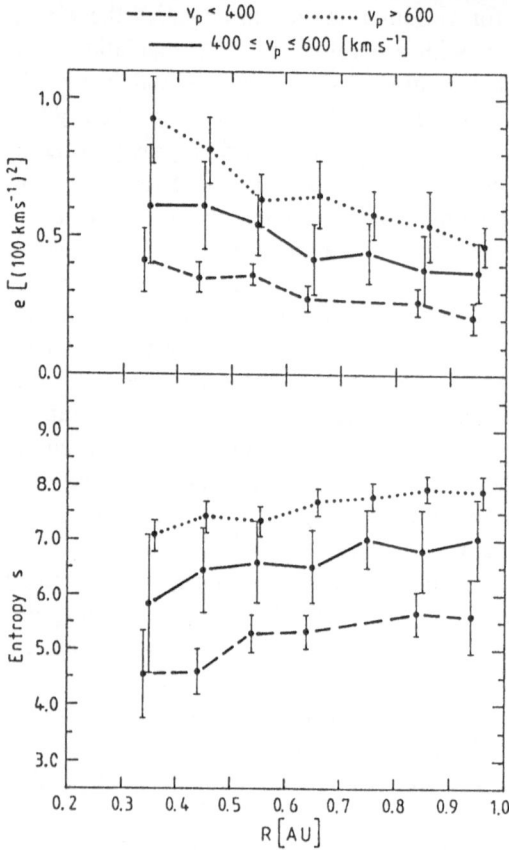

Fig. 8.24. Radial dependence of the solar wind total internal energy (*e*) per unit mass density and of the specific entropy (*s*). Both quantities are based on the combined thermal parameters of electrons, protons, and alpha particles and include the effects of ion differential streaming as well. The entropy has been evaluated with the help of an equivalent bi-Maxwellian based on the total parallel and perpendicular plasma temperatures. Curves are given for the three solar wind speed classes as indicated, and the individual points carry an error bar corresponding to the standard deviation from the mean over each distance bin 0.1 AU in width [8.196]

Considering the solar wind as a collisionless fluid one may ask whether the double-adiabatic invariants are conserved [8.190, 193, 196, 253]. It turns out that this is neither the case for the relevant parameters of the individual species nor for the solar wind plasma as a whole. Detailed investigations of the internal energy and the entropy (corresponding to an equivalent bi-Maxwellian based on the overall temperatures $T_\parallel$ and $T_\perp$ of the solar wind plasma) have been carried out. The result is presented in Fig. 8.24, which demonstrates that there is entropy production in the interplanetary medium, while the solar wind expands radially, and that the specific total energy (*e*) of a fluid parcel [8.196] declines less steeply than in proportion to $R^{-4/3}$ as expected for an adiabatic and isotropic expansion. It is the combined action of all the microphysical processes previously discussed that causes this dissipation and the resultant increase in total entropy.

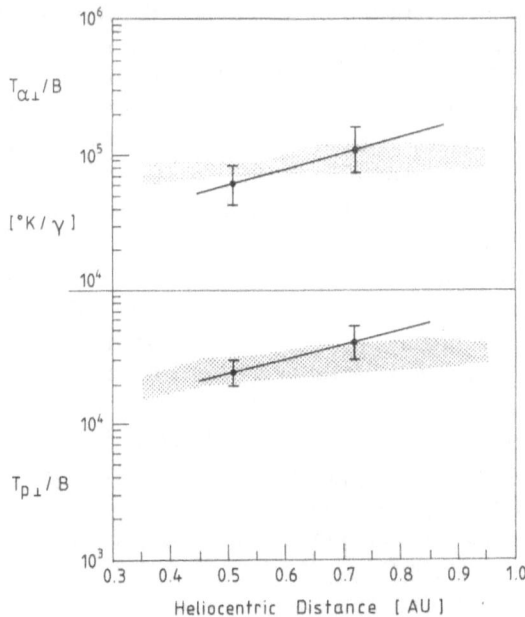

Fig. 8.25. First adiabatic invariants of alpha particles (*top*) and of protons versus heliocentric distance for solar wind high speed streams with speeds above 500 km s^{-1}. The *gray shaded background* relates to statistical results, and the *points* refer to a particular lineup constellation where the two *Helios* spacecraft were able to measure the same plasma parcel at two heliocentric distances indicated by the dots bearing an error bar. The increase of the ion magnetic moments with heliocentric distance is clearly visible, indicating ongoing perpendicular ion heating in the interplanetary medium [8.253]

It is not sufficient, however, to redistribute internal energy between the species and kinetic degrees of freedom, but one needs to invoke heat sources external to the particles in order to explain the radial course of various parameters characterizing the thermal state of the electrons, protons, and alpha particles. For example, the existence of ion heating in the interplanetary medium, and concurrent violation of adiabatic invariance, is most clearly demonstrated by the radial dependence of the magnetic moment μ, which should be strictly conserved in a dissipationless plasma. Statistical studies as well as detailed investigations of special radial line-up configurations of the two *Helios* spacecraft confirm that the ion adiabatic invariants are not conserved. The bottom part of Fig. 8.25 displays $\mu_p = T_{p\perp}/B$ for high-speed solar wind (gray-shaded area) and for a line-up constellation of the two *Helios* spacecraft (heavy line between the two points bearing an error bar) [8.253]. The top panel shows the alpha-particle magnetic moment μ_α. The average increase of μ_p with radial distance indicates extended perpendicular ion heating in high-speed streams throughout the radial distance regime between 0.3 and 1 AU. The corresponding microscopic feature in the distribution function is the large core-temperature anisotropy (see again Figs. 8.1 and 8.4) discussed earlier.

A similar study of the second proton invariant, $T_{p\parallel}(B/n_p)^2$, points to some cooling in the parallel temperature during the radial expansion of a plasma parcel.

The energy source represented by the divergence of the ion heat fluxes was estimated to influence the temperature profiles only marginally [8.190]. As a result, wave–particle processes are most likely responsible for breaking adiabaticity, and Coulomb collisions also play a role of varying importance. In low-speed wind (cold and dense plasma) they are certainly capable of violating double-adiabatic invariance [8.173, 191]. However, in high-speed flows theoretical efforts have to be concentrated primarily on various types of wave–particle interaction in order to explain the observed non-adiabatic behavior [8.191, 196].

In [8.196] an effective "collision time" has been inferred from the *Helios* plasma observations. It is equivalent to the free expansion length of the solar wind that is required to reconcile the predictions from the collisionally modified double-adiabatic theory with the observed behavior of the solar wind conceived as a single-fluid plasma. The inferred collision or total pressure isotropization time (τ) ranges between 10^3 and 10^4 seconds. This is well below the average Coulomb time scales but also still orders of magnitude above the gyration periods of the ions. This estimate of τ compares favorably with the typical growth times of cyclotron-resonant processes at the margin of stability of the underlying nonthermal distributions (see again Sect. 8.4).

8.5.4 Radial Gradients of Heat Fluxes

Besides wave dissipation the degradation of the intrinsic heat fluxes of the particles has also been investigated as a potential source of interplanetary heating and possible cause for the violation of adiabatic invariance. Based on the *Helios* observations several studies have been carried out in which the amount of interplanetary heating associated with the divergences of the ion and electron heat fluxes was estimated. Within the *Helios* orbital range of 0.3 to 1 AU, the proton and alpha-particle heat fluxes, associated with the skewness or double peaks in their distributions, were found to be insufficient to influence the parallel and perpendicular temperature profiles significantly [8.177, 190, 196, 253]. The observed changes in the individual species' adiabatic invariants or entropies could not be accounted for by the heat fluxes as actually observed between 0.3 and 1 AU.

Figure 8.26 summarizes the average radial profiles of the total solar wind heat flux (top curves) and, separately, of the total ion heat flux (bottom part of the figure), both of which are evaluated in the center-of-mass frame of the electron–proton–alpha-particle plasma and thus also comprise the contributions from ion differential streaming. As indicated, these average radial gradients relate to three solar wind speed classes. Apparently, the heat fluxes all increase considerably with decreasing distance from the sun. Whereas the ion contribution amounts to only a few percent of the total heat flux near 1 AU, it can be as high as 30% at 0.3 AU in high-speed solar wind streams. This notion emphasizes the increasing importance of ion differential streaming for the internal energy budget of the plasma. On the average, however, heat conduction is dominated, by far, by the electrons. As can be seen from Fig. 8.7, the electron heat flux Q_e is highly variable, and because of the large electron thermal speed it is less strongly

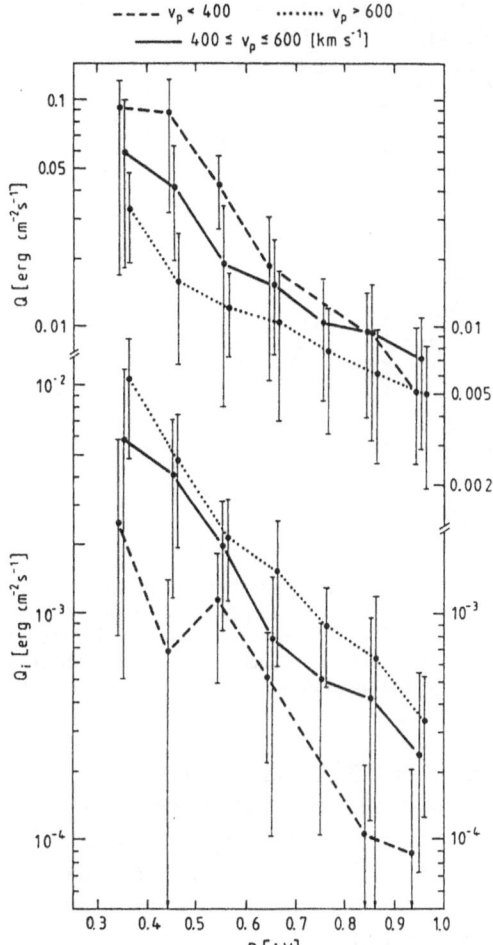

Fig. 8.26. Radial distance variation of the total heat flux Q of the multicomponent solar wind plasma evaluated in the center-of-mass frame (*top*) and of the total ion heat flux Q_i evaluated in the ion center-of-mass frame (virtually identical with the solar wind bulk frame). The three curves refer to high-speed, intermediate-speed, and low-speed wind. Heat fluxes are given on logarithmic scales in units of erg cm^{-2}s^{-1} [8.193]

dependent on the flow speed but reflects more directly the variations of the coronal temperatures.

This fact is substantiated in more detail by Fig. 8.27, which displays the radial gradients of the electron heat flux as measured by *Helios 2* [8.235] in low-speed streams, at sector boundaries, in compression regions, and finally in high-speed streams. Power-law fits are drawn to outline the average trends as indicated in the figure. Sizable differences in the magnitude of Q_e and the steepness of its radial profiles are readily apparent. The heat flux descends most steeply in the slow plasma containing the heliospheric current sheet. This may in part be due to a true decrease in the relative skewness of the underlying electron velocity distribution, but it may also be caused by the fact that at sector boundaries the plasma expands significantly faster than spherically symmetric ($-3.1 \leq \alpha_{NV} \leq -2.4$, for $N_p V_p \propto R^{\alpha_{NV}}$), whereas in fast streams one finds $\alpha_{NV} \approx -2$ [8.146, 257]. Therefore, the radial variation of Q_e is essentially related to the radial variation of the average density profile.

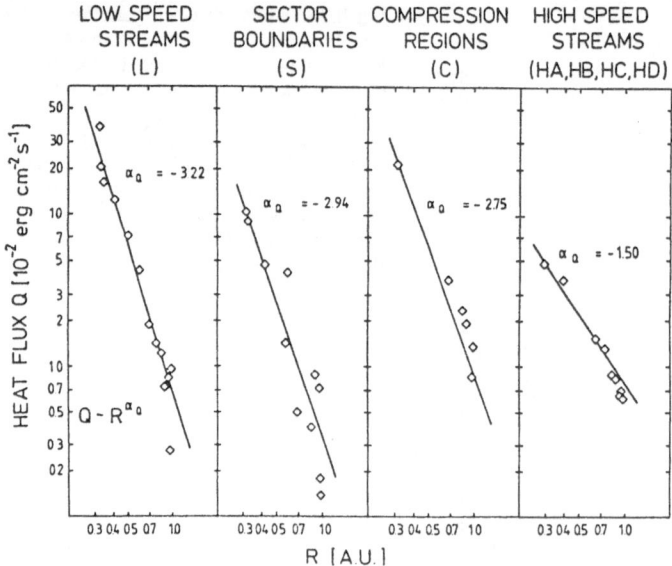

Fig. 8.27. Electron heat flux versus heliocentric distance for various structures of the solar wind as indicated at the top of each panel. The straight lines represent least-squares fits to the measured points by employing a dependence of the form $Q_e \propto R^{\alpha_Q}$. The respective power-law indices are also given. The highest values of Q_e are observed in low-speed wind near 0.3 AU owing to the high number density of the particles there. With increasing flow speed the heat flux profiles become flatter [8.235]

This also explains why Q_e is larger near 0.3 AU in low-speed rather than in high-speed flows. Fast streams are simply less dense. If the density trend is subtracted, the remaining radial trend is characteristic of the normalized heat flux or of the intrinsic skewness and the electron strahl itself. By comparing the first and last columns of Fig. 8.27 it becomes obvious that at 1 AU the values of Q_e are comparable in low- and high-speed streams. This means that, because of lower densities in the high-speed electron distributions, the strahl is maintained out to larger distances, whereas in slow solar wind Coulomb or other scattering processes seem effectively to limit the heat-flux tails (see again Figs. 8.7 and 8.8).

The radial evolution of the internal energy of a single solar wind fluid parcel has been studied thoroughly [8.253] for the rare occasions when the same piece of plasma passed both *Helios* spacecraft at different heliocentric distances. Clear indications were found that the ions did not conserve their adiabatic invariants during the transit through a radial distance interval between 0.507 and 0.72 AU. Some heating was found to be required in the perpendicular temperature (see again Fig. 8.25 and also Fig. 3 in [8.253]) and some cooling needed in the parallel temperature. Dissipation of Alfvénic fluctuations was suggested as one possible cause of nonadiabaticity. From an energy point of view the electrons could be regarded as a potentially important source/sink of internal energy for the ions via coupling through plasma-microinstabilities, as discussed in the previous section.

However, the electron heat flux is a highly variable quantity and is correlated closely with the interplanetary magnetic field sector and the stream structure, as has been emphasized in the discussion of Sect. 8.2 and illustrated in Figs. 8.6–9. Therefore, it remains unclear how effectively this energy reservoir can be tapped by the ions in order to maintain their temperatures above adiabatic values. Moreover, as pointed out in [8.77], it would be striking and "difficult to understand how a cooler gas (the electrons) can heat a hotter gas (the protons) within a region so far away from the sun, which is the ultimate source of most of the energy in interplanetary space".

8.5.5 Radial Gradients of Ion and Electron Temperatures

The radial trends of solar wind temperatures are illustrated in Fig. 8.28, which is taken from [8.186, 235] and shows (left-hand side) the electron and (right-hand side) the proton temperature profiles between 0.3 and 1 AU, as measured by *Helios* for high-speed flows ($V_p > 600 \, \mathrm{km \, s^{-1}}$, open circles) and in low-speed streams ($V_p < 400 \, \mathrm{km \, s^{-1}}$, open boxes). The lines are drawn to guide the eyes. They are based on least-squares fits to the measured points and relate to a temperature dependence of $T_{\parallel, \perp} \propto R^{-\alpha_{\parallel, \perp}}$ on heliocentric distance R. For the electrons (protons) we find $\alpha_{\parallel} \approx 0.23$ (0.72), $\alpha_{\perp} \approx 0.25$ (1.12) for fast wind and $\alpha_{\parallel} \approx 0.67$ (1.0) and $\alpha_{\perp} \approx 0.5$ (0.9) for slow wind. These values are gross

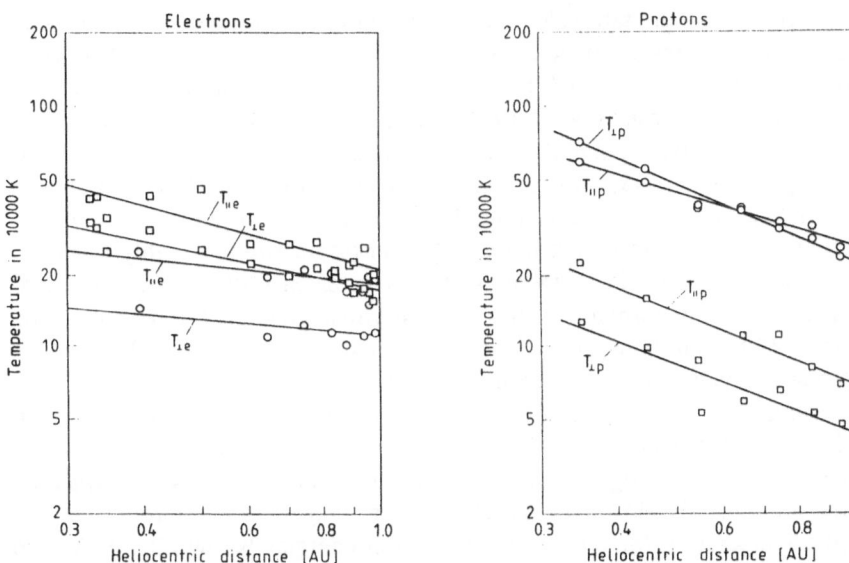

Fig. 8.28. Radial temperature profiles for solar wind electrons (*left*) and protons (*right*). *Circles* refer to high-speed wind ($V_p > 600 \, \mathrm{km \, s^{-1}}$) and *boxes* to low-speed wind ($V_p < 400 \, \mathrm{km \, s^{-1}}$). Protons are much hotter than the electrons in the average high-speed stream whereas the opposite holds in low-speed flows. Temperature anisotropies with $T_{\parallel} > T_{\perp}$ are most distinct in slow wind for the protons and in fast wind for the electrons. Slow (in the sense of the mean flow speed) protons cool off most rapidly, fast electrons most slowly [8.186, 235]

averages with large standard deviations. For a more detailed discussion and for a differentiation of the data according to low-speed, intermediate-speed, and high-speed streams and to sector boundaries and compression regions we must refer to the literature [8.185, 186, 190, 193, 235, 253, 254, 259].

These studies, in qualitative agreement with earlier work [8.77, 80], have clearly established that the thermal properties of electrons and ions correlate strongly with the interplanetary magnetic field and the plasma stream structure, and consequently with the associated coronal source regions, e.g. streamers and coronal holes. An overview of these distinct correlations has been given already in Figs. 8.4, 5, 9, and 23 and in Sect. 8.4. At sector boundaries the electron and ion temperatures show sharp minima (pressure tensors are nearly isotropic). Still close to, but outside of, the current sheet the electron temperature is relatively high, whereas the ions are still fairly cold. All species show the anisotropy signature $T_\parallel > T_\perp$, whereby protons often exhibit resolved double streams (Fig. 8.1). Within the interior of magnetic sectors, connected to open coronal field lines, and in the associated high-speed streams the electron temperature is relatively low, with $T_{e\parallel} \approx 1.2\, T_{e\perp}$, and the protons are much hotter with $T_{p\perp} \geq T_{p\parallel}$. The relative skewness of all particle distributions and their normalized heat fluxes attain maximum values in the middle of fast solar wind streams. But note the opposite trends in the electron temperature in Fig. 8.9 as a function of the solar wind speed.

Detailed studies [8.235] of the electron temperature profiles modeled by a polytropic equation of state (polytrope index β and $T_e \propto N_e^{-\beta}$) have been made. These works follow the rationale that by such an approach the temperature variations due to different expansions of the various plasma parcels could be eliminated, and therefore the "true" radial trend could be determined more clearly compared to a trend based upon fitting the available data directly with a law like $T_e \propto R^{-\alpha}$. As has been pointed out in [8.235] and in similar studies on ion temperature gradients [8.185, 186], all these analyses suffer from limited statistics. Simply increasing the amount of data, however, does not solve the problem, as this implies that even more plasma of different origin and presumably related to different flow tubes would be mixed and perhaps erroneously compared. Therefore, it is extremely difficult to determine the real gradient of any solar wind parameter within an "identifiable" flow tube, and thus the statistical errors and the standard deviations of any mean thermal parameter are large.

Keeping these limitations in mind we may briefly summarize the current knowledge of the temperature gradients in the inner heliosphere, of which Fig. 8.28 gives only a rough overview. The electrons cool down, while expanding radially, significantly slower than expected for adiabatic and isotropic cooling. The core temperature (the core having electron energies below 20–30 eV), not shown here, tends to decrease more steeply than the overall temperature based on the entire velocity distribution including the strahl. The T_e-profiles (as well as Q_e-profiles) are flatter in high-speed streams than in low-speed flows at and neighboring the heliospheric current sheet.

Analogous trends are inferred by analyzing the polytrope indices. They turn out to be smallest in fast streams and indicate possible electron heating, besides that from heat-flux divergency, by other, as yet unspecified interplanetary heat sources, presumably in relation to wave–particle processes or bulk dynamics. The interpretation of the derived gradients is rendered difficult by the variability of these sources. Perhaps it is partly for this reason that the *Helios* electron temperature gradients differ somewhat from earlier evaluations based on data of the *Mariner* missions [8.78, 79, 219, 268] and of the *Voyager* spacecraft from the outer heliosphere, where gradients are of course more strongly influenced by bulk dynamics owing to an increasing Parker spiral angle and ever increasing stream interactions.

The ion temperature profiles are generally steeper than those of the electrons (Fig. 8.28). The protons are considerably hotter than the electrons in fast streams, but much colder than the electrons in slow wind, in particular at 1 AU. Also, the ions tend to cool off more rapidly and exhibit larger overall pressure anisotropies. In our opinion, an explanation of these striking experimental results holds the key to an understanding of the nonequilibrium thermodynamics and to the solution of the acceleration problem of the solar wind.

Voyager observations [8.98] have extended our knowledge of the temperature profiles in the inner heliosphere out to much larger solar distances. Proton temperatures have been derived between 1 and 10 AU and are observed to decrease on the average like $T_p \propto R^{(-0.7\pm0.2)}$, which is also significantly flatter than expected for an adiabatic expansion. The solar wind appears anomalously hot at the leading edges of high-speed streams, presumably by heating due to direct conversion of bulk kinetic into thermal energy by compression regions near stream interfaces [8.106]. This process is still less important within the *Helios* orbital range, but it is likely to dominate the maintenance of thermal energy against adiabatic cooling in the outer realms of the heliosphere. *Pioneer 10* and *11* data [8.145] were used to establish temperature profiles in the far reaches of the solar wind out to 20 AU, which were again found to be weaker than adiabatic with $T_p \propto R^{-(0.6\pm0.1)}$ and most probably produced by conversion of directed into random kinetic energy. *Pioneer 10* data [8.204] resulted in a weighted least-squares fit of the average temperature between 1 and 12.2 AU, yielding $T_p \propto R^{-0.52}$ grossly averaged over the stream structure, which is apparently smeared out and gradually dissolves at larger heliocentric distances.

All these proton temperature profiles are much flatter than those observed in the inner heliosphere [8.85, 175, 185, 186, 254]. Here the ion temperatures in the body of fast, in particular recurrent, streams are still much less influenced by bulk dynamics and stream interactions but are most likely determined by heat conduction and dissipation of convected wave energy. In addition, they still reflect very much the solar boundary conditions of the plasma streams and their inital coronal temperatures. This conclusion is suggested by the strong T_p–V_p-relationship [8.33], which has also been established recently for a large part of the *Helios* data set [8.175]. Whereas it appears that the current-sheet plasma embedded in low-speed wind cools off almost adiabatically (the "cold" solar wind [8.85]), the fast streams exhibit a much higher intrinsic temperature and do cool off more slowly. It seems that the solar wind expansion begins in a turbulent coronal envelope with extremely high initial temperatures ($T_p > 10^6$K), which are needed as a reference if the gradients of Fig. 8.28 are extrapolated back to the sun. It is tempting to speculate that there in fact exists an ion temperature maximum beyond several solar radii. This may perhaps be related to the locations where earlier models of high-speed flows assumed extended ion heating to occur in association with large-amplitude MHD-wave damping (see the review in [8.15] and references therein).

One could also speculate on how proton temperature profiles had to be conditioned to match coronal values and *Helios* observational constraints at 0.3 AU and to behave smoothly and decline monotonously throughout the whole of the inner heliosphere. As a possible process of estimating the close-in heating of the solar wind it was proposed [8.182] to normalize the proton temperature data to 0.3 AU and to establish a temperature–velocity relationship, which could then be used to explore what heating rates are required closer to the sun to predict the empirical T_p–V_p-curve. Thereby the perhaps questionable assumption was made that all the different speed classes have the same coronal temperature, and that this is equal for ions and electrons. Some results of such a study are presented in Fig. 8.29. The related empirical numbers are contained in Tables 8.2 to 8.4, which are taken from [8.182, 198, 271]. They indicate that for high-speed streams the slope of the solar wind proton temperature gradient inside 0.3 AU should be about half of that found beyond 0.3 AU (see again Fig. 8.28

Table 8.2. Gradients in the solar wind electron temperature assuming $T \propto R^{-\gamma}$ [8.198]

Velocity range [km s^{-1}]	γ 0.3 to 1.0 AU	γ 0.014 to 0.3 AU
300 – 400	0.527 ± 0.130	0.650
400 – 500	0.394 ± 0.102	0.680
500 – 600	0.200 ± 0.063	0.767
600 – 700	0.226 ± 0.079	0.805
700 – 800	0.296 ± 0.066	0.812
800 – 900	0.389 ± 0.092	0.825

Table 8.3. Gradients in the solar wind α-particle temperature assuming $T \propto R^{-\gamma}$ [8.271]

Velocity range [km s^{-1}]	γ 0.3 to 1.0 AU	γ 0.014 to 0.3 AU
300 – 400	0.963 ± 0.301	0.577
400 – 500	0.794 ± 0.230	0.322
500 – 600	0.770 ± 0.154	0.136
600 – 700	1.053 ± 0.095	−0.036
700 – 800	1.052 ± 0.067	−0.101
800 – 900	0.922 ± 0.169	−0.151

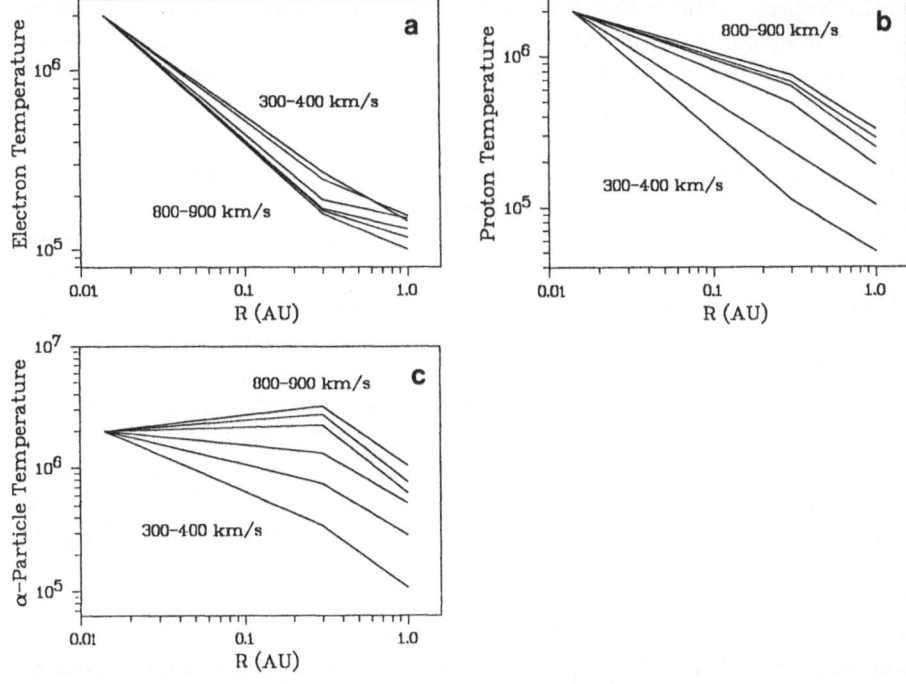

Fig. 8.29a–c. Solar wind mean temperature profiles of the electrons (**a**), protons (**b**) and alpha particles (**c**) as observed by *Helios* (beyond 0.3 AU) and extrapolated to inside 0.3 AU under the constraint to meet a coronal reference temperature of 2×10^6K. The curves relate to particle mean flow speeds ranging between 300 and 400 km s^{-1}, etc. The power-law indices γ for a temperature dependence of the form $T \propto R^{-\gamma}$ are contained in Tables 8.2–4. Notice the distinctly different radial trends in the temperature profiles and the need for extended ion heating closer to the sun in order to comply with the *in situ* measurements of *Helios* [8.182, 198, 271]

Table 8.4. Gradients in the solar wind proton temperature assuming $T \propto R^{-\gamma}$ [8.182]

Velocity range [km s^{-1}]	γ 0.3 to 1.0 AU	γ 0.014 to 0.3 AU
<300	1.331 ± 0.129	1.33
300 – 400	1.223 ± 0.087	0.90
400 – 500	1.033 ± 0.095	0.60
500 – 600	0.826 ± 0.099	0.50
600 – 700	0.762 ± 0.092	0.42
700 – 800	0.808 ± 0.169	0.37

and [8.185, 186, 190, 193, 196, 253, 254, 259]). In contrast, the very low-speed solar wind was found to expand adiabatically all the way out from the corona to 1 AU.

This notion seems to suggest that the solar wind with speeds below about 300 km s^{-1}, as is typical for the heliospheric current sheet, somehow escapes wave heating by originating from sources with closed (probably streamer-belt-related) magnetic fields without intense MHD-wave emission. On the other hand

101

the high-speed ion flows experience considerable interplanetary heating which extends throughout the entire inner heliosphere. This heating is presumably closely connected to damping of Alfvénic fluctuations, as is discussed in more detail below.

A comparison of panels a, b, and c of Fig. 8.29 shows that the electrons, protons, and alpha particles exhibit quite different radial temperature gradients between 0.014 and 1 AU as a function of solar wind flow speed. They are all found not to expand adiabatically, but some not yet well understood interplanetary heating mechanism keeps their mean temperatures above adiabatic values. Whereas electron temperatures are inversely correlated with flow speed, the ion temperatures increase with increasing flow speed. It is most obvious for He^{2+} ions that an extended heating source must be invoked to explain the temperature plateau between several solar radii and 0.3 AU in high-speed wind, provided the temperature profile is to be connected to a coronal temperature of $T_\alpha = 2 \times 10^6 K$.

In fact, the coronal temperatures of minor ions may be much higher. Conclusive evidence for the true minor ion temperatures in the corona is to be expected from the analysis of EUV spectral lines emitted by these ions [8.62]. Compared to He^{2+} the situation is even more dramatic for $O^{6+,7+}$ and Fe^{n+}, for example, which is known to be as hot as $10^7 K$ at just 1 AU [8.205]. In any case, according to our present results protons also need to be more strongly heated closer to the sun than beyond the *Helios* perihelion. In contrast, electrons tend to cool off more strongly closer to the sun than in the outer heliosphere.

8.6 Dissipation of MHD Turbulence and Particle Heating

8.6.1 Kinetic Damping of MHD Waves

The solar wind is a turbulent medium permeated by kinetic and hydromagnetic waves of varying intensity on all scales relevant to the solar wind as a plasma (see e.g. the reviews [8.15, 109, 110, 182]). The problem of the dissipation of kinetic waves and the concurrent transfer of energy and momentum to the particles has been addressed in the previous sections. The collisionless damping of hydromagnetic waves has been discussed, generally as well as for the solar wind in particular, in a series of early pioneering papers by Barnes [8.10–13] and has been reviewed by the same author [8.15]. This work therefore need not be repeated here. It has also found recent interest again in the context of studies on fast-mode magnetohydrodynamic wave-driven coronal hole flows [8.83, 113].

Unfortunately, the theoretical concepts of kinetic MHD-wave damping as developed in [8.15] have not been tested experimentally to the same extent using measured plasma data or distribution functions as has been done for the high-frequency waves. Furthermore, the quasi linear theory may be put into question, since the observations do not yield the small amplitudes required to apply this theory to the interplanetary MHD fluctuations. Density fluctuations are

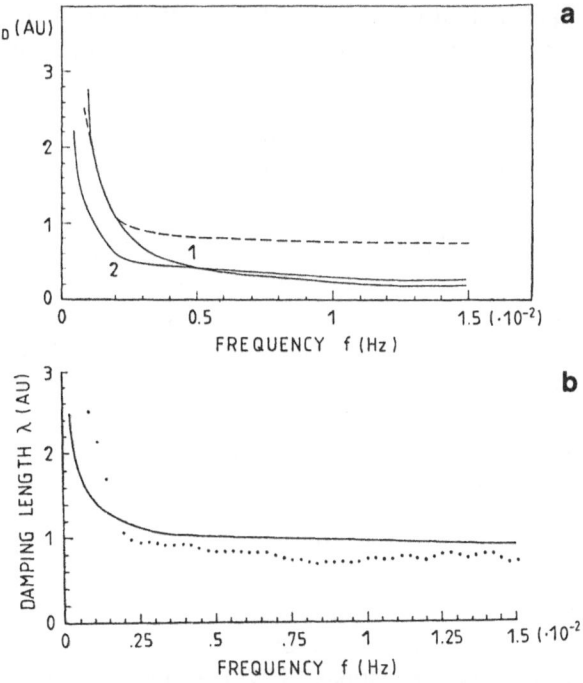

Fig. 8.30. (a) Damping length L_D in AU of a single wave as a function of frequency. Curve 1 for $\theta = 0°$ and curve 2 for $\theta = -45°$ propagation angle with respect to the magnetic field [8.52]. The *dashed line* presents the result inferred from the radial decline of the power spectral density of observed Alfvénic fluctuations [8.16] **(b)** Damping length λ in AU as calculated from the turbulence model of [8.275]. The *dots* again represent the observational results from [8.16]

usually much lower in amplitude than incompressible fluctuations in the inner heliosphere [8.16, 48, 216]. However, it is not clear whether they are related to small-amplitude slow and fast magnetoacoustic waves as is assumed in the theory. It is more likely [8.258] that they are nonlinearly produced as a by-product and additional result of the predominant, large-amplitude "Alfvénic" turbulence.

Direct dissipation of Alfvén waves by Landau damping is possible if the waves propagate obliquely at an angle with respect to the magnetic field. This mechanism has not been considered in the older literature [8.15] because the minimum variance direction of the fluctuations was usually taken to be also the k-vector direction. A linear, fully kinetic treatment [8.52] for an Alfvén wave, e.g. propagating at 45° with respect to B, yields the damping length $L_D \approx 100$ AU and $L_D \approx 0.1$ AU for $\omega/\Omega_p \approx 10^{-2}$ and 10^{-1}, respectively. Thus the damping characteristics exhibit a rather strong dependence on frequency. The results of a detailed numerical calculation shown in Fig. 8.30(a) seem qualitatively to compare favorably with experimental estimates [8.16]. The effects of oblique Landau damping could be shown simply to modify the WKB theory by multiplying the power spectral density of the waves with an exponential decrement factor containing the inverse damping length integrated along the ray path of the wave. It should be mentioned, however, that the experimental and theoretical definitions of L_D in [8.16, 52] are not fully consistent. Unfortunately, no attempts have been made to estimate also the possible particle heating associated with this linear damping mechanism.

8.6.2 Damping of Alfvénic Turbulence by a Nonlinear Cascade

Since the discovery of Alfvénic fluctuations in the solar wind and of the fact that their spectral densities often obey power laws over many decades in frequency, innumerable experimental and theoretical discussions and papers have been devoted to explain their origin, their dynamical spectral and radial evolution, and their final dissipation on kinetic scales. This field of research of solar wind physics is still quite active. Recent works have again emphasized the turbulence aspects of the fluctuations and introduced increasingly more concepts of turbulence theory into the data analysis (see e.g. the reviews [8.200, 206]).

In a series of papers Tu and coworkers [8.274–278] have developed a new theory and a solar wind model that attempt to explain the power spectra of Alfvénic fluctuations and their damping from a unifying point of view, in which the resulting and associated ion heating is also accounted for. Their derivation of governing equations follows the traditional WKB theory, as far as large-scale spatial evolution is concerned, but they also incorporate nonlinear interactions arising from triple correlations of the fluctuations [8.275], which then tend to establish by a cascading process the power-law spectra. It is assumed that the observed fluctuations represent an asymmetric state of MHD turbulence, in which most of the fluctuations are in the Alfvénic mode propagating away from the sun with a small fraction of waves in the inward propagating mode. Together the cascading process and the slow radial variation of the ambient solar wind determine the radial evolution of the power spectra. The theoretical results derived with this model seem to compare favorably with the measurements and are encouraging for further progress with even more realistic modifications of the model. For comparison the dissipation length obtained in [8.275] is plotted in Fig. 8.30(b). Again, the agreement of λ with experimental estimates [8.16] is quite satisfactory.

As was discussed in Sect. 8.5, the relevance of Alfvénic waves for the solar wind dynamics stems from the fact that, once having reached a large enough amplitude, they will strongly accelerate and heat the plasma by dissipation. Since the detailed dissipation mechanism is not known, most models incorporated the assumption of saturated, relative wave amplitudes in association with an unspecified channeling of the energy dissipated by this limiting mechanism into internal energy of the particles [8.119, 123, 128, 142, 143]. However, besides the weaknesses of such an approach lying in the crude treatment of the microscopic dissipation process, this theory is also not self-consistent in so far as it has not been proved that the saturated waves still have the properties of Alfvén waves obeying a WKB theory. The novel approach taken in [8.274, 275, 278] is based on the equation

$$\nabla \cdot \left(\frac{3}{2} (V + V_A) \frac{P(f, r)}{4\pi} \right) - V \cdot \nabla \left(\frac{P(f, r)}{8\pi} \right) = -\frac{\partial}{\partial f} \frac{F(f, r)}{4\pi} , \qquad (8.19)$$

where f is the frequency, V the solar wind velocity, and V_A the local Alfvén velocity, and $P(f, r)$ the power-spectrum density. The right-hand side of (8.19) is the frequency derivative of the wave-energy spectral flux function $F(f, r)$, which describes the turbulent cascade and is given in [8.274] by

$$F(f, r) = C \frac{(4\pi \varrho)^{-1/2}}{V + V_A} f^{5/2} [P(f, r)]^{3/2} \qquad (8.20)$$

with a simple model constant C to be determined self-consistently by the measurements. Here ϱ is the total mass density. Another possible choice for the cascading function may be made which corresponds to the $-3/2$ spectral index predicted for stationary homogeneous MHD turbulence in [8.153]. If we consider a stationary cascade with a frequency-independent flux, (8.20) yields a Kolmogoroff spectrum proportional to $f^{-5/3}$. Without this cascade function, the WKB theory is retained from (8.19). In the case of spherical geometry, analytical solutions for the spectral density are obtained by integration of the partial differential equation along characteristics in r and f [8.274, 275].

Not only does the above theory present a first step towards a fully consistent theoretical description of MHD turbulence in the solar wind, but the cascading function also determines the dissipation rate of the turbulence if the constant C is uniquely determined. At a frequency f_H of the order of the proton-cyclotron frequency $f_p = \Omega_p/2\pi$, multiplied by the factor $1 + M_A$ arising from the Doppler shift, we expect the waves to become strongly damped by cyclotron resonance as discussed extensively in Sect. 8.4. Therefore the heating rate for the protons [8.274] may be estimated as

$$H(r) = \frac{1}{4\pi} F(f_H, r) , \qquad (8.21)$$

a formula allowing us to link the MHD turbulence theory with the nonequilibrium thermodynamics of the particles. It is far beyond the scope of this article to discuss more extensively the merits and advances of the many recent proposals concerned with the evolution of magnetohydrodynamic fluctuations in the solar wind [8.182, 200]. Here we have mainly addressed some basic new ideas and their relevance for the particles' kinetic physics.

8.6.3 Particle Heating by Wave Dissipation

The discussion in Sect. 8.5 resulted in the conclusion that other heat sources needed to be invoked than those related to heat flux degradation in order to explain the nonadiabatic temperature profiles of ions as well as electrons in intermediate- and high-speed solar wind streams. Above we have presented a heating rate in the ion-cyclotron frequency regime, which can be used in the

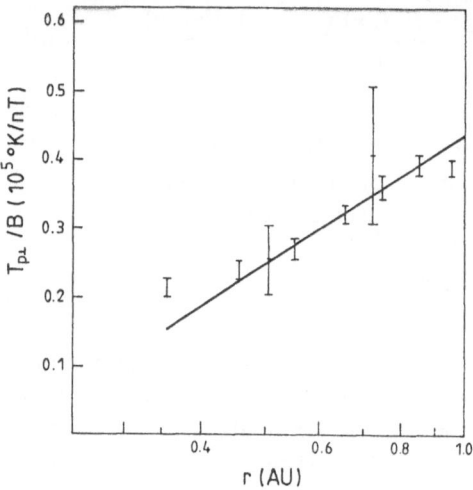

Fig. 8.31. The radial variation of the proton magnetic moment as a result of the heating via cascading of Alfvénic fluctuations in the cyclotron-resonance regime. Observations are indicated by points bearing a standard deviation bar [8.190, 253, 274]. Compare also with Fig. 8.25

energy equations of the ions to calculate the net interplanetary heating. This has been done in [8.274] in a self-consistent fashion and in [8.174] by using the wave power spectra adopted from the observations. The radial variation of the proton magnetic moment is calculated from

$$V \frac{d}{dr} \ln (T_{p\perp}/B) = H(r)/ \left(n_p k_B T_{p\perp} \right) \ . \tag{8.22}$$

If this is integrated with the heating function of the model in [8.274], the resulting radial variation looks like shown in Fig. 8.31. In this case all the wave energy was fed into the perpendicular degree of freedom as expected for a cyclotron-resonant process [8.188]. The straight line gives the model result whereas the small vertical bars refer to the mean magnetic moments found in the high-speed streams with speeds between 600 to 700 km s^{-1} observed during the *Helios* mission in 1976 [8.186]. The two big bars correspond to the detailed investigation of a single fluid parcel which was observed twice at different heliocentric distances by *Helios 1* and *2* in a plasma line-up constellation [8.253]. Apparently, the variation of the proton magnetic moments observed between 0.3 and 1 AU can be well explained by the heating rate determined from the energy cascade and the final dissipation of Alfvénic fluctuations.

Furthermore, a relationship was predicted between the rms value of the velocity fluctuations and the thermal speed of the protons based on the observed mean temperature gradient. For the high-speed range above 500 km s^{-1} the theoretical predictions and the empirical statistical results for the total variance of the magnetic field components were found to be in good agreement and consistent with the proton temperature gradients of Fig. 8.28. Thus the required interplanetary heating of protons can seemingly be provided by the energy extracted via a nonlinear cascade of the Alfvénic turbulence. This dissipation results in a stronger than WKB-like radial decline of the variance of the magnetic field [8.277]. In the models the detailed microscopic dissipation mechanism has not been dealt with explicitly. Ion-cyclotron resonant processes [8.60, 188] appear to be the most likely candidates ultimately to convert the cascaded wave energy into ion internal energy. In view of these recent advances in turbulence theory for the solar wind, it appears promising to resume the work that has already been done on the proton heating and preferential acceleration of heavy ions, and to improve the early model calculations [8.138] by a more sophisticated treatment of the wave dissipation process.

8.7 Energy, Momentum, and Mass Input into the Solar Wind

8.7.1 Coronal Boundary Conditions

In solar wind models one has usually accepted the hot corona as a given fact and has not been concerned with the processes that heat the corona itself [8.226, 228]. However, more recent research suggests that these two problems may be intimately connected [8.123]. Modern coronal heating concepts crucially hinge on the assumed magnetic field structure and topology [8.152], and they invoke direct magnetic mechanisms as the predominant ones instead of sound and shock wave dissipation. Modern fluid concepts of solar wind acceleration also emphasize the important role played by the highly structured coronal magnetic field [8.133, 150, 151, 211] in channeling the plasma flow and guiding the energy and momentum deposition.

As a consequence of the observed inhomogeneity of the solar wind source regions, there appear to exist no unique and global boundary values for the ion and electron velocities, temperatures, and densities, or for the particles' heat fluxes and the wave energy fluxes, for example of Alfvén waves. The determination of the ionic charge states of heavy minor ions in energetic particle events and the analysis of the *in situ* chemical composition [8.6, 31, 99, 100, 102, 130] allow inferences of the coronal "freezing-in" temperatures. These measurements yield different values for the various ionic species, varying typically about several 10^6K, and indicate a high spatial variability of the electron temperature or possibly hint at the existence of non-Maxwellian electron distribution functions in the coronal source regions [8.223].

On the other hand, it has been theoretically shown [8.160] that the simultaneous consideration of the observed proton flux (measured in the narrow range between 2×10^8 and 4×10^8 particles $cm^{-2}s^{-1}$) and the observed electron pressure at the coronal base ($10^{14} \lesssim n_{eo}T_{eo} \lesssim 10^{15}$K cm^{-3}, [8.54]) yield a relatively strong constraint on the coronal hole temperature, if evaluated by a single fluid model. This estimated upper limit is about $2.0-2.6 \times 10^6$K pertaining to the plasma as a whole. However, collisional scales, ionization and recombination rates are such [8.100] that temperature equilibrium between various ions and electrons cannot be expected. Therefore, it has been suggested [8.31, 151] that we seriously consider thermal diffusion, differential mass outflow, and departures from ionization equilibrium in coronal-hole models.

The most reliable data on the plasma density profiles are obtained from the photospheric light scattered by coronal electrons. Their density has also been modeled in [8.211] for a large observed polar hole and was found to decrease rapidly from 5×10^5 to $1.2 \times 10^4 (\mathrm{cm}^{-3})$ between 2 and 5 $R_\odot$. Correspondingly, the radial flow velocity was inferred to increase steeply from 86 to 451 km s^{-1} with a transonic point between 2.2 and 3 $R_\odot$ for coronal temperatures between 1 to 2.5×10^6 K. Outflow velocities of the plasma in the solar wind acceleration region, as well as electron densities and temperatures, can best be obtained by spectroscopic techniques particularly designed for probing the physical state of the transition zone and the outer corona [8.26, 286, 288]. Extreme ultraviolet (EUV) spectrometers have detected a systematic plasma velocity at the base of a coronal hole (for example by Doppler blue shifts in coronal and transition region MgX and OV emission lines) of the order of 5–10 km s^{-1} [8.243]. Similar measurements related to flow in closed field structures are desirable but still missing. These observations are strongly complemented by the indirect method [8.288, 289] of inferring outflow velocities at heights out to 3–5 $R_\odot$ through application of the Doppler dimming technique, in particular for the EUV spectral lines of Fe XII with good sensitivity in the velocity range 20–70 km s^{-1}.

Extensive future use of these methods appears promising, since they seem to serve well as tools for probing the bulk plasma using heavy ions as tracers [8.62]. Consequently, it is of particular theoretical interest to investigate how these minor ions couple to the dominant electrons, protons, and alpha particles under coronal conditions, and if they reliably indicate bulk characteristics. More work needs to be done to answer these crucial questions, which are closely related to the basic problem of whether and where ionization equilibrium actually applies to the corona [8.31, 99, 223].

Analyses of coronal plasma observations acquired in rockets flights [8.288] yield the following empirical constraints: a nearly constant hydrogen kinetic temperature of about 10^6 K between 1.5 and 4 $R_\odot$, subsonic flow for $r < 4 R_\odot$, and an upper limit of 140 km s^{-1} for rms velocities, corresponding to a Doppler broadening of L_α lines of H I at a distance of 4 $R_\odot$. This value poses an upper bound to possible wave velocity amplitudes, for example Alfvén waves. Radio scintillation and scattering measurements performed on various spacecraft [8.22, 290] provide a means of probing the solar wind electron density and flow speed and the magnetic field strength. For example, the average flow velocities inferred from density fluctuation spectra were found to be 30, 200, and 350 km s^{-1} at a distance of 2.17, 10.5, and 30.5 $R_\odot$, respectively ([8.290] and references therein). Radio observations using spacecraft signals can thus complement other measurements, and future efforts should be made to exploit this means fully by conducting more observations in the source region of the solar wind.

The observational evidence reported so far is consistent with the traditional-model idea of a gradually accelerating, fairly homogeneous solar wind flow, start-

Table 8.5. Physical characteristics of jets [8.30]

Parameter	
Maximum velocity [km s^{-1}]	400
Density [cm^{-3}]	7×10^9
Size, large event [km]	3000
Kinetic energy, large event [erg]	3×10^{26}
Mass, large event [g]	3×10^{11}
Birthrate [s^{-1}]	5 – 370
Power over whole solar surface [erg s^{-1}]	7×10^{27} *
Energy flux [erg cm^{-2} s^{-1}]	1×10^5
Maximum lifetime [s]	80
Mass flux into the corona [g s^{-1}]	6×10^{12} *

* Assuming a birthrate of 24 large events per second

ing at some km s^{-1} and reaching its sonic point after few solar radii. However, spectacular measurements of high-energy ion jets in the low corona [8.30, 49] suggest a reconsideration of the coronal boundary values of the solar wind velocity. These turbulent events and jets have been revealed by high-spatial-resolution observations of the ultraviolet solar spectrum of the sun. Their basic characteristics are compiled in Table 8.5, which is taken from [8.30]. Detailed correlations of these jets with other solar phenomena on the disk are as yet very speculative and need further clarification. A connection between jets and spicules, however, can already be excluded. Most likely they originate from small-scale (≈ 1000 km) active magnetic regions, such as exploding loops or possibly X-ray bright points.

Given the features in Table 8.5, it has been argued in [8.30] that jets may account for the entire mass and energy flux required to drive the solar wind. If so, considerable work remains to be done in order to incorporate these new boundary conditions into the conventional solar wind fluid models. In particular, how do jets fill the wide area of, e.g., a coronal hole? Do constraints other than zero velocity at the coronal base have to be imposed, and which ones? Are there any remnants of this jet structure in a highly inhomogeneous primordial solar wind detectable in interplanetary space? What do jets mean for the problem of coronal heating? Do we possibly need a new concept of an ion heat flux due to jets in the acceleration region? It appears that jets represent a challenging problem for any future solar wind modeling.

8.7.2 A Brief Account of Coronal Structures and Associated Interplanetary Phenomena

The solar corona is now known to be magnetically structured in the form of open field regions usually related to coronal holes and of closed regions and intertwined loops of multiple sizes. Whereas the association of holes with recurrent solar wind high-speed streams is well established, the coronal sources of low-speed solar wind, in which magnetic sector boundaries and current sheets are embedded [8.34, 35], remain less clear and demand further studies. Understanding of the dynamics of small-scale magnetic structures in the low corona is prerequisite for a theoretical explanation of the plasma phenomena associated with solar activity, such as solar flares, prominences, and coronal transients [8.53, 238, 264] which manifest themselves in various forms of interplanetary disturbances, such as mass ejections and shock waves ([8.240, 255, 256, 263, 269] and references therein) or noncompressive density enhancements [8.105]. Some typical energy and mass fluxes in these events are listed in [8.38, 257]. In the past few years, magnetic loops or clouds have been observed or inferred to exist in the solar wind [8.36–38, 148], often in close connection with shocks, stream interfaces, or coronal mass ejections. It has been suggested that they originate from coronal transients [8.255]. Many basic features of magnetic structures such as prominences, sun spots, transients, and flares are still not well understood. Neither are their associated plasma phenomena such as mass ejections, shocks, energetic particles and solar cosmic rays, and plasma jets, although research in solar magnetohydrodynamics [8.238] is steadily advancing.

The close correlation between plasma phenomena and flow patterns in interplanetary space with the magnetic geometry of the coronal sources, as described briefly above, leads to many critical questions concerning future solar wind modeling. Certainly, a "quiet" solar wind [8.131, 213] does not exist. During their close approach to the sun the *Helios* solar probes revealed that even recurrent fast streams are highly variable in themselves and seem to be built up of smaller-scale stream tubes with enormous relative velocity differences ($> 100\,\mathrm{km\,s^{-1}}$). These bulk velocity fluctuations are partly caused by Alfvén waves [8.15, 47, 182]. However, the fuzzy plateaus of high-speed streams as visible in Fig. 8.4 suggest a composition by many small-scale flux tubes and also possibly indicate the existence of similar structures in the coronal source regions of the fast flow [8.272]. The inhomogeneity and intermittency of the slow solar wind has long been recognized [8.38, 76, 213, 215, 257]. It may be naturally explained by the inhomogeneity and time-variability of that part of the solar atmosphere seated in mostly closed magnetic field regions which are magnetically open only transiently [8.5].

Generally, mapping of miniature interplanetary solar wind structures back to their coronal counterparts encounters fundamental problems, such as how flux tube areas at the two heliocentric distances correspond to each other. Possibly various flux tubes interact dynamically during their radial evolution such that their areas change differently than for a spherical expansion [8.272]. Modeling of solar wind flow patterns on small scales ($\ll 1$ AU) has not yet been carried out, but it might be required in order to meet observational constraints on the flow dynamics, energetics, and mass load correctly. The governing equations for solar wind expansion along individual stream tubes are known in principle ([8.282], and references therein), but they have not been applied for detailed interplanetary modeling except for the evolution of corotating shocks. It should be pointed out, however, that considerable progress has been made in understanding the large-scale dynamics of quasi-steady, corotating solar wind structures beyond 0.3 AU [8.236].

To conclude, it appears necessary to repeat the "search for a structureless solar wind", which can be observationally well defined on certain spatial and temporal scales and with respect to its coronal boundaries, and thus be well prepared and adopted for a description by comparatively simple theoretical models. At least three types of flow must be accounted for: "quiet" high-speed and "quiet" low-speed wind originating from stationary large-scale coronal sources and "intermittent" or "disturbed" flows associated with local impulsive plasma ejections from the sun. Typical parameters for these three types of flow are compared in [8.254, 257] and contained in the reviews [8.53, 255].

8.7.3 Interplanetary and Coronal Energy Fluxes and Solar Wind Constants of Motion

The current knowledge of the coronal plasma boundary conditions and the magnetic structure of the corona has been briefly discussed in the previous sections. The existing information on the coronal plasma state is still rather sparse, in particular with regard to the energy input in terms of various possible energy fluxes at the coronal base (for a good recent review see [8.287]). These are the fluxes of bulk kinetic energy, the convected enthalpy flux, the electron and ion heat fluxes, and the energy fluxes carried by various waves (MHD or kinetic), which are believed, though as yet not directly observed, to emanate from the solar surface and to penetrate the different solar atmospheric layers [8.122]. Some of these waves may be readily damped and heat the corona [8.15, 119], others can make it through out into the solar wind where they appear as remnants of coronal wave turbulence. However, to this day none of these energy fluxes has been directly measured. The first solid experimental data have been obtained beyond 0.3 AU from the *Helios* twin probes and other space probes. Observational constraints on solar wind acceleration and expansion can thus be derived from the energy fluxes as determined by *in situ* plasma and magnetic field measurements [8.76, 193].

At larger heliocentric distances we may disregard the complex magnetic field topology and assume the simple law $B_r \propto r^{-2}$ for the radial magnetic field component. The large-scale, average field in the ecliptic plane is organized in the standard spiral pattern. For a steady-state, spherically symmetric, single-fluid solar wind model there exist five global constants of motion [8.180, 193, 228, 280]: the radial magnetic flux

$$F_B = r^2 B_r \qquad (8.23)$$

and the out-of-ecliptic component of the electric field obtained from Faraday's law yielding

$$-F_B \Omega_\odot = r(u_r B_\phi - u_\phi B_r) , \qquad (8.24)$$

where $\Omega_\odot$ is the solar angular rotation frequency. The flow velocity is $u = (u_r, u_\phi)$ and the magnetic field $B = (B_r, B_\phi)$. The equation of continuity implies that the mass flux

$$\dot{M} = r^2 \varrho u_r \tag{8.25}$$

is conserved. Balance of mechanical and magnetic stresses requires $\dot{M}L$ to be a constant of motion, whereby

$$L = r \left\{ u_\phi - \frac{B_\phi}{4\pi} \frac{F_B}{\dot{M}} \left(1 - \frac{4\pi(p_\parallel - p_\perp)}{B^2} \right) \right\} \tag{8.26}$$

is the specific angular momentum per proton mass. This quantity is related to the Alfvén critical radius r_A [8.192, 237] by the simple relation $L = \Omega_\odot r_A^2$. Here $p_{\parallel,\perp}$ are the parallel and perpendicular total plasma pressures. Finally, the total energy flux $S_r = \dot{M}\varepsilon$ is conserved during the radial expansion, whereby ε, as the specific energy per amu of the fluid parcel, is most relevant for a determination of the radial bulk flow speed and is given by

$$\varepsilon = \tfrac{1}{2}u_r^2 + \phi_G + \phi_\Omega + \phi_T . \tag{8.27}$$

Since all potentials vanish at infinity, the streamline constant ε is directly related to the asymptotic flow speed by $\varepsilon = \tfrac{1}{2}u_\infty^2$. The "potentials" are defined as gravitational, $\phi_G = -\tfrac{1}{2}v_\infty^2 (R_\odot/r)$, where $v_\infty = 618\,\mathrm{km\,s^{-1}}$ is the escape speed from the solar surface, and as "magnetorotational",

$$\phi_\Omega = \frac{1}{2}u_\phi^2 + \frac{B_\phi^2}{4\pi\varrho} - \frac{u_\phi B_\phi F_B}{4\pi\dot{M}} , \tag{8.28}$$

combining the terms associated with the azimuthal kinetic energy and the Poynting flux. The "thermal" potential is given by

$$\phi_T = (\tfrac{1}{2}(\mathrm{Tr}\,\mathsf{P})\boldsymbol{u} + \mathsf{P} \cdot \boldsymbol{u} + \boldsymbol{Q})_r / \varrho u_r , \tag{8.29}$$

which is the enthalpy plus heat flux divided by the mass flux. More explicit algebraic details can be found in [8.193]. In (8.29) the pressure tensor P and heat flux vector $\boldsymbol{Q}$ pertain to the plasma as a whole and have been obtained by summing over the individual contributions of electrons, protons and alpha particles. In a polytropic single-fluid model [8.228] one would simply have

$$\phi_T(\varrho) = \frac{\gamma}{\gamma - 1} c_0^2 \left(\frac{\varrho}{\varrho_0} \right)^{\gamma-1} \tag{8.30}$$

where c_0 and ϱ_0 are the thermal speed and mass density at some reference level r_0, and γ is the constant polytropic index. Inserting (8.30) into ε then closes the system of equations. Generally, however, an additional equation for ϕ_T or the pressure is required which involves the full complexity of the internal energy of the multicomponent solar wind as discussed in the previous sections.

The set (8.23–27) are the solar wind model equations of motions in an integrated form and implicitly determine $\boldsymbol{u}, \boldsymbol{B}$, and ϱ as functions of r. In the case of an azimuthally structured solar wind the above, comparatively simple, streamline constants no longer apply because parameters will then explicitly depend on the ϕ-coordinate in addition to the radial distance r. As pointed out already, in the corona the assumption of spherical symmetry certainly breaks down. This fact casts doubts on a straightforward extrapolation of $L, \varepsilon, \dot{M}$, and F_B from the interplanetary medium into the corona. However, in view of the lack of more realistic theoretical models with which to compare, this extrapolation presently remains the only possibility of placing empirical constraints on coronal expansion. The rapid divergence of coronal-hole magnetic fields can be incorporated into our scheme by replacing r^2 in (8.23) and (8.25) by a radially dependent area function $A(r)$ [8.126, 133]. This replacement has an essential impact on ϕ_T, since rapid flow-tube divergence allows for an effective enhancement of the heat conduction energy supply [8.127] by a factor of $A(r)/r^2$ compared to spherical expansion. Analysis of observed coronal holes [8.133] indicates that this area ratio ranges from about 3 to 7, based on estimates of the spatial angular extent of a hole at 1 $R_\odot$ and the related solar wind stream at 1 AU.

The constants of motion associated with the solar wind energy and mass flux (equivalent to the energy and mass-loss rate of the sun) are plotted versus heliocentric radial distance between 0.3 and 1 AU in Fig. 8.32. Average parameters for three solar wind speed classes are contained in Table 8.6, taken from [8.193], which presents a more detailed discussion of these results. The dispersion of ε and $\dot{M}$ with the proton flow speed is obvious. All quantities are fairly constant and support the assumption of a spherical expansion. Truly remarkable is the fact that the total energy flux $\dot{M}(\varepsilon + \frac{1}{2}v_\infty^2)$ appears to be a global constant ($\approx 3.6 \times 10^{26}$ erg s^{-1}sr^{-1}) and does not reflect the stream structure of the solar wind. A similar result has been found in other data sets [8.254, 257] with a much larger statistical basis. The origin of this phenomenon is not explained. A resolution of this puzzle presumably provides a key for the understanding of the solar wind acceleration process. Similar plots for $L, \dot{M}L$, and F_B, as in Fig. 8.32, can be found in [8.193], where ϕ_Ω is also discussed more extensively.

Table 8.6. Conserved quantities averaged over heliocentric distance [8.193]

Parameter	$v_p < 400$ km s^{-1}	400 km s^{-1} $\leq v_p$ $v_p \leq 600$ km s^{-1}	600 km s$^{-1} < v_p$	Total
$\dot{M}$, 10^{11}g s^{-1} sr^{-1}	1.49 ± 0.61	1.09 ± 0.62	0.81 ± 0.22	1.02 ± 0.53
L, AU km s^{-1}	10.6 ± 16.0	2.34 ± 17.0	-1.55 ± 20.85	1.71 ± 19.11
$\dot{M}L$, 10^{30}dyne cm sr^{-1}	2.59 ± 4.66	0.56 ± 4.36	-0.37 ± 2.91	0.42 ± 3.95
ε, $(100$ km s$^{-1})^2$	6.86 ± 1.08	13.65 ± 3.49	24.18 ± 3.49	17.4 ± 7.2
$\dot{M}\varepsilon$, 10^{26}erg s^{-1} sr^{-1}	1.03 ± 0.47	1.42 ± 0.68	1.94 ± 0.56	1.60 ± 0.68
F_B, nT AU^{2}sr^{-1}	2.86 ± 1.87	3.54 ± 1.73	3.15 ± 1.49	3.28 ± 1.67
$\dot{M}(\varepsilon + v_\infty^2/2)$, 10^{26}erg s^{-1} sr^{-1}	3.88 ± 1.61	3.51 ± 1.79	3.48 ± 0.94	3.55 ± 1.45

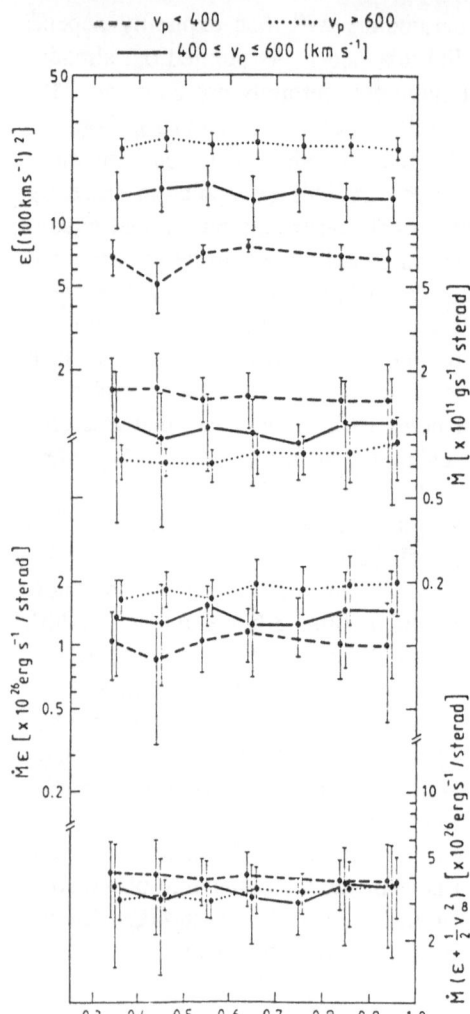

Fig. 8.32. Constants of motion conserved for solar wind flow in the ecliptic plane with spherical symmetry [8.193]

The distribution of solar wind angular momentum between particles and magnetic field, as observed by the *Helios* solar probes, has been exploited to infer parameters at the Alfvén critical point [8.180, 192, 237], which was estimated to be located at $r_A = 14$–17 $(34$–$49)$ $R_\odot$ for high-speed (low-speed) wind. The average critical mass density is readily derived as $\varrho_A = 4\pi(\dot{M}/F_B)^2$ and yields $n_A = 978$ $(4017)\,\text{cm}^{-3}$ for fast (slow) solar wind flow. The average radial acceleration was found to be $\approx 50\,\text{km s}^{-1}/R_\odot$ at the Alfvén point of high-velocity streams. In general, the concept of angular momentum transport as developed in [8.280] was observationally well confirmed. In slow wind the particles appear to dominate the angular momentum loss of the sun, whereas in fast streams the magnetic field stresses are equally important or slightly predominant.

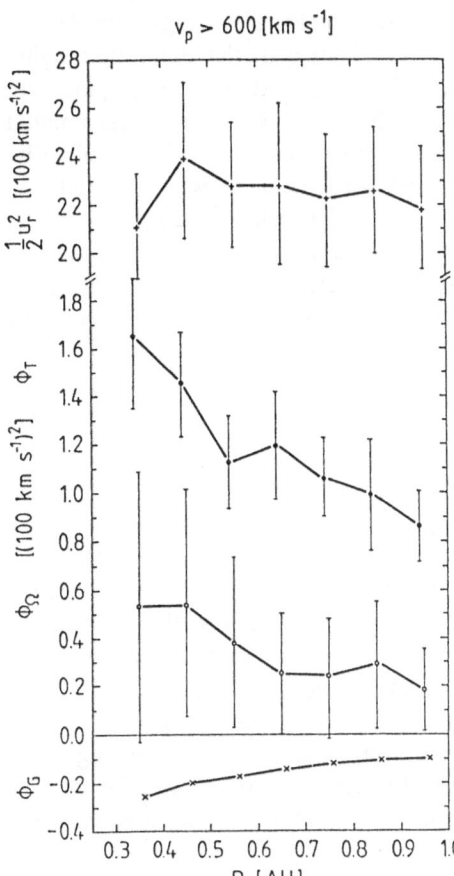

$v_p > 600\,[\mathrm{km\ s^{-1}}]$

Fig. 8.33. Kinetic energy per amu and potentials for high-speed flows with $V_p > 600\,\mathrm{km\,s^{-1}}$ [8.193]

The specific energy ε can be broken into kinetic energy and potentials according to (8.27). Radial profiles of these quantities are displayed in Fig. 8.33 for solar wind streams with a proton speed larger than $600\,\mathrm{km\,s^{-1}}$. Apparently, beyond 0.3 AU the high-speed wind has almost reached its asymptotic state, since only about ten percent of ε is due to potential energy. The potential ϕ_T decreases by about a factor of two between 0.3 and 1 AU. The essence of the concept of a thermally driven wind is that at the coronal base (where $u_r \approx 0$) with radius r_0 we have

$$\varepsilon + \frac{1}{2}v_\infty^2 \frac{R_\odot}{r_0} = \phi_T(r_0) \ . \tag{8.31}$$

Thus the thermal potential at r_0 must be equal to the asymptotic bulk kinetic plus gravitational escape energy. This requirement imposes a rather severe constraint on the microphysics of the coronal enthalpy and heat flux.

According to Table 8.6, bottom row, we find the corresponding average energy flux at 1 AU to be 3.6×10^{26} erg s^{-1}, independent of the stream structure. Spherical mapping back to the solar surface yields the energy flux density $F_\odot = 7.3 \times 10^4$ erg cm^{-2}s^{-1}. This value could be larger by the area ratio ranging from 3 to 7 for some observed coronal holes. Consequently the flux requirement for a thermally driven fast wind out of rapidly diverging holes would amount to 2–5 $\times 10^5$ erg cm^{-2}s^{-1}. This energy flux is then believed to be supplied by enthalpy and heat conduction alone. However, it has been shown by various authors [8.162] that, given these constraints, conductive solar wind models cannot explain the existence of the observed high-speed streams. Therefore, the addition of energy and momentum has been considered [8.125, 161] as a possible remedy to the acceleration problem. Formally, we can incorporate these sources by adding another potential to ε in the form

$$\phi_{ADD} = \int_{r_0}^{r} dr' \left(D(r') + A(r')H(r')/\dot{M} \right) , \qquad (8.32)$$

where $D(r)$ and $H(r)$ are the volumetric rates at which momentum and heat are added to the plasma. It was found [8.161, 162] that heat addition in the region of supersonic flow invariably leads to an increased flow speed at 1 AU. Similarly, addition of momentum or direct acceleration in the supersonic regime again produces larger flow speeds at 1 AU in better agreement with observations. These phenomenological studies undoubtedly have their merits and help to build up intuition on the physics of solar wind acceleration. On the other hand, the source terms D and H still need to be specified by real physical processes. For example, fast and slow magnetoacoustic wave damping has been suggested to provide the predominant ion heating in the corona and acceleration region. Much work remains to be done so as to justify the existence of an energy flux like $\dot{M}\phi_{ADD}$ in the corona and to clarify the associated physics.

Alfvén waves have received most attention as a means of achieving high-speed flows. First, they are ubiquitously observed in high-speed streams, and second, they were shown to provide a body acceleration force proportional to the gradient of their energy density. The energy flux associated with undamped Alfvén waves has been thoroughly studied. For outward-propagating waves one can write [8.15, 119, 120]

$$\boldsymbol{F}_A = \varrho(\delta v)^2 \left\{ \frac{3}{2} \boldsymbol{u} + \boldsymbol{B} \frac{1}{\sqrt{4\pi\varrho}} \left(1 - \frac{4\pi(p_\parallel - p_\perp)}{B^2} \right)^{1/2} \right\} . \qquad (8.33)$$

The Alfvén wave potential may be defined as $\phi_A = \boldsymbol{F}_A \cdot \hat{\boldsymbol{r}}/\varrho u_r$ and the Alfvénic Mach number as $M_A = u_r/v_{A,r}$, where $v_{A,r}$ is the radial component of the Alfvén velocity. The velocity-fluctuation amplitude [8.19] varies radially as

$$(\delta v)^2 \sim \frac{F_B}{\dot{M}} \frac{M_A}{(1 + M_A)^2} . \qquad (8.34)$$

Therefore the potential can be written as

$$\phi_A = \phi_{A0} \left(\frac{2 + 3M_A}{2 + 3M_{A0}} \right) \left(\frac{1 + M_{A0}}{1 + M_A} \right)^2 , \qquad (8.35)$$

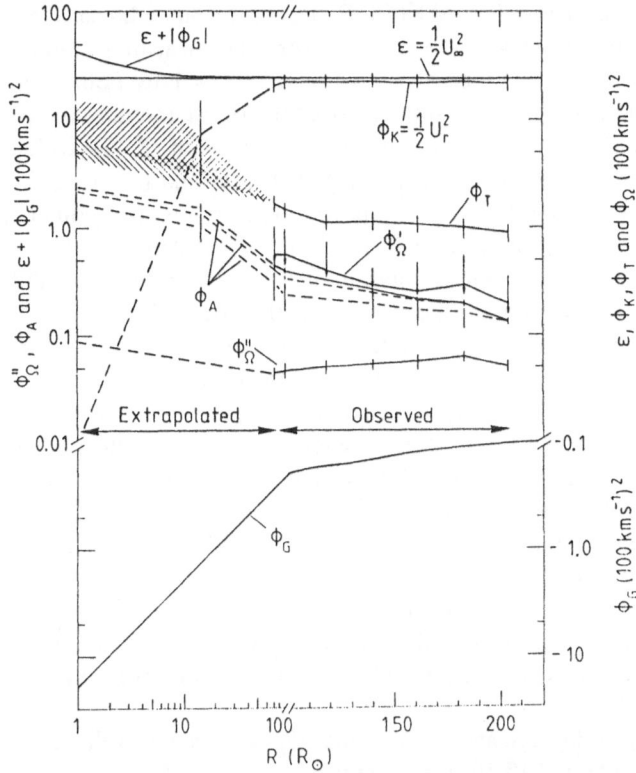

Fig. 8.34. Streamline constant ε, the various potentials, and the kinetic energy as observed *in situ* and theoretically extrapolated (*broken line*) as a function of heliocentric distance [8.178]

where the index zero refers to any coronal reference level r_0. Beyond 0.3 AU the Alfvén wave potential can be experimentally determined from the relation $F_{A,r} r^2 / \dot{M} = \phi_A(r)$. Actual numbers, published first in [8.20], are $F_{A,r} = 0.013$ erg cm^{-2}s^{-1} at 1 AU. *Helios* measurements [8.47, 212] yielded maximum values as high as $F_{A,r} = 0.44$ erg cm^{-2}s^{-1} in high-speed streams at 0.3 AU, corresponding to an rms velocity amplitude of $\delta v \approx 77$ km s^{-1} for one-hour averages and to a potential of $\phi_A \approx (105$ km s$^{-1})^2$. Extrapolating this value back to the coronal base with the help of (8.35) with $M_A = 4$ gives $\phi_{A0} \approx (200$ km s$^{-1})^2$, which may be considered a typical value. According to Table 8.6 this number corresponds to about 17 % of ε in high-speed streams. In interplanetary space one finds the mean of ϕ_A/ε to be even smaller and to decrease from about 3 % to 1 % between 0.3 and 1 AU.

Using these numbers the radial course of $\phi_A(r)$ was calculated in [8.178]; it is plotted together with all the other constituents of ε in Fig. 8.34. Note that the radial scale is logarithmic below 100 $R_\odot$ and linear above. All potentials are plotted in units of $(100$ km s$^{-1})^2$ on a logarithmic scale extending over four orders of magnitude. The constant of motion $\varepsilon = 2.42 \times 10^5$ (km s$^{-1})^2$ corresponds to an asymptotic average flow speed $u_\infty = 695$ km s^{-1}. The kinetic energy is denoted by $\phi_K = \frac{1}{2} u_r^2$. Heavy lines correspond to *Helios* measurements, and broken

117

lines indicate extrapolations below 0.3 AU (63 $R_\odot$) back to the solar surface. Estimates of the Alfvén radius yield $r_A = 14\ R_\odot$ from the measured angular momentum in high-speed streams. Since at r_A we have $M_A = 1$ by definition, the flow speed at that point and ϕ_A can be inferred from the *in situ* measurements. The potential ϕ_A has then been extrapolated inwards by means of (8.35), which describes undamped wave propagation with the WKB theory, and by using the measured values at 0.95, 0.85, and 0.35 AU, respectively, as reference numbers. The three corresponding curves provide a measure of the typical variations in ϕ_A and are compatible with each other within measurement uncertainties (standard deviation bars are given for each radial distance bin of 0.1 AU width).

Apparently, the observed Alfvén wave potential ϕ_A as based on one-hour-variances, if extrapolated from 0.3 AU back to the sun, is insufficient to exert a strong influence on solar wind dynamics unless appreciable dissipation occurs inside 0.3 AU. Therefore in all Alfvén-wave-driven solar wind models [8.61, 120, 123, 124, 126, 128, 162, 264, 265, 273], which claim a reasonably good agreement with *in situ* constraints, the authors have to make assumptions about the dissipative input of waves into the nascent solar wind and the outer corona that cannot be easily confirmed by interplanetary measurements but must be taken for granted and be accepted by plausible theoretical arguments. This issue may be resolved if better coronal observations become available that allow us to draw detailed inferences about the plasma state and acceleration processes in the outer solar corona [8.62].

The two curves in Fig. 8.34 indicated by ϕ'_Ω and ϕ''_Ω correspond to the measured magnetorotational potential [8.193], whereby

$$\phi''_\Omega = \frac{1}{2r^2} L^2 \left\{ 1 + (2M_A^2 - 1)\left(1 - \left(\frac{r}{r_A}\right)^2\right)^2 \Big/ (1 - M_A^2)^2 \right\} \qquad (8.36)$$

is the theoretical potential depending on M_A. The potential ϕ''_Ω has been determined from the measurements by help of (8.28) and also includes, by time averaging, contributions from Alfvénic fluctuations. This fact explains the quite similar course of ϕ_A and ϕ''_Ω. This potential can be evaluated at the solar surface giving $\phi'_\Omega = (\Omega_\odot r_A)^2 \approx (26\,\mathrm{km\,s^{-1}})^2$, which is rather small. Consequently, the solar wind is not rotationally driven [8.14], and ϕ_Ω may be entirely neglected.

The effects of finite-amplitude compressional magnetohydrodynamic waves on the acceleration and angular momentum transport of the solar wind have been investigated theoretically in [8.180], by using a second-order perturbation scheme for the general stress tensor and by employing the WKB theory. In the Bernoulli equation (8.27) the fast and slow magnetoacoustic waves then appear through an additional acceleration potential of the form

$$\phi_w = \phi_{w0}\sqrt{\alpha}\, \frac{\left[\left(M_A \cdot (1 + \hat{P}_w) + \hat{V}_g\right) \cdot \hat{r}\right]}{\left[(M_A + \hat{V}_g) \cdot \hat{r} \left(\sqrt{\alpha} + M_A \cdot \hat{k}\right)\right]}, \qquad (8.37)$$

where the reference value ϕ_{w0} is again the wave energy flux divided by the mass flux $\dot{M}$ at the coronal reference level r_0. The other symbols are $M_A = u/V_A$, the Alfvénic Mach vector; $\hat{k}$, the unit wave vector; $\hat{r}$, the unit vector in the radial direction; $\alpha = C_{F,S}/V_A$, the normalized phase speed $C_{F,S}$ for the fast and slow wave; $\hat{V}_g$, the corresponding normalized group velocity; and finally $\hat{P}_w$, the dimensionless wave stress tensor [8.180]. At the solar surface the value ϕ_{w0} is attained for $M_A = 0$. Unfortunately, no firm empirical estimates exist of the energy flux of compressional waves in the solar atmosphere. Compressional waves are also attenuated faster than Alfvén waves and are probably completely damped out at 0.3 AU. Therefore, (8.37) could play the role of a "hidden" acceleration potential. In conclusion, the relevance of the potential ϕ_w for the acceleration of the solar wind still remains to be explored in theoretical models.

Figure 8.34 quite clearly illustrates the observational energy constraints to be placed on high-speed solar wind expansion. In order to create the solar wind, first a fluid parcel has to be given the specific escape energy $\phi_G(R_\odot) = \frac{1}{2}v_\infty^2$ so as to surmount the gravitational barrier. Second the coronal energy reservoir must also supply the asymptotic kinetic energy $\varepsilon = \frac{1}{2}u_\infty^2$. Their sum in equation (8.27) must then equal the sum of all the potentials involved in the source region, if the solar wind is assumed to start with zero velocity. Unfortunately, the relevance of the heuristic models, which solve the acceleration problem phenomenologically by energy and momentum addition as described by ϕ_{ADD} in (8.32), cannot be empirically confirmed or denied, because the region where these processes may operate has not yet been accessible to any *in situ* measurements.

8.7.4 Physics of the Thermal Potential

The physics of the thermal potential ϕ_T defined by (8.29) is very complicated and can be dealt with on various levels of sophistication. The simplest expression is (8.30) which was extensively exploited in the early days of solar wind modeling. The shortcomings of the polytropic equation of state have been discussed at length in Sect. 8.5. A constant polytropic index is not acceptable. If the full energy equations are considered instead, the basic question concerns the heat flux laws. Classical heat conduction and transport coefficients are not appropriate, neither in the interplanetary medium nor in coronal holes, as we have extensively elaborated in previous sections.

In Fig. 8.26 the observed total (ion) heat flux $Q(Q_i)$ has been logarithmically plotted versus heliocentric distance for various solar wind velocities. The radial decline is apparently rather steep, and it is difficult to extrapolate Q into the corona. One can derive from Table 8.6 that for spherical expansion a coronal energy flux of $F_\odot = 7.3 \times 10^4$ erg cm^{-2}s^{-1} is required in high-speed flows. Thus Q must increase by more than six orders of magnitude between 64 and 1 $R_\odot$ in order to match this value. By assuming, for the time being and the sake of simplicity, a single-fluid temperature $T_e = T_p$ and $T_e(r) = T_0(r/r_0)^{-\delta}$ we can, though admittedly in a crude and non-self-consistent fashion, estimate the Coulomb-collision-based thermal potential in the corona. In units of $(100\,\mathrm{km\,s^{-1}})^2$ and with $\tilde{r} = r/R_\odot$, and $\tilde{T}_0 = T_0/10^6 \mathrm{K}$ this is obtained using (8.29) as follows:

$$\phi_T(\tilde{r}) = 4.6\,\tilde{T}_0\tilde{r}^{-\delta} + 6.8\,\delta\tilde{T}_0^{7/2}\tilde{r}^{(1-7/2\delta)} . \tag{8.38}$$

Here Spitzer's law, $Q_e = -\kappa T_e^{5/2} dT_e/dr$, with $\kappa = 7.8 \times 10^{-7}$ erg cm^{-1}s^{-1} K$^{-7/2}$, has been employed [8.131], and $\dot{M} = 0.8 \times 10^{11}$ g s^{-1} from Table 8.6 with a relative alpha-particle density of $n_\alpha/n_p = 5$ % has been used. At $\tilde{r} = 1$ we should have $\phi_T(1) = 43.27$ according to (8.31). The required coronal temperatures are $\tilde{T}_0 = 2.25, 1.61, 1.48$ for $\delta = 2/7, 1, 4/3$. These values appear to be somewhat high for coronal holes. The gray shaded part of the ϕ_T curve in Fig. 8.34 corresponds to an interpolation as in (8.38) between an assumed coronal ϕ_T and the firm, measured value of $\phi_T = 1.67$ (100 km s^{-1})^2 at $\tilde{r} = 64$. With a slope of $\delta = 2/7$ and $\tilde{T}_0 = 1$ (1.5) one obtains the potential $\phi_T = 6.54$ (14.93) and 4.32 (11.60) (100 km s^{-1})^2 at $\tilde{r} = 1$ and 64. As a result, classical electron heat conduction yields much too low coronal and too high interplanetary thermal potentials for realistic temperatures in the corona. With steep initial temperature gradients ($\delta > 1$) the energy flux requirement on fast solar wind expansion in the corona may be met, but if the same conduction law is subsequently applied to the solar wind expansion out to 0.3 AU (*Helios* perihelion) the theoretical heat flux falls entirely below the *in situ* constraint.

Consequently, the role of the various particle populations for the evolution of internal plasma energy has to be reexamined, in particular the role of suprathermal electrons [8.221, 222] which carry most of the interplanetary heat flux. The skewness (double beams) of proton distributions and the differential streaming between alpha particles and protons at a sizable fraction of the local Alfvén speed gives rise to an ion heat flux of about 10^{-2} erg cm^{-2}s^{-1} at 0.3 AU. The interplanetary observations discussed before yield different temperatures for the various ions and electrons and the kinetic degrees of freedom, which indicates considerable deviations from thermodynamic equilibrium. From these remarks it is clear that the calculation of ϕ_T demands a detailed multifluid treatment, or, even better, a fully kinetic treatment.

There are probably several ways to deal with this problem. We have pointed out already that high-frequency turbulence may significantly modify the true transport compared to purely collisional transport. However, such a theory becomes rather complicated and requires self-consistent calculations also to be done for the temporal and spatial evolution of the spectral energy densities of the waves involved in the particle-scattering process. But even if wave–particle interactions were absent, the effects of Coulomb collisions must be reexamined within a local–global transport theory [8.221, 260, 261], which is more appropiate for the tenuous solar wind plasma than the classical, local theory. A complete model should also correctly incorporate the high-energy strahl electrons, which are particularly sensitive to the coronal boundaries and the interplanetary magnetic field topology. The transport theory yet to be developed must pay special attention to the problem of multiple spatial scales. Since collisional free paths are highly velocity dependent, this fact has to be taken care of in a Vlasov–

Boltzmann approach. In this way one may hope to treat the tail electrons and their free paths adequately in comparison with the continuously changing fluid scale lengths in rapidly diverging coronal-hole geometries. Similar comments apply to the collisional transport of the ions.

It is generally believed, probably owing to a biased view from the early *in situ* observations at 1 AU, that for solar wind dynamics the ion heat flux is entirely negligible. Figure 8.26 tells us, however, that the total ion heat flux at 0.3 AU, caused by pronounced differential ion streaming, can already amount to forty percent of the electron heat flux. Extrapolating this trend into the corona suggests that a sizable fraction of the total heat flux in coronal holes might be due to the skewness of the overall ion distribution function in terms of H^+–He^{2+} differential motion or ion double beams. As discussed already, the drifts at 0.3 AU are typically of the order of the Alfvén speed. Extrapolated to the corona this scaling would imply rather large drifts there. If these invoked double-beams really exist in the corona, perhaps as the remnants of, or associated with, the observed jets discussed before, then sizable coronal ion heat fluxes may not be unlikely. Clearly, considerable work remains to be done to confirm this conjecture and the feasibility of such a new heat flux concept.

8.8 Conclusions

In the previous sections the impasse of current solar wind kinetic and transport theory has been extensively described. As the fast solar wind plasma may already become collisionless deep in coronal holes, the particle distributions in the solar wind source regions are expected to develop non-Maxwellian features and velocity-space anisotropies which then will drive various plasma instabilities. Our present-day understanding of the role of microinstabilities in the interplanetary plasma is still incomplete and far from being satisfying. A new transport concept, revising the classical Coulomb-collision picture, may finally evolve from future studies. A kinetic description of the transition from locally collision-dominated to collisionless behavior is urgently needed for the lower solar corona. Yet Coulomb dissipation may always play an important role in the global scale of the whole solar system.

The reaction of distributions to the observed high-frequency wave turbulence ought to be better understood. Quasi-linear concepts of wave–particle interactions should be applied to the inhomogeneous solar wind plasma in order to achieve consistency between models and real distributions, which are usually found to be at the margin of stability. It appears that the new transport theory and the resulting physics of the thermal potential ϕ_T will be rather complicated and may differ radically from the older concepts. Ultimately, further advances must also come from improved observations of the coronal plasma state, and from better and more detailed measurements of the relevant plasma parameters, energy fluxes, and distribution functions.

Closely connected to the above problems are the questions of slow- and fast-mode wave damping, and nonlinear processes leading to damping of Alfvén waves. These mechanisms have frequently been invoked so as to achieve extended heating of, predominantly, ions in the corona and the solar wind, where violation of adiabatic invariance indeed suggests ion heating by waves. A theory of fast- and slow-wave generation at the sun is needed. The effects of nonlinearity and spatial inhomogeneity, as well as Landau damping of these modes, must be studied more thoroughly. A rigorous theory of transport and dissipation of energy and momentum by coronal magnetosonic waves is desirable and necessary to specify the physics of the phenomenological potential ϕ_{ADD} as well as for ϕ_w. All these questions are intimately related to the very basic field of research concerned with the spatial and temporal evolution of MHD turbulence in the solar wind. In order to make further progress here, better *in situ* or indirect measurements of coronal MHD fluctuations and their associated energy fluxes are urgently needed. These observations are, to our mind, prerequisite to a solution of the solar wind acceleration problem.

It cannot be excluded that, in the past, a possible mechanism for solar wind propulsion was entirely overlooked. Figure 8.34 may suggest that there is a "missing potential problem", unless we invoke the existence of an appropriate ϕ_{ADD} or wave potential or alter the boundary condition for the radial flow speed at the solar surface. If jets can indeed be shown to contribute significantly to the coronal ion heat flux, the role of ion heat conduction has to be considered and evaluated anew and the physics of ϕ_T modified accordingly. All the theoretical tasks mentioned are rendered even more difficult when the complex coronal magnetic field topology is fully accounted for. This appears, however, to be unavoidable, since active phenomena on the sun reveal the prominent role played by the magnetic field. Similarly, the coronal magnetic field is believed to channel the initial solar wind plasma flow and to strongly influence its acceleration in the solar source regions.

Acknowledgments. I thank P.W. Daly, R. Hernández, B. Inhester, R. Schwenn, and C.Y. Tu for a careful reading of the manuscript and many valuable comments and suggestions. I am also grateful to S.J. Schwartz and W.C. Feldman for their constructive review of the manuscript and many helpful suggestions for improvement.

References

8.1   Abraham-Shrauner, B., W.C. Feldman, Whistler heat flux instability in the solar wind with bi-Lorentzian velocity distributions, J. Geophys. Res., **82**, 1889–1892, 1977.

8.2   Abraham-Shrauner, B., W.C. Feldman, Nonlinear Alfvén waves in high-speed solar wind streams, J. Geophys. Res., **82**, 618–624, 1977.

8.3   Abraham-Shrauner, B., J.R. Asbridge, S.J. Bame, W.C. Feldman, Proton-driven electromagnetic instabilities in high-speed solar wind streams, J. Geophys. Res., **84**, 553–559, 1979.

8.4   Asbridge, J.R., S.J. Bame, W.C. Feldman, M.D. Montgomery, Helium and hydrogen velocity differences in the solar wind, J. Geophys. Res., **81**, 2719–2727, 1976.

8.5 Axford, W.I., The solar wind, Solar Phys., **100**, 575–586, 1985.

8.6 Bame, S.J., J.R. Asbridge, W.C. Feldman, P.D. Kearny, The quiet corona: temperature and temperature gradient, Solar Phys., **35**, 137–142, 1974.

8.7 Bame, S.J., J.R. Asbridge, W.C. Feldman, M.D. Montgomery, P.D. Kearney, Solar wind heavy ion abundances, Solar Phys., **43**, 463–473, 1975.

8.8 Bame, S.J., J.R. Asbridge, W.C. Feldman, S.P. Gary, M.D. Montgomery, Evidence for local ion heating in solar wind high speed streams, Geophys. Res. Lett., **2**, 373–375, 1975.

8.9 Barakat, A.R., R.W. Schunk, Transport equations for multicomponent anisotropic space plasmas: A review, Plasma Physics, **24**, 389–418, 1982.

8.10 Barnes, A., Collisionless damping of hydromagnetic waves, Phys. Fluids, **9**, e.g. 1483, 1966.

8.11 Barnes, A., Stochastic electron heating and hydromagnetic wave damping, Phys. Fluids, **9**, e.g. 2427, 1967.

8.12 Barnes, A., Quasi linear theory of hydromagnetic waves in collisionless plasma, Phys. Fluids, **11**, 2644–2654, 1968.

8.13 Barnes, A., Collisionless heating of the solar wind plasma, 2. Application of the theory of plasma heating by hydromagnetic waves, Astrophys. J., **155**, 311–321, 1969.

8.14 Barnes, A., Acceleration of the solar wind by the interplanetary magnetic field, Astrophys. J., **188**, 645–648, 1974.

8.15 Barnes, A., Hydromagnetic waves and turbulence and in the solar wind, in *Solar System Plasma Physics*, Vol I, ed. by E.N. Parker, C.F. Kennel, L.J. Lanzerotti, North-Holland, Amsterdam, 249–319, 1979.

8.16 Bavassano, B., M. Dobrowolny, F. Mariani, N.F. Ness, Radial evolution of power spectra of interplanetary Alfvénic turbulence, J. Geophys. Res., **87**, 3617–3622, 1982.

8.17 Behannon, K.W., Observations of the interplanetary magnetic field between 0.41 and 1 AU by the Mariner 10 spacecraft, Doc. 692-76-2, Goddard Space Flight Center, Greenbelt, Md., 1976.

8.18 Beinroth, H.J., F.M. Neubauer, Properties of whistler-mode waves between 0.3 and 1 AU from Helios observations, J. Geophys. Res., **86**, 7755–7760, 1981.

8.19 Belcher, J.W., Alfvénic wave pressure and the solar wind, Astrophys. J., **168**, 509–524, 1971.

8.20 Belcher, J.W., L. Davis, Large-amplitude Alfvén waves in the interplanetary medium, J. Geophys. Res., **76**, 3534–3563, 1971.

8.21 Belcher, J.W., H.S. Bridge, A.J. Lazarus, J.D. Sullivan, Preliminary results from the Voyager solar wind experiment, in *Solar Wind Four*, ed. by H. Rosenbauer, Report MPAE-W-100-81-31, Max-Planck-Institut für Aeronomie, Katlenburg-Lindau, F.R.Germany, 131–142, 1981.

8.22 Bird, M.K., Coronal investigations with occulted spacecraft signals, Space Sci. Rev., **33**, 99–126, 1982.

8.23 Bird, M.K., P. Edenhofer, Remote sensing of the solar corona, in *Physics of the Inner Heliosphere*, Vol 1, ed. by R. Schwenn and E. Marsch, Springer-Verlag, Berlin, Heidelberg, New York, 13–97, 1990.

8.24 Bochsler, P., J. Geiss, R. Joos, Kinetic temperatures of heavy ions in the solar wind, J. Geophys. Res., **90**, 10779–10789, 1985.

8.25 Bochsler, P., J. Geiss, S. Kunz, Abundances of carbon, oxygen, neon in the solar wind during the period from August 1978 to June 1982. Solar Phys., **103**, 177–201, 1986.

8.26 Bohlin, J.D., E.O. Hulbert, An observational definition of coronal holes, in *Coronal Holes and High Speed Streams*, ed. by J.B. Zirker, Colorado Associated University Press, Boulder, Colorado, USA, 27–70, 1977.

8.27 Borrini, G., J.T. Gosling, S.J. Bame, W.C. Feldman, J.M. Wilcox, Solar wind helium and hydrogen structure near the heliospheric current sheet: a signal of coronal streamers at 1 AU, J. Geophys. Res., **86**, 4565–4573, 1981.

8.28 Braginskii, S.I., Transport processes in plasma, in *Review of Plasma Physics*, Vol 1, ed. by M.A. Leontovich, Consultants Bureau, New York, 205–311, 1966.

8.29 Brandt, J.C., J.P. Cassinelli, Interplanetary gas, Icarus, **5**, 47–63, 1966.

8.30 Brückner, G.E., J.D.F. Bartoe, Observations of high energy jets in the corona above the quiet sun, the heating of the corona and the acceleration of the solar wind, Astrophys. J., **272**, 329–348, 1983.

8.31 Bürgi, A., J. Geiss, Helium and minor ions in the corona and solar wind: Dynamics and charge states, Solar Phys., **103**, 347–383, 1986.

8.32 Burlaga, L.F., Hydromagnetic waves and discontinuities in the solar wind, Space Sci. Rev., **12**, 600–657, 1971.

8.33 Burlaga, L.F., K.W. Ogilvie, Heating of the solar wind, Astrophys. J., **159**, 659–670, 1970.

8.34 Burlaga, L.F., J.F. Lemaire, J.M. Turner, Interplanetary current sheets at 1 AU, J. Geophys. Res., **82**, 3191–3200, 1977.

8.35 Burlaga, L.F., K.W. Behannon, S.F. Hansen, G.W. Pneuman, W.C. Feldman, Sources of magnetic fields in recurrent interplanetary streams, J. Geophys. Res., **83**, 4177–4185, 1978.

8.36 Burlaga, L., E. Sittler, F. Mariani, R. Schwenn, Magnetic loop behind an interplanetary shock: Voyager, Helios, IMP8 observations, J. Geophys. Res., **86**, 6673–6684, 1981.

8.37 Burlaga, L.F., L. Klein, A magnetic cloud and a coronal mass ejection, Geophys. Res. Lett., **9**, 1317–1320, 1982.

8.38 Burlaga, L., Magnetic clouds, in *Physics of the Inner Heliosphere*, Vol 1, ed. by R. Schwenn and E. Marsch, Springer-Verlag, Berlin, Heidelberg, New York, 1990.

8.39 Chew, G.F., M.L. Goldberger, F.E. Low, The Boltzmann equation and the one-fluid hydromagnetic equations in the absence of particle collisions. Proc. Roy. Soc. London, **236A**, e.g. 112, 1956.

8.40 Coroniti, F.V., C.F. Kennel, F.L. Scarf, E.J. Smith, Whistler mode turbulence in the disturbed solar wind, J. Geophys. Res., **87**, 6029–6044, 1982.

8.41 Cuperman, S., Electromagnetic kinetic instabilities in multicomponent space plasmas: Theoretical predictions and computer simulation experiments, Rev. Geophys. Space Sci., **19**, 307–343, 1981.

8.42 Cuperman, S., Solar wind models, in *Solar Wind Four*, ed. by H. Rosenbauer, Report MPAE-W-100-81-31, Max-Planck-Institut für Aeronomie, Katlenburg-Lindau, F.R. Germany, 13–27, 1981.

8.43 Cuperman, S., R.W. Landau, Ion cyclotron resonant instability of RH waves propagating at an angle to the interplanetary magnetic field, Astrophys. Space Sci., **5**, 333–341, 1969.

8.44 Cuperman, S., I. Weiss, M. Dryer, Higher order fluid equations for multicomponent nonequilibrium stellar (plasma) atmospheres and star clusters, Astrophys. J., **239**, 345–359, 1980.

8.45 Cuperman, S., I. Weiss, M. Dryer, Theoretical non-Maxwellian particle velocity distribution functions for spherically-symmetric solar wind-like plasma systems and consequences, Astrophys. J., **273**, 363–373, 1983.

8.46 Cuperman, S., L. Ofman, M. Dryer, On the dispersion of ion-cyclotron waves in the H^+-He^{++} solar wind-like magnetized plasma, J. Geophys. Res., **93**, 2533–2538, 1988.

8.47 Denskat, K.U., F.M. Neubauer, R. Schwenn, Properties of "Alfvénic" fluctuations near the sun: Helios 1 and Helios 2, in *Solar Wind Four*, ed. by H. Rosenbauer, Report MPAE-W-100-81-31, Max-Planck-Institut für Aeronomie, Katlenburg-Lindau, F.R.Germany, 392–397, 1981.

8.48 Denskat, K.U., F.M. Neubauer, Statistical properties of low frequency magnetic field fluctuations in the solar wind from 0.29 to 1 AU during solar minimum conditions: Helios 1 and Helios 2, J. Geophys. Res., **87**, 2215–2223, 1982.

8.49 Dere, K.P., HTRS observations of the fine structure and dynamics of the solar chromosphere and transition zone, in *Solar Wind Five*, ed. by M. Neugebauer, NASA Conference Publication 2280, Washington, USA, 34–43, 1983.

8.50 Dobrowolny, M., G. Moreno, Plasma kinetics in the solar wind, Space Sci. Rev., **20**, 577–620, 1977.

8.51 Dobrowolny, M., M. Tessarotto, Electron kinetic instabilities in the solar wind, Astrophys. Space Sci., **57**, 153–162, 1978.

8.52 Dobrowolny, M., G. Torricelli-Ciamponi, Astron. Astrophys., **142**, 404–410, 1985.

8.53 Dryer, M., Coronal transient phenomena, Space Sci. Rev., **33**, 233–275, 1983.

8.54 Dulk, G.A., K.V. Sheridan, S.F. Smerd, G.L. Withbroe, Radio and EUV observations of a coronal hole, Solar Phys., **52**, 349–367, 1977.

8.55 Dum, C.T., Electrostatic waves and anomalous transport in the solar wind, in *Solar Wind Five*, ed. by M. Neugebauer, NASA Conference Publication 2280, Washington, USA, 369–376, 1983.

8.56 Dum, C.T., E. Marsch, W. Pilipp, Analysis of electromagnetic instabilities using measured solar wind distribution functions, paper presented at the European Geophysical Society Meeting, Vienna, 1979.

8.57 Dum, C.T., E. Marsch, W.G. Pilipp, Determination of wave growth from measured distribution functions and transport theory, J. Plasma Phys., **23**, 91–113, 1980.

8.58 Dum, C.T., E. Marsch, W.G. Pilipp, D.A. Gurnett, Ion sound turbulence in the solar wind, in *Solar Wind Four*, ed. by H. Rosenbauer, Report MPAE-W-100-81-31, Max-Planck-Institut für Aeronomie, Katlenburg-Lindau, F.R.Germany, 299–304, 1981.

124

8.59   Dum, C.T., E. Marsch, W.G. Pilipp, D.A. Gurnett, Ion sound wave instability in the solar wind, unpublished manuscript, 1988.

8.60   Dusenbery, P.B., J.V. Hollweg, Ion cyclotron heating and acceleration of solar wind minor ions, J. Geophys. Res., **86**, 153–164, 1981.

8.61   Esser, R., E. Leer, S. Habbal, G.L. Withbroe, A two-fluid solar wind model with Alfvén waves: Parameter study and application to observations, J. Geophys. Res., **91**, 2950–2960, 1986.

8.62   Esser, R., T.E. Holzer, E. Leer, Drawing inferences about solar wind acceleration from coronal minor ion observations, J. Geophys. Res., **92**, 13377–13389, 1987.

8.63   Eviatar, A., M. Schulz, Ion temperature anisotropies and structure of the solar wind, Planet. Space Sci., **18**, 321–332, 1970.

8.64   Eyni, M., A.S. Kaufman, The adiabatic cooling of the protons in the solar wind: The case where the interplanetary magnetic field is of spiral form, Astrophys. Space Sci., **28**, 177–183, 1974.

8.65   Fahr, H.J., B. Shizgal, Modern exospheric theories and their observational relevance, Rev. Geophys., Space Phys., **21**, 75–124, 1983.

8.66   Feldman, W.C., Solar wind plasma processes and transport, Rev. Geophys. Space Phys., **7**, 1743–1751, 1979.

8.67   Feldman, W.C., Kinetic processes in the solar wind, in *Solar System Plasma Physics*, Vol I, ed. by E.N. Parker, C.F. Kennel, L.J. Lanzerotti, North-Holland, Amsterdam, 331–344, 1979.

8.68   Feldman, W.C., J.R. Asbridge, S.J. Bame, M.D. Montgomery, Double ion streams in the solar wind, J. Geophys. Res., **78**, 2017–2027, 1973.

8.69   Feldman, W.C., J.R. Asbridge, S.J. Bame, M.D. Montgomery, On the origin of solar wind proton thermal anisotropy, J. Geophys. Res., **78**, 6451–6468, 1973.

8.70   Feldman, W.C., J.R. Asbridge, S.J. Bame, M.D. Montgomery, Interpenetrating solar wind streams, Rev. Geophys. Space Phys., **4**, 715–723, 1974.

8.71   Feldman, W.C., J.R. Asbridge, S.J. Bame, The solar wind He^{2+} to H^+ temperature ratio, J. Geophys. Res., **79**, 2319–2323, 1974.

8.72   Feldman, W.C., J.R. Asbridge, S.J. Bame, M.D. Montgomery, S.P. Gary, Solar wind electrons, J. Geophys. Res., **80**, 4181–4196, 1975.

8.73   Feldman, W.C., J.R. Asbridge, S.J. Bame, J.T. Gosling, High–speed solar wind flow parameters at 1 AU, J. Geophys. Res., **81**, 5054–5060, 1976.

8.74   Feldman, W.C., J.R. Asbridge, S.J. Bame, S.P. Gary, M.D. Montgomery, Electron parameter correlations in high-speed streams and heat flux instabilities, J. Geophys. Res., **81**, 2377–2382, 1976.

8.75   Feldman, W.C., J.R. Asbridge, S.J. Bame, S.P. Gary, M.D. Montgomery, S.M. Zink, Evidence for the regulation of solar wind heat flux at 1 AU, J. Geophys. Res., **81**, 5207–5211, 1976.

8.76   Feldman, W.C., J.R. Asbridge, S.J. Bame, J.T. Gosling, Plasma and magnetic fields from the sun, in *The Solar Output and its Variations*, ed. by O.R. White, Colorado Associated University Press, Boulder, Colorado, 351–382, 1977.

8.77   Feldman, W.C., J.R. Asbridge, S.J. Bame, J.T. Gosling, D.S. Lemons, Characteristics electron variations across simple high-speed solar wind streams, J. Geophys. Res., **83**, 5285–5295, 1978.

8.78   Feldman, W.C., J.R. Asbridge, S.J. Bame, J.T. Gosling, D.S. Lemons, Electron heating within interaction zones of simple high-speed solar wind streams, J. Geophys. Res., **83**, 5297–5303, 1978.

8.79   Feldman, W.C., J.R. Asbridge, S.J. Bame, J.T. Gosling, D.S. Lemons, The core electron temperature profile between 0.5 and 1 AU in the steady-state high speed solar wind, J. Geophys. Res., **84**, 4463–4467, 1979.

8.80   Feldman, W.C., J.R. Asbridge, S.J. Bame, E.E. Fenimore, J.T. Gosling, The origins of solar wind interstream flows: near-equatorial coronal streamers, J. Geophys. Res., **86**, 5408–5416, 1981.

8.81   Feldman, W.C., J.R. Asbridge, S.J. Bame, J.T. Gosling, Quantitative tests of a steady state theory of solar wind electrons, J. Geophys. Res., **87**, 7355–7362, 1982.

8.82   Feynman, J., On solar wind helium and heavy ion temperatures, Solar Phys., **43**, 249–252, 1975.

8.83   Flå, T., S.R. Habbal, T.E. Holzer, E. Leer, Fast-mode magnetohydrodynamic waves in coronal holes and the solar wind, Astrophys. J., **280**, 382–390, 1984.

8.84   Forslund, D.W., Instabilities associated with heat conduction in the solar wind and their consequences, J. Geophys. Res., **75**, 17–28, 1970.

8.85   Freeman, J.W., R.W. Lopez, The cold solar wind, J. Geophys. Res., **90** 9885–9887, 1985.

8.86   Freeman, J.W., Estimates of solar wind heating inside 0.3 AU, Geophys. Res. Lett., **15**, 88–91, 1988.

8.87   Fuselier, S.A., D.A. Gurnett, R.J. Fitzenreiter, The downshift of electron plasma oscillations in the electron foreshock region, J. Geophs. Res., **90**, 3935–3946, 1985.

8.88   Gary, S.P., Ion-acoustic-like instabilities in the solar wind, J. Geophys. Res., **83**, 2504–2510, 1978.

8.89   Gary, S.P., Electrostatic heat flux instabilities, J. Plasma Phys., **20**, 47–60, 1978.

8.90   Gary, S.P., Wave–particle transport from electrostatic instabilities, Phys. Fluids, **23**, 1193–1204, 1980.

8.91   Gary, S.P., Electrostatic instabilities in plasmas with two electron components, J. Geophys. Res., **90**, 8213–8219, 1985.

8.92   Gary, S.P., The electron/electron acoustic instability, Phys. Fluids, **30**, 2745–2749, 1987.

8.93   Gary, S.P., W.C. Feldman, D.W. Forslund, M.D. Montgomery, Electron heat flux instabilities in the solar wind, Geophys. Res. Lett., **2**, 79–82, 1975.

8.94   Gary, S.P., W.C. Feldman, D.W. Forslund, M.D. Montgomery, Heat flux instabilities in the solar wind, J. Geophys. Res., **80**, 4197–4203, 1975.

8.95   Gary, S.P., W.C. Feldman, A second order theory for $k \| B_0$ electromagnetic instabilities, Phys. Fluids, **21**, 72–80, 1978.

8.96   Gary, S.P., J.T. Gosling, D.W. Forslund, The electromagnetic beam instability upstream of the earth's bow shock, J. Geophys. Res., **86**, 6691–6696, 1981.

8.97   Gary, S.P., C.D. Madland, B.T. Tsurutani, Electromagnetic ion beam instabilities: II, Phys. Fluids, **28**, 3691–3695, 1985.

8.98   Gazis, P.R., A.J. Lazarus, Voyager observations of solar wind proton temperature: 1–10 AU, Geophys. Res. Lett., **9**, 431–434, 1982.

8.99   Geiss, J., Processes affecting abundances in the solar wind, Space Science Reviews, **33**, 201–217, 1982.

8.100 Geiss, J., Diagnostics of corona by in-situ composition measurements at 1 AU, Proceedings of an ESA workshop on "Future missions in solar, heliospheric and space plasma physics", ESA SP-235, 37–50, 1985.

8.101 Geiss, J., P. Bochsler, Ion composition in the solar wind in relation to solar abundances, in *Rapports Isotopiques dans le Systeme Solaire*, Cepadues-Editions, Paris, France, 1–16, 1985.

8.102 Geiss, J., P. Bochsler, Solar wind composition and what we expect to learn from out-of-ecliptic measurements, in *The Sun and the Heliosphere in Three Dimensions*, ed. by R.G. Marsden, D. Reidel Publishing Company, Dordrecht, 173–186, 1986.

8.103 Gendrin, R., Wave particle interactions as an energy transfer mechanism between different particle species, Space Sci. Rev., **34**, 271–287, 1983.

8.104 Goodrich, C.C., A.J. Lazarus, Suprathermal protons in the interplanetary solar wind, J. Geophys. Res., **81**, 2750–2754, 1976.

8.105 Gosling, J.T., E. Hildner, S.R. Asbridge, S.J. Bame, W.C. Feldman, Noncompressive density enhancements in the solar wind, J. Geophys. Res., **82**, 5005–5010, 1977.

8.106 Gosling, J.T., J.R. Asbridge, S.J. Bame, W.C. Feldman, Solar wind stream interfaces, J. Geophys. Res., **83**, 1401–1412, 1978.

8.107 Gosling, J.T., G. Borrini, J.R. Asbridge, S.J. Bame, W.C. Feldman, R.T. Hansen, Coronal streamers in the solar wind at 1 AU, J. Geophys. Res., **86**, 5438–5448, 1981.

8.108 Griffel, D.H., L. Davis, The anisotropy of the solar wind, Planet. Space Sci., **17**, 1009–1020, 1969.

8.109 Gurnett, D.A., Plasma waves in the solar wind: A review of observations, in *Solar Wind Four*, ed. by H. Rosenbauer, Report MPAE-W-100-81-31, Max-Planck-Institut für Aeronomie, Katlenburg-Lindau, F.R.Germany, 286–298, 1981.

8.110 Gurnett, D.A., Waves and Instabilities, in *Physics of the Inner Heliosphere* (this volume).

8.111 Gurnett, D.A., L.A. Frank, Ion-acoustic waves in the solar wind, J. Geophys. Res., **83**, 58–74, 1978.

8.112 Gurnett, D.A., E. Marsch, W.G. Pilipp, R. Schwenn, H. Rosenbauer, Ion-acoustic waves and related plasma observations in the solar wind, J. Geophys. Res., **84**, 2029–2038, 1979.

8.113 Habbal, S.R., E. Leer, Electron heating by Fast-mode magnetohydrodynamic waves in the solar wind emanating from coronal holes, Astrophys. J., **253**, 318–322, 1982.

8.114 Hernández, R., E. Marsch, Collisional time scales for temperature and velocity exchange between drifting Maxwellians, J. Geophys. Res., **90**, 11062–11066, 1985.

8.115 Hernández, R., S. Livi, E. Marsch, On the He^{2+} to H^+ temperature ratio in slow solar wind, J. Geophys. Res., **92**, 7723–7727, 1987.

8.116 Hinton, F.L., Collisional transport in plasma, in *Basic Plasma Physics I*, ed. by A.A. Galeev and R.N. Sudan, North-Holland Publishing Company, Amsterdam, 147–197, 1983.

8.117 Hollweg, J.V., Collisionless solar wind. 1. Constant electron temperature, J. Geophys. Res., **75**, 2403–2418, 1970.

8.118 Hollweg, J.V., On electron heat conduction in the solar wind, J. Geophys. Res., **79**, 3845–3850, 1974.

8.119 Hollweg, J.V., Waves and instabilities in the solar wind, Rev. Geophys. Space Phys., **13**, 263–289, 1975.

8.120 Hollweg, J.V., Some physical processes in the solar wind, Rev. Geophys. Space Phys., **16**, 689–720, 1978.

8.121 Hollweg, J.V., Helium and heavy ions, in *Solar Wind Four*, ed. by H. Rosenbauer, Report MPAE-W-100-81-31, Max-Planck-Institut für Aeronmie, Katlenburg-Lindau, F.R.Germany, 414–424, 1981.

8.122 Hollweg J.V., Energy and momentum transport by waves in the solar atmosphere, in *Proceedings of the 1985 Trieste Summer College on Plasma Physics*, ed. by B. Buti, Advances in Space Plasma Physics, World Scientific, Singapore, 77, 1985.

8.123 Hollweg, J.V., Transition region, corona, solar wind in coronal holes, J. Geophys. Res., **91**, 4111–4125, 1986.

8.124 Hollweg, J.V., W. Johnsen, Transition region, corona, solar wind in coronal holes: some two-fluid models, J. Geophys. Res., **93**, 9547–9554, 1988.

8.125 Holzer, T.E., Effects of rapidly diverging flow, heat addition and momentum addition in the solar wind and stellar winds, J. Geophys. Res., **82**, 23–35, 1977.

8.126 Holzer, T.E., The solar wind and related astrophysical phenomena, in *Solar System Plasma Physics*, Vol I, ed. by E.N. Parker, C.F. Kennel, L.J. Lanzerotti, 101–176, North-Holland, Amsterdam, 1979.

8.127 Holzer, T.E., E. Leer, Conductive solar wind models in rapidly diverging flow geometries, J. Geophys. Res., **85**, 4665–4679, 1980.

8.128 Holzer, T.E., T. Flå, E. Leer, Alfvén waves in stellar winds, Astrophys. J., **275**, 808–835, 1983.

8.129 Holzer, E.T., E. Leer, Xue-Pu Zhao, Viscosity in the solar wind, J. Geophys. Res., **91**, 4126–4132, 1986.

8.130 Hundhausen, A.J., Composition and dynamics of the solar wind plasma, Rev. Geophys. Space Phys., **8**, 729–811, 1970.

8.131 Hundhausen, A.J., *Coronal Expansion and Solar Wind*, Springer-Verlag, New York, Berlin, Heidelberg, 1972.

8.132 Hundhausen, A.J., Solar wind stream interactions and interplanetary heat conduction, J. Geophys. Res., **78**, 7996–8010, 1973.

8.133 Hundhausen, A.J., An interplanetary view of coronal holes, in *Coronal Holes and High Speed Streams*, ed. by J.B. Zirker, Colorado Associated University Press, Boulder, Colorado, USA, 223–329, 1977.

8.134 Hundhausen, A.J., Solar activity and the solar wind, Rev. Geophys. Space Phys., **17**, 2034–2048, 1979.

8.135 Hundhausen, A.J., S.J. Bame, N.F. Ness, Solar wind thermal anisotropies: Vela 3 and IMP 3, J. Geophys. Res., **72**, 5265–5274, 1967.

8.136 Hundhausen, A.J., M.D. Montgomery, Heat conduction and nonsteady phenomena in the solar wind, J. Geophys. Res., **76**, 2236–2244, 1971.

8.137 Ipavich, F.M., A.B. Calvin, G. Gloeckler, D. Hovestadt, S.J. Bame, B. Klecker, M. Scholer, L.A. Fisk, C.Y. Fan, Solar wind Fe and CNO measurements in high-speed flows, J. Geophys. Res., **91**, 4133–4141, 1986.

8.138 Isenberg, P.A., Acceleration of heavy ions in the solar wind, in *Solar Wind Five*, ed. by M. Neugebauer, NASA Conference Publication 2280, Washington, USA, 655–661, 1983.

8.139 Isenberg, P.A., The ion-cyclotron dispersion relation in a proton–alpha solar wind, J. Geophys. Res., **89**, 2133–2141, 1984.

8.140 Isenberg, P.A., J.V. Hollweg, Finite amplitude Alfvén waves in a multi-ion plasma: Propagation, acceleration, heating, J. Geophys. Res., **87**, 5023–5029, 1982.

8.141 Isenberg, P.A., J.V. Hollweg, On the preferential acceleration and heating of solar wind heavy ions, J. Geophys. Res., **88**, 3923–3935, 1983.

8.142 Jacques, S.A., Momentum and energy transport by waves in the solar atmosphere and solar wind, Astrophys. J., **215**, 942–951, 1977.

8.143 Jacques, S.A., Solar wind models with Alfvén waves, Astrophys. J., **226**, 632–649, 1978.

8.144 Jockers, K., Solar wind models based on exospheric theory, Astron. Astrophys., **6**, 219–239, 1970.

8.145 Kayser, S.E., A. Barnes, J.D. Mihalov, The far reaches of the solar wind: *Pioneer 16* and *Pioneer 11* plasma results, Astrophys. J., **285**, 339–346, 1984.

8.146 Kennel, C.F., F.L. Scarf, F.V. Coroniti, R.W. Fredericks, D.A. Gurnett, E.J. Smith, Correlated whistler and electron plasma oscillation bursts detected on ISEE 3, Geophys. Res. Lett.,7, 129–132, 1980.

8.147 Kennel, C.F., F.L. Scarf, F.V. Coroniti, E.J. Smith, D.A. Gurnett, Nonlocal plasma turbulence associated with interplanetary shocks, J. Geophys. Res., **87**, 17–34, 1982.

8.148 Klein, L.W., L.F. Burlaga, Interplanetary magnetic clouds at 1 AU, J. Geophys. Res., **87**, 613–624, 1982.

8.149 Klein, L.W., K.W. Ogilvie, L.F. Burlaga, Coulomb collisions in the solar wind, J. Geophys. Res., **90**, 7389–7395, 1985.

8.150 Kopp, R.A., T.E. Holzer, Dynamics of coronal hole regions, I. Steady polytropic flows with multiple critical points, Solar Phys., **49**, 43–56, 1976.

8.151 Kopp, R.A., F.Q. Orrall, Models of coronal holes above the transition region, in *Coronal Holes and High Speed Streams*, ed. by J.B. Zirker, Colorado Associated University Press, Boulder, Colorado, USA, 179–224, 1977.

8.152 Kuperus, M., Heating processes of the solar corona, in *Plasma Astrophysics*, ed. by T.D. Guyenne and G. Levy, ESA-SP 161, Noordwijk, Netherlands, 113–128, 1981.

8.153 Kraichnan, R.H., Inertial-range spectrum of hydromagnetic turbulence, Phys. Fluids **8**, 1385–1387, 1965.

8.154 Lakhina, G.S., Regulation of solar wind heat flux by ordinary mode instability, Solar Phys., **52**, 153–162, 1977.

8.155 Lakhina, G.S., Ion cyclotron instability in the solar wind, Solar Phys., **57**, 467–473, 1978.

8.156 Lakhina, G.S., B. Buti, Stability of solar wind double ion streams, J. Geophys. Res., **81**, 2135–2139, 1976.

8.157 Lee, M.A., I. Lerche, Waves and irregularities in the solar wind, Rev. Geophys. Space Phys., **12**, 671–687, 1974.

8.158 Leer, E., W.I. Axford, A two fluid model with anisotropic proton temperature, Solar Phys., **23**, 238–250, 1972.

8.159 Leer, E., T.E. Holzer, Collisionless solar wind protons: A comparison of kinetic and hydro-dynamic descriptions, J. Geophys. Res., 77, 4035–4041, 1972.

8.160 Leer, E., T.E. Holzer, Constraints on the solar coronal temperature in regions of open magnetic field, Solar Phys., **63**, 143–156, 1979.

8.161 Leer, E., T.E. Holzer, Energy addition to the solar wind, J. Geophys. Res., **85**, 4681–4688, 1980.

8.162 Leer, E., T.E. Holzer, T. Flå, Acceleration of the solar wind, Space Sci. Rev., **33**, 161–200, 1982.

8.163 Lemaire, J., M. Scherer, Kinetic models of the solar wind, J. Geophys. Res., **76**, 7479–7490, 1971.

8.164 Lemaire, J., M. Scherer, Kinetic models of the solar and polar wind, Rev. Geophys. Space Phys., **11**, 427–468, 1972.

8.165 Lemons, D.S., S.P. Gary, Temperature anisotropy instability in a plasma of two ion components, J. Plasma Phys, **15**, 83–89, 1976.

8.166 Lemons, D.S., W.C. Feldman, Collisional modification to the exospheric theory of solar wind halo electron pitch angle distributions, J. Geophys. Res., **88**, 6881–6687, 1983.

8.167 Leubner, M.P., Influence of non-bi-Maxwellian distribution function of solar wind protons on the ion cyclotron instability, J. Geophys. Res., **83**, 3900–3902, 1978.

8.168 Leubner, M.P., Velocity distribution function and cyclotron wave growth in a modified bi-Maxwellian two-ion-component solar wind plasma, J. Geophys. Res., **84**, 2661–2665, 1979.

8.169 Leubner, M.P., A.F. Viñas, Stability analysis of double peaked proton distribution functions in the solar wind, J. Geophys. Res., **91**, 13366–13372, 1986.

8.170 Livi, S., E. Marsch, On the collisional relaxation of solar wind velocity distributions, Ann. Geophys., **4A**, 333–340, 1986.

8.171 Livi, S., E. Marsch, H. Rosenbauer, Coulomb collisional domains in the solar wind, J. Geophys., **91**, 8045–8050, 1986.

8.172 Livi, S., E. Marsch, Comparison of the Bhatnagar–Gross–Krook–approximation with the exact Coulomb collision operator, Phys. Rev. A, **34**, 533–540, 1986.

8.173   Livi, S., E. Marsch, Generation of solar wind proton tails and double beams by Coulomb collisions, J. Geophys. Res., **92**, 7255–7261, 1987.

8.174   Lomonosov, V.N., The heating of solar-wind protons by Alfvén waves in the inner heliosphere, Geomagnetism and Aeronomy, **27**, 313–316, 1987.

8.175   Lopez, R.E., J.W. Freeman, Solar wind proton temperature-velocity relationship, J. Geophys. Res., **91**, 1701.–1705, 1986.

8.176   Mariani, F., F.M. Neubauer, The interplanetary magnetic field, *Physics of the Inner Heliosphere*, Vol 1, ed. by R. Schwenn and E. Marsch, Springer-Verlag, Berlin, Heidelberg, New York, 183–206, 1990.

8.177   Marsch, E., Velocity distributions of solar wind ions and electrons, in Proceedings of a Course & Workshop on Plasma Astrophysics, Varenna, Italy, 28 Aug–7 Sept 1984, ESA SP-207, 33–40, 1984.

8.178   Marsch, E., Energy input into the solar wind, in Proceedings of an ESA workshop on "Future missions in solar, heliospheric and space plasma physics", ESA SP-235, 11–21, 1985.

8.179   Marsch, E., Beam-driven electron acoustic waves upstream of the earth's bow shock, J. Geophys. Res., **90**, 6327–6336, 1985.

8.180   Marsch, E., Acceleration potential and angular momentum of undamped MHD waves in stellar winds, Astron. Astrophys., **164**, 77–85, 1986.

8.181   Marsch, E., Wave-particle interactions in the solar wind, in *Proceedings of the workshop on "Nonlinear phenomena in Vlasov plasmas"*, ed. by F. Doveil, l'Institut d'Etudes Scientifiques de Cargèse, Corsica, France, 145–162, 1988.

8.182   Marsch, E., MHD turbulence in the solar wind, in *Physics of the Inner Heliosphere*, (this volume).

8.183   Marsch, E., K.-H. Mühlhäuser, W.G. Pilipp, R. Schwenn, H. Rosenbauer, Initial results on solar wind alpha particle distributions as measured by Helios between 0.3 and 1 AU, in *Solar Wind Four*, ed. by H. Rosenbauer, Report MPAE-W-100-81-31, Max-Planck-Institut für Aeronomie, Katlenburg-Lindau, F.R.Germany, 443–449, 1981.

8.184   Marsch, E., K.-H. Mühlhäuser, H. Rosenbauer, R. Schwenn, K.U. Denskat, Pronounced proton core temperature anisotropy, ion differential speed, and simultaneous Alfvén wave activity in slow solar wind at 0.3 AU, J. Geophys. Res., **86**, 9199–9203, 1981.

8.185   Marsch, E., K.-H. Mühlhäuser, H. Rosenbauer, R. Schwenn, F.M. Neubauer, Solar wind helium ions: Observations of the Helios solar probes between 0.3 and 1 AU, J. Geophys. Res., **87**, 35–51, 1982.

8.186   Marsch, E., K.-H. Mühlhäuser, R. Schwenn, H. Rosenbauer, W.G. Pilipp, F.M. Neubauer, Solar wind protons: Three-dimensional velocity distributions and derived plasma parameters measured between 0.3 and 1 AU, J. Geophys. Res., **87**, 52–72, 1982.

8.187   Marsch, E., T. Chang, Lower hybrid waves in the solar wind, Geophys. Res. Lett., **9**, 1155–1158, 1982.

8.188   Marsch, E., C.K. Goertz, K. Richter, Wave heating and acceleration of solar wind ions by cyclotron resonance, J. Geophys. Res., **87**, 5030–5044, 1982.

8.189   Marsch, E., T. Chang, Electromagnetic lower hybrid waves in the solar wind, J. Geophys. Res., **88**, 6869–6880, 1983.

8.190   Marsch, E., K.-H. Mühlhäuser, H. Rosenbauer, R. Schwenn, On the equation of state of solar wind ions derived from Helios measurements, J. Geophys. Res., **88**, 2982–2992, 1983.

8.191   Marsch E., H. Goldstein, The effects of Coulomb collisions on solar wind ion velocity distributions, J. Geophys. Res., **88**, 9933–9940, 1983.

8.192   Marsch, E., A.K. Richter, Distribution of solar wind angular momentum between particles and magnetic field: Inferences about the Alfvén critical point from Helios observations, J. Geophys. Res., **89**, 5386–5394, 1984.

8.193   Marsch, E., A.K. Richter, Helios observational constraints on solar wind expansion, J. Geophys. Res., **89**, 6599–6612, 1984.

8.194   Marsch, E., S. Livi, Coulomb collision rates for the self-similar and kappa distributions, Phys. Fluids, **28**, 1379–1386, 1985.

8.195   Marsch, E., S. Livi, Coulomb self-collision frequencies for nonthermal velocity distributions in the solar wind, Annales Geophysicae, **3**, 545–556, 1985.

8.196   Marsch, E., A.K. Richter, On the equation of state and collision time for a multicomponent, anisotropic solar wind, Ann. Geophys., **5A**, 71–82, 1987.

8.197   Marsch, E., S. Livi, Observational evidence for marginal stability of solar wind ion beams, J. Geophys. Res., **92**, 7263–7268, 1987.

8.198 Marsch, E., W.G. Pilipp, K.M. Thieme, H. Rosenbauer, Cooling of solar wind electron inside 0.3 AU, J. Geophys. Res., **94**, 6893–6898, 1989.

8.199 Marsden, R.G., K.P. Wenzel, The international solar polar mission (ISPM), in *Plasma Astrophysics*, ed. by T.D. Guyenne and G. Levy, ESA-SP 161, Noordwijk, Netherlands, 167–175, 1981.

8.200 Matthaeus, W.H., M.L. Goldstein, Magnetohydrodynamic turbulence in the solar wind, in *Solar Wind Five*, ed. by M. Neugebauer, NASA Conference Publication 2280, Washington, USA, 73–80, 1983.

8.201 McKenzie, J.F., W.H. Ip, W.I. Axford, The acceleration of minor ion species in the solar wind, Astrophys. Space Sci., **64**, 183–211, 1979.

8.202 McKenzie, J.F., E. Marsch, Resonant wave acceleration of minor ions in the solar wind, Astrophys. Space Sci., **81**, 295–314, 1982.

8.203 Melrose, D.B., *Instabilities in Space and Laboratory Plasmas*, Cambridge University Press, Cambridge, 1986.

8.204 Mihalov, J.D., J.H. Wolfe, *Pioneer-10* observation of the solar wind proton temperature heliocentric gradient, Solar Phys. **60**, 399–406, 1978.

8.205 Mitchell, D.G., E.C. Roelof, W.C. Feldman, S.J. Bame, D.J. Williams, Thermal iron ions in high-speed solar wind streams, 2. Temperatures and bulk velocities, Geophys, Res. Lett., **8**, 827–830, 1981.

8.206 Montgomery, D.M., Theory of hydromagnetic turbulence, in *Solar Wind Five*, ed. by M. Neugebauer, NASA Conference Publication 2280, Washington, USA, 107–130, 1983.

8.207 Montgomery, D., M.R. Brown, W.H. Matthaeus, J. Geophys. Res., **92**, 282–284, 1987.

8.208 Montgomery, M.D., S.J. Bame, A.J. Hundhausen, Solar wind electrons: Vela 4 measurements, J. Geophys. Res., **73**, 4999–5003, 1968.

8.209 Montgomery, M.D., S.P. Gary, D.W. Forslund, W.C. Feldman, Electromagnetic ion-beam instabilities in the solar wind, Phys. Rev. Lett., **35**, 667–670, 1975.

8.210 Montgomery, M.D., S.P. Gary, W.C. Feldman, D.W. Forslund, Electromagnetic instabilities driven by unequal proton beams in the solar wind, J. Geophys. Res., **81**, 2743–2749, 1976.

8.211 Munro, R.H., B.V. Jackson, Physical properties of a coronal hole from 2 to 5 $R_\odot$, Astrophys. J., **213**, 874–886, 1977.

8.212 Neubauer, F.M., G. Musmann, G. Dehmel, Fast magnetic fluctuations in the solar wind: Helios 1, J. Geophys. Res., **82**, 3201–3212, 1977.

8.213 Neugebauer, M., The quiet solar wind, J. Geophys. Res., **81**, 4664–4670, 1976.

8.214 Neugebauer, M., The role of Coulomb collisions in limiting differential flow and temperature differences in the solar wind, J. Geophys. Res., **81**, 78–82, 1976.

8.215 Neugebauer, M., Observations of solar wind helium, Fundam. Cosmic Phys., **7**, 131–199, 1981.

8.216 Neugebauer, M., C.W. Wu, J.D. Huba, Plasma fluctuations in the solar wind, J. Geophys. Res., **83**, 1027–1033, 1978.

8.217 Neugebauer, M., W.C. Feldman, Relation between superheating and superacceleration of helium in the solar wind, Solar Phys., **63**, 201–205, 1979.

8.218 Neupert, W.M., V. Pizzo, Solar coronal holes as sources of recurrent geomagnetic disturbances, J. Geophys. Res., **79**, 3701–3709, 1974.

8.219 Ogilvie, K.W., J.D. Scudder, The radial gradient and collisionless properties of solar wind electrons, J. Geophys. Res., **83**, 3776–3782, 1978.

8.220 Ogilvie, K.W., P. Bochsler, M.A. Coplan, J. Geiss, Observations of the velocity distribution of solar wind ions, J. Geophys. Res., **85**, 6069–6074, 1980.

8.221 Olbert, S., Inferences about the solar wind dynamics from observed distributions of electrons and ions, in Proceedings of an International School and Workshop on Plasma Astrophysics, Varenna, Como, Italy, Eur. Space Agency Spec. Publ., ESA SP–161, 135–144, 1981.

8.222 Olbert, S., The role of thermal conduction in the acceleration of the solar wind, in *Solar Wind Five*, ed. by M. Neugebauer, NASA Conference Publication 2280, Washington, USA, 149–162, 1983.

8.223 Owocki, S.P., J.D. Scudder, The effect of a non-Maxwellian electron distribution on oxygen and iron ionization balances in the solar corona, Astrophys. J., **270**, 758–768, 1983.

8.224 Owocki, S.P., R.C. Canfield, The role of nonclassical electron transport in the lower solar transition region, Astrophys. J., **300**, 420–427, 1986.

8.225 Papadopoulus, K., Electrostatic turbulence at colliding plasma streams as the source of ion heating in the solar wind, Astrophys. J., **179**, 931–938, 1973.

130

8.226 Parker, E.N., Dynamics of the interplanetary gas and magnetic fields, Astrophys. J., **128**, 664–684, 1958.

8.227 Parker, E.N., Dynamical instability of an anisotropic gas of low density, Phys. Rev., **109**, 1874–1876, 1958.

8.228 Parker, E.N., *Interplanetary Dynamical Processes*, Interscience, New York, 1963.

8.229 Parker, E.N., Dynamical theory of the solar wind, Space Sci. Rev., 4, 666–708, 1965.

8.230 Pilipp, W.G., R. Schwenn, E. Marsch, K.-H. Mühlhäuser, H. Rosenbauer, Electron characteristics in the solar wind as deduced from Helios observations in *Solar Wind Four*, ed. by H. Rosenbauer, Report MPAE-W-100-81-31, Max-Planck-Institut für Aeronomie, Katlenburg-Lindau, F.R. Germany, 241–249, 1981.

8.231 Pilipp, W.G., Solar wind electrons as a probe for the global structure of the interplanetary magnetic field, in *Topics in Plasma-, Astro-, and Space Physics*, ed. by G. Haerendel and B. Battrick, Max-Planck-Institut für Physik und Astrophysik, Institut für Extraterrestrische Physik, Garching bei München, F.R.Germany, 91–107, 1983.

8.232 Pilipp, W.G., H. Miggenrieder, K.-H. Mühlhäuser, H. Rosenbauer, R. Schwenn, F.M. Neubauer, Variations of electron distribution functions in the solar wind, J. Geophys. Res., **92**, 1103–1118, 1987.

8.233 Pilipp, W.G., H. Miggenrieder, M.D. Montgomery, K.-H. Mühlhäuser, H. Rosenbauer, R. Schwenn, Characteristics of electron velocity distribution functions in the solar wind derived from the Helios plasma experiment, J. Geophys. Res., **92**, 1075–1092, 1987.

8.234 Pilipp, W.G., H. Miggenrieder, M.D. Montgomery, K.-H. Mühlhäuser, H. Rosenbauer, R. Schwenn, Unusual electron distribution functions in the solar wind derived from the Helios plasma experiment: Double-strahl distributions and distributions with an extremely anisotropic core, J. Geophys. Res., **92**, 1093–1101, 1987.

8.235 Pilipp, W.G., H. Miggenrieder, K.H. Mühlhäuser, H. Rosenbauer, R. Schwenn, Large scale variations of thermal electron parameters in the solar wind, J. Geophys. Res., in press, 1988.

8.236 Pizzo, V.J., Quasi-steady solar wind dynamics, in *Solar Wind Five*, ed. by M. Neugebauer, NASA Conference Publication 2280, Washington, USA, 675–691, 1983.

8.237 Pizzo, V., R. Schwenn, E. Marsch, H. Rosenbauer, K.H. Mühlhäuser, F.M. Neubauer, Determination of the solar wind angular momentum flux from Helios data–An observational test of the Weber and Davis theory, Astrophys. J., **271**, 335–354, 1983.

8.238 Priest, E.R., *Solar Magnetohydrodynamics*, D. Reidel Publishing Company, Dordrecht, The Netherlands, 1982.

8.239 Randolph, J.E., Solar probe study, in *A Close Up of the Sun*, ed. by M. Neugebauer and R.W. Davies, JPL Publication 78-70, Jet Propulsion Laboratory, Pasadena, California, USA, 521–534, 1978.

8.240 Richter, A.K., K.C. Hsieh, A.H. Luttrell, E. Marsch, R. Schwenn, Review of interplanetary shock phenomena near and within 1 AU, in *Collisionless Shocks in the Heliosphere: Reviews of Current Research*, ed. by B.T. Tsurutani and R.G. Stone, Geophysical Monograph, **35**, 33–50, 1985.

8.241 Rosenbauer, H., R. Schwenn, E. Marsch, B. Meyer, H. Miggenrider, M.D. Montgomery, K.-H. Mühlhäuser, W.G. Pilipp, W. Voges, S.M. Zink, A survey of initial results of the Helios plasma experiment, J. Geophys., **42**, 561–580, 1977.

8.242 Rossi, P.B., S. Olbert, *Introduction to the Physics of Space*, McGraw-Hill, New York, 1970.

8.243 Rottman, G.J., F.Q. Orrall, Observational evidence for solar wind acceleration at the base of coronal holes, in *Solar Wind Five*. ed. by M. Neugebauer, NASA Conference Publication 2280, Washington, USA, 199–210, 1983.

8.244 Scarf, F.L., Microscopic structure of the solar wind, Space Sci. Rev., **11**, 234–270, 1970.

8.245 Scarf, F.L., D.A. Gurnett, W.S. Kurth, The first year of Voyager plasma wave observations in the solar wind, in *Solar Wind Four*, ed. by H. Rosenbauer, Report MPAE-W-100-81-31, Max-Planck-Institut für Aeronomie, Katlenburg-Lindau, F.R.Germany, 305–316, 1981.

8.246 Scarf, F.L., J.H. Wolfe, R.W. Silva, A plasma instability associated with thermal anisotropies in the solar wind, J. Geophys. Res., **72**, 993–1005, 1967.

8.247 Schmidt, J., P. Bochsler, J. Geiss, Velocity of iron ions in the solar wind, J. Geophys. Res., **92**, 9901–9906, 1987.

8.248 Schmidt, W.K.H., H. Rosenbauer, E.G. Shelley, J. Geiss, On temperature and speed of He^{2+} and O^{6+} ions in the solar wind, Geophys. Res. Lett., **7**, 697–700, 1980.

8.249 Schwartz, S.J., Microturbulence of the solar wind, 1. Analytical results for fast mode instability growth rates, J. Geophys. Res., **83**, 3745–3752, 1978.

8.250 Schwartz, S.J., Plasma instabilities in the solar wind: A theoretical review, Rev. Geophys. Space Phys., **18**, 313–336, 1980.

8.251 Schwartz, S.J., W.C. Feldman, S.P. Gary, The source of proton anisotropy in the high-speed solar wind, J. Geophys. Res., **86**, 541–546, 1981.

8.252 Schwartz, S.J., W.C. Feldman, S.P. Gary, Wave–electron interactions in the high speed solar wind, J. Geophys. Res., **86**, 4574–4578, 1981.

8.253 Schwartz, S.J., E. Marsch, The radial evolution of a single solar wind plasma parcel, J. Geophys. Res., **88**, 9919–9932, 1983.

8.254 Schwenn, R., The "average" solar wind in the inner heliosphere: Structures and slow variations, in *Solar Wind Five*, ed. by M. Neugebauer, NASA Conference Publication 2280, Washington, USA, 489–508, 1983.

8.255 Schwenn, R., Direct correlations between coronal transients and interplanetary disturbances, Space Sci. Rev., **34**, 85–99, 1983.

8.256 Schwenn, R., Relationship of coronal transients to interplanetary shocks: 3-D aspects, Space Sci. Rev., **44**, 139–168, 1986.

8.257 Schwenn, R., Large scale structure of the interplanetary medium, in *Physics of the Inner Heliosphere*, Vol 1, ed. by R. Schwenn and E. Marsch, Springer-Verlag, Berlin, Heidelberg, New York, 99–181, 1990.

8.258 Schwenn, R., M.D. Montgomery, H. Rosenbauer, H. Miggenrieder, K.H. Mühlhäuser, S.J. Bame, W.C. Feldman, R.T. Hansen, Direct observations of the latitudinal extent of a high speed stream in the solar wind, J. Geophys. Res., **83**, 1011–1018, 1978.

8.259 Schwenn, R., K.-H. Mühlhäuser, E. Marsch, and H. Rosenbauer, Two states of the solar wind at the time of solar activity minimum, II, Radial gradients of plasma parameters in fast and slow streams, in *Solar Wind Four*, ed. by H. Rosenbauer, Report MPAE-W-100-81-31, Max-Planck-Institut für Aeronomie, Katlenburg-Lindau, F.R.Germany, 126–130, 1981.

8.260 Scudder, J.D., S. Olbert, A theory of local and global processes which affect solar wind electrons, 1. The origin of typical 1 AU velocity distribution functions – Steady state theory, J. Geophys. Res., **84**, 2755–2772, 1979.

8.261 Scudder, J.D., S. Olbert, A theory of local and global processes which affect solar wind electrons, 2. Experimental support, J. Geophys. Res., **84**, 6603–6620, 1979.

8.262 Sheeley, N.R., J.W. Harvey, W.C. Feldman, Coronal holes, solar wind streams, recurrent geomagnetic disturbances: 1973–1976, Solar Phys., **49**, 271–278, 1976.

8.263 Sheeley, N.R., Jr., R.A. Howard, M.J. Koomen, D.J. Michels, R. Schwenn, K.H. Mühlhäuser, H. Rosenbauer, Association between coronal mass ejections and interplanetary shocks, in *Solar Wind Five*, ed. by M. Neugebauer, NASA Conference Publication 2280, Washington, USA, 693–702, 1983.

8.264 Sheeley, N.R., R.A. Howard, M.J. Koomen, D.J. Michels, K.L. Harvey, J.W. Harvey, Observations of coronal structure during sunspot maximum, Space Sci. Rev., **33**, 219–231, 1983.

8.265 Shoub, E.C., Invalidity of local thermodynamic equilibrium for electrons in the solar transition region. I. Fokker–Planck results, Astrophys. J., **266**, 339–369, 1983.

8.266 Shoub, E.C., Failure of the Fokker–Planck approximation to the Boltzmann integral for $1/r$ potentials, Phys. Fluids, **30**, 1340–1352, 1987.

8.267 Singer, C.E., I.W. Roxburgh, The onset of microinstabilities and its consequences in the solar wind, J. Geophys. Res., **82**, 2677–2685, 1977.

8.268 Sittler, E.C., Jr., J.D. Scudder, An empirical polytrope law for solar wind thermal electrons between 0.45 and 4.76 AU: Voyager 2 and Mariner 10, J. Geophys. Res., **85**, 5131–5137, 1980.

8.269 Smith, E.J., Observations of interplanetary shocks: Recent progress, Space Sci. Rev., **34**, 101–110, 1984.

8.270 Spitzer, L., *Physics of Fully Ionized Gases*, Interscience Publ., New York, 1962.

8.271 Thieme, K.M., E. Marsch, H. Rosenbauer, Estimates of alpha particle heating in the solar wind inside 0.3 AU, J. Geophys. Res., **94**, 2673–2676, 1988.

8.272 Thieme, K.M., R. Schwenn, E. Marsch, Are structures in high-speed streams signatures of coronal fine structures?, Advances in Space Research, **9**, (4), 127–130, 1989.

8.273 Tu, C.-Y., A solar wind model with the power spectrum of Alfvénic fluctuations, Solar Phys., **109**, 149–186, 1987.

8.274 Tu, C.-Y., The damping of interplanetary Alfvénic fluctuations and the heating of the solar wind, J. Geophys. Res., **93**, 7–20, 1988.

8.275 Tu, C.-Y., Z.-Y. Pu, F.-S. Wei, The power spectrum of interplanetary Alfvénic fluctuations: Derivation of the governing equation and its solution, J. Geophys. Res., **89**, 9695–9702, 1984.

8.276   Tu, C.-Y., L. Dong, Dissipation mechanism of interplanetary Alfvénic fluctuations, Chin. Astron. Astrophys., **9**, 60–65, 1985.

8.277   Tu, C.-Y., J.W. Freeman, R.E. Lopez, The proton temperature and the total hourly variance of the magnetic field components in different solar wind speed regions, Solar Phys., **119**, 197–206, 1988.

8.278   Tu, C.-Y., D.A. Roberts, M.L. Goldstein, Determination of the spectral evolution and cascade constant in solar wind Alfvénic turbulence observed cross helicity values, J. Geophys. Res., in print, 1989.

8.279   Völk, H.J., Microstructure of the solar wind, Space Sci. Rev., **17**, 255–276, 1975.

8.280   Weber, E.J., L. Davis, The angular momentum of the solar wind, Astrophys. J., **148**, 217–227, 1967.

8.281   Whang, Y.C., Higher moment equations and the distribution function of the solar wind plasma, J. Geophys. Res., **76**, 7503–7507, 1971.

8.282   Whang, Y.C., T.H. Chien, Magnetohydrodynamic interaction of high-speed streams, J. Geophys. Res., **86**, 3263–3272, 1981.

8.283   Winske, D., M.M. Leroy, Diffuse ions produced by electromagnetic ion beam instabilities, J. Geophys. Res., **89**, 2673–2688, 1984.

8.284   Winske, D., C.S. Wu, Y.Y. Li, Z.Z. Mou, J.Y. Guo, Coupling of newborn ions to the solar wind by electromagnetic instabilities and their interaction with the bow shock, J. Geophys. Res., **90**, 2713–2726, 1985.

8.285   Winske, D., S.P. Gary, Electromagnetic instabilities driven by cool heavy ion beams, J. Geophys. Res., **91**, 6825–6832, 1986.

8.286   Withbroe, G.L., The chromospheric and transition layers in coronal holes, in *Coronal Holes and High Speed Streams*, ed. by J.B. Zirker, Colorado Associated University Press, Boulder, Colorado, USA, 145–177, 1977.

8.287   Withbroe, G.L., The temperature structure, mass and energy flow in the corona and inner solar wind, Astrophys. J., **325**, 442–467, 1988.

8.288   Withbroe, G.L., J.L. Kohl, H. Weiser, R.H. Munro, Probing the solar wind acceleration region using spectroscopic techniques, Space Sci. Rev., **33**, 17–52, 1982.

8.289   Withbroe, G.L., J.C. Raymond, Plasma diagnostics for the outer solar corona: UV and XUV Fe XII lines, Astrophys. J., **285**, 347–353, 1984.

8.290   Woo, R., Spacecraft radio scintillation and scattering measurements of the solar wind, in *Solar Wind Four*, ed. by H. Rosenbauer, Report MPAE-W-100-81-31, Max-Planck-Institut für Aeronomie, Katlenburg-Lindau, F.R. Germany, 66–77, 1981.

9. Waves and Instabilities

Donald A. Gurnett

9.1 Introduction

It is widely recognized that even though the solar wind is essentially collisionless, to a good approximation it can be treated as a fluid. To understand this behavior, it is first necessary to understand the microscopic processes that control the macroscopic properties of the plasma. In a collisionless plasma such as the solar wind, it is now widely recognized that waves play a role similar to collisions in an ordinary fluid. As the plasma flows outward from the sun, dynamical changes cause the velocity distribution function to deviate from an equilibrium thermal distribution. In the absence of collisions these deviations continue to grow until the velocity-space gradients parallel and perpendicular to the magnetic field, $\partial f / \partial v_\parallel$ and $\partial f / \partial v_\perp$, become so large that plasma instabilities start to occur. These instabilities lead to the growth of waves. As the waves grow to large amplitudes, wave–particle interactions eventually act to eliminate the velocity-space gradients that cause the instability. Waves and instabilities thereby play a crucial role in preventing large deviations from thermal equilibrium. For a review of the types of plasma instabilities and wave–particle interactions that can occur in a plasma see, for example, Hasegawa [9.34], or Melrose [9.47].

For reference, the primary plasma wave modes of importance in the solar wind are summarized in Fig. 9.1, which shows a plot of the phase velocity, ω/k, as a function of frequency for typical solar wind conditions at 1 AU. To simplify the presentation, only waves propagating parallel to the magnetic field are shown. At low frequencies the relevant characteristic speeds of the plasma are the Alfvén speed, $V_A = B/\sqrt{\mu_0 \varrho_m}$, and the sound speed, $V_s = \sqrt{kT/m}$, where B is the magnetic field strength, ϱ_m is the mass density, T is the temperature, and m is the ion mass. Except very close to the sun the solar wind speed, V_{sw}, is substantially greater than either of these speeds.

At low frequencies three modes exist. These modes are called the fast, intermediate, and slow magnetohydrodynamic (MHD) waves. At higher frequencies these three modes merge into the whistler mode, the ion cyclotron mode, and the ion acoustic mode. The whistler mode has a resonance (point of zero phase velocity) at the electron cyclotron frequency, f_c^-, and the ion cyclotron mode has a resonance at the ion cyclotron frequency, f_c^+. The upper frequency limits of these two modes are at f_c^- and f_c^+, respectively. Both the whistler mode and the ion cyclotron mode are electromagnetic, since the waves have both electric

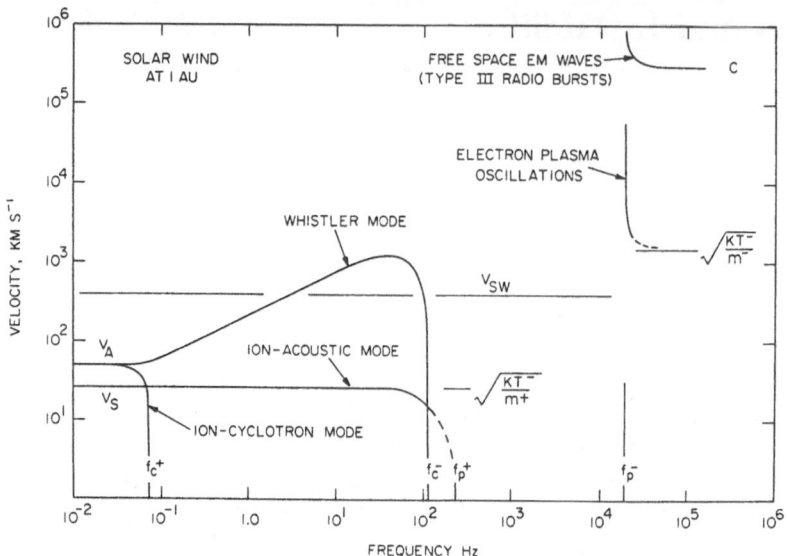

Fig. 9.1. A plot showing the phase velocity of selected plasma wave modes in the solar wind for propagation parallel to the magnetic field. Modes with phase velocities below the solar wind velocity, V_{sw}, are strongly Doppler shifted

and magnetic fields. These modes can be driven unstable by a variety of free energy sources, including currents, anisotropies and particle beams [9.47]. At frequencies above the ion cyclotron frequency the magnetic field of the ion acoustic mode becomes negligibly small compared to the electric field. This type of wave, with no magnetic field, is called an electrostatic wave, since the electric field can be derived from a potential. The ion acoustic mode has many properties similar to a sound wave in an ordinary gas and is strongly damped by Landau damping unless the electron temperature, T_e, is much greater than the ion temperature, T_i. The region of strong damping is indicated by a dashed line in Fig. 9.1. The upper frequency limit of the ion acoustic mode is the ion plasma frequency, f_p^+. If $T_e \gg T_i$, then the ion acoustic mode can be driven unstable by a number of free-energy sources, including currents [9.57], electron heat conduction [9.14], and ion beams [9.44].

At high frequencies, near the electron plasma frequency, f_p^-, three additional modes appear. Two of these modes are the free space electromagnetic modes, one of which is right-hand polarized, and the other of which is left-hand polarized. The phase velocities go to infinity as the wave frequency approaches f_p^-. At frequencies well above the plasma frequency the phase velocities of the two free space modes approach the speed of light, c. The free space modes do not propagate at frequencies below the electron plasma frequency.

Slightly above the plasma frequency a third purely electrostatic mode also occurs. This mode is called a Langmuir wave or an electron plasma oscillation. For phase velocities well above the electron thermal speed, the electron plasma oscillation represents an almost purely electrostatic oscillation at the electron

136

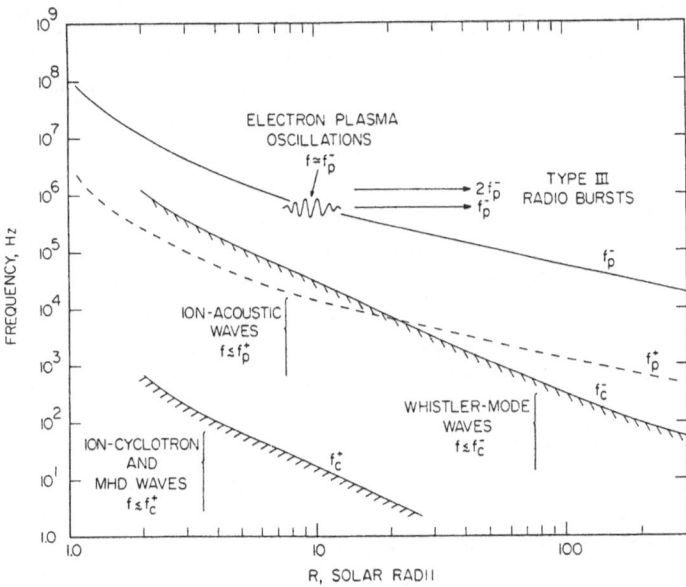

Fig. 9.2. The characteristic frequencies of the plasma as a function of radial distance from the sun

plasma frequency. As the phase velocity decreases, approaching the electron thermal speed, the frequency begins to increase. However, in this region the oscillations are strongly damped by Landau damping. The region of strong damping is indicated by a dashed line. Electron plasma oscillations are driven unstable whenever a region of positive slope, $\partial f / \partial v_{\parallel} > 0$, occurs in the electron distribution function. These distribution functions are characteristic of beams, such as those injected into the solar wind by solar flares [9.45].

When only small amplitude waves are considered, the two free space electromagnetic modes propagate completely independent of the other plasma wave modes. However, as the amplitude increases, nonlinear effects cause the two free space modes to be coupled to the plasma wave modes. Plasma waves can then transfer energy to the free space modes, thereby producing radio emissions that can be detected at large distances from the coupling region. This nonlinear mode conversion process is the mechanism by which most solar radio emissions detected at the earth are believed to be produced [9.42].

To illustrate the typical radial variations that occur in the solar wind, Fig. 9.2 shows the characteristic frequencies of the solar wind plasma as a function of heliocentric radial distance, R. At the orbit of the earth the electron plasma frequency is typically about 20 kHz, and the ion plasma frequency is about 450 Hz. The electron and ion cyclotron frequencies are typically about 140 Hz and 0.07 Hz, respectively. At large distances from the sun, where the solar wind velocity is essentially constant, conservation of particles implies that the solar wind electron density must vary as $1/R^2$. Since the electron plasma frequency depends on the square root of the electron density, $f_{\mathrm{p}}^- = 9\sqrt{n_{\mathrm{e}}}$ kHz, where n_{e} is in units of cm^{-3}, the electron plasma frequency must vary as $1/R$. As the solar wind velocity decreases near the sun, the electron plasma frequency decreases more rapidly than $1/R$. The approximate radial variation based on existing models of the solar wind velocity [9.35] is shown in Fig. 9.2. The ion plasma frequency varies like the electron plasma frequency but for protons is a factor of $\sqrt{m^-/m^+} = \frac{1}{43}$ lower than the electron plasma frequency. The radial variations of the electron and ion cyclotron frequencies are more complicated since the magnetic field makes a transition from a radial, $1/R^2$, configuration to an Archimedian spiral, $1/R$, configuration near the earth's orbit. Near the sun even more complicated and less predictable variations, generally steeper than $1/R^2$, occur because of coronal loops and other complicated magnetic field structures. The radial variations of f_c^- and f_c^+ shown in Fig. 9.2 are based on magnetic field models discussed by Hundhausen [9.35]. The ion cyclotron frequency shown is for protons and is a factor of $m^-/m^+ = \frac{1}{1836}$ lower than the electron cyclotron frequency. Other ions, such as He^{++}, also exist in the solar wind. The cyclotron frequencies of these heavier ions, which are normally much less abundant than protons, are not shown.

For the purposes of this review, only waves with frequencies near and above the proton cyclotron frequency are considered. Waves and fluctuations at frequencies below the ion cyclotron frequency are normally studied by static-field magnetometers whose characteristics are quite different from instruments designed to detect plasma waves at frequencies above the ion cyclotron frequency. Numerous spacecraft have carried instruments to study plasma waves in the solar wind. These include *OGO 3* and *5, Pioneer 8* and *9, IMP 6, 7* and *8, Hawkeye 1, Helios 1* and *2, ISEE 1, 2* and *3*, and *Voyager 1* and *2*. Of these, only *Helios 1* and *2* provided measurements significantly inside of 1 AU. Except for *Voyager 1* and *2*, which provided measurements beyond 1 AU, the remaining spacecraft provided measurements near the earth's orbit. Since this review concentrates mainly on the inner heliosphere, inside of 1 AU, the results presented are mainly from *Helios 1* and *2*, although results from other spacecraft are discussed when appropriate.

The plasma wave instrumentation on *Helios 1* and *2* is typical of the types of instrumentation used to detect plasma waves in the solar wind. Two types of sensors are used, one to detect electric fields and the other to detect magnetic fields. The electric field sensor consists of an electric dipole antenna, 32 meters tip-to-tip, mounted perpendicular to the spacecraft spin axis as shown in Fig. 9.3. Electric fields are detected by measuring the voltage difference between the two antenna elements using a sensitive differential amplifier. For a description of the electric field sensor and associated instrumentation on *Helios*, see Gurnett et al. [9.27]. The magnetic field sensor consists of a triaxial search coil magnetometer mounted on the end of a boom as shown in Fig. 9.3. The search coil magnetometer consists of a high permeability iron rod wound with a large number of turns of fine wire. Magnetic fields are measured by detecting the voltage induced in the winding with a sensitive preamplifier. Since only time-varying magnetic fields cause an induced voltage, this type of magnetometer responds only to wave magnetic fields, and not to static fields. For a description of the search coil

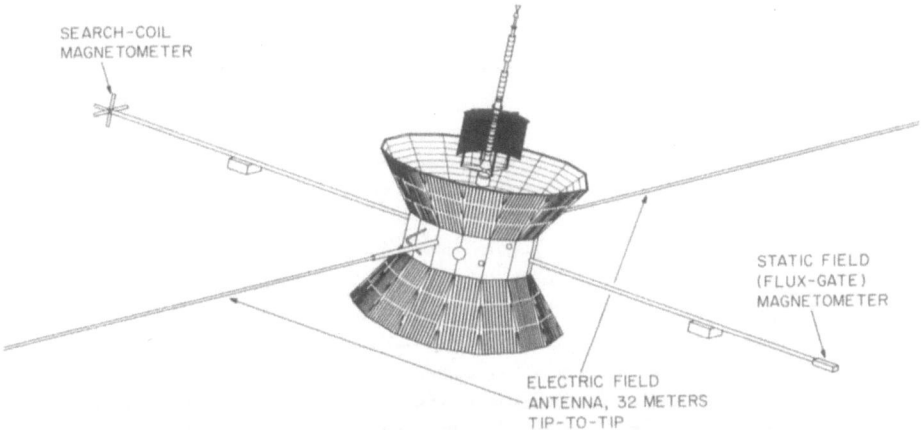

Fig. 9.3. A sketch of the *Helios* spacecraft showing the electric dipole antenna used for detecting wave electric fields and the search-coil magnetometer used for detecting wave magnetic fields

magnetometer and associated instrumentation on *Helios*, see Dehmel et al. [9.7], and Gliem et al. [9.22].

Since the types of plasma wave occurring in the solar wind are divided conveniently into electrostatic and electromagnetic waves, this review is organized by separately considering electrostatic and electromagnetic waves.

9.2 Electrostatic Waves

The *Helios* plasma wave instrument detected two primary types of electrostatic wave: electron plasma oscillations and ion acoustic waves. Examples of both types of wave are illustrated in Fig. 9.4 which shows a 16-channel plot of the electric field intensities detected by *Helios 2* at a heliocentric radial distance of 0.45 AU. The solid line in each frequency channel gives the peak electric field strength and the upper edge of the solid black band gives the average field strength. The time resolution for the peak and average field strength measurements varies with bit rate and in this case is 40 seconds. The electron plasma oscillations occur in the 56.2 kHz channel during the interval from about 0650 to 0740 UT, and the ion acoustic waves occur in the frequency range from about 1.0 to 17.8 kHz for the entire 8-hour duration of the plot. Because these two types of emission usually occur independently and have different characteristics, they will be discussed separately.

9.2.1 Electron Plasma Oscillations

Electron plasma oscillations are easily identified in the *Helios* electric field data because they usually occur in only one or two frequency channels near the local

139

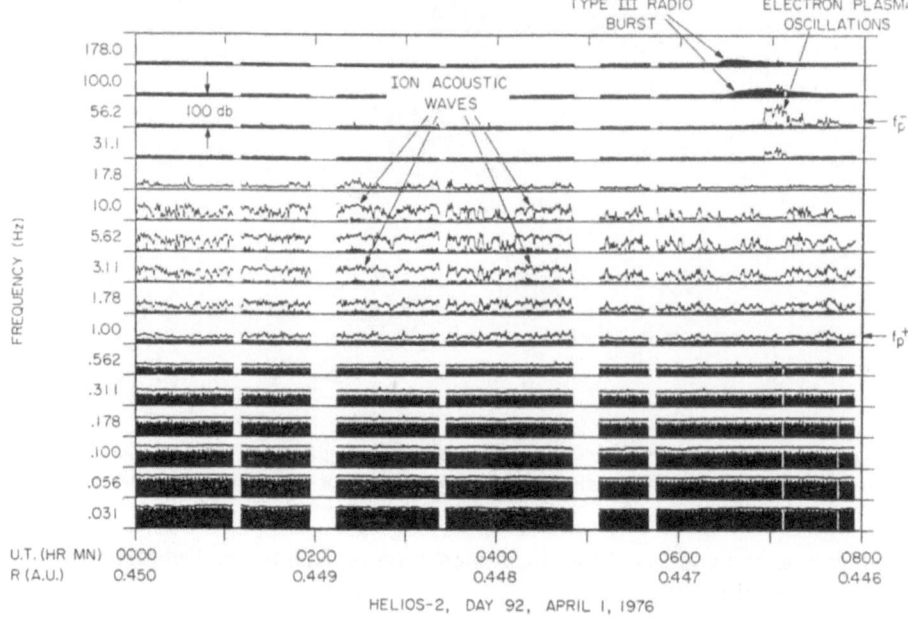

Fig. 9.4. A representative plot of the electric field intensities from *Helios 2* showing an electron plasma oscillation event associated with a type III solar radio burst and a period of enhanced ion acoustic wave activity

electron plasma frequency, f_p^-. Electron plasma oscillations are often associated with type III solar radio bursts of the type shown in Fig. 9.4, although they can also occur separately. Type III radio bursts are produced by flares and active regions on the sun and have a characteristic frequency–time variation, decreasing monotonically in frequency with increasing time [9.59]. The *Helios* observations of electron plasma oscillations occurring in association with type III radio bursts [9.24] confirm a long-standing theory for the origin of these radio bursts. According to this theory, first proposed by Ginzburg and Zheleznakov [9.21], the generation of type III radio bursts is a two-step process in which (1) electron plasma oscillations are first produced by energetic electrons from a solar flare, and (2) the energy in the plasma oscillations is converted to electromagnetic radiation via coupling to the free space electromagnetic modes. Because electrons are impulsively released by the flare, a beam-like region of positive slope, $\partial f / \partial v_{\parallel} > 0$, is generated near the leading edge of the energetic electron stream owing to time-of-flight considerations. The temporal evolution of the reduced one-dimensional distribution function, $f(v_{\parallel})$, is illustrated in Fig. 9.5 for an electron plasma oscillation event detected by the *ISEE 3* spacecraft at 1 AU [9.45]. Since the growth rate of the plasma oscillations is proportional to $\partial f / \partial v_{\parallel}$ [9.57, 41], waves are only generated during the interval when $\partial f / \partial v_{\parallel}$ is positive. This interval is usually about 1 hour at 1 AU, decreasing to about 20 minutes at 0.3 AU. Because the region of positive slope is only present near the leading edge of the electron stream, the plasma oscillations are confined to a localized region that moves outward from the sun at a speed comparable to the energetic

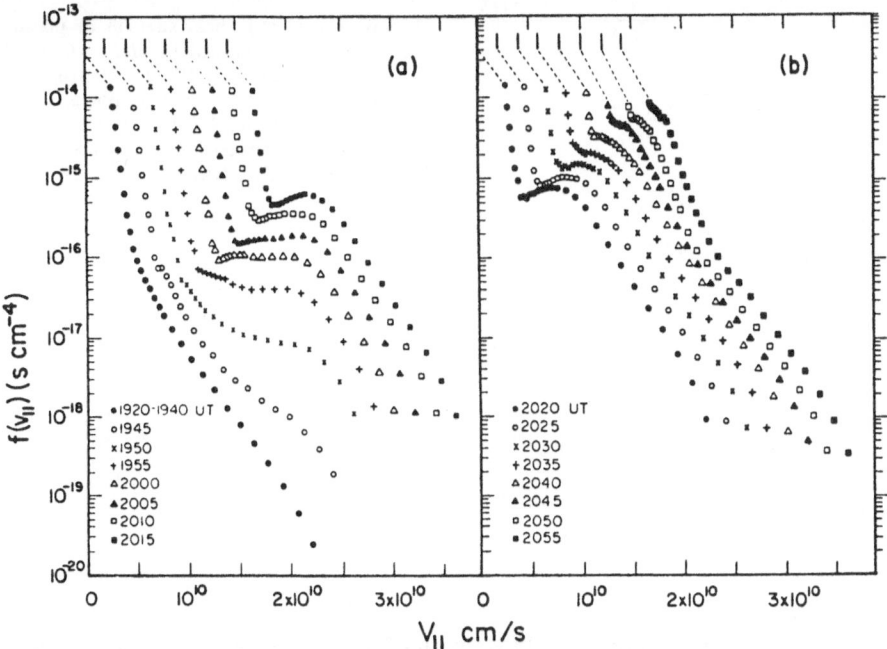

Fig. 9.5a,b. A sequence of reduced one-dimensional electron distribution functions from *ISEE 3* [9.45] showing energetic electrons arriving from a solar flare. Electron plasma oscillations occur during the times when the distribution function has a region of positive slope, $\partial f / \partial v_{\parallel} > 0$

electron stream, typically $0.1c$ to $0.5c$. As the region of intense plasma oscillations moves outward from the sun, electromagnetic radiation is generated at f_p^- and $2f_p^-$, producing the decreasing emission frequency with increasing time that is characteristic of type III radio bursts. This radio emission process is illustrated in Fig. 9.2. At low frequencies, below about 1 MHz, the strongest radiation is usually at the second harmonic, $2f_p^-$, [9.11, 36, 28], although evidence also exists for radiation at the fundamental [9.37, 38]. Because the electromagnetic radiation propagates freely away from the source, the type III radiation is typically detected 10 to 20 minutes before the region of intense plasma oscillations arrives at the spacecraft.

By using a large number of electron plasma oscillation events detected over a several-year period, the variation of the electric field strength of the plasma oscillations can be determined as a function of radial distance from the sun [9.29, 32]. This variation is shown in Fig. 9.6. As can be seen, the electric field strength tends to decrease with increasing heliocentric radial distance, varying approximately as $R^{-1.4\pm0.5}$. This decrease is expected because the electric field to plasma energy density ratio, $E^2/8\pi nkT$, which controls the saturation intensity, is approximately constant. Since the number density n varies as R^{-2} and the temperature T varies as $R^{-2/7}$ [9.35], for a constant energy density ratio, the maximum electric field E is expected to vary approximately as $R^{-8/7}$. The observed radial variation is seen to be reasonably consistent with the expected radial variation. Typical maximum values for $E^2/8\pi nkT$ are about 2×10^{-5}.

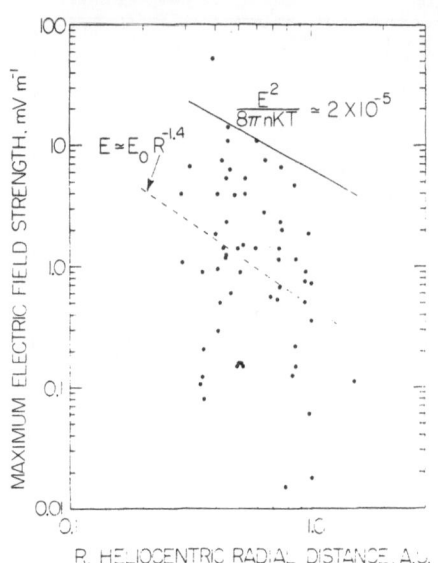

Fig. 9.6. The electric field strength of plasma oscillation events detected by *Helios 1* and 2 in association with type III radio bursts. The best-fit power law through these points varies as $R^{-1.4\pm0.5}$. The *solid line* shows the electric field intensity corresponding to an electric field to plasma energy density ratio of $E^2/8\pi nkT \simeq 2 \times 10^{-5}$

The radial variation of the field strength of the plasma oscillations can also be compared with the emissivity of the type III radiation, which varies as approximately R^{-6} [9.58]. For second-harmonic emission mechanisms, the emissivity should vary as E^4. This dependence occurs because the radiating current at the second harmonic is proportional to the product of the field strength of two interacting plasma oscillations. Since the power radiated is proportional to the square of the current, a fourth-power dependence on the electric field strength is expected. Since the emissivity varies approximately as R^{-6}, the electric field intensity should then vary as $R^{-1.5}$. This radial variation is seen to be in reasonable quantitative agreement with the observed radial variation.

High-time-resolution measurements of electron plasma oscillations detected by *Helios* [9.25] show that the electric field intensity is very spiky, with rapid intensity variations down to the time resolution of the instrument, which is about 50 ms. This very spiky electric field structure accounts for the very large peak to average field strength ratios evident in Fig. 9.4. Consideration of the nonlinear effects that stabilize the electron beam has led a number of investigators to predict that the electric field structure should be very spiky [9.52, 17, 2, 51, 50, 15, 23]. The origin of the spiky structure involves a focusing effect that causes the plasma oscillation intensity to increase in regions of decreased plasma density. If the wave intensity becomes sufficiently large, the electric field pressure causes the plasma density to decrease, further increasing the focusing and eventually forming a soliton-like structure that collapses down to a spatial scale of only a few Debye lengths. Collapsed plasma oscillation structures of this type, sometimes referred to as "spiky turbulence", have been observed in the laboratory [9.60]. For the receiver averaging time constant of the *Helios* instrument, ~ 50 msec, it is not possible to resolve intensity fluctuations at the time scales, hundreds of μs, required to resolve soliton-like structures. The maximum plasma oscillation field strengths measured by *Helios*, typically only 5 to 10 mV/m, are too small to reach the threshold electric field strength, $E^2/8\pi nkT \gtrsim (k\lambda_D)^2 \simeq 10^{-2}$ to 10^{-3}, required for soliton collapse. However, because of the relatively long, 50 ms, averaging interval, it is possible that field strengths much larger than 5 to 10 mV/m could be occurring on short time scales. Evidence of soliton-like intensity fluctuation has been reported by Gurnett et al. [9.33] for electron plasma oscillations detected in the solar wind upstream of Jupiter's bow shock.

9.2.2 Ion Acoustic Waves

The second basic type of electrostatic noise detected in the solar wind by *Helios* is a band of sporadic emissions between the proton and electron plasma frequency, f_p^+ and f_p^-. Because of the frequency range involved, these emissions were initially referred to as $f_p^+ < f < f_p^-$ noise [9.25]. In their initial interpretation of this noise, Gurnett and Anderson [9.25] suggested that the emissions are ion acoustic waves which are Doppler shifted into the frequency range $f_p^+ < f < f_p^-$ by the motion of the solar wind. As will be discussed shortly, considerable evidence now exists that these waves are ion acoustic waves, or a closely related ion-acoustic-like mode. For this reason, we will refer to the emissions as ion acoustic waves.

A typical example of the ion acoustic waves detected by *Helios* is shown in Fig. 9.4. The noise does not occur continuously but rather in episodes lasting for periods ranging from a few hours to several days. Episodes of enhanced ion acoustic wave activity usually occur a few times per month and tend to be more frequent and more intense closer to the sun. Periods of enhanced ion acoustic wave activity often recur on successive solar rotations, particularly ahead of high-speed solar wind streams [9.31] and in association with interplanetary shocks [9.30, 40]. An example of an ion acoustic wave event associated with an interplanetary shock is shown in Fig. 9.7. The ratio of the peak to average electric field strength is usually very large, typically greater than 40 dB, indicating that the noise is impulsive and sporadic on time scales of a few seconds or less. When the intensities are averaged over several minutes the spectrum appears very broad, usually extending from near f_p^+ to slightly below f_p^-. The cutoffs at f_p^+ and f_p^- are not rigid cutoffs, and, as will be discussed shortly, it is probably coincidental that the spectrum falls in this frequency range. As shown by Gurnett and Frank [9.26] and Gurnett et al. [9.31] both the intensity and frequency of the noise tend to increase with decreasing radial distance. This tendency is illustrated in Fig. 9.8, which shows representative electric field spectrums of the ion acoustic noise at three heliocentric radial distances: 1.73 AU from *Voyager 1*, 0.98 AU from *Helios 1*, and 0.47 AU from *Helios 1*. A statistical study of the radial dependence shows that the broad-band electric field strength varies approximately as $1/R$, varying from about 1 mV/m (10 % quartile) at 0.33 AU to about 0.3 mV/m at 1 AU. Although *Helios* had no capability for obtaining electric field waveforms, wide-band measurements from *Voyager 2* [9.43] show that the ion acoustic waves

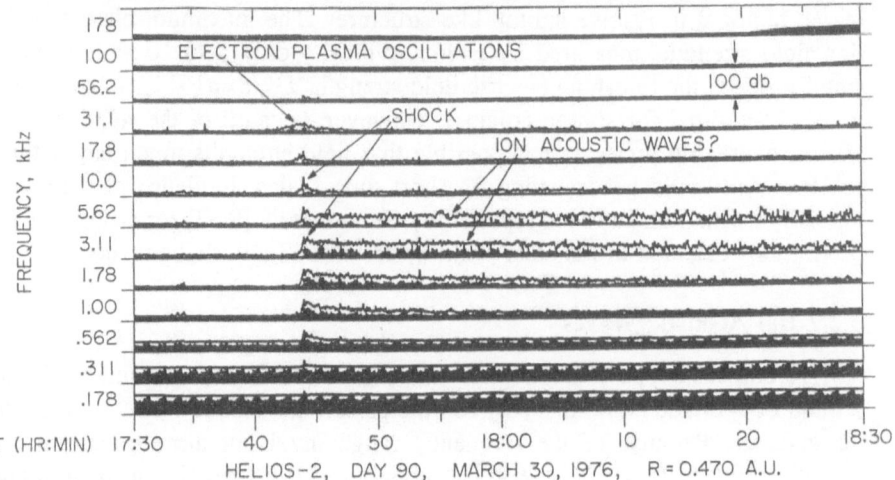

Fig. 9.7. An example of an ion acoustic wave event detected by *Helios 2* in association with an interplanetary shock. A region of weak electron plasma oscillations also occurs near the shock. For an analysis of this event, see Gurnett et al. [9.30]

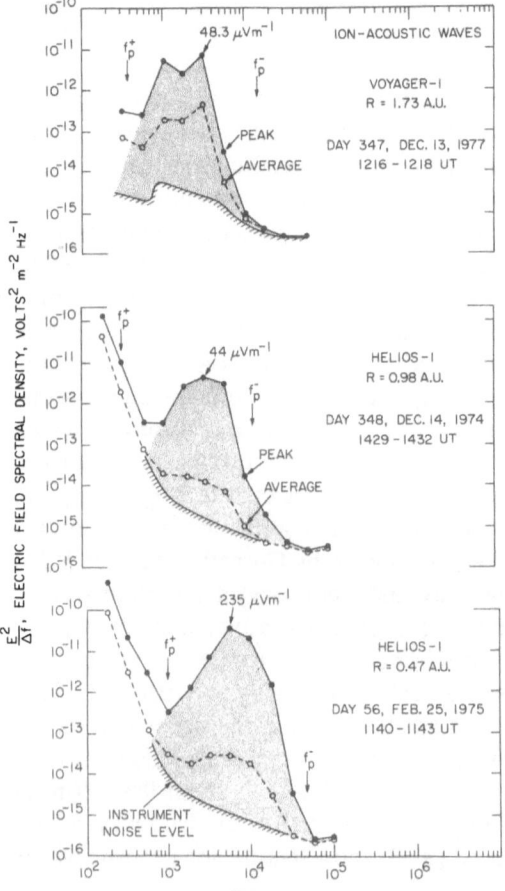

Fig. 9.8. Peak and average electric field spectral densities for ion acoustic waves at three representative radial distances from the sun. The spectrum increases in intensity and moves to higher frequencies with decreasing radial distance, usually staying in the range $f_p^+ < f < f_p^-$

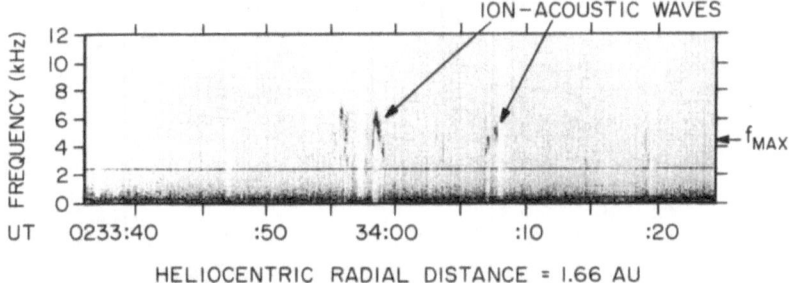

Fig. 9.9. A high-resolution frequency–time spectrogram of ion acoustic waves detected by *Voyager 2* at 1.66 AU. The ion acoustic emissions consist of a nearly monochromatic tone with a rapidly varying center frequency

consist of nearly monochromatic emissions whose center frequencies vary over a wide range on a time scale of a few seconds. A typical high-resolution frequency–time spectrogram of a burst of ion acoustic waves detected by *Voyager 2* is shown in Fig. 9.9. Because of rapid fluctuations in the center frequency of the emission, the spectrum appears to be very broad when averaged over longer intervals, as in Fig. 9.8, even though it is nearly monochromatic on short time scales.

The evidence that the $f_p^+ < f < f_p^-$ noise detected by *Helios* is caused by ion acoustic waves is somewhat indirect and is based mainly on comparisons with very similar waves observed upstream of the earth's bow shock. Because ion acoustic waves have very short wavelengths, the Doppler shift caused by the motion of the solar wind is substantial. When the Doppler shift is included, the frequency of an ion acoustic wave in the spacecraft frame of rest is given by

$$\omega = \frac{C_S k}{\sqrt{1 + k^2 \lambda_D^2}} + V_{sw} k \cos \theta_{kv} , \tag{9.1}$$

where $C_S = \sqrt{kT_e/m_i}$ is the ion acoustic speed, k is the wave number, θ_{vk} is the angle between the wave vector and the solar wind velocity, and λ_D is the Debye length. The rest-frame frequency of the ion acoustic wave is given by the first term, and the Doppler shift is given by the second term. For typical solar wind parameters, the Doppler shift is much larger than the rest-frame frequency. Therefore, to a good approximation $\omega \simeq V_{sw} k \cos \theta_{kv}$. Because of the onset of strong Landau damping at short wavelengths, k has a maximum value of $k\lambda_D \simeq 1$. The maximum frequency is then given by $\omega_{max} = V_{sw}/\lambda_D$. For typical solar wind parameters at 1 AU ($n = 5$ cm^{-3}, $T = 1.5 \times 10^5$ K, $V_{sw} = 400$ km/s) the maximum frequency is approximately $f_{max} \simeq 8.0$ kHz, which is in good agreement with the observed upper frequency limit (see Fig. 9.8). Since the electron temperature is nearly independent of radial distance [9.35] and since both V_{sw}/λ_D and f_p^- vary as the square root of the electron density, it is easy to verify that f_{max} is proportional to f_p^-. This relationship explains why the frequency of the ion acoustic waves appears to vary in direct proportion to f_p^- (see Fig. 9.8).

145

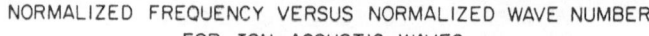

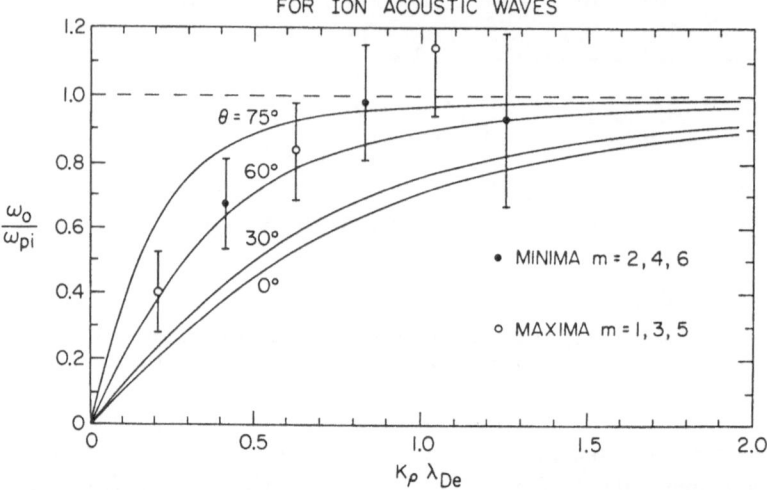

Fig. 9.10. Measurements of the frequency and wave number of ion acoustic noise detected by *ISEE 1* upstream of the earth's bow shock. The *solid curves* show the dispersion relation for the ion acoustic mode as a function of the polar angle θ of the wave vector, which is unknown

However, in this interpretation the upper cutoff frequency actually has no direct relationship to f_p^-. It is only coincidental that f_{max} is near f_p^-.

Although the upper cutoff frequency of the $f_p^+ < f < f_p^-$ noise is consistent with an ion acoustic wave interpretation, it does not uniquely identify the mode because other short wavelength modes also exist with cutoffs near $k\lambda_D \simeq 1$. These modes would also have an upper frequency cutoff at $\omega_{max} = V_{sw}/\lambda_D$. The evidence that the noise is due to ion acoustic waves rests on other observations. First, a well-known characteristic of the ion acoustic mode is that the damping is very sensitive to the electron to ion temperature ratio, T_e/T_i. Ion acoustic waves are weakly damped if $T_e/T_i \gg 1$, and strongly damped if $T_e/T_i = 1$. Using solar wind plasma measurements from *Helios*, Gurnett et al. [9.31] have shown that the electric field intensity of the $f_p^+ < f < f_p^-$ noise increases as T_e/T_i increases, as would be expected for an ion acoustic instability. Second, Gurnett and Frank [9.26] have pointed out that the ion acoustic noise detected by *Helios* has essentially identical characteristics to a band of low-frequency electric field noise detected upstream of the earth's bow shock between about 1 to 10 kHz. By using techniques not available on *Helios*, Fuselier and Gurnett [9.16] have been able to measure the wavelength of the waves. By independently measuring both the wavelength and the frequency, the dispersion relation can be determined. An example of such dispersion-relation measurements is shown in Fig. 9.10. Although the polar angle θ of the wave vector is a free parameter, the basic shape of the dispersion curve, including the break at $k\lambda_{De} \simeq 1$, and the upper frequency limit at ω_{pi}, is in excellent agreement with the ion acoustic dispersion relation. Although these measurement are not available on *Helios*, they strongly suggest that the waves detected by *Helios* are also ion acoustic waves.

Given that the waves detected by *Helios* are ion acoustic waves, the question arises as to how these waves are generated. This is a difficult question that has not been completely answered. Upstream of the earth's bow shock it is now well established [9.54, 26, 1] that the comparable types of waves are closely associated with energetic, 1 to 10 keV, proton beams arriving from the bow shock. Since ion acoustic waves are also observed ahead of interplanetary shocks [9.40], one might think that the ion acoustic waves detected by *Helios* far from the earth are produced by an ion beam in the solar wind. In fact, Lemons et al. [9.44], have suggested that the waves are produced by an ion-beam instability. Despite the reasonableness of this line of argument, to date no definite relationship has been established between ion beams in the interplanetary medium and ion acoustic waves. This situation is in part due to the absence of suitable energetic ion measurements on *Helios*. However, attempts to establish a relationship using the more comprehensive ion measurements available on *ISEE 3* at 1 AU have also not been successful. Finally, it should be pointed out that energetic protons with energies of 1 to 10 keV have velocities too high to resonate with the ion acoustic mode. Therefore, it is unlikely that these ions are the direct source of excitation of the ion acoustic waves, if indeed they are ion acoustic waves.

Many years ago, even before the discovery of the solar wind ion acoustic waves, it was suggested by Forslund [9.14] that the conduction of heat away from the sun could excite ion acoustic waves in the solar wind. This instability occurs because if no net current is allowed to flow in the plasma then the presence of a third moment (heat flux) in the electron distribution function produces a shift between the peaks of the electron and ion distribution functions. If the heat flux is sufficiently large, this shift produces a double hump in the reduced distribution function that makes the ion acoustic mode unstable. Comparative studies of the *Helios* plasma and plasma wave data [9.31] show that the ion acoustic wave intensities are closely correlated with the electron heat flux. This correlation is shown in Fig. 9.11. Detailed studies of electron and ion distribution functions by Dum et al. [9.10] also show that the electron heat flux can account for the ion acoustic waves observed by *Helios*. The electron heat flux instability therefore seems to provide the most reasonable basis for understanding how these waves are generated. However, there are still unresolved difficulties. In some cases ion acoustic waves are observed when T_e/T_i is near one. Under these conditions the ion acoustic mode should be strongly damped. Also, in a few of the cases analyzed by Dum et al. [9.10] the ion acoustic mode was found to be stable even though waves were present. The origin of the instability under these unusual conditions remains unresolved. Marsch and Chang [9.46] have suggested that under these conditions another instability involving lower hybrid waves may be required to explain the noise.

The possible macroscopic consequences of the solar wind ion acoustic waves detected by *Helios* has not been established. Forslund [9.14] suggested that ion acoustic waves in the solar wind could have important consequences for controlling heat conduction in the solar wind. Usually the ion acoustic waves detected by *Helios* are very weak. The ratio of the electric field energy density to the plasma

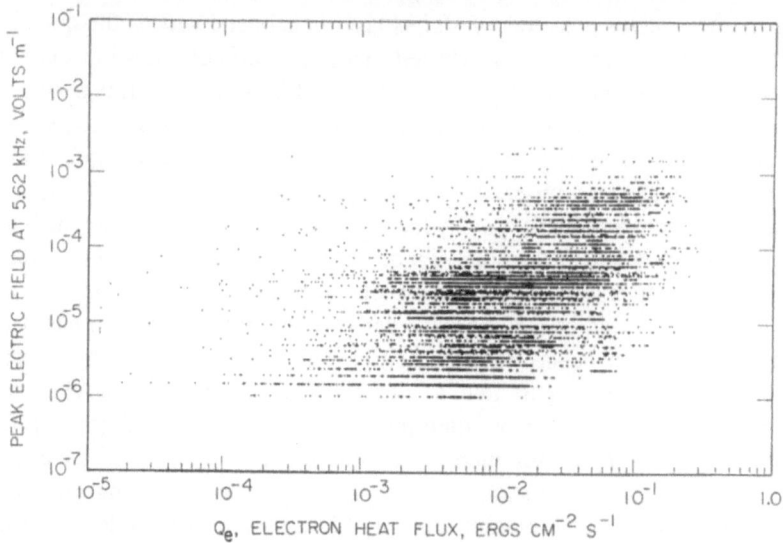

Fig. 9.11. A plot of simultaneous measurements of the electric field intensity of ion acoustic waves and the electron heat flux. The field intensity clearly tends to increase as the heat flux increases, suggesting that the ion acoustic waves may be driven by an electron heat-flux instability

energy density, $E^2/8\pi nkT$, is only about 10^{-5} to 10^{-7} during intense bursts, and much smaller on the average. Whether such low intensities can have significant macroscopic effects on the solar wind has not been adequately explored.

9.3 Electromagnetic Waves

Three primary types of electromagnetic waves are observed in the solar wind: MHD waves, ion cyclotron waves, and whistler-mode waves. Since electromagnetic waves always have a magnetic field component, these waves can be most easily identified by using magnetic field measurements. As can be seen from Fig. 9.1, at 1 AU MHD waves and ion cyclotron waves occur at frequencies below 10^{-1} Hz, in the frequency range where static-field magnetometers provide the best sensitivity. Whistler-mode waves on the other hand occur at much higher frequencies, up to 10^2 Hz, in the frequency range where search-coil magnetometers provide the best sensitivity. Therefore, both types of measurements must be used to study the spectrums of these waves.

9.3.1 MHD Waves and Ion Cyclotron Waves

An example of a magnetic field spectrum obtained by combining measurements from a static-field (flux-gate) magnetometer and a search-coil magnetometer is shown in Fig. 9.12 (from Denskat et al. [9.9]). These data were obtained from *Helios 2* at a radial distance of 0.30 AU. The magnetic field spectral densities

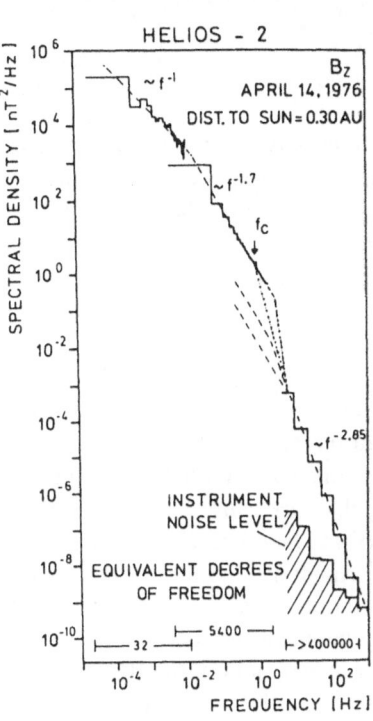

Fig. 9.12. A representative magnetic field spectrum from *Helios* 2 at 0.3 AU. The abrupt decrease in amplitude at about 2 Hz is believed to be due to cyclotron damping of electromagnetic ion cyclotron waves propagating outward from the sun. The spectrum above 2 Hz is mainly due to whistler-mode noise

below 2 Hz were computed by performing a power spectral analysis of a time series of flux-gate magnetometer measurements. The magnetic field spectral densities at frequencies above 4.7 Hz were obtained using a set of eight bandpass filters from the search-coil magnetometer.

As can be seen, the spectrum decreases monotonically with increasing frequencies, varying approximately as $f^{-1.55}$ at low frequencies, steepening to approximately $f^{-2.9}$ at high frequencies. A discontinuity, or inflection point, in the spectrum occurs from 2 Hz to 4.7 Hz. This portion of the spectrum is shown by dashed lines in Fig. 9.12. Although the inflection point corresponds almost exactly to the gap between the two instruments, Denskat et al. [9.9] argue that the effect is real and is caused by a change in the plasma wave spectrum. Further evidence for an important change in the spectrum in this frequency range is given by the slope, which shifts from a $f^{-1.7}$ variation below 2 Hz to a $f^{-2.8}$ variation above 4.7 Hz.

In the frequency range around a few Hz, the only resonance or propagation cutoff that could account for a change in the spectrum is the proton cyclotron frequency. The measured proton cyclotron frequency, f_c^+, is indicated in Fig. 9.12. As can be seen, the proton cyclotron frequency is near, but somewhat below, the point where the inflection occurs in the spectrum. This offset is believed to be caused by a Doppler shift. Because the solar wind is moving relative to the spacecraft, the frequency, f', detected in the spacecraft frame of reference is shifted from the rest frame frequency, f, by the Doppler effect. The frequency in the spacecraft frame of reference is given by

$$f' = f\left[1 + (V_{sw}/V_p)\cos\theta_{kv}\right] , \qquad (9.2)$$

where V_p is the phase velocity and θ_{kv} is the angle between the solar wind velocity and the direction of propagation. Since the solar wind velocity is typically a factor of five or more greater than the phase velocity of the ion cyclotron mode, the Doppler shift could be easily a factor of five, or more. Doppler shifts of this magnitude could easily explain the offset between the inflection point in the spectrum and the proton cyclotron frequency. The offset is positive, toward higher frequencies, for waves propagating away from the sun ($\theta_{kv} < 90°$).

Next we consider the origin of the MHD waves. For many years [9.5,6] it has been known that intense low-frequency MHD wave fluctuations exist in the solar wind. These waves propagate outward from the sun, apparently originating from turbulent solar wind heating processes in the solar corona. The intense low-frequency magnetic field fluctuations in Fig. 9.12 are caused by these waves. These fluctuations are probably a superposition of the fast, intermediate and slow MHD waves described in the introduction. Unfortunately, the three MHD wave modes cannot be uniquely resolved by measurements from a single spacecraft. Therefore, the exact distribution of wave energy between the three modes is not known. For a review of the observations and their interpretation, see Barnes [9.3]. Usually, it is assumed that some energy exists in the MHD mode that connects to the ion cyclotron mode. As the waves propagate outward from the sun the ion cyclotron frequency gradually decreases (see Fig. 9.1). As the ion cyclotron frequency approaches the wave frequency, the waves are strongly damped by ion cyclotron damping. This damping eventually absorbs all of the energy in the ion cyclotron mode. Denskat et al. [9.9] propose that absorption of wave energy by cyclotron damping causes the abrupt decrease in the magnetic field spectral density near the ion cyclotron frequency. Studies of the spatial variation in the magnetic field fluctuations by Denskat and Neubauer [9.8] show that the fluctuations decrease rapidly with increasing radial distance from the sun, as would be expected if the waves are being absorbed.

9.3.2 Whistler-Mode Waves

It is evident from the magnetic field spectrum in Fig. 9.12 that a significant amount of wave energy extends up to frequencies as high as 300 Hz, above the proton cyclotron frequency. These frequencies are much too high to be propagating in the ion cyclotron mode. Since the ion acoustic mode is electrostatic and has no magnetic field, the only electromagnetic mode that can account for magnetic field fluctuations in this frequency range is the whistler mode. For this reason Denskat et al. [9.9] conclude that this turbulence must consist of whistler-mode waves.

Whistler-mode turbulence comparable to that shown in Fig. 9.12 is observed at all heliocentric radial distances sampled by *Helios 1* and *2*, and appears to

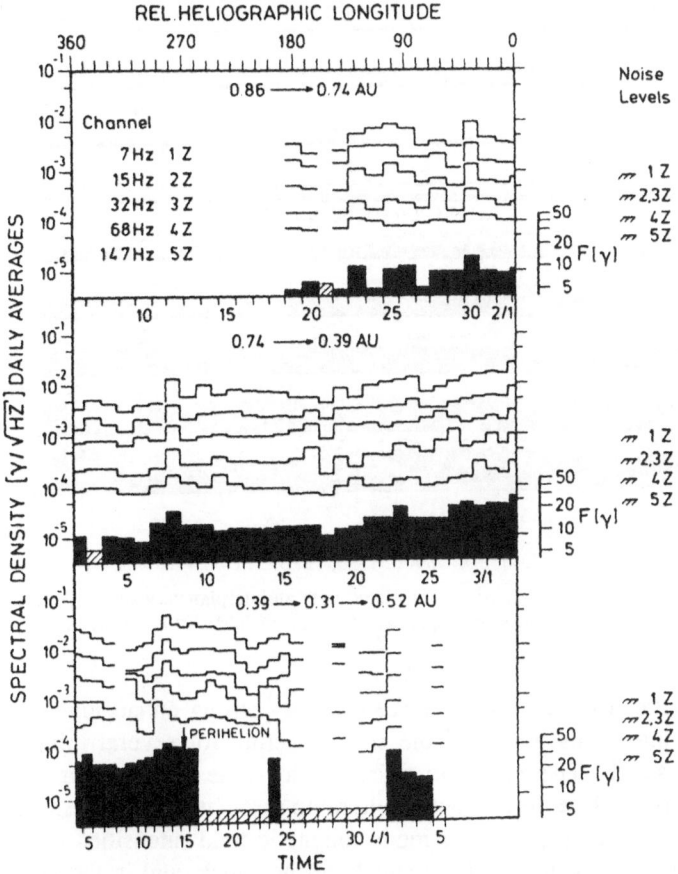

Fig. 9.13. Representative whistler-mode magnetic field intensities for a *Helios 1* pass from 0.86 to 0.31 AU. Note the increase in intensity as the spacecraft approaches closer to the sun

be a permanent feature of the solar wind plasma. Figure 9.13 shows a plot of daily averages of the magnetic field intensities in five frequency channels (7 Hz to 147 Hz) for a period in which *Helios 1* covers the radial distance range from near aphelion (0.86 AU) to perihelion (0.31 AU). These frequency channels all respond exclusively to the whistler-mode portion of the solar wind magnetic field spectrum. As can be seen, substantial fluctuations in the whistler-mode magnetic field intensities occur from day to day. However, a general trend can also be seen for the intensities to increase with decreasing heliocentric radial distance. This trend, for the whistler-mode intensities to increase closer to the sun, has been confirmed on a statistical basis by Beinroth and Neubauer [9.4] using several years of data from *Helios 1* and 2.

Enhanced whistler-mode wave intensities are also observed in association with interplanetary shocks. Figure 9.14 shows the magnetic field intensities in eight channels from 6.8 Hz to 1.47 kHz for an interplanetary shock that occurred on 30 March 1976. This is the same shock that was described earlier in the

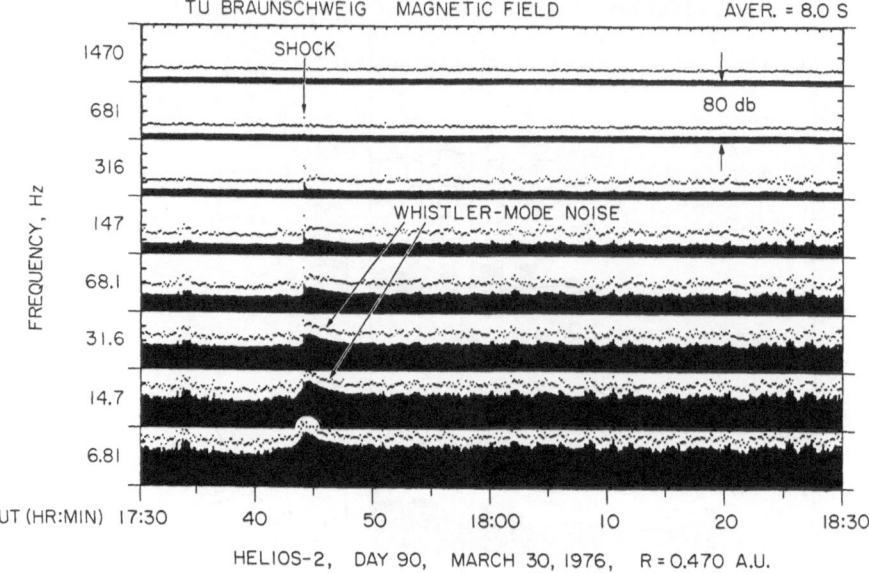

Fig. 9.14. Whistler-mode magnetic field intensities associated with an interplanetary shock. Note the abrupt burst of noise at the shock, followed by a gradual decrease in intensity downstream of the shock

section on ion acoustic waves. As can be seen, a very strong enhancement of the magnetic noise intensities occurs at the shock, lasting for several minutes downstream of the shock. The magnetic noise in this case actually starts to increase several seconds before the magnetic field ramp associated with the shock. Representative spectrums of the whistler-mode magnetic field intensities in the precursor region immediately ahead of the shock, at the shock, and in the wake region downstream of the shock are shown in Fig. 9.15. For a further analysis of this shock, see Gurnett et al. [9.30]. These magnetic field intensities are typical of the other shocks detected by *Helios*.

Short bursts of whistler-mode noise have also been observed in association with other types of discontinuities in the solar wind. These include tangential discontinuities, rotational discontinuities, and reversible magnetic field variation. For a discussion of these events, see Neubauer et al. [9.49]. In some of these cases the whistler-mode noise is produced by instabilities generated in the current sheet associated with the discontinuity, whereas in others the discontinuity itself may act to duct the waves to the spacecraft from a distant source.

Although it is possible that the background of whistler- mode noise in the solar wind may be propagating outward from a source near the sun, the occurrence of enhanced whistler-mode intensities at shocks and other local discontinuities demonstrates that a local whistler-mode instability must be operative in the solar wind. As discussed by Kennel and Petschek [9.39] a Doppler-shifted cyclotron resonance interaction provides the most likely mechanism for converting free energy in the particle distribution into the whistler-mode waves. The normal cyclotron resonance interaction of whistler-mode waves is with electrons, since

152

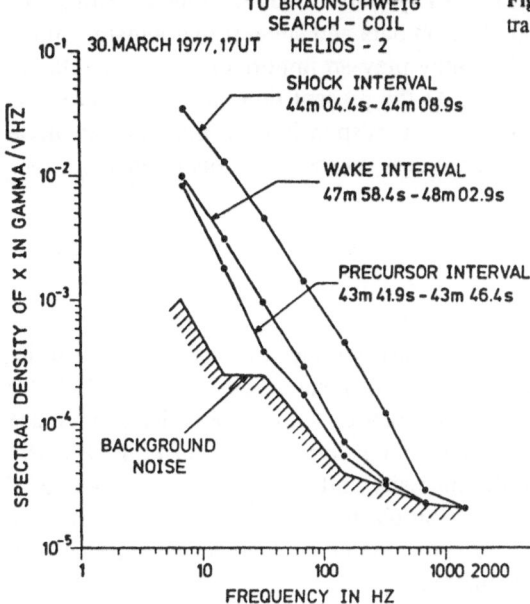

both the waves and electrons rotate in the right-hand sense with respect to the static magnetic field. However, whistler-mode waves can also resonate with ions if the component of particle velocity along the magnetic field, $v_{\parallel}$, exceeds the parallel component of the phase velocity, $\omega/k_{\parallel}$. This type of interaction is called the anomalous cyclotron resonance.

A wide variety of free energy sources can give rise to whistler-mode wave growth via cyclotron resonance interactions. For the normal cyclotron resonance interaction, electron anisotropies with perpendicular temperature greater than the parallel temperature, $T_{\perp} > T_{\parallel}$, provide a well-known free energy source for producing whistler-mode noise. This free energy source occurs whenever a loss-cone is present and is responsible for the intense whistler-mode noise commonly observed in planetary magnetospheres. However, in the solar wind the plasma expands into a region of decreasing magnetic field, which because of conservation of the first adiabatic invariant tends to make $T_{\parallel} > T_{\perp}$, thereby inhibiting the loss-cone type of instability. When $T_{\parallel} > T_{\perp}$, the ions have the proper anisotropy to produce whistler-mode wave growth via the anomalous cyclotron resonance interaction [9.53, 20].

Other free energy sources that could produce whistler-mode waves via a cyclotron resonance interaction include currents and the electron heat flux. Except for certain isolated conditions that occur near magnetic field discontinuities, currents are not expected to be an important source of whistler-mode noise. However, the electron heat flux remains a viable possibility. As discussed by Feldman et al. [9.12], the electron heat flux produces a shift between the velocities of the cool "core" electron distribution and the hot "halo" electron distribution. Gary et al. [9.18, 19] have shown that this shift can lead to the growth of whistler-mode waves in the solar wind. Feldman et al. [9.13] have further examined various correlations that exist in the solar wind electron data and suggest that whistler-mode waves driven by the electron heat flux may play an important role in regulating the electron heat flux in the solar wind. At the present time, no clear consensus has emerged as to which of these processes is responsible for the whistler-mode noise in the solar wind. For further comments on the theoretical considerations involved, see Schwartz [9.56].

9.4 Conclusion

In this paper we have reviewed the basic characteristics of plasma waves in the inner heliosphere, with particular emphasis on the results from *Helios 1* and 2. Of the various observations presented, the tendency for wave intensities to increase with decreasing radial distance from the sun is particularly striking. Electric field measurements show that both the intensity and frequency of occurrence of electrostatic ion acoustic waves and electron plasma oscillations increase rapidly with decreasing heliocentric radial distance. Magnetic field measurements also show that the intensity of the electromagnetic ion cyclotron and whistler-mode turbulence increases rapidly with decreasing radial distance from the sun.

The general tendency for plasma wave intensities to increase near the sun is also confirmed by the *Voyagers 1* and 2 measurements beyond 1 AU. As discussed by Gurnett et al. [9.29], Kurth et al. [9.43], and Scarf et al. [9.55], the intensities of electron plasma oscillations, ion acoustic waves, and type III solar radio bursts all decrease rapidly with increasing heliocentric radial distance. In fact, beyond about 2 AU, solar wind plasma waves are seldom detected by *Voyager*. Even at the orbit of the earth, some of these waves, such as the electron plasma oscillations associated with type III radio bursts, are very rare and seldom detectable.

Although the precise reasons for the strong radial dependence has not been analyzed in detail, the fundamental principles involved are obvious. As described in Sect. 9.1, plasma instabilities are driven by deviations from thermal equilibrium. Since the strongest spatial gradients occur in the region near the sun, and since the sun provides the source of the deviations from thermal equilibrium, it is to be expected that the strongest wave intensities should be observed near the sun. What is not presently understood is the role of the observed waves in determining the steady-state properties of the solar wind. Future theoretical effort is needed to understand the effect of plasma waves on the equilibrium state of the solar wind. Considerable room also exists for improvements in the observations. *Helios 1* and 2 only provided measurements over a little more than a factor of three in heliocentric radial distance. Given the rapid increase in plasma wave intensities as one proceeds from 1.0 to 0.29 AU, one naturally wonders how much stronger the plasma wave intensities become closer to the sun. The answer will have to await future missions that approach closer to the sun, such as the solar probe that is currently being studied.

Acknowledgements. The research at the University of Iowa has been supported by NASA through contract NAS5-11279 with Goddard Space Flight Center and grant NAGW-1488 with NASA Headquarters.

References

9.1 Anderson, R.R., C.C. Harvey, M.M. Hoppe, B.T. Tsurutani, T.E. Eastman, J. Etcheto, Plasma waves near the magnetopause, J. Geophys. Res., **87**, 2087, 1982.

9.2 Bardwell, S., M.V. Goldman, Three-dimensional Langmuir wave instabilities in type III solar radio bursts, Astrophys. J., **209**, 912, 1976.

9.3 Barnes, A., Hydromagnetic waves and turbulence in the solar wind, *Solar System Plasma Physics,* Vol. 1, ed. by E.N. Parker, C.F. Kennel, and L.J. Lanzerotti, North-Holland, Amsterdam, 249, 1979.

9.4 Beinroth, H.J., F.M. Neubauer, Properties of whistler-mode waves from 0.3 and 1.0 AU from Helios observations, J. Geophys. Res., **86**, 7755, 1981.

9.5 Coleman, P.J., Jr., Hydromagnetic waves in the interplanetary plasma, Phys. Rev. Lett., **17**, 207, 1966.

9.6 Coleman, P.J., Jr., Turbulence, viscosity, and dissipation in the solar wind plasma, Astrophys. J., **153**, 371, 1968.

9.7 Dehmel, G., F.M. Neubauer, D. Lukoschus, J. Wawretzko, E. Lammers, Das Induktionsspulen-Magnetometer-Experiment (E4), Raumfahrtforschung, **19**, 241, 1975.

9.8 Denskat, K.U., F.M. Neubauer, Statistical properties of low-frequency magnetic fluctuations in the solar wind from 0.29 to 1.0 AU during solar minimum conditions: Helios 1 and 2, J. Geophys. Res., **87**, 2215, 1982.

9.9 Denskat, K.U., H.J. Beinroth, F.M. Neubauer, Interplanetary magnetic field power spectra with frequencies from 2.4 x 10-5 Hz to 470 Hz from Helios-observations during solar minimum conditions, J. Geophys. Res., **54**, 60, 1983.

9.10 Dum, C.T., E. Marsch, W. Pilipp, D.A. Gurnett, Ion sound turbulence in the solar wind, in *Solar Wind Four,* ed. by H. Rosenbauer, Max-Planck-Institut Report MPAE W100-81-31, Lindau, Germany, 299, 1981.

9.11 Fainberg, J., R.G. Stone, Satellite observations of type III solar radio bursts at low frequencies, Space Sci. Rev., **16**, 145, 1974.

9.12 Feldman, W.C., J.R. Asbridge, S.J. Bame, M.D. Montgomery, S.P. Gary, Solar wind electrons, J. Geophys. Res., **80**, 4181, 1975.

9.13 Feldman, W.C., J.R. Asbridge, S.J. Bame, S.P. Gary, M.D. Montgomery, Electron parameter correlations in high-speed streams and heat flux instabilities, J. Geophys. Res., **81**, 2377, 1976.

9.14 Forslund, D.S., Instabilities associated with heat conduction in the solar wind and their consequences, J. Geophys. Res., **75**, 17, 1970.

9.15 Freud, H.P., K. Papadopoulos, Oscillating two-stream and parametric decay instability in a weakly magnetized plasma, Phys. Fluids, **23**, 139, 1980.

9.16 Fuselier, S.A., D.A. Gurnett, Short wavelength ion waves upstream of the Earth's bow shock, J. Geophys. Res., **89**, 91, 1984.

9.17 Galeev, A.A., R.Z. Sagdeev, Yu.S. Sigov, V.D. Shapiro, V.I. Shevchenko, Nonlinear theory for the modulation instability of plasma waves, Sov. J. Plasma Phys. Engl. Transl., **1**, 5, 1975.

9.18 Gary, S.P., W.C. Feldman, D.W. Forslund, M.D. Montgomery, Electron heat flux instabilities in the solar wind, Geophys. Res. Lett., **2**, 79, 1975a.

9.19 Gary, S.P., W.C. Feldman, D.W. Forslund, M.D. Montgomery, Heat flux instabilities in the solar wind, J. Geophys. Res., **80**, 4197, 1975b.

9.20 Gary, S.P., M.D. Montgomery, W.C. Feldman, D.W. Forslund, Proton temperature anisotrophy instabilities in the solar wind, J. Geophys. Res., **81**, 1241, 1976.

9.21 Ginzburg, V.L., V.V. Zheleznyakov, On the possible mechanism of sporadic radio emission (radiation in an isotropic plasma), Sov. Astron., **AJ 2**, 653, 1958.

9.22 Gliem, F., G. Dehmel, G. Musmann, C. Turke, U. Krupstedt, R.P. Kugel, Die Bordrechner der Helios-Magnetometer- Experiment E2 and E4, Raumfahrtforschung, **20**, 16, 1976.

9.23 Goldman, M.V., G.F. Reiter, D.R. Nicholson, Radiation from a strongly turbulent plasma: Application to electron beam-excited solar emission, Phys. Fluids, **23**, 388, 1980.

9.24   Gurnett, D.A., R.R. Anderson, Electron plasma oscillations associated with type III radio bursts, Science, **194**, 1159, 1976.

9.25   Gurnett, D.A., R.R. Anderson, Plasma wave electric fields in the solar wind: Initial results from Helios 1, J. Geophys. Res., **82**, 632, 1977.

9.26   Gurnett, D.A., L.A. Frank, Ion acoustic waves in the solar wind, J. Geophys. Res., **83**, 58, 1978.

9.27   Gurnett, D.A., R.R. Anderson, D.L. Odem, The University of Iowa, HELIOS solar wind plasma wave experiment, [E5a], Raumfahrtforschung, **5**, 245, 1975.

9.28   Gurnett, D.A., M.M. Baumback, H. Rosenbauer, Stereoscopic direction finding analysis of a type III solar radio burst: Evidence for emission at 2fp-, J. Geophys. Res., **83**, 616, 1978.

9.29   Gurnett, D.A., R.R. Anderson, F.L. Scarf, W.S. Kurth, The heliocentric radial variation of plasma oscillations associated with type III radio bursts, J. Geophys. Res., **83**, 4147, 1978.

9.30   Gurnett, D.A., F.M. Neubauer, R. Schwenn, Plasma wave turbulence associated with an interplanetary shock, J. Geophys. Res., **84**, 541, 1979.

9.31   Gurnett, D.A., E. Marsch, W. Pilipp, R. Schwenn, H. Rosenbauer, Ion acoustic waves and related plasma observations in the solar wind, J. Geophys. Res., **84**, 2029, 1979.

9.32   Gurnett, D.A., R.R. Anderson, R.L. Tokar, Plasma oscillations and the emissivity of type III radio bursts, in *Radio Physics of the Sun*, ed. by M. Kundu and T. Gergely, IAU, 369, 1980.

9.33   Gurnett, D.A., J.E. Maggs, D.L. Gallagher, W.S. Kurth, F.L. Scarf, Parametric interaction and spatial collapse of beam-driven Langmuir waves in the solar wind, J. Geophys. Res., **86**, 8833, 1981.

9.34   Hasegawa, A., *Plasma Instabilities and Nonlinear Effects*, Springer-Verlag, Berlin, Heidelberg, New York, 1975.

9.35   Hundhausen, A.J., *Coronal Expansion and Solar Wind*, Springer-Verlag, Berlin, Heidelberg, New York, 1972.

9.36   Kaiser, M.L., The solar elongation distribution of low frequency radio bursts, Solar Physics, **45**, 181, 1975.

9.37   Kellogg, P.J., Fundamental emission in three type III solar bursts, Astrophys. J., **236**, 696, 1980.

9.38   Kellogg, P.J., Observations concerning the generation and propagation of type III solar bursts, Astron. Astrophys., **169**, 329, 1986.

9.39   Kennel, C.F., H.E. Petschek, Limit on stably trapped particle fluxes, J. Geophys. Res., **71**, 1–28, 1966.

9.40   Kennel, C.F., F.L. Scarf, F.V. Coroniti, E.J. Smith, D.A. Gurnett, J. Geophys. Res., **87**, 17, 1982.

9.41   Krall, N.A., A.W. Trivelpiece, *Principles of Plasma Physics*, McGraw-Hill, 1973.

9.42   Kundu, M.R., *Solar Radio Astronomy*, Interscience, New York, 1965.

9.43   Kurth, W.S., D.A. Gurnett, F.L. Scarf, High-resolution spectrograms of ion acoustic waves in the solar wind, J. Geophys. Res., **84**, 3413, 1979.

9.44   Lemons, D.S., J.R. Asbridge, S.J. Bame, W.C. Feldman, S.P. Gary, J.T. Gosling, The source of electrostatic fluctuations in the solar wind, J. Geophys. Res., **84**, 2135, 1979.

9.45   Lin, R.P., D.W. Potter, D.A. Gurnett, F.L. Scarf, Energetic electrons and plasma waves associated with a solar type III radio burst, Astrophys. J., **251**, 364, 1981.

9.46   Marsch, E., T. Chang, Lower hybrid waves in the solar wind, J. Geophys. Res., **88**, 6869, 1983.

9.47   Melrose, D.B., *Instabilities in Space and Laboratory Plasmas*, Cambridge University Press, Cambridge, 1986.

9.48   Musmann, G., F.M. Neubauer, A. Maier, E. Lammers, Das Förstersonden-Magneticfeld-experiment (E2), Raumfahrtforschung, **19**, 232, 1975.

9.49   Neubauer, F.M., G. Musmann, G. Dehmel, Fast magnetic fluctuations in the solar wind: Helios 1, J. Geophys. Res., **82**, 3201, 1977.

9.50   Nicholson, D.R., M.V. Goldman, P. Hoyng, J.S. Weatherall, Nonlinear langmuir waves during type III solar radio bursts, Ap. J., **223**, 605, 1978.

9.51   Papadopoulos, K., On the physics of strong turbulence for electron plasma waves, *Proc. Varenna School on Plasma Physics*, Pergamon, New York, 355, 1978.

9.52   Papadopoulos, K., M.L. Goldstein, R.A. Smith, Stabilization of electron streams in the type III solar radio bursts, Astrophys. J., **190**, 175, 1974.

9.53   Scarf, F.L., J.H. Wolfe, R.W. Silva, A plasma instability associated with thermal anisotropies in the solar wind, J. Geophys. Res., **72**, 993, 1967.

9.54   Scarf, F.L., R.W. Fredricks, L.A. Frank, C.T. Russell, P.J. Coleman, Jr., M. Neugebauer, Direct correlations of large amplitude waves with suprathermal protons in the upstream solar wind, J. Geophys. Res., **75**, 7316, 1970.

9.55   Scarf, F.L., E. Marsch, W. Pilipp, D.A. Gurnett, Ion sound turbulence in the solar wind, in *Solar Wind Four*, ed. by H. Rosenbauer, Max-Planck-Institut Report MPAE W100-81-31, Lindau, Germany, 299, 1981.

9.56   Schwartz, S.J., Plasma instabilities in the solar wind: A theoretical review, Rev. Geophys. and Space Phys., **18**, 313, 1980.

9.57   Stix, T., *The Theory of Plasma Waves*, McGraw-Hill, New York, 1962.

9.58   Tokar, R.L., D.A. Gurnett, The volume emissivity of type III radio bursts, J. Geophys. Res., **85**, 2353, 1980.

9.59   Wild, J.P., Observations of the spectrum of high- intensity solar radiation at meter wavelengths, Aust. J. Sci. Res., **3**, 541, 1950.

9.60   Wong, A.Y., B.H. Quon, Spatial collapse of beam-driven plasma waves, Phys. Rev. Lett., **34**, 1499, 1975.

10. MHD Turbulence in the Solar Wind

Eckart Marsch

10.1 Introduction

10.1.1 General and Historical Remarks

From the very beginning of *in situ* observations of the solar wind it was realized that the interplanetary medium by all appearances was usually not quiet but rather turbulent and visibly permeated by sizable fluctuations of the plasma flow velocity and density and of the magnetic field. Fluctuations occurred on all observed spatial and temporal scales, extending from the vast dimensions of the inner heliosphere and the corresponding solar wind transit time, or from the solar rotation period, down to the minute kinetic scales associated with the particles' gyromotion, where the dissipation was assumed finally to occur. The observational studies often revealed random and nonreproducible behavior of solar wind parameters as a function of time, thus indicating properties typical of a turbulent magnetofluid. The measured fractional variances of the magnetic field components, when normalized to the mean intensity, turned out to be large, suggesting the importance of nonlinear processes that couple a large number of degrees of freedom and turbulent "eddies" of disparate scales.

In a series of seminal papers Coleman [10.39–41] firmly established the turbulent character of solar wind fluctuations by showing that power-law spectra aptly fitted the data and that existing theoretical concepts, pertaining to the inertial range, as advanced by Kraichnan [10.93] seemed to be applicable. However, it also occurred some years later in studies that would have a similar impact on future research by Belcher and coworkers [10.19–21], that examples were found in the measurements of what resembled pure magnetohydrodynamic waves, notably Alfvén waves, stressing the possible persistence and coherence of the fluctuations.

Ever since, solar wind research has been perhaps somewhat misled by the fruitless controversy of "waves versus turbulence", which dominated interplanetary studies in the seventies and yielded many unsuccessful attempts to identify idealized wave properties in the data [10.3, 16, 44, 159]. In the past decade we have witnessed a period of new and fruitful approaches aimed at reconciling and unifying the two opposite points of view by means of modern concepts of magnetohydrodynamic turbulence, which provide valuable research tools for guiding solar wind studies, though they need to be adopted with caution, as discussed in detail in Montgomery's [10.128, 129] excellent survey of hydromagnetic tur-

Physics and Chemistry in Space - Space and Solar Physics, Vol. 21
Physics of the Inner Heliosphere II Editors: R. Schwenn · E. Marsch
© Springer-Verlag Berlin Heidelberg 1991

bulence theory and its relevance to the solar wind plasma. We recommend as a more general account of the statistical theory of fully developed turbulence the review by Rose and Sulem [10.156]. Magnetohydrodynamic turbulence has been recently reviewed by Pouquet [10.142].

It now appears that solar wind plasma and magnetic field fluctuations may be meaningfully viewed as representing MHD turbulence. In a set of pioneering papers Matthaeus and Goldstein [10.118, 119, 121, 123] convincingly showed that the observed small-scale fluctuations are adequately described as examples of time-stationary, and to some degree spatially homogeneous, turbulence as theoreticians liked to conceive it. In this new kind of data analysis the principal quantities used to characterize the observed plasma state were the so-called rugged invariants of incompressible ideal MHD [10.200], the spectra of which suitably describe the overall turbulent energy, as well as the degree of Alfvénic correlations and helical winding of the magnetic fluctuations [10.60, 127, 143, 144]. It was by means of these quantities that long-standing problems concerning the origin and nature, and the spectral and spatial evolution of interplanetary fluctuations could be freshly tackled, and perhaps be solved.

The current prospect for future research in solar wind turbulence may even be more attractive, since the first adaptation of phenomenological concepts of turbulence [10.91, 93] to the solar wind by Dobrowolny, Mangeney and Veltri [10.52, 53] and their theoretical rephrasing of turbulence problems in terms of Elsässer variables turned out to be equally fruitful for data analyses and numerical simulations of turbulent processes. At present it seems as if turbulence studies in the solar wind have entered another active epoch, and the previews of the future suggested by the ongoing research reviewed below appear promising. More importantly, the prodigious amount of solar wind data acquired by many spacecraft are readily available and include all the information needed to test the advancing turbulence theory. The data sets are also found, to a certain extent, to comply with the idealizations of homogeneity and stationarity as assumed in most of the theories. All these virtues make the interplanetary medium a natural laboratory, being ideally suited to studying MHD turbulence and to providing empirical results to foster basic theoretical developments.

10.1.2 The Importance of MHD Turbulence for Space Plasma Physics

As mentioned above, the solar wind satisfies the needs for turbulence studies excellently, which are of fundamental importance in basic plasma physics and which may prove to be equally relevant to plasma astrophysics. Studying solar wind turbulence also provides essential knowledge to be transferred to other astrophysical systems, notably to stellar winds of sun-like stars [10.86] and to the outer turbulent envelopes of other stars. It is well known that hydromagnetic waves and turbulence carry energy and momentum [10.50, 78, 85, 86] and therefore exert forces on the ambient flow or via dissipation contribute to heating of the plasma [10.179]. Therefore, an understanding of MHD turbulence is prerequisite for solar and stellar wind acceleration theories.

The dissipation mechanism in turbulent plasmas is not well understood [10.4, 106, 128]. By enabling us to measure the detailed underlying velocity distributions of the particles in the plasma, solar wind studies are hoped to shed further light on the microphysics of the dissipation process of MHD turbulence. An accompanying review in this volume by Marsch [10.106] addresses this unresolved issue. Studies of hydromagnetic fluctuations provide important clues to the plasma state of the solar wind and its source regions in the corona, and thus deserve to be carried out in their own right as an essential part of space plasma research.

Among the topics of most interest are the morphology of interplanetary turbulence and of the resulting large-scale magnetic field that emerges out of the fast fluctuations by long-enough averaging of the data. A companion review [10.104] and other papers [10.18, 33, 102, 131] are exclusively devoted to this large-scale magnetic field, apparently ordered in the form of the Parker spiral [10.139]. As detailed later on, the occurrence and intensity of MHD turbulence is intimately connected to the stream structure of the solar wind, particularly during solar activity minimum, which is extensively reviewed for the *Helios* epoch in Schwenn's [10.162] paper in this book. The interaction of predominantly Alfvénic fluctuations with the large streams and their intrinsic velocity shear is one of the principal subjects of current turbulence research.

A fundamental topic of space physics and astrophysics is to understand the effects interplanetary and interstellar magnetohydrodynamic fluctuations have on solar and cosmic rays of various energies and origins. Detailed knowledge of the spectral and spatial characteristics of the turbulence is indispensable if the diffusive propagation of energetic particles in any kind of natural space plasma is to be fully comprehended. A companion paper reviews the current state of this field concerning the inner solar system [10.94]. It is only in the solar wind that the interaction between fluctuations and particles can be thoroughly studied, notably at interplanetary shocks and their associated turbulence [10.64, 98, 99, 175]. Further references on this topic can be found in [10.147, 176]. Finally, waves and turbulence may heat and accelerate protons and solar wind minor ions [10.89, 108] and thereby influence the nonequilibrium thermodynamics of the plasma [10.106].

10.1.3 Outline and Scope of the Article and Its Relation to Previous Reviews

This presentation of MHD turbulence in the solar wind is meant to be a continuation of earlier reviews on this subject, but it is more limited in scope. Throughout the whole article the discussion of new experimental results from the *Helios* mission will be in the foreground, whereby the emphasis is placed on the observations and basic physical concepts. Extensive mathematical derivations are avoided and the reader will be referred to the original literature for algebraic details. Admittedly, leaving out a detailed discussion of *Voyager* results represents a shortcoming of this article. Certainly, neither the properties of the fluctuations nor the physics depend on the fortuitous orbital characteristics of spacecraft. However, the subject of this book is the inner heliosphere and we will largely comply with this editorial constraint.

Yet there are some good physical reasons to distinguish turbulence in the outer and inner heliosphere. Within 1 AU the interplanetary medium by and large is still strongly influenced by coronal boundary conditions and reflects the large-scale coronal magnetic field and solar magnetic-activity-related phenomena. These imprints get increasingly lost as the plasma expands to several AU, because the streams collide and interact owing to an increasingly bent magnetic field spiral, and because shocks produced through this interaction may largely reprocess the plasma and thus change its properties. The associated dynamic phenomena have been reviewed for the outer heliosphere by Burlaga [10.34]. The references found therein and the more recent papers [10.90, 151] give a lively account of physical processes occurring in the outer interplanetary medium.

Several reviews [10.1, 4, 32, 77, 195] on MHD waves and turbulence are available in the enormous literature on this subject, which up to the year 1979 is almost completely covered by the excellent reviews by Barnes [10.3, 6], which are highly recommended to the reader. These papers contain lookup tables of typical lengths and times of the interplanetary plasma, and define characteristic parameters and the basic sets of kinetic and fluid equations to describe magnetohydrodynamic waves at various levels of sophistication. Special emphasis is placed on MHD waves, their dissipation and propagation, and their interaction with the wind. However, these early reviews did not include the plentiful *Helios* and *Voyager* results. One main task of our paper is to remedy this lack of documentation.

In the older literature most of the attempts to understand the nature and origin of the fluctuations used the theoretical concepts of linear MHD, classifying fluctuations as Alfvén, fast and slow magnetoacoustic waves, plus the nonpropagating rotational and tangential discontinuities (Burlaga [10.30]). Interplanetary fluctuations were conceived as a superposition of these modes [10.3, 77] or also considered as pure examples of large-amplitude simple waves [10.2, 67, 76]. However, the failure of the superposition principle for waves at large amplitudes rendered weak turbulence concepts rather questionable [10.4].

The crucial difference between the wave and turbulence pictures lies in the nonlinear evolution of the fluctuations. As pointed out in [10.128], by any conceivable definition of strong turbulence the solar wind fulfills the requirements and looks unmistakably turbulent. But the nonlinear interactions building power-law spectra by an energy cascade are substantially weakened by high Alfvénic correlations, as first emphasized by Dobrowolny, Mangeney, and Veltri [10.53], and they entirely cease for ideal correlations. So how can spectra develop and persist in such conditions, and what is the evolution time in the presence of strong wave-like correlations? This key problem has been studied in numerical calculations by Grappin et al. [10.69], who found that the higher the correlations the longer the interaction times were. Thus it seems that MHD turbulence and waves are not mutually exclusive, but are inextricably linked.

The present work reviews a turbulence-style description of solar wind fluctuations and experimental results collected over the past decade. It also addresses theoretical issues related to a proper description of the turbulence for the inhomogeneous wind by means of recent two-scale dynamical theories in terms of correlation functions for the evolution of turbulence [10.113, 125, 178, 181, 202]. We start out with a description of the large-scale fluctuations and their relation to the solar wind stream structure and average magnetic field morphology. Then the MHD equations, terminology, and some concepts of turbulence theory are discussed. Polarization state and magnetic helicity are briefly addressed. Subsequently, the basic properties of the observed turbulent spectra are presented by means of the invariants of ideal MHD [10.119, 200] and the so-called Elsässer variables [10.52, 53, 56, 65, 69, 70, 93, 112]. Pressure-balanced structures are discussed and the role of compressive fluctuations is considered.

The following section is concerned with the origin and evolution of solar wind MHD turbulence. The sources of interplanetary fluctuations, the observed radial and spectral evolution, and the possible dissipation of turbulence are studied. A subsequent section is devoted to theoretical concepts of turbulence generation and evolution. Some numerical simulations are evaluated and exploited as a guide to interpret the observations. Models for the dynamical evolution of spectral density functions by the effects of nonlinearities and cascading are discussed. The importance of velocity shear and of compressibility in turbulence generation is evaluated. A final section summarizes our results and conclusions and outlines some future research perspectives.

10.2 Structures and Fluctuations in the Inner Heliosphere

10.2.1 Large-Scale Solar Wind Structures

The interplanetary medium is continuously permeated by fluctuations of the magnetic field, particle density, and flow velocity, which occur on all time scales characterizing the solar wind plasma as a kinetic entity and as a fluid. Fluctuations with tiny amplitudes are recorded in the frequency domains determined by the particles' plasma and gyro-frequencies (for an up-to-date review see the companion paper by Gurnett [10.73]), but they tend to have ever increasing amplitudes for longer periods. Finally at scales beyond half a day or so the fluctuating fields or flows are nearly as large in the solar wind rest frame as anything that might be defined as their long-time averages, thus indicating "strong turbulence". Some relevant typical scales in space and time are contained in Table 10.1.

The spatial inhomogeneity of the solar wind due to the expansion from a point source, $R_\odot \ll 1\,\mathrm{AU}$, the variability of its coronal boundaries and sources (in time as well as in space and due to solar rotation), and the resulting complex topology of the interplanetary field and plasma stream structure all make the interplanetary medium a rather inhomogeneous physical system, highly dynamic

Table 10.1. Relevant scales in space and time

Phenomenon	Frequency [s^{-1}]	Time [Day]	Speed [km s^{-1}]	Turbulence
Solar rotation	$\Omega_\odot = 2.7 \times 10^{-6}$ $\nu_\odot = 4.3 \times 10^{-7}$	27	$\Omega_\odot R_\odot$ 2	Generation
Solar wind expansion	$\nu_{exp} = 5.1\text{--}1.9 \times 10^{-6}$	2–6	V_{sw} 900–300	Generation
Alfvén waves $\lambda \approx 10^{-3}$ AU	$\nu_A = 2.8 \times 10^{-4}$	1/24 (1 hour)	V_A 50 (1 AU)	Inertial range
Ion-cyclotron waves $\lambda \approx 50$ km	$\Omega_p = 6.3$ $\nu_p = 1$	1.2×10^{-5} 1 s	V_A 50	Dissipation

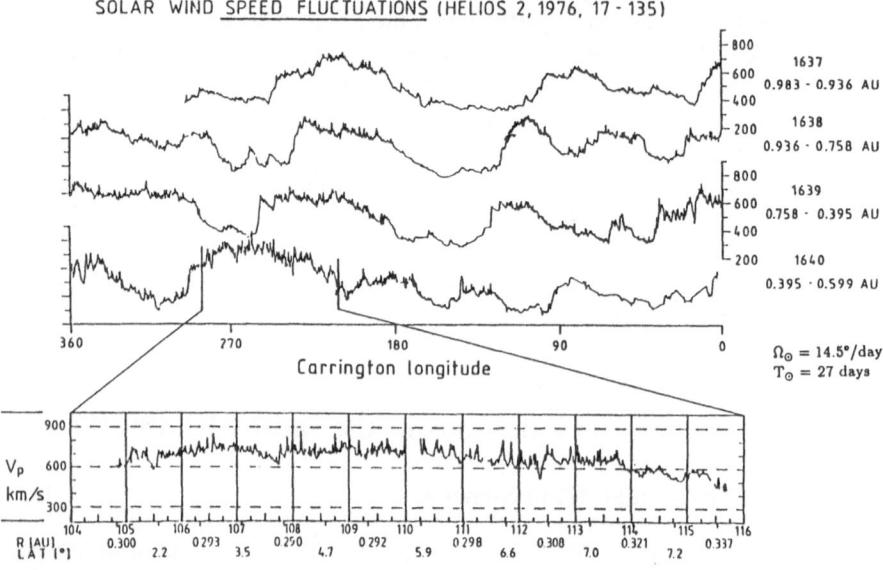

Fig. 10.1. Solar wind streams and speed variations measured by *Helios* 2 in 1976 between 0.29 and 1 AU. Four subsequent solar rotations are shown with speed displayed versus Carrington longitude. The *blow-up* shows the detailed structure of a large corotating stream, consisting of flow tubes and Alfvénic fluctuations. The plot covers twelve days of the perihelion passage of the probe [10.109]

and nonstationary. The stream structure, as it appeared typically near and during the solar activity minimum of the *Helios* epoch in the years 1974–1976, is illustrated in Fig. 10.1, which shows four subsequent solar rotations versus Carrington longitude [10.109, 157]. The heliocentric distances covered range from 0.29 to 1.0 AU, corresponding to the perihelion and aphelion of the *Helios* 2 spacecraft. Below the main frame a blow-up shows a single high-speed stream enlarged over a time period of twelve days.

Several basic features are obvious. Fairly steady plasma sources on the sun emit high- and low-speed flows that recur about every 27 days as seen from the earth. The enlargement of Fig. 10.1 reveals that fast streams are fairly random

164

in appearance, indicating by their fuzzy plateaus sizable fluctuations or that they are possibly composed of many mesoscale flow tubes, which add up to form the entire stream. The overall turbulence level seems to increase as the space probe approaches the sun, which is true in all types of flows observed (with speeds ranging between about 250 and 800 km/s). However, by and large one can also say that slow-speed profiles are less noisy than the fast ones, which thus indicates a direct correlation between stream structure and turbulence level during this time period. Undoubtedly the kinetic energy contained in the large streams and represented by their velocity differences is the only major interplanetary energy reservoir available for the stirring of smaller-scale fluctuations and the excitation of fresh turbulence, i.e. by shear-induced speed equalization as the plasma radially expands. This point of view is later substantiated by detailed references.

10.2.2 Mesoscale Structures and Fluctuations

It is now fairly well established [10.172–174] that high-speed streams are partly composed of mesoscale flow tubes which are related to coronal fine structures and seem to reflect in their angular scales of 2°–4° in Carrington longitude the size of single or several supergranular cells of the chromospheric network on the sun. The flow-tube boundaries and structures can be identified by characteristic concurrent variations in the proton and alpha-particle parameters. That these variations are not of a turbulent nature but signify stable spatial structures was corroborated by an investigation of a "plasma line-up", whereby the *Helios* probes happened to be positioned such that the same plasma parcel that had passed the inner probe at 0.5 AU could later be observed again by the outer probe at 0.7 AU.

The results of this study are summarized in Fig. 10.2. Part (a) gives the trace of the proton and alpha-particle speed (in the second panel the proton speed was "corrected" for Alfvénic fluctuations [10.173, 174]), the densities, and finally the proton temperature. Various "flow tubes" are marked and delineated by dotted lines. As can be seen from part (b), in case of a line-up the flow tubes can be identified by the remarkably similar traces of the plasma parameters, and they obviously remain stable while the plasma expands through a distance of 0.2 AU. Yet at larger solar distance these structures tend gradually to disappear, and at 1 AU they are usually "washed out" by dynamical processes. To conclude, evidence was found for flow tubes to exist in the solar wind close to the sun which appear to be relics of plasma structures in the corona and tend to dissolve during the expansion, just as the streams themselves do in the outer heliosphere [10.34].

However, the picture is usually not as clear as that shown in Fig. 10.2. There are contenders for an explanation of some of the variations on time scales of several hours, which are genuine fluctuations. In the inner heliosphere [10.26] clear evidence has been obtained for long-period Alfvén waves, which can be persistent with periods of up to 15 hours or longer [10.119, 152]. In the trailing edges of high-speed streams Alfvénic fluctuations with periods of several hours are

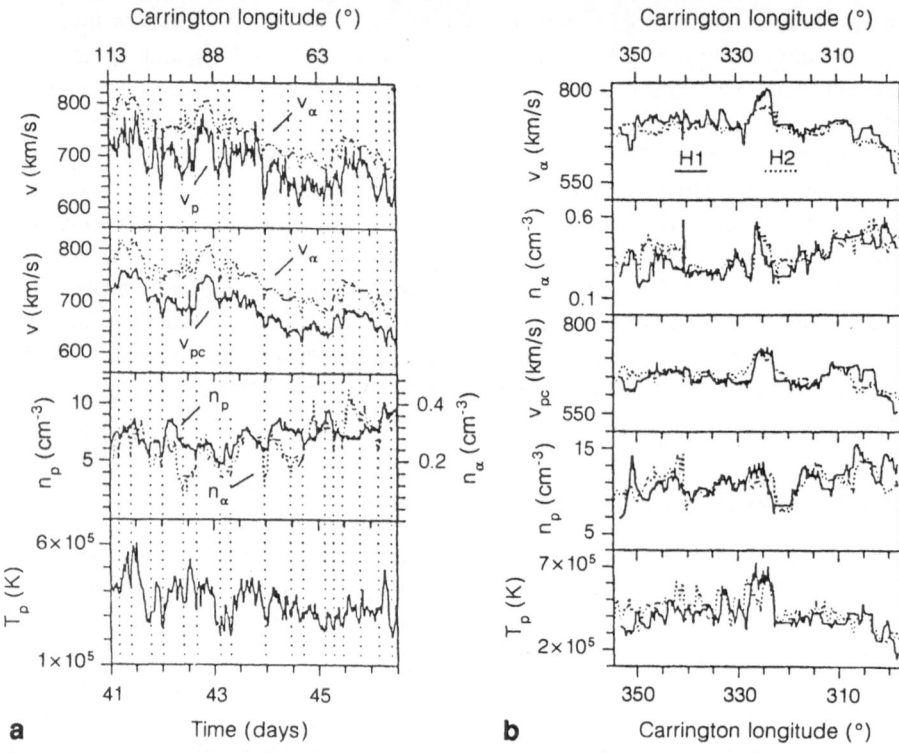

Fig. 10.2. (a) Velocities, densities, and temperatures of protons and α-particles from day 41 until day 46.5 in 1975, as measured by *Helios 1* between 0.6 and 0.66 AU. This fast stream shows significant related variations of the plasma parameters. The boundaries of the mesoscale flow tubes are marked by *dotted vertical lines*. (b) The plasma line-up around day 69.5 in 1976. The data of *Helios 2* (*dotted curves*) have been projected on the orbit of *Helios 1* by taking radial gradients into account. Radially invariant spatial structures are clearly revealed [10.173, 174]

ubiquitously observed. The well-known theoretical correlation in Alfvén waves (see e.g. Barnes' reviews [10.3, 6])

$$\delta V = \pm \frac{1}{\sqrt{4\pi\varrho}} \delta B \qquad (10.1)$$

for the fluctuations of velocity V and magnetic field vector B (note that ϱ denotes as usual the plasma mass density) is nicely illustrated in Fig. 10.3 after [10.26]. For the details of the data analysis see the figure caption. Even at distances as large as 2.5 and 5.0 AU, large-amplitude Alfvén waves have been seen [10.126]. They seemed to propagate with little dissipation between the two spacecraft observation points. As a result, one may expect that flow tubes and long-wavelength fluctuations in the frequency domain from 10^{-4} down to 10^{-5} Hz are sometimes intermingled and interact strongly. Clearly, variations with frequencies below 10^{-5} Hz or periods above a day are associated with the large-scale stream structure itself as shown in Fig. 10.1 and with smaller-sized streams or transient flows and structures. The long-living corotating streams and their intrinsic scales de-

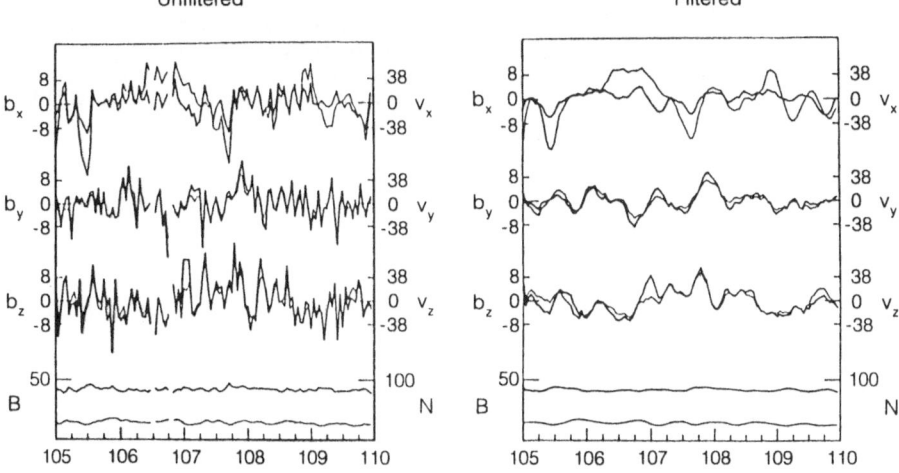

Fig. 10.3. (a) Five days of *Helios 2* magnetic field and plasma velocity components given in nT and km s^{-1}, respectively. Data points are 1-hour averages and refer to minimum-variance (along x) coordinates. Sectors are rectified such that a positive $b - v$ correlation implies an outward sense of propagation for the Alfvénic fluctuations. The two lower curves are the field strength and plasma density (in cm^{-3}). (b) Filtered data in the same format using a low-pass filter with a cutoff frequency of 1.7×10^{-5} Hz [10.26]

fine what might be considered the "energy-containing eddies" in the language of turbulence.

The spectrum of large-scale structures illustrated in Fig. 10.4 is taken from *Voyager 1* data by Burlaga and Mish [10.37]. It shows the power of fluctuations in the strength of the interplanetary magnetic field as a function of heliocentric distance from the sun, as is indicated at each spectrum. For comparison an f^{-2} spectrum is also shown, which can be produced by a sequence of jumps. The jumps were suggested as a model for shocks and boundaries of stream interaction regions or merged stream interfaces as they typically occur in the outer heliosphere [10.34]. The speed fluctuation spectra look very similar to the ones for the magnetic field in Fig. 10.4 and have equal slopes. Note from Table 10.1 that the solar rotation frequency corresponds to 4.3×10^{-7} Hz for a 27-day period as seen from the earth. Peaks are seen at the subharmonics of this basic period. The magnetic intensity spectra were normalized to remove the systematic large-scale radial trends of B.

It was concluded in [10.37, 150] that the spectra such as those in Fig. 10.4 resulted from a superposition of a large number of steps, of varying sizes and scales, in the plasma parameters rather than from a few large discontinuities. The -2 power law describing the large-scale speed fluctuations was found to extend from periods of a few hours to 13 days at 1 AU and to 26 days beyond a few AU. This observed period-doubling was attributed to the formation of compound streams, for example as the result of entrainment [10.34] of slow and

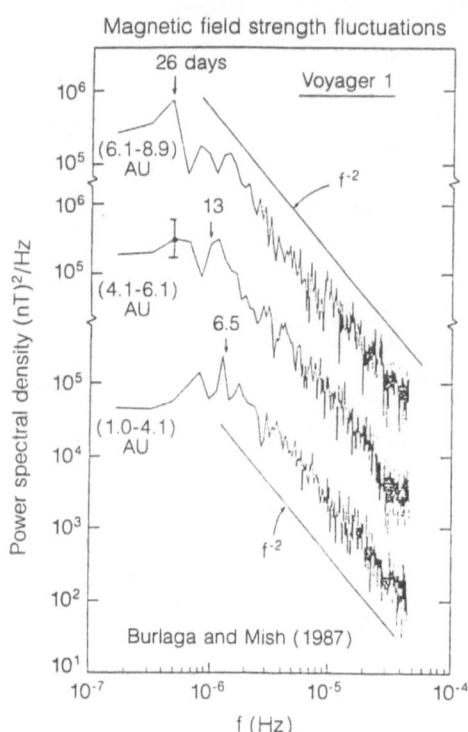

Fig. 10.4. Spectra of fluctuations in the magnetic field strength as a function of heliocentric distance. These *Voyager 1* data cover a distance range from 1 AU to 8.9 AU. The single error bar indicates the 90% confidence level [10.37]

fast streams. The overall power level of long-period speed fluctuations decreased radially, signifying the ultimate damping of the stream modulation and the total disappearance of such primordial stream patterns as displayed in Fig. 10.1, which were observed close to the sun inside 1 AU. These results put strong constraints on the evolution of smaller-scale fluctuations and provide an empirical framework within which their dynamics and spatial evolution has to be conceived.

In a study [10.150] on spectral signatures of jumps and turbulence in interplanetary speed and magnetic field data it was demonstrated that f^{-2} spectra such as those in Fig. 10.4 largely arose from phase-coherent structures that could be distinguished clearly from incoherent turbulent fluctuations. Moreover, if the original time series were purged from jumps, the remaining time series of V and B exhibited $f^{-5/3}$ spectra characteristic of fluid turbulence [10.91]. Jumps, by definition, comprised coherent MHD structures such as shocks and discontinuities [10.30], and many of the observed jumps actually appeared as such upon examination of high-resolution data, including complete plasma parameter sets.

Since the propagation and interaction of interplanetary shocks is well described by nonturbulent MHD theory, the interesting question was raised in [10.150] "to what extent the jumps and fluctuations are genuinely independent". Whether preexisting fluctuations are affected by shocks remains an open question. Similarly one might ask whether, and if so how, fluctuations are affected by the large-scale stream structure discussed above. Apparently, the coherent large

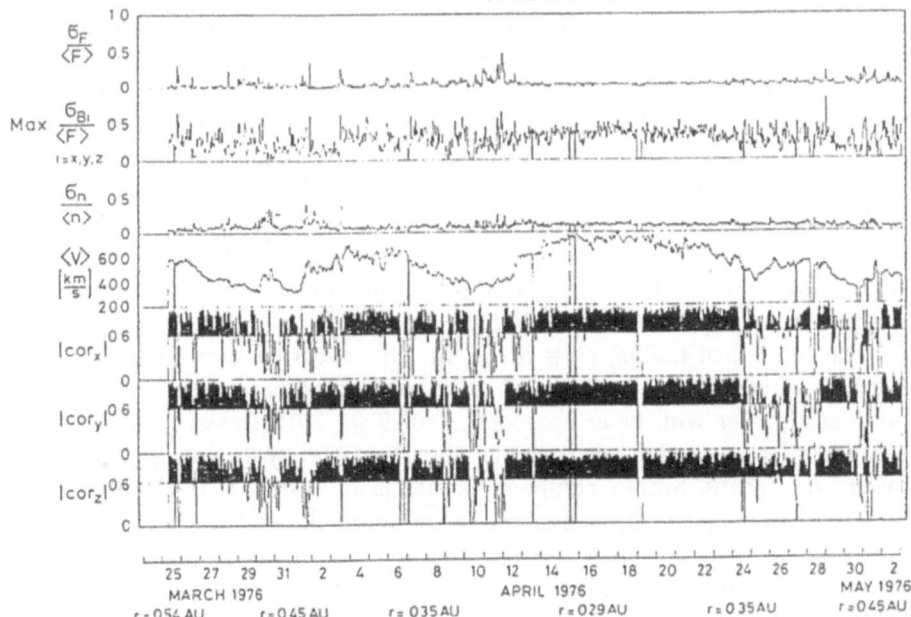

Fig. 10.5. General properties of solar wind fluctuations in relation to the speed profile. *Helios 2* data of the first perihelion passage between 0.29 and 0.54 AU are shown versus time in days. From the top the panels show normalized standard deviations (1-hour rms values) of the fluctuations in magnetic field magnitude and components (only the maximum is displayed), and in proton number density. Below the average bulk speed panel the absolute values of the correlation coefficients between δB and δV in mean-field coordinates are given, where values above 0.6 are in black [10.48]

streams do not simply act as waveguides for the small-scale turbulence, but they take an active part in turbulence generation and evolution, as will be shown in subsequent sections.

Long-term trends in large-scale fluctuations (in the solar wind at 1 AU during the years 1978–1982) of V and B were further analyzed in much detail in [10.38]. In the frequency range 3×10^{-6} to 3×10^{-5} Hz the fluctuations, persistently over many years, revealed an f^{-2} spectral shape within measurement uncertainties. The jumps in B and V were found not to occur simultaneously always. The typical jump width was 2 hours, and 13 percent (22 percent) of the jumps in V (B) had the signature for a forward or reverse shock. Their frequency of occurrence was found to be about 2 jumps per day. To conclude, the solar wind contains fairly coherent MHD structures giving rise to sizable power in the low-frequency domain, in fact even dominating the power density there. Beyond 10^{-5} Hz, however, "genuine" turbulence becomes increasingly prominent and without doubt governs the higher-frequency domains.

General properties of these MHD fluctuations [10.46, 48] are presented in the survey plots of Fig. 10.5, which shows data obtained by *Helios 2* during its first perihelion passage in spring 1976. The Carrington rotation plot corresponds to heliocentric distances between 0.20 AU and 0.54 AU; data for the same streams were shown earlier in the bottom panel of Fig. 10.1. From top to bottom we have the fluctuations (in terms of normalized one-hour standard deviations) of the magnetic field magnitude and its components (only the maximum is shown), the proton density n, the averaged proton speed, and the absolute values of the correlation coefficients between δB and δV (evaluated in mean-field coordinates). The symbols σ_F, σ_{B_i}, σ_n denote rms values of the field magnitude $F(B)$, and of B_i and n, respectively. Correlation coefficients larger than (0.6) are indicated in black. About 75% of the data exceed this threshold, signifying high Alfvénic correlations (with an outward sense of propagation) that occur predominantly in fast plasma during near-solar-minimum conditions.

The center and trailing parts of the streams reveal the clearest examples of Alfvénic turbulence coincident with a lower level of compressive fluctuations (as found earlier with other spacecraft [10.19–21, 36]). Density and magnetic field intensity fluctuations are variable, but are apparently increasingly present at lower flow speeds. Sizable compressive effects are visible at the leading edges of fast streams and in the neighboring low-speed flows. However, field magnitude fluctuations are much smaller than directional fluctuations. The vector component fluctuations are big but if normalized hardly ever exceed 0.5. They therefore reveal signatures of amplitude saturation, and indeed exhibit few systematic changes as a function of heliocentric distance [10.46, 48, 192].

It should be kept in mind, however, that the apparent stream-structure dependence of turbulence properties as illustrated in Fig. 10.5 is typical for near-solar-minimum conditions and not characteristic of the whole solar cycle. Large recurrent high-speed streams do not occur near activity maximum, and therefore flow speed is of limited use as a parameter to order generally turbulence patterns in the solar wind. In fact, in [10.107] a clear example of Alfvén-wave activity was found in slow solar wind, which, however, strongly resembled, in its microscopic plasma features, the fast streams of Fig. 10.1. Other examples of Alfvénic turbulence in all kinds of flow will be given below. Having discussed the basic features of observed MHD structures and fluctuations in the solar wind we now turn our attention to the theoretical description of the turbulence.

10.3 MHD Theory and Turbulence Diagnostics

10.3.1 Basic Equations

The applicability of ideal MHD theory to a great number of interplanetary phenomena observationally seems unquestionable. The large-scale flows reviewed in the previous section, hydromagnetic waves, and discontinuities in the solar wind are all successfully described by magnetohydrodynamics. The early classic review by Burlaga [10.30] amply proves this, and other reviews [10.1, 3, 4, 6, 31, 32, 77, 195] do as well. However, since the solar wind plasma is only weakly collisional, the direct application of MHD remains theoretically questionable. Its low collisionality renders classical transport results meaningless to the solar wind. In particular, a Coulomb-collision-based scalar kinematic viscosity ν and magnetic diffusivity η appear obsolete according to [10.128] in view of the fact that particles' free paths are evaluated to be comparable to the size of the entire inner heliosphere. Instead, it appears that microscopic wave–particle interactions and plasma microinstabilities govern the transport and dissipation properties [10.106]. How to incorporate these in a manageable fashion into the MHD picture remains an open problem. Moreover, a simple equation of state for the solar wind remains elusive, and approximations involving the energy equations of ions and electrons are inevitably complicated.

Facing these and other problems summarized in [10.3, 4, 106], we are inclined to follow the arguments in Montgomery's review [10.128] and are ready to accept the following set of equations as the simplest model for a description of dissipative MHD turbulence, including compressibility, in the solar wind:

$$\frac{\partial}{\partial t}\varrho + \nabla \cdot (\varrho V) = 0 , \tag{10.2}$$

$$\varrho \left(\frac{\partial}{\partial t} + V \cdot \nabla \right) V = -\nabla \left(p + \frac{B^2}{8\pi} \right) + \frac{1}{4\pi}(B \cdot \nabla)B + \varrho\nu\nabla^2 V , \tag{10.3}$$

$$\left(\frac{\partial}{\partial t} + V \cdot \nabla \right) B = (B \cdot \nabla)V - B(\nabla \cdot V) + \eta\nabla^2 B , \tag{10.4}$$

of course complemented by the subsidiary condition that $\nabla \cdot B = 0$. The viscous and ohmic terms are included here to provide the set of equations needed in later sections on numerical simulations and to signify and simply model possible dissipation at wave numbers much higher than those in the inertial range, which we are mainly concerned with in what follows. For the sake of simplicity we may assume that the thermal pressure obeys a polytropic equation of state with index γ:

$$p = p_0(\varrho/\varrho_0)^\gamma , \tag{10.5}$$

implying that the sound speed can be expressed by

$$c_S^2 = \frac{\partial p}{\partial \varrho} = \gamma\frac{p}{\varrho} . \tag{10.6}$$

Instead of the magnetic field vector B the Alfvén velocity

$$V_A = B/\sqrt{4\pi\varrho} \tag{10.7}$$

may be used as a variable. Because it has the same dimension as the flow velocity V, the Alfvén velocity is useful and allows us to introduce a set of new variables after Elsässer [10.56] defined as

$$Z^\pm = V \pm V_A . \tag{10.8}$$

Dobrowolny, Mangeney, and Veltri [10.52, 53] first made use of $Z^\pm$ in the solar wind context. The equations of motion for Alfvénic fluctuations can be expressed by means of $Z^\pm$ in a very transparent manner. Marsch and Mangeney [10.112] have also cast the ideal MHD equations in terms of compressive Elsässer variables including density variations. By taking in addition the dissipation terms in (10.3) and (10.4) into account, we may rewrite their basic equations as follows:

$$\frac{D^\pm}{Dt} \ln \varrho = -\frac{1}{2} \nabla \cdot (3Z^\pm - Z^\mp) , \qquad (10.9)$$

$$\frac{D^\mp}{Dt} Z^\pm = \pm \frac{1}{4}(Z^+ - Z^-)\frac{D^\pm}{Dt} \ln \varrho - \left(\frac{1}{8}(Z^+ - Z^-)^2 + c_s^2\right) \nabla \ln \varrho$$

$$+ D_\eta + \nu^\pm \nabla^2 Z^+ + \nu^\mp \nabla^2 Z^- - \frac{1}{8}\nabla(Z^+ - Z^-)^2 . \qquad (10.10a)$$

The dissipative force related to density variations can be expressed as

$$D_\eta = \eta\{\tfrac{1}{2}\nabla \ln \varrho \cdot \nabla(Z^+ - Z^-)$$

$$+ \tfrac{1}{8}(Z^+ - Z^-)(2\nabla^2 \ln \varrho + (\nabla \ln \varrho)^2)\} . \qquad (10.10b)$$

Here the advective derivative along Alfvénic characteristics is defined by

$$\frac{D^\pm}{Dt} = \frac{\partial}{\partial t} + Z^\mp \cdot \nabla \qquad (10.11)$$

and the viscosities $\nu^\pm = \frac{1}{2}(\nu \pm \eta)$ have been employed. Frequently the assumption of $\eta = \nu$, i.e. $\nu^- = 0$, is made in numerical simulations for simplification [10.70].

In the literature the limit of incompressible, nondissipative MHD is often considered, in which case the set of equations [10.52] reduces to

$$\nabla \cdot Z^\pm = 0 , \qquad (10.12)$$

$$\frac{D^\mp}{Dt} Z^\pm = -\frac{1}{\varrho}\nabla p_T , \qquad (10.13a)$$

where the total plasma pressure is given by

$$p_T = p + \frac{1}{8\pi}B^2 \qquad (10.13b)$$

as the sum of the thermal pressure and magnetic energy density. Of course, the sets of equations (10.2–4) and (10.9, 10) are entirely equivalent and it is largely a matter of taste or convenience which of them is used. Elsässer variables are advantageous if Alfvénic fluctuations are dealt with.

10.3.2 Alfvén Waves and Magnetosonic Waves

Consider a constant background fluid described by $Z_0^\pm$ and superimposed fluctuations $\delta Z^\pm$ of small amplitude such that $Z^\pm = Z_0^\pm + \delta Z^\pm$. In the absence of nonlinearities the incompressible transverse fluctuations obey a simple linear

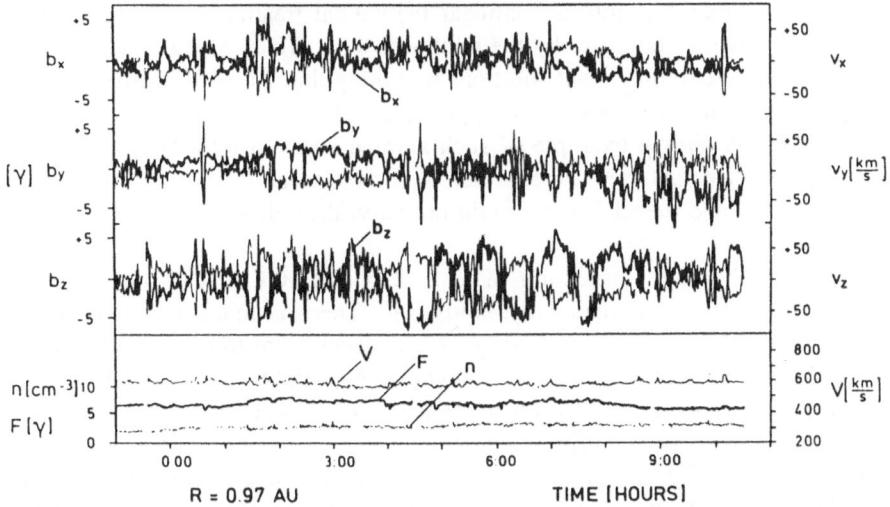

R = 0.97 AU   TIME [HOURS]

Fig. 10.6. *Helios* 2 data from 0.97 AU, showing for half a day the components of the magnetic field vector and flow velocity fluctuations in nT and km s^{-1}, respectively. Solar ecliptic coordinates were used. The *bottom panel* given density, magnetic field intensity and solar wind speed. Note the high anticorrelations between δV and δB, which are typical for Alfvénic fluctuations propagating along an outward-oriented mean field [10.43]

wave equation, which is just derived from a perturbation of (10.13):

$$\frac{d^{\mp}}{dt}\delta Z^{\pm} = \left(\frac{\partial}{\partial t} + Z_0^{\mp} \cdot \nabla\right)\delta Z^{\pm} = 0 , \tag{10.14}$$

which after Fourier transformation yields the dispersion relation

$$(\omega - k \cdot Z_0^{\mp})\delta Z_k^{\pm} = 0 \tag{10.15}$$

implying the well-known relations for Alfvén waves

$$\omega - k \cdot V_0 = \omega' = \pm k \cdot V_{A0} , \tag{10.16a}$$

$$\delta Z^{\pm} = 0 \Rightarrow \delta V = \mp \delta V_A , \tag{10.16b}$$

where ω' is the frequency in the plasma rest frame and the $\pm$ sign refers to parallel and antiparallel propagation with respect to the mean field. The early work by Belcher et al. [10.19, 20] is given the credit for having first established that extended periods of solar wind measurements existed where the Alfvénic correlations (10.16b) were in fact observed. Figures 10.3 and 10.5 confirm these observations. Further *Helios* examples of Alfvénic fluctuations are explicitly given in Fig. 10.6 taken from [10.46, 48]. In this case the magnetic field pointed outwards away from the sun, and its intensity B and the number density n were particularly steady, such that the incompressibility condition was fulfilled. However, the amplitudes were observed to have variances of 20–30 percent of the mean field, and therefore certainly a nonlinear theoretical treatment was required. If (10.14) is supplemented again by the nonlinear term $(\delta Z^{\mp} \cdot \nabla) \delta Z^{\pm}$ left off previously, then we can find exact solutions by noting that the nonlinearity actually disappears for pure solutions of the type either $\delta Z^{+} = 0$ and $\delta Z^{-} \neq 0$ or vice versa [10.52]. A simple example of such a wave is given by a large-amplitude circularly polarized Alfvén wave, which keeps the total B constant and, given that the density and thermal pressure do not vary, thus also maintains a constant total pressure.

However, thorough inspection of Figs. 10.5 and 10.6 reveals that none of the idealized wave properties fully holds true. There is much less than 100% correlation, implying that even if Alfvénic properties seem to prevail, we always deal with a state where a given δZ^{+} dominates a slight admixture of δZ^{-} is present and vice versa. As a result, nonlinear effects described by (10.13a) are unavoidably to be expected. The assumption of incompressibility does not strictly hold. Observationally, a simple transverse or circular polarization is not found either [10.16]. These experimental findings stimulated research on what might be called generalized "Alfvénic waves", which have characteristics in common with the ideal waves discussed above but must be modified to comply with the observations. Barnes [10.5] proposed a stochastic model for Alfvénic fluctuations, where the field vector, while keeping its length constant, was assumed to wander around in a random fashion with its tip on a sphere. In an experimental check of this model [10.12] partial agreement between predicted and observed directional variations of the field was found. However, no further attempts have since been made to incorporate more realistic features of the fluctuations into the model, such as weak compressive effects.

Behannon and Burlaga [10.16] have also discussed a similar concept of a "general Alfvén wave", where the fluctuations δB are not restricted to a line or plane but may move around in three dimensions such that the tips of δB and B move together on the surface of a sphere while $|\delta B + B_0| = |B| = \mathrm{const.}$ Thus, although incompressible, these waves would have a fluctuating component $\delta B_{\parallel}$ along the background field B_0 in accord with many observations [10.9, 12, 19, 20, 36]. Dobrowolny et al. [10.52] identified this parallel component as corresponding to the incompressible limit of the slow magnetoacoustic mode [10.3, 30]. In the absence of nonlinear interactions the observed Alfvénic fluctuations were argued to be then composed of two types of wave, one with transverse Alfvén and one with slow magnetoacoustic polarization. In such a case the assumption that the minimum variance direction determined the wave propagation direction was found to be no longer tenable. Likewise, a simple purely transverse or circular polarization state would not describe the data well. Predictions were then made in [10.52], by considering a homogeneous and "Gaussian turbulence" model, on the structure of the variance matrix of the magnetic field fluctuations and their eigenvalues in qualitative accord with some of the observations.

Inspired by these theoretical considerations Bavassano et al. [10.9] set out to verify their validity empirically by means of a variance analysis of *Helios 2* magnetic field data. Incompressible fluctuations with a constant total field amplitude should theoretically obey the relation

$$B^2 = (B_0 + \delta B_{\parallel} + \delta B_{\perp})^2 = \text{const} , \tag{10.17a}$$

implying (with a constant C) that

$$\frac{\delta B_{\parallel}}{B_0} = C - \frac{1}{2} \frac{\delta B_{\perp}^2}{B_0^2} , \tag{10.17b}$$

which could easily be tested against *in situ* observations. Here $\parallel$ and $\perp$ indicate the parallel and perpendicular components of δB with respect to B_0. Scatter plots in the fluctuation variables of (10.17b) showed this relation to be extremely well fulfilled, with time series of 22.5 min and individual field vectors sampled at 6 s. This confirmed the constancy of B, and further that the fluctuations could in fact be decomposed in two incompressible polarization components, obeying the inequality $|\delta B_{\parallel}| \ll |\delta B_{\perp}|$. The time interval of 22.5 min used corresponds to a frequency of 7.4×10^{-4} Hz, falling right into the range where most analyses have revealed the purest examples of Alfvénic fluctuations.

10.3.3 Analysis of Random MHD Fields

We discussed in the previous section the virtues and limitations of the wave picture in describing the observed MHD fluctuations in the solar wind. There exist some clean examples of Alfvén waves, but usually the correlation is not ideal, indicating that nonlinear interactions are active to a varying degree. In the following the wave picture is therefore entirely abandoned and replaced by the turbulence approach. In a fully turbulent magnetofluid the detailed behavior of the fields $V(x,t)$ and $V_A(x,t)$ becomes analytically inaccessible because of their random nature (illustrated in Fig. 10.6), and the only adequate description is a statistical one. Our subsequent presentation closely follows the basic references [10.52, 53, 112–114, 119, 121, 123, 125, 128, 202] for the solar wind.

In the turbulence picture it is assumed that the velocity and magnetic field fluctuations can be decomposed into rapid fluctuations and slowly varying means, described by an ensemble average, usually indicated by brackets, $\langle ... \rangle$, that separates the large-scale, slowly varying from the small-scale, rapidly varying coordinates. Accordingly we may adopt, for example, the decomposition

$$V = \langle V \rangle + \delta V , \quad B = \langle B \rangle + \delta B , \tag{10.18}$$

$$Z^{\pm} = \langle Z^{\pm} \rangle + \delta Z^{\pm} , \tag{10.19}$$

where by definition the fluctuations have a zero mean, e.g. $\langle \delta Z^{\pm} \rangle = 0$. The ensemble averages are often indicated by the index nought, i.e. $Z_0^\pm = \langle Z^\pm \rangle$, and they may yet be variable on the large spatial and temporal scales. In accord with the observations the fast and small fluctuations are often assumed to be incompressible, whereas compressibility prevails on the large scales.

The ensemble-averaging operator deserves some further discussion. What the brackets $\langle ... \rangle$ mean from the *in situ* measurement point of view is a difficult problem. In practice most researchers have equated their ensemble averages by time averages over a space–time trajectory of the spacecraft. Of course, the means represented by $\langle ... \rangle$ need be shown to be relatively smooth and reproducible. If these averages are independent of the origin of time, the fluctuations are said to be stationary. Thorough stationarity tests have been applied [10.118, 119, 123] to a number of magnetic field data sets covering heliocentric distances of 1–10 AU and data record lengths from about 10 to 100 days. These analyses indicated good convergence of the estimates of $\langle ... \rangle$, provided data sets were long enough and did not include only a few coherent MHD structures such as those discussed in Sect. 10.2. Therefore Matthaeus and Goldstein [10.121, 123] suggested that "the interplanetary magnetic field can be meaningfully viewed as a *weakly* stationary random function".

The quantities of main theoretical interest in the analysis of random MHD fields are the double-correlation functions or covariances of the field variables, which represent second-rank tensors depending on two points in space and time, and which have the form

$$R^{AA'}(\boldsymbol{x}, t; \boldsymbol{x}', t') = \langle A(\boldsymbol{x}, t) A'(\boldsymbol{x}', t') \rangle \tag{10.20}$$

for any field vectors A and A', which may be identical (autocorrelation function) or different, such as the flow velocity $\boldsymbol{V}(\boldsymbol{x}, t)$ and the Alfvén velocity $\boldsymbol{V}_A(\boldsymbol{x}, t)$ or the magnetic field $\boldsymbol{B}(\boldsymbol{x}, t)$. Much of the existing turbulence analysis, as summarized in the monographs by Batchelor [10.7] or by Hinze [10.75] for Navier–Stokes fluids, is concerned with homogeneous turbulence, in which case the covariances only depend on the relative distance $\boldsymbol{r} = \boldsymbol{x}' - \boldsymbol{x}$. If stationarity can also be assumed, then the covariance is only a function of the time difference $\tau = t' - t$. As space and time are independent coordinates, time stationarity and spatial homogeneity are usually separate issues. If both these properties can be assumed, we have the simplification that

$$R^{AA'}(\boldsymbol{r}, \tau) = \langle A(\boldsymbol{x}, t) A'(\boldsymbol{x} + \boldsymbol{r}, t + \tau) \rangle . \tag{10.21}$$

In the super-Alfvénic solar wind we may further assume that

$$R^{AA}(\boldsymbol{0}, \tau) = R^{AA}(-V_{\mathrm{sw}}\hat{\boldsymbol{r}}, 0), \tag{10.22}$$

e.g. for the autocorrelation function of A, where V_{sw} is the solar wind speed. This relation is know as the Taylor frozen-in-flow assumption [10.170], which is usually valid (cf. [10.65]) for phenomena occurring on MHD time scales. As a consequence of (10.22), we find that if $A(\boldsymbol{x}, t)$ is a time-stationary random process it would also be spatially homogeneous. Of course, these conclusions only apply to scales such that $\tau \ll T$ and $r \ll R$, where T is the solar wind transit time through heliocentric distance R. More detailed discussion of the property (10.22) is found in [10.119], which also thoroughly analyzes the so-called "rugged" invariants of ideal MHD, which is our next topic.

10.3.4 Integral Invariants of Incompressible MHD

In a statistical description of a turbulent magnetofluid the detailed behavior of the field variables (flow or magnetic) is not of interest. Key quantities are those which are integral invariants of the ideal model equations (10.2–4) without dissipation, i.e. for $\nu = \eta = 0$. The general invariants for MHD were first discussed by Woltjer [10.199, 200]. For an incompressible magnetofluid these are the total energy,

$$E = \frac{1}{2} \int d^3 x (V^2 + V_A^2) = \int d^3 x \; e(\boldsymbol{x}) \;, \tag{10.23}$$

per unit mass density and the integral cross-helicity

$$H_c = \int d^3 x (V \cdot V_A) = \int d^3 x \; h_c(\boldsymbol{x}) \;, \tag{10.24}$$

which we here define as differing by a factor of 2 from the definition in [10.119]. The third quantity is the magnetic helicity

$$H_m = \int d^3 x (A \cdot B) \;, \tag{10.25}$$

where A here denotes the magnetic vector potential defining the field B through its curl. The physical meaning of H_m is extensively discussed in Moffatt's [10.127] monograph. It may be considered a measure of topological linkage or "knottedness" of magnetic field lines. A nonvanishing H_m can be looked upon as measuring the polarization state of wave-like magnetic fields. An appropriately normalized H_m (e.g. by the magnetic field energy density) yields $+1$ or -1 for circularly polarized waves, while H_m ranges between these extremes for complex fluctuations with mixed polarization.

If use is made of the Elsässer variables $Z^{\pm} = V \pm V_A$, the two invariants corresponding to E and H_c are given by the specific energies (of the "oppositely propagating" Alfvénic flows)

$$E^{\pm} = \frac{1}{2} \int d^3 x (Z^{\pm})^2 = \int d^3 x \; e^{\pm}(\boldsymbol{x}) \tag{10.26}$$

The densities $e^{\pm}(\boldsymbol{x})$ have the advantage of being positive-definite quantities. Of course, there exist simple relations between these two pictures by way of

$$e = \tfrac{1}{2}(e^+ + e^-) \;, \tag{10.27}$$

$$h_c = \tfrac{1}{2}(e^+ - e^-) \;. \tag{10.28}$$

These linear relations between the densities are a natural outcome of the definition of $Z^\pm$. By exploiting the basic equations (10.8) through (10.10), the conservation equation for the energy density $e^\pm$ [10.112] in the incompressible limit attains the instructive form

$$\frac{\partial}{\partial t} e^\pm = -\nabla \cdot \left(e^\pm Z^\mp + \frac{1}{\varrho} p_T Z^\pm \right) . \tag{10.29}$$

This equation explicitly demonstrates, by means of spatial integration over the volume of the system under consideration, that $e^\pm$ are integral invariants, provided the fluxes contained under the divergence of (10.29) are everywhere normal to the surface bounding the plasma-containing region. The proof of invariance therefore depends crucially on the nature of the boundary conditions. These two fluxes represent the transport of $e^\pm$ by the "opposite" velocity $Z^\mp$ and of the total pressure (10.13b) by the velocity $Z^\pm$ into the volume considered. Generalizations of (10.29) to compressible MHD have been discussed in [10.112]. It should be noted that H_c also is an invariant for a compressible magnetofluid [10.200].

10.3.5 Turbulence Spectra

The literature on turbulence emphasizes the role of the integral invariants because they cannot be modified by nonlinear effects and give, by their wave-number spectra or spectral decomposition in frequency, valuable clues to the dynamical evolution and turbulent state of the magnetofluid. A proper treatment of this important topic is beyond the scope of this article. We must refer the reader to some general reviews [10.128, 142, 156] and books [10.7, 75, 127] on this subject and seminal papers on MHD turbulence and the dynamo problem [10.60, 143, 144]. Of particular interest are the autocorrelations of the fluctuations, which are directly related to their energy density at zero time lag ($\tau = 0$) and zero relative distance ($r = 0$). Let us first consider, for $r = 0$, the fluctuation energy spectra of Elsässer variables, after [10.114, 184] given by

$$e^\pm(f) = 4 \int_0^\infty d\tau \, \cos(2\pi f \tau) \tfrac{1}{2} \langle \delta Z^\pm(0) \cdot \delta Z^\pm(\tau) \rangle . \tag{10.30}$$

Clearly the spectrum is directly given by the Fourier transform of the autocorrelation function, which is shown to be symmetric in the time variable τ [10.113, 119]. Observationally, the correlation time (i.e. the time for which the correlation functions differ significantly from zero and do not oscillate about zero) turns out to be a few hours for solar wind fluctuations. In deriving (10.30) the assumption of stationarity had to be made. Another spectrum of interest is that of the symmetric cross-correlations

$$e^R(f) = 4 \int_0^\infty d\tau \, \cos(2\pi f \tau) \tfrac{1}{4} \langle \delta Z^+ \cdot \delta Z^-(\tau) + \delta Z^- \cdot \delta Z^+(\tau) \rangle , \tag{10.31}$$

or "residual energy" spectrum, which is the difference between the kinetic and magnetic energy spectrum [10.69, 70] and which allows us to calculate the spectrum of the Alfvén ratio of kinetic to magnetic energy as

$$r_A(f) = \frac{e(f) + e^R(f)}{e(f) - e^R(f)} ,$$ (10.32)

where the total energy spectrum of the fluctuations is given by the sum

$$e(f) = \tfrac{1}{2}(e^+(f) + e^-(f))$$ (10.33)

and corresponding to (10.24) and (10.28) the cross-helicity spectrum by

$$h_c(f) = \tfrac{1}{2}(e^+(f) - e^-(f)) .$$ (10.34)

It is convenient to normalize some of the spectra such that they range between -1 and $+1$. For example the normalized cross-helicity reads

$$\sigma_c(f) = h_c(f)/e(f) .$$ (10.35)

The normalized residual spectrum is similarly given by $\sigma_R(f) = e^R(f)/e(f)$. The above-defined spectra are readily derivable from the time series of observed fluctuations. They may then be transformed into wave-number space by using the frozen-in-flow assumption (10.22) relating frequency f and reduced wave number $k^* = k/2\pi$ by the Doppler-shift formula $f = k^* V_{sw}$. A thorough discussion concerning the validity of this relation for evaluating wave-number spectra can be found in [10.65, 119].

Traditional turbulence analysis is concerned with equal-time covariances, the evaluation of which demands, of course, the measurement of fluctuations at two points in space at a distance r apart. Such a measurement cannot be provided by a single spacecraft and therefore the frozen-in-flow assumption is at present the only and crucial means by which k-spectra can be constructed. For the sake of completeness we quote a few such covariances by using the traditional variables V and V_A. The kinetic fluctuation energy spectrum is given after [10.7, 119, 128] by the expression

$$e_V(k) = \int d^3k \, e^{-ik \cdot r} \tfrac{1}{2}\langle \delta V(0) \cdot \delta V(r) \rangle ,$$ (10.36)

and similarly the magnetic energy spectrum per unit mass density by

$$e_{V_A}(k) = \int d^3k \, e^{-ik \cdot r} \tfrac{1}{2}\langle \delta V_A(0) \cdot \delta V_A(r) \rangle .$$ (10.37)

The cross-helicity spectrum is defined to be symmetric:

$$h_c(k) = \int d^3k \, e^{-ik \cdot r} \tfrac{1}{2}\langle \delta V(0) \cdot \delta V_A(r) + \delta V_A(0) \cdot \delta V(r) \rangle .$$ (10.38)

Again, $h_c(k)$ may also be normalized to the total energy spectrum:

179

$$e(\boldsymbol{k}) = e_V(\boldsymbol{k}) + e_{V_A}(\boldsymbol{k}) \,. \tag{10.39}$$

Another quantity of key interest for the dynamical evolution of the turbulence is the Alfvén-ratio spectrum,

$$r_A(\boldsymbol{k}) = e_V(\boldsymbol{k})/e_{V_A}(\boldsymbol{k}) \,, \tag{10.40}$$

which measures the spectral distribution of kinetic versus magnetic energy. The analogue in Elsässer variables is what might be coined the "Elsässer ratio" [10.113],

$$r_E(\boldsymbol{k}) = e^-(\boldsymbol{k})/e^+(\boldsymbol{k}) \,, \tag{10.41}$$

which compares, in the language of wave theory, the energy of inward- and outward-propagating Alfvén waves against each other, and is closely connected to the cross-helicity spectrum by the equation

$$r_E = \frac{1 - \sigma_c}{1 + \sigma_c} \,. \tag{10.42}$$

All of these spectra have now been empirically established by a great number of solar wind fluctuation studies to be discussed in the subsequent sections.

The study of the dynamical or time evolution of the spectra (10.36–38) is at the heart of MHD turbulence theory. We shall not pursue this issue any further at present but postpone it to a later section. Dimensional analyses [10.91, 93] have led to predictions of power-law spectra in wave-number space, for fully developed homogeneous turbulence, applying to energy spectra such as $e_V(\boldsymbol{k})$ and $e_{V_A}(\boldsymbol{k})$ or $e(\boldsymbol{k})$. Recently, theoretical suggestions have also been made for the solar wind spectra of $e^{\pm}(\boldsymbol{k})$ [10.187, 189] by adapting the Kolmogorov phenomenology. The predicted spectral laws depend on the dimensions of the magnetofluid (see e.g. Table 1 in [10.128]) and usually assume symmetry properties of the correlation functions, which often are not well substantiated by the observations. A comprehensive analysis of various possible symmetries that a function like (10.21) might have can be found in [10.117, 119]. Phenomenology of turbulence essentially provides two spectral indices for developed turbulence, against which measured spectra might be compared: $-\frac{5}{3}$ or $-\frac{3}{2}$, depending on the type of assumption on the nonlinear interaction times going into the dimensional analysis. These two indices will be frequently referred to while judging observed spectra.

The paper by Grappin et al. [10.70] generalizes the Kraichnan phenomenology of homogeneous MHD turbulence to the situation with high velocity–magnetic-field correlations and predicts the sum of the slopes for the e^+ and e^- spectra to be equal to -3. This phenomenology seems to receive substantial confirmation by direct numerical simulation [10.146]. However, it is not clear whether these results are relevant to the inhomogeneous solar wind, which as shown below does not exhibit the predicted spectral characteristics. A slope of -1 has been obtained for the magnetic energy spectrum [10.143], resulting from an inverse cascade of the magnetic helicity spectrum (with a -2 slope) as studied by means of numerical computations. Spectral measurements of the magnetic helicity from *Helios* data have been presented by Bruno and Dobrowolny [10.27] and from *Voyager* data by Matthaeus and Goldstein [10.119]. The magnetic helicity will not be further discussed in this review.

10.3.6 Analysis Techniques in the Time Domain

Spatial and temporal correlation functions, as they occur as integrands in formulae (10.30, 36, 37), are the basic ingredients for establishing power spectra from time series of random fluctuations. From them the correlation time τ_c can be directly obtained [10.114]. Other methods to determine τ_c explicitly rely on the spectra or special properties of the correlation function. Various methods have been discussed by Matthaeus and Goldstein [10.119], who also were the first to give empirical estimates of correlation times and lengths relating to the three "rugged" invariants of Sect. 10.3.4. Correlation lengths in the solar wind are basically estimated by means of $\lambda_c = V_{sw}\tau_c$. Typically τ_c is a few hours, and, corresponding to the solar wind flow speed, the correlation length amounts to about 10^6 km or a few percent of 1 AU.

The detailed values and the variation of λ_c with heliocentric distance are presented in Fig. 10.7, which combines *Helios* and *Voyager* data from [10.27, 119]. A clear radial trend for all lengths to increase with distance is visible. The various symbols refer to different calculations for λ_c, which are based on the magnetic correlation function (L1), or on normalized moments of k; on the first (L4) and second (L2) moment of k evaluated with the magnetic field intensity; or (L3) with the helicity spectrum. Similar studies have been carried out in [10.114, 184] by means of Elsässer variables . They revealed a distinct difference between the autocorrelation of δZ^+ and δZ^-. A general property found was that correla-

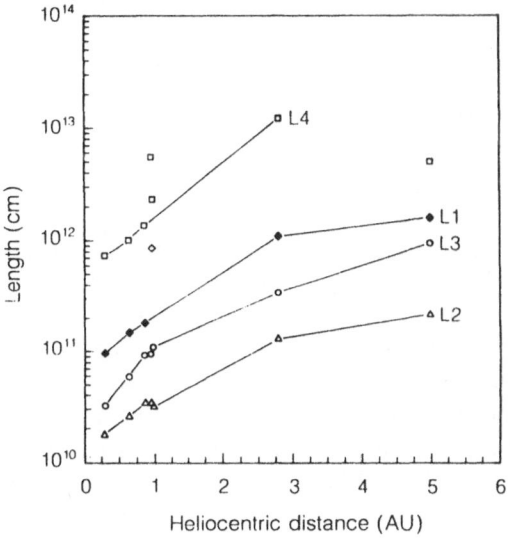

Fig. 10.7. Various estimates of the turbulence correlation length as a function of heliocentric distance. Individual points obtained by *Helios* [10.27] and *Voyager* [10.119] were connected to visualize better the radial trends

tions between inward-oriented fluctuations tended to last longer than those of the outward-propagating fluctuations (by a factor of 2–3). This result applied to both high- and low-speed solar wind flows.

Finally, the technique of Roberts et al. [10.151, 152] of studying the properties of fluctuations by directly using the ensemble averages in the time domain is worth mentioning. This method is straightforward and complements the spectral methods discussed before. The normalized cross-helicity may thus be simply defined by

$$\sigma_c(t) = \frac{h_c(t)}{e(t)} = \frac{2\langle \delta V \cdot \delta V_A \rangle}{\langle \delta V^2 + \delta V_A^2 \rangle} ,$$ (10.43)

where the slow time t is meant to characterize the start or center time of the interval used to evaluate the ensemble average. The absolute value of σ_c is always less than or equal to unity. The number ± 1 indicates perfect Alfvénic correlations, and a number in between these extremes signifies a mixed state, in which both "propagation" directions in the wave jargon or non-Alfvénic fluctuations are present. The same property is of course shared by the spectral definition (10.35), as becomes immediately apparent if the properties of Alfvén waves in (10.16a,b) are exploited.

The same technique can be applied to study the Alfvén ratio (10.32) if the cross-correlation function in terms of $\delta Z^\pm$ is used. After normalization, we find

$$r_A(t) = \frac{1 + \sigma_z(t)}{1 - \sigma_z(t)} ,$$ (10.44a)

where the symbol σ_z denotes

$$\sigma_z(t) = \frac{2\langle \delta Z^+ \cdot \delta Z^- \rangle}{\langle (\delta Z^+)^2 + (\delta Z^-)^2 \rangle} .$$ (10.44b)

This quantity is readily obtained from the measurements.

10.4 Nature and Morphology of MHD Fluctuations

10.4.1 Power Spectra in Terms of Elsässer Variables

In this section we first investigate high-resolution spectral properties of energy spectra of the Elsässer velocities $Z^\pm$. In subsequent sections spectra of the magnetic field components and intensity will also be shown. There exists a variety of techniques to construct power spectra from the time series of observed fluctuations. For this purpose researchers traditionally use fast-Fourier-transform (FFT) methods assisted by various other practices to prepare and smooth the raw data. Classic references in this field are the monographs by Bendat and Piersol [10.22], Blackman and Tuckey [10.24], and Panchev [10.138]. We will not dwell further on these topics here but refer the reader to [10.119] for a detailed description of how to apply various mathematical techniques. Consequently, technical aspects concerning the reliability of constructed power spectra are not discussed here, but the published spectra are taken for granted in the following.

The potential of $e^{\pm}$ spectra for analysis of the turbulent state of the solar wind was first recognized and explored by Grappin, Mangeney, and Marsch [10.71, 72], who were inspired by the results of numerical simulations [10.68, 69] and wanted to investigate some theoretical predictions on separate cascades for e^+ and e^- made in [10.53, 70]. Extensive use of Elsässer variables in data analyses was then made by Tu et al. [10.184] and Marsch and Tu [10.114]. The key point is to decompose the fluctuations according to (10.8) into their two components, which would correspond in the linear small-amplitude limit to Alfvén waves traveling inwards and outwards with respect to the sun, along the average interplanetary magnetic field. Of course, this suggestive picture does not apply to the nonlinear situation or in case of equipartition between e^+ and e^-. We will still keep the terminology of inward and outward Alfvén modes for semantic simplicity. It should be kept in mind, however, that it is often not even clear whether these modes really propagate, since in the case of strong interactions there is nothing like a dispersion relation (10.16a,b). Analysis in terms of "Alfvén modes" has the advantage that their energies are both positive definite, and if they are considered per unit mass density, the main radial trends related to the decline of the plasma density in proportion to R^{-2} are then eliminated. As e^+ and e^- are the inviscid invariants (see again Sect. 10.3.4) they are expected to exhibit spectral shapes of their own (see e.g. Grappin et al. [10.69, 70]).

That this is in fact true was first found in [10.71, 72]. The basic spectral features are illustrated in Fig. 10.8 taken from [10.184]. The left-hand box (a) refers to fast wind and the right-hand box (b) to slow wind; solid lines represent $e^+(f)$ and dotted lines $e^-(f)$. These data were obtained near 0.3 AU and are characteristic for the turbulent state as encountered by the *Helios* spacecraft in their perihelion passages near solar minimum. (See again Figs. 10.1, 3, 5.) The detailed plasma parameters, spectral slopes, and a possible explanation for the various frequency regimes indicated in Fig. 10.8 can be found in [10.184]. The following basic features are obvious. There is an excess of e^+ over e^- in both types of flow, with the outward Alfvén mode dominating the inward one, which is clearly visible in the middle frequency range between 10^{-4} and 10^{-3} Hz and in the fast stream. Spectra in (b) are more developed towards an extended inertial regime with an overall spectral index ranging between $-\frac{3}{2}$ and $-\frac{5}{3}$. In contrast, spectra in (a) are badly fitted by a single power law, reveal distinct kinks at middle frequencies, have a flat part at low (e^+) and high (e^-) frequencies, and, if looked at together, form a characteristic rhomboid shape. The spectra probably join each other again at the highest and lowest frequencies of the domain considered here (6×10^{-6} up to 6×10^{-3} Hz). The reliability of these spectral characteristics is confirmed also by applying the Blackman and Tuckey [10.24] techniques yielding the dashed lines which nicely coincide with the FFT spectra.

The general spectral pattern of e^+ and e^-, reminding one of a systematic "breathing" in the sense of opening and closing the gap between both spectra as

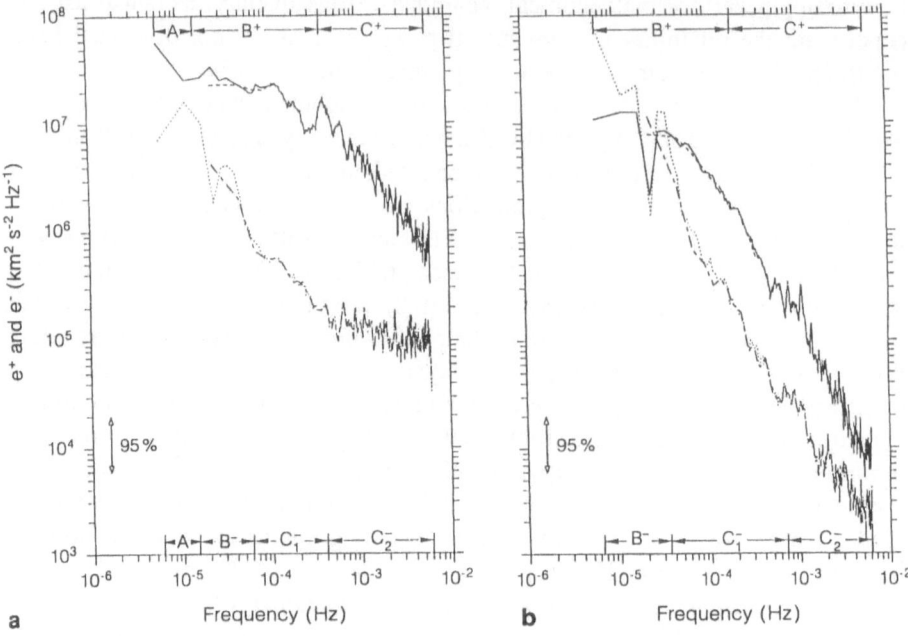

Fig. 10.8a,b. Power spectra of Elsässer variables for high-speed (**a**) and low-speed (**b**) wind derived from *Helios* data at 0.3 AU. *Solid lines* give e^+ and *dotted lines* e^-. The *dashed* and *dot-dashed lines* are calculated by the Blackman–Tuckey technique. The error bar is based on a fit with 22 degrees of freedom [10.184]

first revealed in Fig. 10.4 of [10.72], is typical of the period near solar minimum and repeats itself as the spacecraft successively traverses the recurrent stream pattern displayed in Fig. 10.1. This morphological pattern of $e^\pm$ spectra is displayed in Fig. 10.9 taken from [10.186], showing the passage through a typical high-speed stream and its adjacent low-speed flows observed near 0.3 AU. Each spectrum covers one day, which is indicated on the top panel, which displays the solar wind flow speed, by a hatched bar. For reference Kolmogorov's spectral law is given with a slope of $-\frac{5}{3}$, which is apparently obeyed fairly well by the $e^\pm$ spectra in the low-speed flow neighboring the heliospheric current sheet crossed during day 101. As one goes away from this region the spectral appearance changes through a sequence of intermediate stages finally to reach the rhombus shape discussed in Fig. 10.8 in the body of the fast stream (c). The bottom panels are not of interest at the moment and may be disregarded here. Clearly, there is a systematic change in the $e^\pm$ spectra as a function of flow speed, which indicates more developed turbulence in slow wind and a sizable excess of e^+ over e^- in fast streams, signifying the dominance of outward Alfvén modes (or for $e^+ \gg e^-$, of outward-propagating Alfvén waves in the wave picture) of near-solar origin. A similar pattern is observed at 1 AU (see Fig. 2 of [10.186]), but all spectra there appear to be further evolved towards equipartition between e^+ and e^- and steeper slopes.

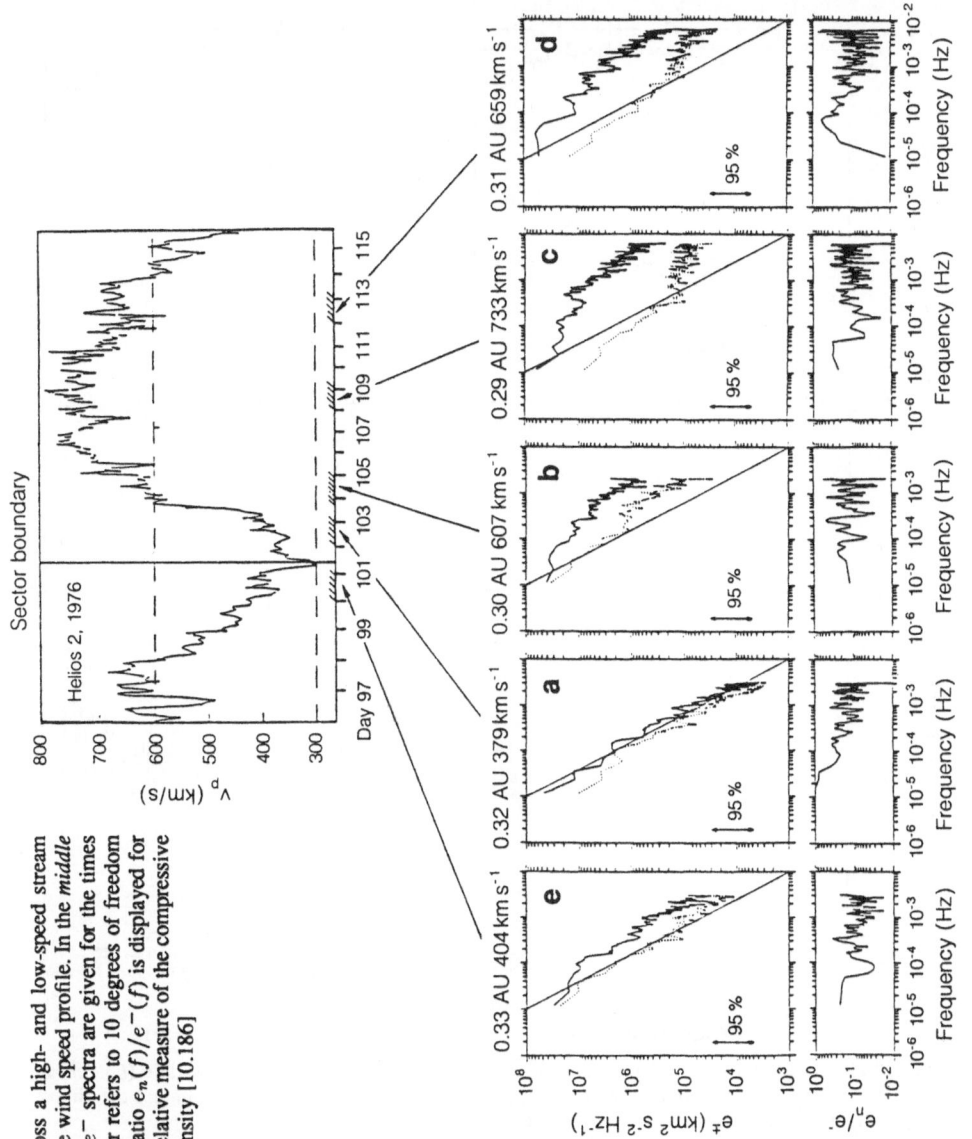

Fig. 10.9. Sequence of $e^{\pm}$ spectra across a high- and low-speed stream near 0.3 AU. The *upper panel* shows the wind speed profile. In the *middle panel* five pairs of e^{+} (*solid lines*) and e^{-} spectra are given for the times indicated by *shaded bars*. The error bar refers to 10 degrees of freedom for the spectral fits. At the bottom the ratio $e_n(f)/e^{-}(f)$ is displayed for each time period in order to provide a relative measure of the compressive contribution to the total turbulence intensity [10.186]

185

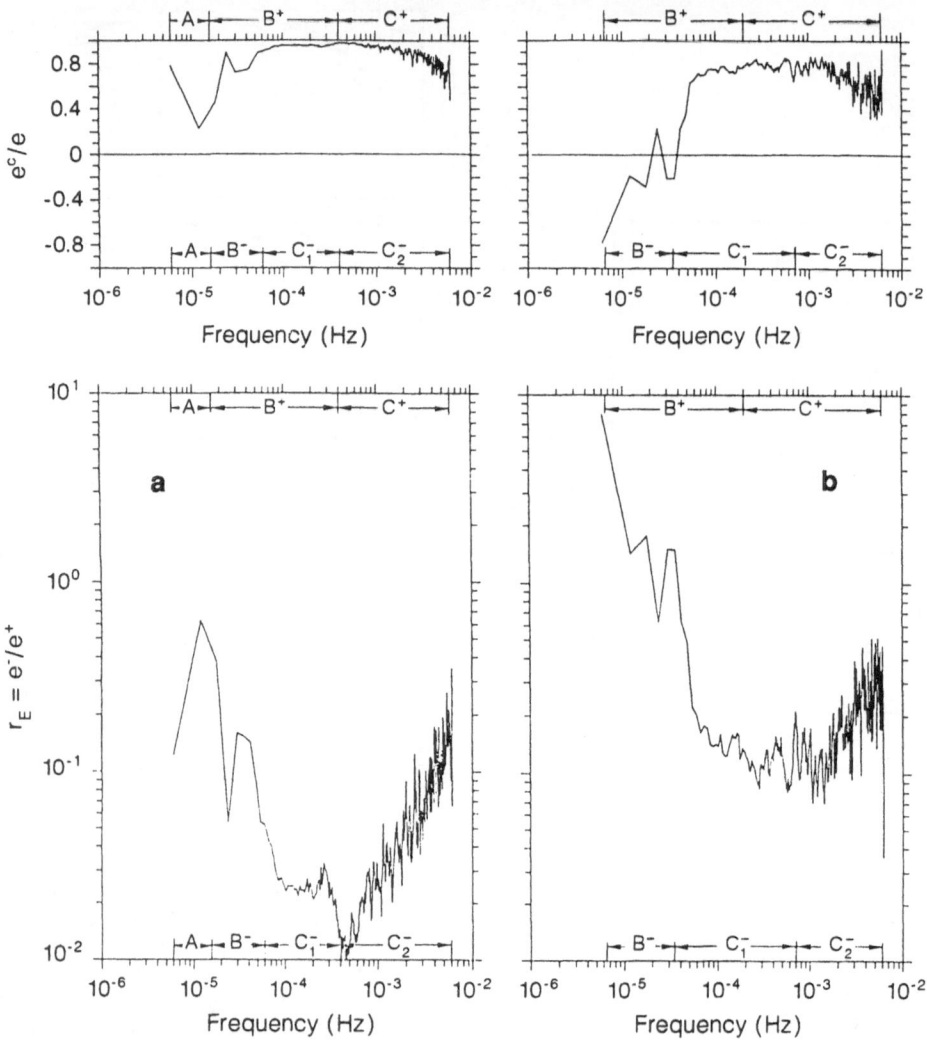

Fig. 10.10a, b. Spectra of normalized cross-helicity (*top panels*) and of the Elsässer ratio (*bottom*) in the same format as in Fig. 10.8. Note the V-shaped form of $r_E(f)$ in the fast stream and the almost ideal Alfvénic correlations about 5×10^{-4} Hz [10.184]

The degree of Alfvénicity of the fluctuations is indicated by the normalized cross-helicity and the Elsässer ratio. Figure 10.10 presents high-resolution measurements of their spectra in the range 6×10^{-6}–6×10^{-3} Hz for the same data as displayed in Fig. 10.8. The top part gives $\sigma_c = h_c/e$ for high (a) and low (b) solar wind speed, and the bottom part the corresponding ratios of $r_E = e^-/e^+$. Of course, both quantities are intimately related through (10.42) and the various formats just emphasize in different ways the same features, of which the prevailing ones are that in (a) σ_c is very close to unity in the range 10^{-4}–10^{-3} Hz and drops to lower values at both the lower- and higher-frequency ends, and, correspondingly, the spectrum of r_E is typically V-shaped with a significant minimum near 5×10^{-4} Hz or 33 min, where the cleanest "Alfvén waves" are found.

This overall shape of the σ_c-spectrum resembles traditional coherence spectra between, for example, B_x and V_x as evaluated by Denskat and Neubauer [10.46] and other researchers [10.14, 26, 43, 46, 48]. Comparison of frames (a) and (b) shows that the spectral course of σ_c and r_E is similar in high- and low-speed wind, however, two major differences are apparent. First, in slow wind the minimum of r_E is higher and ≈ 0.1, indicating a more strongly mixed state in the Alfvénic regime, and concurrently σ_c never goes beyond 0.8. Second, below 2×10^{-5} Hz, we have $r_E > 1$ and σ_c negative, indicating the dominance of large-scale ingoing fluctuations, a phenomenon revealed also in the *Voyager* data [10.119].

All intermediate evolutionary stages of the turbulence between state (a) and (b) of Figs. 10.8 and 10.10 can be found by sampling data across the stream structure of Figs. 10.1 and 10.9. It seems reasonable, therefore, to average over the whole period of four solar rotations covered in Fig. 10.1. The result [10.71, 72] is that in the Alfvénic regime between 10^{-4} and 10^{-3} Hz there always is a distinct excess (more than 45%) of the outward mode, which is generally more abundant than the inward mode at all frequencies. This distinct asymmetry in e^+ and e^- strongly supports a coronal origin of the fluctuations, but the mere existence of e^- also indicates nonlinear couplings and local generation of the fluctuations, topics to be discussed below in more depth.

10.4.2 Stream-Structure Dependence of Alfvénic Fluctuations

On discussing the stream-structure dependence of solar wind fluctuations, we should be aware of the fact that, for the near-solar-minimum epoch of activity cycle 21 under investigation here, the stream structure has a clear association with the interplanetary and coronal magnetic field when expressed in heliomagnetic coordinates. Bruno and coworkers [10.28, 29] investigated the latitudinal gradients of solar wind parameters and magnetic field power spectra. The flow speed minimizes in the near-equatorial region, whereas beyond a latitudinal belt of about ± 25 degrees only fast plasma streams were observed [10.162]. In contrast the magnetic field energy density does not show any significant variation with angular separation from the current sheet or magnetic equator and thus no dependence on the stream structure. This observation is also made for the radial component B_r of B, which according to Gauss' law should obey $F_B = B_r R^2 = \text{const}$. The radial magnetic flux was estimated to be $F_B = (3.28 \pm 1.67)\,\text{nT AU}^2\,\text{sr}^{-1}$, independent of flow speed and heliocentric distance R [10.110].

What varies distinctly, however, is the magnetic field power spectrum defined by the trace of the autocorrelation matrix. The power level was shown to change with angular displacement from the current sheet as illustrated in Fig. 10.11. Wave numbers were used instead of frequency in order to eliminate the latitudinal dependence of flow speed and to get results referring to invariant spatial scales. To achieve k-spectra the frozen-in-flow assumption (see Sect. 10.3.3) had to be employed. Three wave number bands were discriminated. The results for band

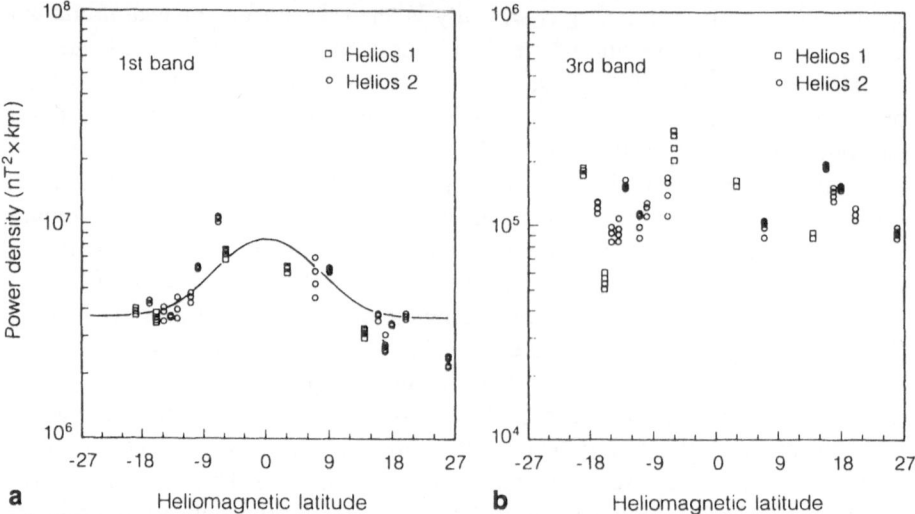

Fig. 10.11a,b. Average magnetic field power density versus displacement from the heliomagnetic equator. All power densities have been extrapolated to their 1 AU values. *Box* (a) refers to the wave vector band 1.36×10^{-7}–2.0×10^{-6} km^{-1} and *box* (b) to higher k in the band 6.0×10^{-6}–1.4×10^{-5} km^{-1}. Note the different behavior between the two bands. The *solid line* in (a) represents a least-mean-square Gaussian fit to the data points and emphasizes the enhanced power about the current sheet within a $\pm 18°$ latitudinal belt [10.29]

1: 1.36×10^{-7}–2.00×10^{-6} km^{-1} and band 3: 6.0×10^{-6}–1.4×10^{-5} km^{-1} are shown in parts (a) and (b) of Fig. 10.11, where the power density is displayed versus displacement of the measurement location from the heliomagnetic equator. The different behavior of the two bands is obvious. The study comprised a large sample of power spectra from various solar distances, which were all corrected for radial trends by normalization to 1 AU. The average power density was calculated to scale radially like $R^{-\alpha}$ with $\alpha = 2.63$ for band 1 and $\alpha = 3.77$ for band 2 [10.11].

In the low-k band the magnetic power clearly decreased while moving away from the equator (Fig. 10.11(a)). For each latitudinal interval four different estimates of the power are given. The solid line represents a mean-square Gaussian fit to the data, with a weak indication of plateaus beyond about $\pm 15°$. In contrast, part (b) of the figure shows no clear trend and scattering of the points between 8×10^4 and 2×10^5 nT^2 km. On the average the power spectrum at high wave numbers ($> 2 \times 10^{-6}$ km^{-1}) did not change with latitude and showed spectral characteristics of fully developed turbulence with an invariant slope close to $\frac{5}{3}$. At first glance these results seem to be at odds with those in Fig. 10.9. However, the discrepancy disappears by noting that the plasma density maximizes at zero latitudes and drops by a factor of 2–3 on the average when going to $\pm 20°$ [10.28], and that the spectra in Fig. 10.11(a,b) need to be divided by the plasma mass density to be comparable with spectra in Fig. 10.9. If this is done and all spectra are evaluated in frequency space then qualitative agreement is achieved.

We briefly summarize the basic features of turbulent spectra (power per unit mass density) of $e^{\pm}$ near solar minimum. There is always an excess of the outward Alfvén mode. In the bulk and trailing edges of fast streams $e^+ \gg e^-$, indicating the emergence of pure "Alfvén wave" patterns in the turbulence with comparatively enhanced power levels between 10^{-4} and 10^{-3} Hz and relatively flat spectra. Plasma neighboring the current sheet exhibits more fully developed turbulence with $e^+ \geq e^-$ and Kolmogorov-type spectra. This morphology of the turbulence appears to hinge on the well-ordered IMF pattern and associated stream structure, as is typical for the near-solar-minimum conditions. The turbulence pattern becomes, however, more complex and less ordered, like the solar wind stream structure itself, during later phases of the solar cycle.

This is illustrated by Figs. 10.12 and 10.13, which display data for time periods in the ascending phase (1978) and near solar maximum (1980). Both figures show magnetic field intensity, solar wind speed, heliocentric distance (normalized radial component of B to indicate the sector pattern), and normalized cross-helicity σ_c, calculated according to (10.43) in the time domain and based on 3-, 27-, 81-, and 243-hour running values. These figures by Roberts et al. [10.151, 152] comprehensively survey the morphology of the solar wind and its turbulence and the degree of Alfvénic correlations of the fluctuations. Note that the intervals covered include the whole distance range from 0.3 to 1 AU traversed by the *Helios* spacecraft and that σ_c is defined such that positive values always mean outward propagation in "wave language".

Figures 10.12 and 10.13 illustrate many important characteristics of the fluctuations. High Alfvénic correlations predominantly occur closest to the sun on all time scales analyzed, and they tend to fade away gradually with increasing heliocentric distance as is most clearly seen by comparison of panels 3 and 5 of Fig. 10.12. Lower and more negative values of σ_c occur more often near 1 AU than 0.3 AU, where σ_c at the 3-hour scale always remains on a high level, yet with some modulation in accord with magnetic sector patterns as evidenced by the B_r trace in Fig. 10.13. Unlike in Fig. 10.9, the purest Alfvénic fluctuations ($\sigma_c > 0.8$) are not observed in association with high-speed streams but in fact were often clearly associated with slow flows, e.g. as around days 50, 75, and 107 of Fig. 10.13. Similar events are documented in the *Helios 2* data [10.43, 107]. The lower correlation for intervals between these days were related to an inward-pointing magnetic sector, thus indicating that turbulent conditions on either side of the current sheet were remarkably different.

Another striking feature evident in Fig. 10.12 around day 10 is the existence of very long-period Alfvén waves near 0.3 AU on the 243-hour scale. Similar situations are visible on day 110 of Fig. 10.13 and in Fig. 10.3. On all these occasions the σ_c-values at smaller scales are even higher. Such very large-scale correlations represent equivalent spatial structures of the size of the entire inner heliosphere (1 AU), and seemingly can only arise as a result of coronal processes. In view of the generally higher correlation during the perihelion passages of *Helios 1* and 2, we are persuaded to follow Roberts et al. [10.151, 152] in concluding

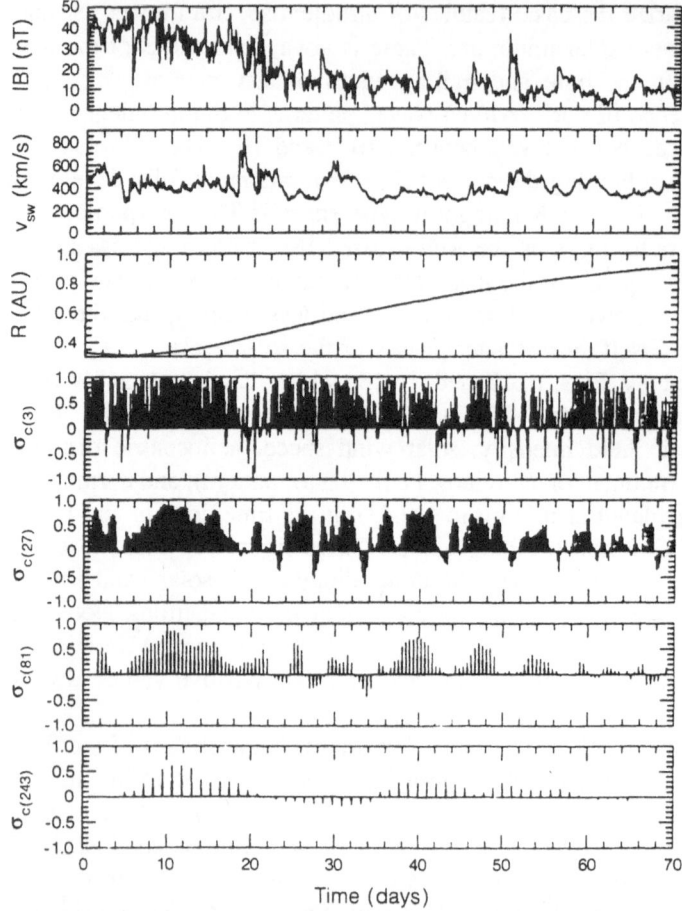

Fig. 10.12. *Helios 1* data for 70 days of the year 1978. *Top* to *bottom*: Magnetic field strength, solar wind speed, solar distance, and the running average values of the cross-helicity σ_c for 3, 27, 81, and 243 hours. The "spikes" are computed at each value, whereas the correlations were only calculated as often as was needed to resolve significant changes [10.152]

that the sun, or its outer corona, is the source of most of these outward-going Alfvénic fluctuations. The enhanced level of Alfvénicity even on the 3-hour scale near 0.3 AU generally points towards a predominantly coronal origin of interplanetary turbulence in the inner heliosphere. This is also suggested by Fig. 10.9, indicating that the enhanced level of e^+ and concurrent depletion of e^- is the result of a long-lasting connection to coronal sources with open magnetic field topology, from which coronal turbulence can freely escape conducted by the large streams acting as "wave guides" for the fluctuations.

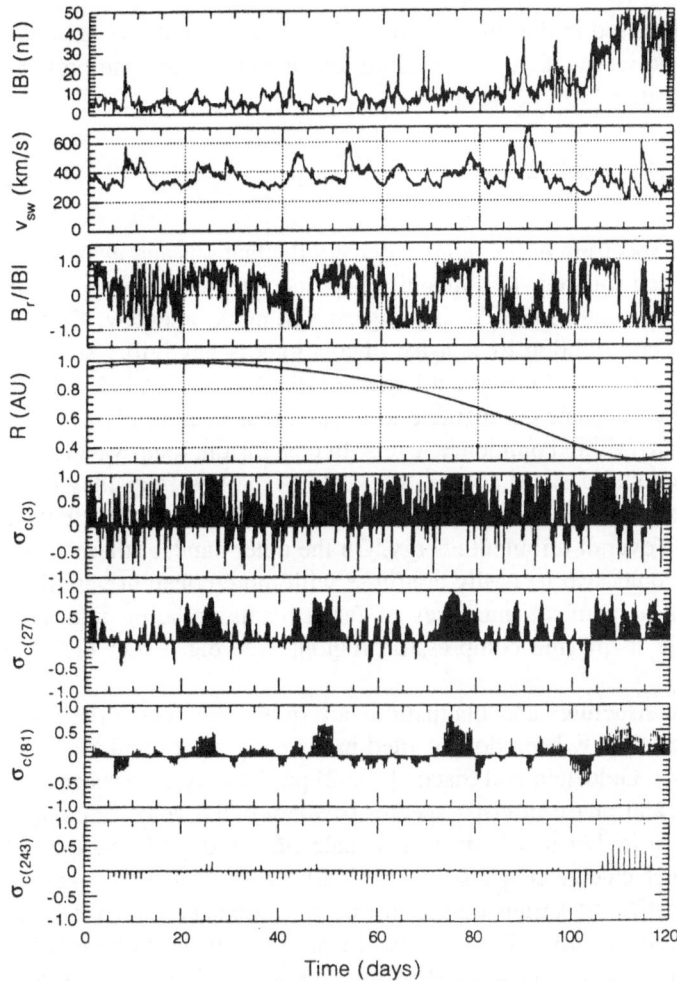

Fig. 10.13. *Helios 1* data for 120 days of the year 1980 near solar activity maximum. The format is the same as in Fig. 10.12. In addition the radial field component B_r is given in the *third panel* to indicate the magnetic sector structure [10.152]

10.4.3 Compressive Fluctuations

Figure 10.5, discussed earlier, indicates the occurrence of sizable compressive fluctuations in the high-density solar wind flows traversed by *Helios 2* during its first perihelion passage. Generally, the predominantly Alfvénic periods also reveal variable compressive noise, with sometimes significant amplitudes of the fluctuations in plasma density and magnetic field. Enhancements in the noise intensity coincide with passages through dense plasma regions often near the current sheet, which are, as previously shown in Fig. 10.9, characterized by low Alfvénic correlations and by turbulence more developed toward equipartition of e^+ and e^-. The true nature of the density fluctuation is not fully understood.

This issue will be addressed in the subsequent section. If we assume we are dealing with magnetosonic waves, then a reasonable upper limit to the compressive contribution is given after [10.112, 186] by

$$ e_n = \left\langle \left(\frac{\delta \varrho}{\varrho} \right)^2 \right\rangle \left(V_A^2 + c_S^2 \right) / 2 \, , \tag{10.45} $$

employing the power spectrum of the relative density fluctuations and the average Alfvén and sound speed squared. This quantity is plotted in the bottom panels of Fig. 10.9, where it is normalized to the ingoing Alfvén mode intensity. Clearly, beyond 10^{-4} Hz the contribution of density fluctuations is negligible since e_n hardly amounts to 10% of e^-. However, at larger scales ($f < 5 \times 10^{-5}$ Hz) and in low-speed flows e_n can become comparable to e^- and thus even to e^+, which is about equal to e^- in (a) within confidence levels. In conclusion, long-wavelength magnetosonic waves could perhaps account for some of the low-frequency noise occurring in the plasma neighboring the current sheet and there contribute a sizable fraction to the overall turbulent energy. On the other hand, at these scales density fluctuations could also be easily confused with more coherent variations in ϱ arising from the dynamical interfaces of fast streams colliding into slow ones, thus giving rise to plasma compression regions (see e.g. Pizzo [10.141] and [10.34, 162]).

That compressive structures and fluctuations are quite common in the interplanetary medium has already been documented in the early solar wind literature [10.20, 36, 62, 88, 134]. Goldstein and Siscoe [10.62] particularly emphasized the change in coherence and phase in cross-spectra between radial velocity component V_r and density ϱ in *Mariner 5* data at a scale of one day. At periods of 12–24 hours a distinct change in phase from $-180°$ to $0°$ was also found in $B - \varrho$ cross-spectra, indicating a transition from anticorrelation at high frequencies to positive correlations at low frequencies. Such a distinct anticorrelation was first found by Burlaga [10.31] when studying "pressure-balanced structures" (PBS). Besides abrupt coherent structures such as tangential discontinuities (TD) wave-like PBS were also detected. A similar transition from negative to positive correlation, when progressing from scales of hours to scales of days, was revealed to exist in $B - p$ correlation studies [10.35, 61], thus establishing the existence of PBS.

More recent studies by Vellante and Lazarus [10.188], Roberts [10.149], and Bavassano and Bruno [10.14] provided a systematic analysis of radial variations in the compressive fluctuations. Roberts et al. [10.152] also studied compressive fluctuations in the inner heliosphere. Figure 10.14 displays some of their results. The lowest two panels show on a 9- and 243-hour scale the correlation coefficient between δn and δb defined in the time domain as

$$ \varrho_{bn} = \langle \delta b \, \delta n \rangle / (\langle \delta b^2 \rangle \langle \delta n^2 \rangle)^{1/2} \, , \tag{10.46a} $$

$$ \delta b = \delta B \, V_A / B_0 \quad \text{with} \quad B_0 = \langle B \rangle \, . \tag{10.46b} $$

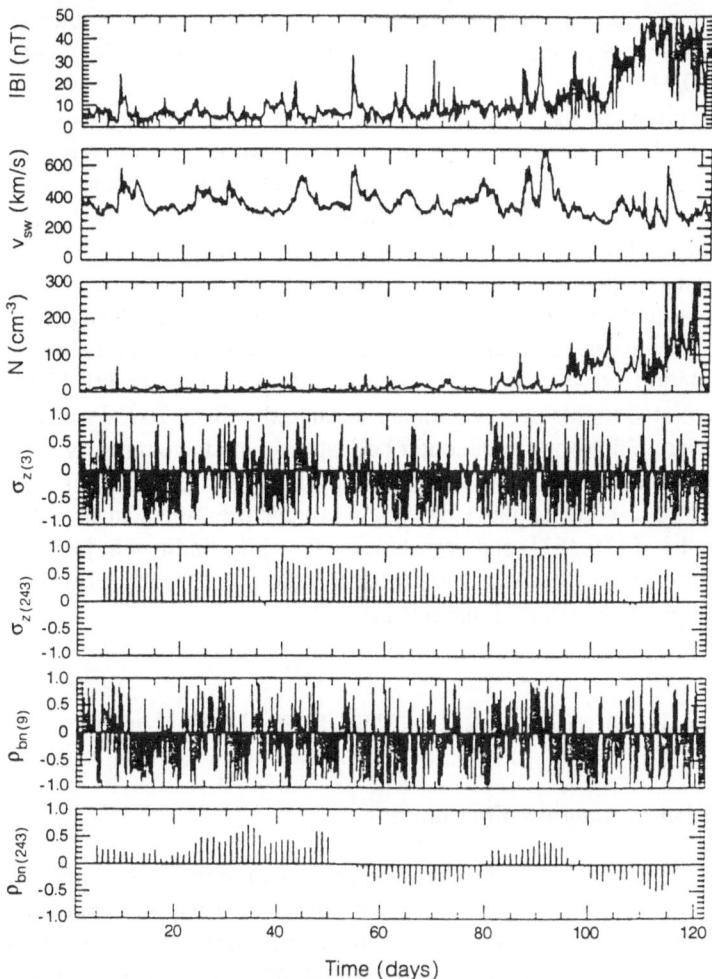

Fig. 10.14. Same time period as in Fig. 10.13. The *third panel* gives the density and the *subsequent panels* show 3- and 243-hour running averages of the normalized cross-correlation σ_z and the correlation coefficient between density and field strength fluctuations [10.152]

The two middle panels show the normalized cross-correlation function between δZ^+ and δZ^- from (10.44a,b). The parameter σ_z is related to the Alfvén ratio r_A in (10.40) and quantifies the relative excess of kinetic over magnetic fluctuation energy. The time period shown is the same as in Fig. 10.13 and corresponds to near- solar-maximum conditions in early 1980. The data used are based on 1-hour averages of *Helios 1* parameters. Several prominent features are apparent. The stream structure comprises intermediate-speed flows (≈ 400 km/s) with intermittent faster flow lasting for a couple of days. Stream kinetic energy dominates magnetic energy of the fluctuations on the ten-day scale ($\sigma_z(243)$); however, the opposite is found on a 3-hour scale, where distinctly negative values of $\sigma_z(3)$ correspond to an Alfvén ratio less than unity, indicating dominance of magnetic energy. Similar behavior is found for the near-minimum period shown in Fig. 10.1, for which a survey plot of the means of σ_z as a function of frequency can be found in [10.71].

Compression regions on the 10-day scale are revealed by a positive ϱ_{bn} in the last panel of Fig. 10.14. However, even here anticorrelations are found during days 55–80 and 110–120, perhaps indicating the presence of PBS. These structures (or fluctuations) dominate the compressional variations seen on the 9-hour scale. Strong anticorrelations were even evident on a 1.8-min scale [10.151] and were shown to exist independently of TDs. High-resolution analysis of PBS in [10.188] with 192-s data strengthen the conclusion that in the inertial range spatial variations of the total pressure are suppressed, i.e. $\delta p_T \approx 0$, which then naturally yields an anticorrelation between magnetic and thermal pressure (see (10.13b)). Theoretical issues related to this observation are discussed in Sect. 10.6. Pressure-balanced structures are ubiquitous in the outer heliosphere and have recently been studied in considerable detail [10.149]. The aforementioned switch from negative to positive values of ϱ_{bn} close to the 1-day scale is very prominent in the *Voyager* data (e.g. Fig. 14 of [10.151]) and corroborates the early findings in [10.61]. On scales larger than 10 days genuine compression regions are clearly visible in association with large-scale stream dynamics [10.34].

10.5 Origin and Evolution of Solar Wind MHD Turbulence

10.5.1 Sources of Interplanetary Fluctuations

Undoubtedly, the ultimate source of interplanetary fluctuations is the sun and its corona. Magnetohydrodynamic fluctuations originate from the turbulent solar atmosphere [10.81, 82] and propagate from there or are convected outward by the expanding wind. Alfvénic fluctuations propagating away from the sun [10.20] prevail in the inner heliosphere and dominate the turbulence at 0.3 AU [10.152]. After the solar wind has passed its Alfvénic critical point, inferred to be located at about $14\,R_\odot$ or beyond [10.111], it is energetically cut off from its sources. For interplanetary turbulence generation it is destined to use up its differential flow kinetic energy associated with the large-scale stream pattern illustrated in Fig. 10.1. The large streams define the energy-containing scales, which of course change as the plasma expands from essentially a point source into the vast interplanetary space. Out at several tens of AU the primordial stream structures are entirely wiped out by coherent large-scale dynamical processes [10.34], which redistribute kinetic energy spatially and replenish and sustain thermal energy above the level for an adiabatic expansion [10.106].

Close to the sun, however, and in the inner heliosphere out to 1 or 2 AU the original streams are well preserved and play the role of "large-scale eddies" that guide and determine the turbulence evolution at smaller scales. The coherent stream structure is permeated by very long-wavelength Alfvénic fluctuations, as evidenced in Fig. 10.12, which may be conceived as being generated by slowly moving the foot points of interplanetary spiral field lines anchored at the bottom of the corona. This low-frequency "ringing" of the spiral field fades away while the "waves" spread over the inhomogeneous stream pattern and lose their energy in multiple interactions with it. This effect is nicely seen in the three bottom panels of Figs. 10.12 and 10.13.

Overlapping in scales with these fluctuations are the mesoscale flow tubes illustrated in Fig. 10.2, which provide a reservoir of kinetic energy in differential flows available for the generations of smaller-scale turbulence. Smoothing out speed differences results in overall stream profiles that appear less noisy at 1 AU than at 0.3 AU, as can be seen in Fig. 10.1. The frequency domain from 10^{-7} to 10^{-4} comprises the energy-containing scales near 0.3 AU. The adjacent domain from 10^{-4} up to 10^{-1} Hz may be identified with the inertial domain, and this is the proper realm of turbulence. Most of the spectral analysis to be discussed below is concerned with this frequency range. As shown below, spectra tend to steepen strongly at even higher frequencies ($> 10^{-1}$ Hz), thus indicating dissipative effects and marking the upper boundary of the inertial range. The picture of fluctuations drawn here combines the wave viewpoint of Belcher and Davis [10.20] with the turbulence concept of Coleman [10.41], which was essentially established more than two decades ago and ever since then has been supported remarkably well by increasingly complex solar wind observations.

Energy contained in the differential flow of macroscale and mesoscale streams is the only major interplanetary source for newly stirring turbulence. This energy is continuously fed into the inertial range, possibly by two major mechanisms. Velocity-shear-driven instabilities in an incompressible magnetofluid are theoretically predicted [10.8, 92] and are numerically demonstrated to be capable of creating energy at scales larger than the exciting ones, but also at much smaller scales owing to purely nonlinear processes [10.154]. In this process an initially Alfvénic shear layer evolves towards a state of overall zero cross-helicity, while both Alfvén modes δZ^+ and δZ^- are excited. A compressive turbulence generation mechanism is related to dynamic equalization of lateral pressure imbalances (of solar origin or in relation to the spiral curvature) between "flow tubes" at all frequencies below about 10^{-4} Hz. That this process works at large scales is amply documented in the solar wind literature (see e.g. [10.34, 141, 162]). Its occurrence at smaller scales is suggested by the increasing appearance of pressure-balanced structures, perhaps as the final outcome of the relaxation of pressure gradients, with increasing heliocentric distance [10.14, 151, 152, 188]. While it is observationally difficult to establish whether these mechanisms are actually at work, there are arguments in favor of the velocity shear as the prevailing generation mechanism within 1 AU.

Roberts et al. [10.151, 152] pointed out that *Voyager* observations in compressive and rarefaction regions did not produce significantly different changes in cross-helicity, thus ruling out large-scale compressive regions as preferred sources of Alfvén modes at mixed helicity. Furthermore, the radial decay of σ_c displayed in Fig. 10.12 and 10.13 seems not to occur preferentially in particular flow regions. Moreover, fading of Alfvénicity (see Sect. 10.5.2) occurs most strongly inside about 0.7 AU, where dynamical effects are less evident. Of course smaller-scale density fluctuations might have an effect, although from the bottom panel of Fig. 10.9 it is hard to believe, because of their low energy content, that these compressive fluctuations have an impact on the cross-helicity in the inertial domain ($f > 10^{-4}$ Hz). We will take up this topic again in a subsequent section.

10.5.2 Radial Evolution of Alfvénicity

The previous discussion has provided ample observational evidence that the Alfvénic turbulence in the inertial domain, notably in the high-speed streams, appears to be predominantly of coronal origin. The sun or its outer corona generates outward-propagating fluctuations over scales covering many orders of magnitude. What is seen as fossils of these primordial fluctuations at 0.3 AU is active turbulence in a dynamical evolutionary state, which develops strongly further as the plasma expands and convects the fluctuations to larger solar distances. We are then witnesses to a rapid decay of the original Alfvénic correlations in the inner heliosphere. This is illustrated in Fig. 10.15 for two separate frequency ranges: "low", 3.9×10^{-4} to 4×10^{-3} Hz, and "high", 4×10^{-3} to 1.2×10^{-2} Hz, corresponding to the high-frequency branches of the spectra shown in Fig. 10.8. The two boxes show percentages of Alfvénic fluctuations, for 1827 coherence and phase spectra (a) and 1138 cross-helicity spectra (b) taken from the *Helios* data

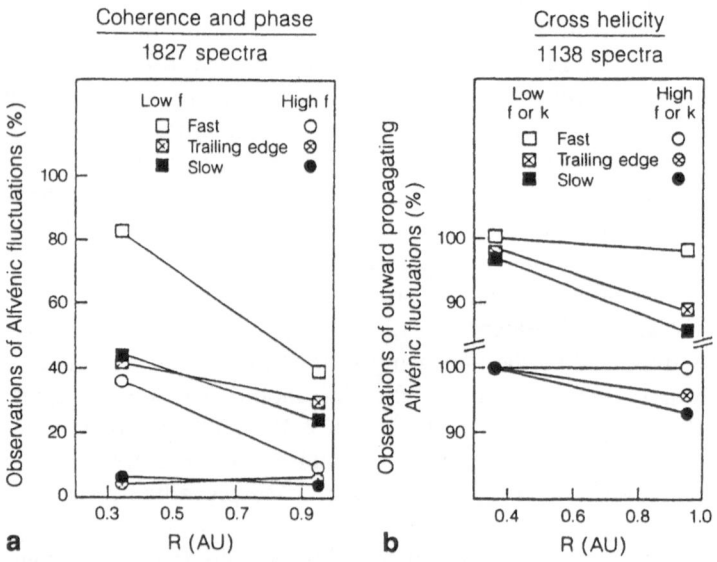

Fig. 10.15. (a) Plot of the percentages of low-frequency (3.9×10^{-4}–4×10^{-3} Hz) and high-frequency (4×10^{-3}–1.2×10^{-2} Hz) fluctuations analyzed (which are predominantly Alfvénic) versus distance from the sun for slow and fast wind and for trailing edges of high-speed streams. (b) Percentages of outward-propagating Alfvénic fluctuations versus heliocentric distance in the same format as in box (a) [10.100]

analysis in [10.100], whereby fluctuations were judged to be Alfvénic if phases ranged within 20° around 0° or 180°, and if the absolute coherence was higher than 0.6. The data were further grouped into fast and slow wind and into flows occurring in the fast-stream trailing edges. The resulting points near 0.35 and 0.95 AU were joined by straight lines to signify radial trends.

From Fig. 10.15(a) it is obvious that the fluctuations at frequencies higher than 4×10^{-3} Hz are considerably less Alfvénic than those below this limit, in general accord with the trends in Fig. 10.10 that indicate that the purest "Alfvén waves" occur near 5×10^{-4} Hz. Clearly the fluctuations tend to become less coherent with increasing radial distance, indicating a rapid decline of Alfvénicity particularly in fast solar wind. As a result, the turbulence in fast streams evolves significantly more between 0.3 and 1 AU than does the turbulence in slow wind, which statistically confirms the conclusions obtained from individual specta in Figs. 10.8 and 10.9. Furthermore, Fig. 10.15(b) shows that those fluctuations which were found to be predominantly Alfvénic nearer the sun radially evolved into a more mixed state. The inward Alfvén modes first appear in slow wind and at low frequencies giving a possible hint of their interplanetary origin. These findings at periods above 42 min are complemented by studies at longer periods, which have established similar evolutionary trends.

Figure 10.16 shows these trends by means of percentage distributions of the normalized cross-helicity σ_c, as calculated in [10.151, 152] from (10.43), on a 3-hour (a), 9-hour (b), and 81-hour (c) scale. The curves are for radial distances at 0.3 AU (circles), 2.0 AU (triangles), and 20 AU (squares), and compare *Helios* and *Voyager* data, thereby revealing the overall large-scale evolution of σ_c. Whereas the circles indicate quite asymmetric distributions, associated with outward "propagating" Alfvén modes near 0.3 AU, the triangles already reveal much less skewed distributions at 2 AU, and ultimately at 20 AU fairly symmetric, though broad, percentage distributions are measured. Apparently, the signature of primordial fluctuations having originated from the sun gradually fades away, and as the wind reaches the outer heliosphere, no memory of them is left, just as there remains little memory of the original large-scale stream structure (see again Fig. 10.1) [10.34].

It occasionally happens that large solar wind streams do not undergo strong interactions with their neighboring streams, in which case they may conduct the smaller-scale turbulence relatively unchanged even out to 8 AU. Figure 7(a,b) in [10.152] presents such a case where outward fluctuations observed by *Voyager 1* are seemingly remnants of solar-generated fluctuations that passed by *Helios 1* earlier on their way into the outer heliosphere. According to [10.198] this plasma region had weakly interacted with ambient material, thus providing an almost laterally undisturbed stream undergoing merely radial evolution.

The radial evolution of solar wind fluctuations at scales beyond 1 hour in the inner heliosphere close to the activity minimum of cycle 21 has been statistically investigated by Bavassano and Bruno [10.14]. This study provides complementary evidence on the evolution of Alfvénicity to the single case studies shown in Figs. 10.8–10 and the subsequent section. It was found that at scales from 1

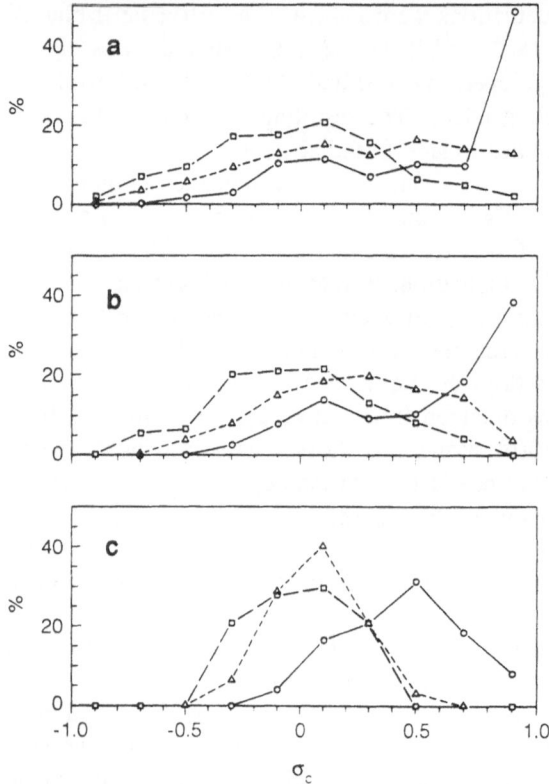

Fig. 10.16a–c. Evolution of normalized cross-helicity from 0.3 to 20 AU as given by percentage distributions from both *Helios* and *Voyager* spacecraft. Curves shown are for 0.3 AU (*circles*), 2 AU (*triangles*), and 20 AU (*squares*), and refer to 3-hour (a), 9-hour (b), and 81-hour (c) running averages [10.152]

to 12 hours Alfvénic fluctuations prevailed in all regions investigated and under all solar wind conditions. The highest degree of Alfvénicity was found in fast streams near the sun. At scales from 0.5 to 3 days a much more mixed state was found in which both types of Alfvén mode or non-Alfvénic fluctuation were present. The central parts of fast streams revealed the purest "Alfvén waves" (see also Fig. 10.9), thus indicating that those regions where compressive and velocity shear effects are weakest have best preserved the Alfvénic fluctuations originating from the sun.

In Sect. 10.3.2 we addressed the polarization state of the fluctuations and provided evidence for the existence of sizable fluctuation components along the mean magnetic field. Figure 10.17 presents the results of an analysis of the relative components of 6-hour running averages of magnetic field fluctuations as functions of Alfvénicity and radial distance after [10.14]. Light lines refer to mixed, and heavy lines to pure, Alfvénic fluctuations. In the mixed or non-Alfvénic state relative $\delta B/B$ and $\delta n/n$ fluctuations larger than 0.1 were allowed. Correlation coefficients for at least two components had to be bigger than 0.8

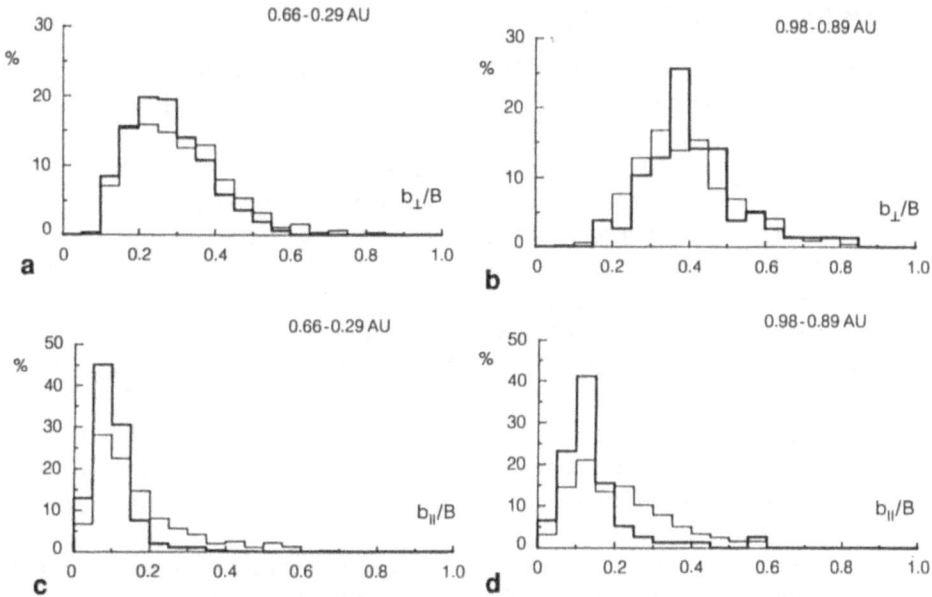

Fig. 10.17. Relative amplitude of the parallel and perpendicular component fluctuations of the magnetic field with respect to the mean field. Parts (a) and (c) refer to a solar rotaion close to the sun, and (b) and (d) to a rotation as observed close to earth orbit. Data are classified according to Alfvénic or non-Alfvénic (*light lines*) periods [10.14]

and compressive effects below the 0.1 threshold for fluctuations to qualify as Alfvénic. The right-hand panels refer to 0.98–0.89 AU and the left-hand panels to 0.29–0.66 AU.

It is apparent that, independent of the Alfvénicity, the fluctuations are predominantly transverse in all cases with $\delta B_\perp$ generally larger than $\delta B_{\parallel}$ by about a factor of 3, corresponding to about one order of magnitude in the power level. Shifts of the maxima of the histograms indicate a slight increase in relative amplitudes, which is due to the fact noticed in [10.27] that at an hourly scale fluctuation amplitudes radially decrease less strongly than the background field. In the non-Alfvénic regions (Fig. 10.17(c,d)) the amplitude $\delta B_{\parallel}/B$ clearly shows larger values than in the Alfvénic intervals, indicating reduced transversality with decreasing cross-helicity and increasing radial distance. This trend is also apparent in detailed individual spectra of component fluctuations in Elsässer variables and their ratios as evidenced in Figs. 13 and 14 of [10.114], which show that, as the turbulence radially evolves or as it transits to a regime of lower Alfvénicity, the transverse character of the fluctuations tends to disappear and the turbulence becomes more isotropic with respect to the three vector components. In a statistical analysis of magnetic field fluctuations Bavassano et al. [10.10] found opposite trends indicating that the degree of anisotropy at high frequencies increases with distance from the sun. This unresolved issue needs further thorough investigation.

10.5.3 Radial Evolution of Spectra

One major issue of solar wind MHD turbulence studies has been the question of how the fluctuations and their spectra radially evolve. Several papers using *Pioneer 10* and *11* data and *Mariner 10* data [10.25] addressed this question in the 1970s and considered how different features of the fluctuations varied with distance from the sun. These results were reviewed by Behannon [10.18]. Notably, the *Helios* and *Voyager* missions allowed us to carry out investigations of the radial gradients of spectral characteristics of the turbulence by means of individual case studies and statistical analyses. Here the emphasis is placed on the distance range between 0.3 and 1 AU corresponding to the *Helios* orbits. Several studies were performed in the early 1980s on magnetic field component and magnitude power spectra [10.10, 11, 13, 17, 43, 46–48, 100, 193]. It was established that their slopes changed significantly, especially between 0.3 and 0.4 AU. Inside 0.4 AU the spectra were generally flatter, particularly in high-speed streams. Also, differences in spectral slope as a function of flow speed were established.

The results of a statistical study by Denskat and Neubauer [10.46, 48] are presented in Fig. 10.18, which shows distributions of the best-fit exponent α for the

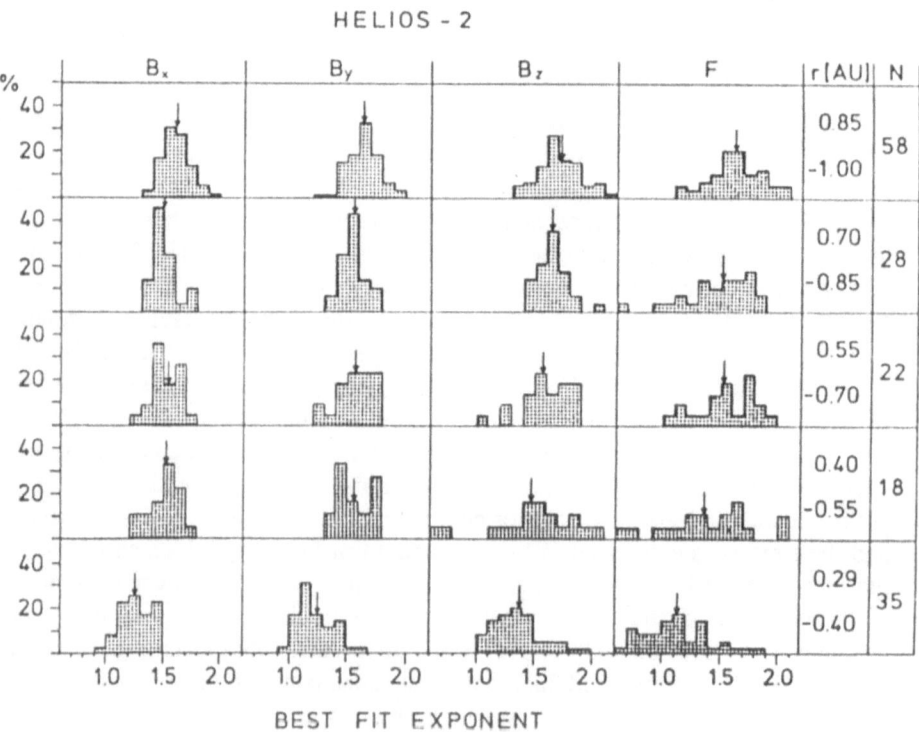

Fig. 10.18. Distributions of best-fit exponents for magnetic field component and intensity power spectra for five heliocentric distance ranges. N gives the number of spectra computed in each distance interval, the *arrows* mark the mean values [10.48]

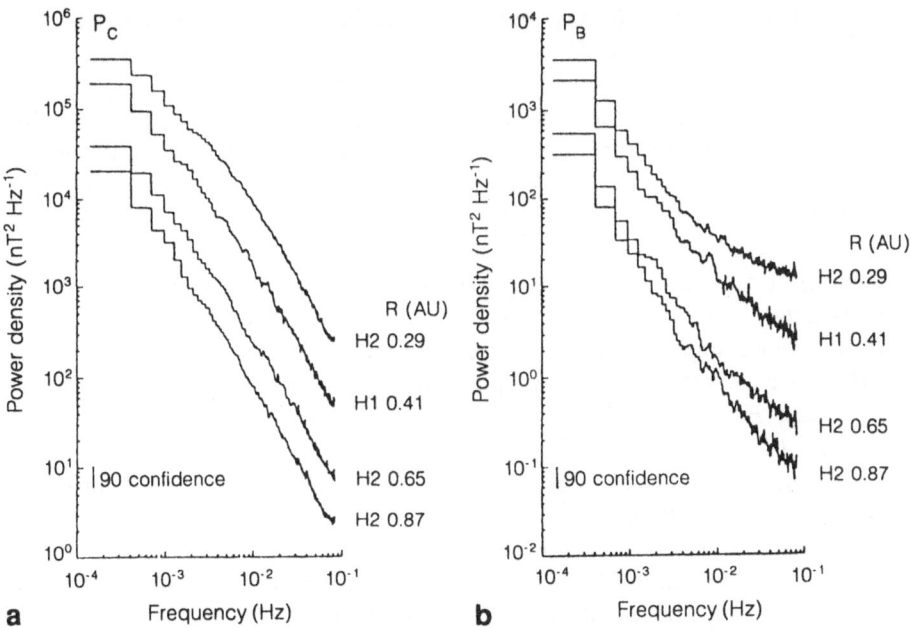

Fig. 10.19a,b. Average spectral densities for four heliocentric distances. The total power (trace of spectral matrix) P_C is shown in (a) and the power P_B in magnitude fluctuations in (b). Note that all curves steepen radially and never cross each other [10.11]

magnetic field power spectra, $P \propto f^{-\alpha}$, for five heliocentric distance intervals. N gives the number of spectra computed (in the range 2.4×10^{-5} to 1.2×10^{-2} Hz), and the arrows indicate the mean values. The index α was found to vary between 1.59 and 1.69 at 0.97 AU and between 0.87 and 1.15 at 0.29 AU with a typical uncertainty of about 10%. The authors emphasized that simple power laws did not fit their spectra well near 0.3 AU, where a distinct flattening with $\alpha \approx 1$ was found below some 10^{-3} Hz. Matthaeus and Goldstein [10.122] also established a $1/f$ spectrum in the range 2.7×10^{-6}–8.0×10^{-5} Hz from selected observations at 1 AU. By and large, however, the spectral index of -1 could be applied only approximately to this frequency regime, because it overlaps with the domain of the structure discussed in Sect. 10.2.2.

Estimation of spectral indices was further improved by considering lineup configurations of the *Helios 1* and 2 spacecraft in order to determine the genuine radial evolution of turbulence within a recurrent stream encountered several times at increasing radial distances. Bavassano et al. [10.11] corroborated the trends shown in Fig. 10.18 and were able to extend the frequency range up to 10^{-1} Hz. Their results are displayed in Fig. 10.19, showing the average smoothed spectral density P_C, evaluated from the trace of the correlation matrix, and the power P_B in field magnitude. Several features are noticeable. At 0.29 AU the aforementioned flattening in the low-frequency region is visible, which fades away with radial distance, while the spectra straighten and are increasingly better fitted by single power laws. The radial decline in intensity of P_C and P_B is obvious. Papers [10.11, 46, 48] give the radial intensity variations in different frequency bands. Whereas P_C revealed a convex shape, P_B showed a distinctly concave shape, indicating a significant flattening between 10^{-2} and 10^{-1} Hz. This signature tends to disappear with larger solar distance. Clearly the compressible turbulence evolves quite differently than the incompressible Alfvénic fluctuations.

The spectral domain was also divided into four ranges, for each of which power-law indices were separately evaluated. They are given in Table 2 of [10.11], which contains spectral indices for each field component. Generally speaking, the radial evolution found could be described as leading to Kolmogorov-type spectra for P_C with index $-\frac{5}{3}$ at 1 AU within measurement uncertainties. The turbulence appeared to be strongly evolving on its way out from 0.3 to 1 AU with a distinct spectral reshaping and radial intensity decline. If the radial scaling $P_C \propto R^{-\beta}$ was assumed, the index turned out to be $\beta = 3.2$ for $f \leq 2.5 \times 10^{-3}$ Hz and $\beta = 4.1 - 4.2$ beyond 10^{-3} Hz, which is compatible with a radial dissipation at an e-folding length of about 1 AU. Similar results on radial dependences were obtained in [10.43, 46–48].

Whereas spectral information about magnetic fluctuations is abundant, similar knowledge of the simultaneous plasma velocity fluctuations in the inner heliosphere became available only quite recently [10.100]. The radial evolution of MHD turbulence between 0.3 and 1 AU was reanalyzed and newly evaluated in a series of papers [10.71, 72, 114, 184, 185, 187] exploiting the Elsässer variables (10.8), which incorporate the wind velocity by definition. Their spectral intensities have been defined and discussed in Sect. 10.3.5, and typical features of spectra were presented in Sect. 10.4.1. The radial evolution of $e^{\pm}$ spectra is illustrated in Fig. 10.20, taken from [10.114], which presents high-speed (right-hand side) and low-speed data for three radial distances as indicated. The time periods are the same as for the recurrent fast stream of Fig. 10.19. Analyzing the same stream repeatedly comes closest to the ideal of studying the true radial evolution within the same plasma fluid element. Certainly, it is even better to study the plasma in a radial line-up [10.160, 163] configuration. However, such favorable conditions are too rare. The low-speed periods shown in Fig. 10.19 are related to the neighboring plasma of the fast streams.

The prevailing features of this figure are the following. All $e^{\pm}$ spectra tend to steepen with heliocentric distance. The most prominent signature is the "rhomboid" shape of e^+ (right side) in perihelion near 0.3 AU, which tends to fade away at larger distances. Generally, the spectra e^+ and e^- (dotted lines) approach each other with increasing distance with a tendency towards equipartition, which is most clearly revealed in solar wind at 0.9 AU (bottom-left frame). The spectral slopes evolve radially, and at the same time an extended inertial regime expands into the low-frequency domain, with ultimate spectral slopes close to the famous $-\frac{5}{3}$ law [10.91]. The distinct plateau in e^- in the top-right frame is gradually bent down and disappears towards 1 AU. The concurrent evolution of the Elsässer ratio r_E and cross-helicity σ_c is illustrated in Figs. 3 and 4 of [10.114], which support the overall trends for the radial evolution of Alfvénicity established in Sect. 10.5.2, namely that the fluctuations increasingly lose their

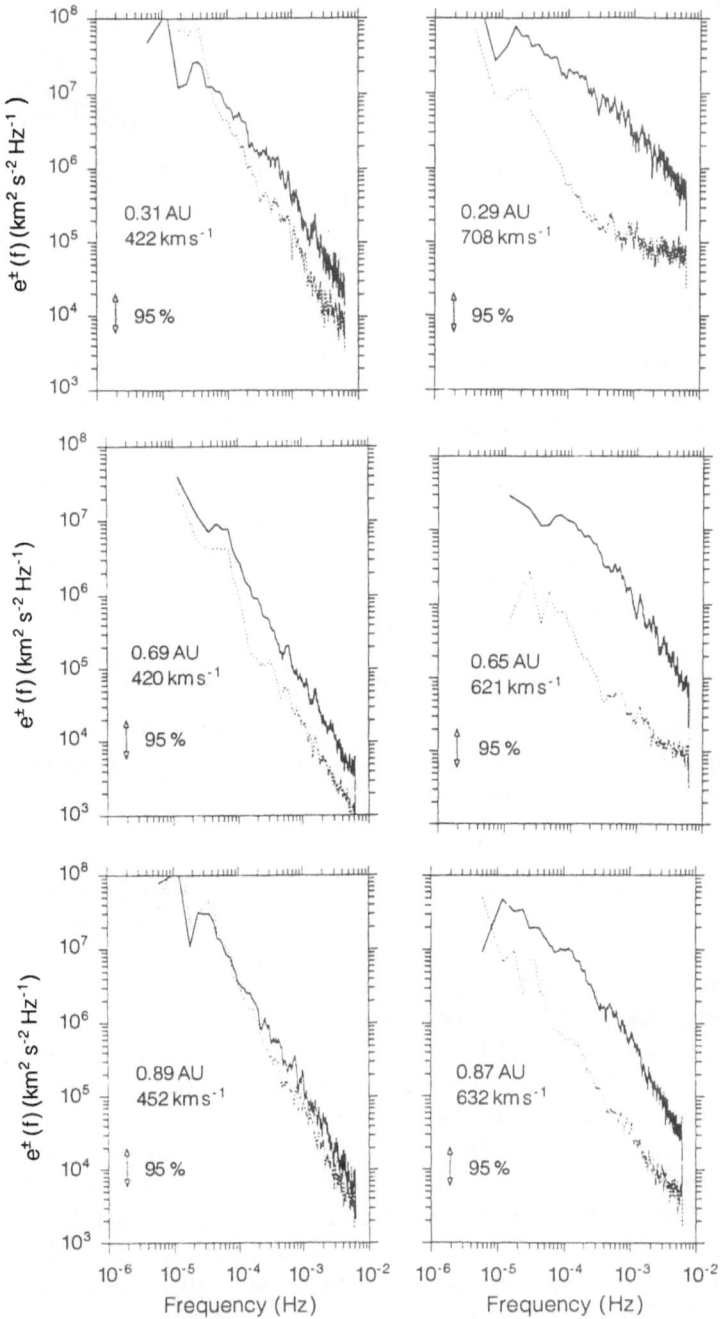

Fig. 10.20. Power spectra $e^{\pm}$ of the Elsässer variables for those heliocentric distances as indicated and for high-speed (*right column*) and low-speed wind. The e^+ spectra are given by continuous traces [10.114]

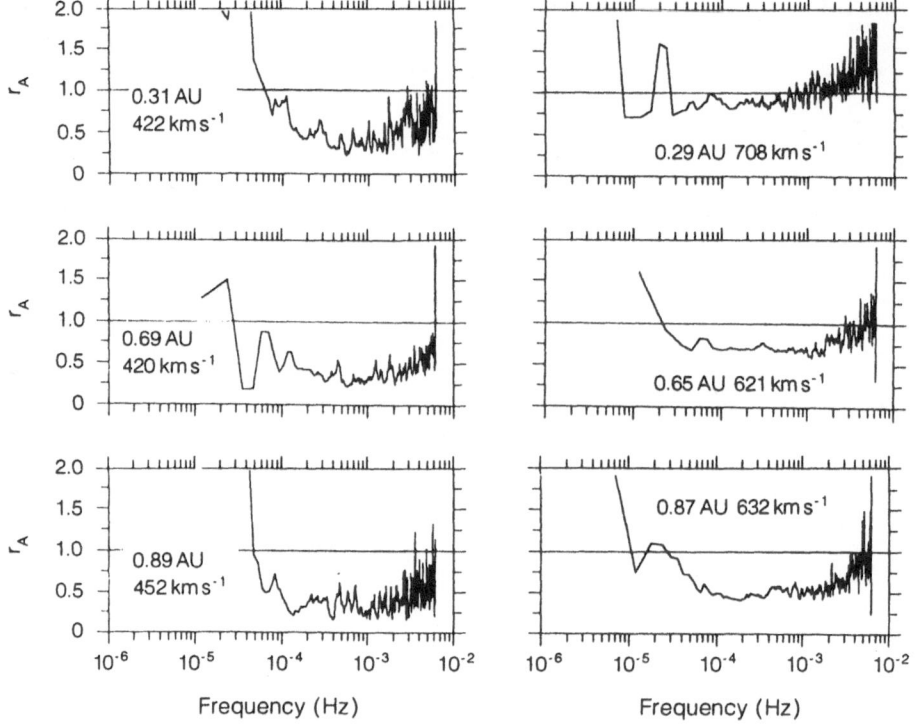

Fig. 10.21. The spectra of the Alfvén ratio $r_A(f)$ in the same format as in Fig. 10.20. Note the radial decline of r_A in the intermediate frequency range between 10^{-4} and 10^{-3} Hz [10.114]

Alfvénic correlations, and, while proceeding through intermediate evolutionary stages, they finally end up in fully developed turbulence with an equipartition in energy between the δZ^+ and δZ^- Alfvén mode.

During this evolution a significant spectral redistribution of turbulent kinetic and magnetic energy takes place. It has been established in [10.114] that the cross-correlation spectrum $|e^R|$ in (10.31) evolves radially towards an extended inertial range (note that $e^R(f)$ itself is a rapidly oscillating function of frequency) with a final slope close to that of the final $e^\pm(f)$ spectra. Another way to illustrate this behavior is by means of the Alfvén ratio (10.32), which provides equivalent spectral information. For the evolutionary sequence of $e^\pm$ spectra of Fig. 10.20 the corresponding Alfvén ratio is shown in Fig. 10.21 in a similar format. Recall that e^R measures the difference between kinetic and magnetic energy, whilst r_A measures their ratio. Furthermore, notice that for a pure Alfvén wave, from (10.16a,b), one finds $e^R = 0$ and $r_A = 1$.

The following features are characteristic of both high- and low-speed flows. Kinetic energy dominates in the low-frequency range, below about 2×10^{-5} Hz corresponding to mesoscale stream structures, whereas magnetic energy is preponderant in the Alfvénic domain with a distinct tendency towards $r_A \approx 1$ at higher frequencies. For the fast stream at 0.29 AU (upper-right box) one finds

204

$r_A \approx 1$ over a broad band in frequency, in accord with the predominance of e^+ and a high degree of Alfvénicity. Generally, $r_A(f)$ tends to decline with radial distance and to approach a value near 0.5. This evolutionary tendency is furthest developed in slow-wind spectra which also reveal fully developed inertial ranges. Similar findings have been established for the outer heliosphere [10.119, 151], although the radial trends are less clear and variations in r_A are larger. An excess of magnetic over kinetic energy has been predicted for developed homogeneous MHD turbulence in numerical computations by Pouquet et al. [10.143].

We now concentrate on the dependence of the spectral indices for $e^\pm$ on the solar distance and wind flow speed. The results are compiled in Fig. 10.22, discriminating between the two frequency bands between 5×10^{-5} and 3×10^{-4} Hz (upper panels) and between 5×10^{-4} and 2×10^{-3} Hz (lower panels). The indices for 22 individual spectra (based on 48-hour-long data records) are explicitly given [10.114, 185]. Respectively, two regression lines are shown for speeds below and above 550 km/s taken as the dividing line between fast streams and low to intermediate-speed plasma flows. These lines and points mainly pertain to distances between 0.72 and 0.98 AU, with an average of 0.87 AU. Correspondingly, the dashed regression lines represent the fits through other points which are not shown individually but are cursorily represented by the gray-shaded clouds. These point clouds belong to the distance range between 0.29 and 0.5 AU, with an average of 0.35 AU.

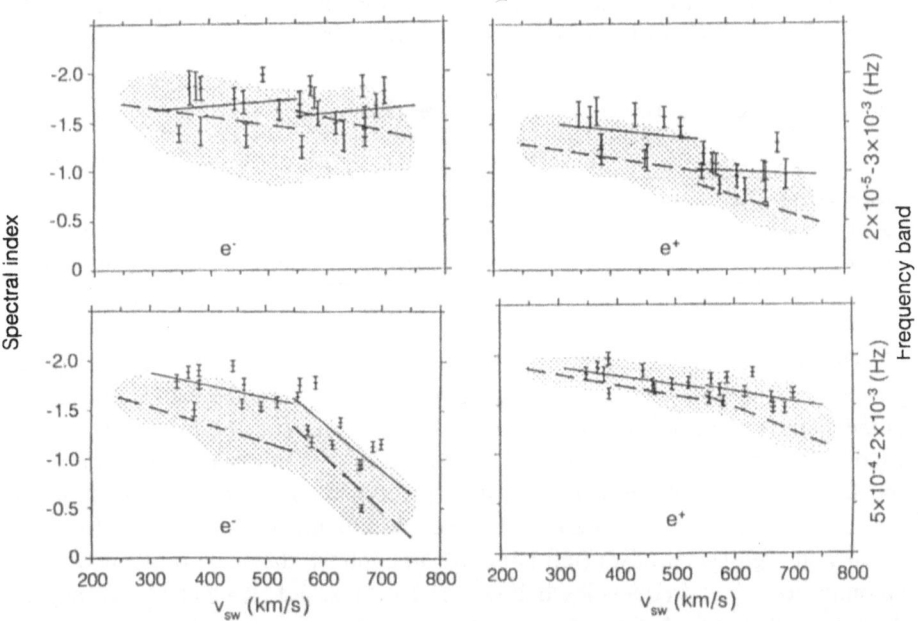

Fig. 10.22. Spectral indices, based on 22 individual spectra, for two different frequency bands as indicated. The indices for $e^\pm$ are given as a function of solar wind flow speed and refer to about 0.35 AU (*broken lines* and *shaded area*) and 0.87 AU. The lines represent least-squares-fits through the measured points [10.114]

As a result, one can infer from Fig. 10.22 the radial trends of the spectral indices in each frequency band and convince oneself of the overall trend towards steepening of the spectra with increasing heliocentric distance. The low-frequency e^- spectrum resists any systematic changes with flow speed and radial distance and exhibits a mean spectral index (represented by the regression lines) which is somewhat above $-\frac{3}{2}$ and close to $-\frac{5}{3}$. This part of the spectrum was first described in [10.71, 72, 184] and then was established as a kind of "universal background" spectrum [10.185]. It appears to be invariant also against solar-cycle variations.

The high-frequency part of the e^+ spectrum also shows an almost stationary index with a mean above 1.5 and a tendency towards smaller values at flow speeds beyond 600 km/s. Most strongly evolving are the low-frequency part of e^+ and the high-frequency part of the e^- spectrum. For these two spectral branches we see a definite steepening with heliocentric distance. Furthermore, there is a whitening in these spectral domains with increasing flow speed. These spectral features are also quite obvious in the evolutionary trends in Fig. 10.20.

We may emphasize that the results in terms of Elsässer variables are consistent with the previous results of references [10.10, 11, 43, 46, 48], using only B-spectra in their high-speed-stream analysis, since e^+ can be considered nearly equivalent to the trace of the magnetic field correlation matrix for predominantly Alfvénic periods. Note, however, that Fig. 10.22 provides further new results in that it exhibits complementary spectral features of the minority wave species δZ^-, which is theoretically assumed to play an essential role in the spectral and radial evolution revealed in Fig. 10.20, if we follow the arguments given in [10.52, 53, 68, 69, 70, 113, 120, 178, 181, 187, 189, 190].

Table 10.2. Average radial gradients of spectral indices (R in AU)

Frequency range [Hz]	$V_{sw} \geq 550 \, \text{km s}^{-1}$		$V_{sw} < 550 \, \text{km s}^{-1}$	
	e^+	e^-	e^+	e^-
2×10^{-5}–3×10^{-4}	$1.04 + 0.56(R-1)$	$1.60 + 0.20(R-1)$	$1.50 + 0.60(R-1)$	$1.66 + 0.08(R-1)$
5×10^{-4}–2×10^{-3}	$1.64 + 0.52(R-1)$	$1.35 + 0.90(R-1)$	$1.70 + 0.12(R-1)$	$1.83 + 0.76(R-1)$

The average radial trends of the spectral indices of $e^\pm$, according to a simple power law $e^\pm(f) \propto f^{-\alpha}$, are given in Table 10.2, in which the data are classified in the same way as in Fig. 10.22. This table has been extracted from Figs. 11 and 18 of Marsch and Tu [10.114], which reveal a wide spread of the individual spectral indices about the regression lines given here. The most significant radial variation occurs at lower frequencies for e^+ and higher frequencies for e^-, notably in high-speed flows (left column). In contrast, the slope of e^- invariably amounts to $-\frac{5}{3}$ at frequencies below 3×10^{-4} Hz, independent of V_{sw} and R. Also note that the e^+ slope at low frequencies is close to -1, as emphasized earlier [10.46, 122]. In conclusion, there appears to be an overall trend towards radial steepening of $e^\pm$ spectra. Single power laws do not fit the data well.

10.5.4 Self-similarity and Origin of Turbulence

The preceding sections have provided ample evidence that solar wind MHD turbulence is radially and spectrally evolving. The observed spectra cover many orders of magnitude in frequency, and there exist distinctly different spectral domains such that a single power law is not appropriate. By decomposing the turbulent energy into the two components of inward and outward Alfvén modes, the existence of separate nonlinear cascades for e^+ and e^- was established, in compliance with $e^{\pm}$ being independent inviscid invariants of ideal MHD (see again Sect. 10.3.4). Figure 10.22 reveals a significant dependence of spectral characteristics upon the solar wind speed for the near-solar-minimum epoch illustrated in Fig. 10.1. Also, the Alfvénicity of the turbulence seemed to vary strongly with the stream structure, thus suggesting two different kinds of turbulence related to the two states of solar wind flow [10.161–163]. Yet there appears to be an underlying common pattern in the turbulence.

In a recent paper Grappin, Mangeney, and Marsch [10.72] showed that the turbulent energy spectra indeed undergo large daily variations in shape and intensity, but despite these variations the outward component e^+ correlates distinctly with the daily rms-value of the proton thermal speed, and the inward (low-frequency) component e^- closely follows the relative density fluctuations $\Delta n/n$ averaged over a day. This is shown in Fig. 10.23, which displays the above quantities (as daily means) for the first 120 days of the *Helios* mission. The spectral intensities $e^{\pm}$ are given for nine frequency bands with $f_m = f_0 2^m/\sqrt{2}$ ($m > 1$) and $f_0 = f_1 = 1/$day. The proton thermal speed c_S is defined as usual as $c_S = (k_B T_p/m_p)^{1/2}$, with the proton temperature T_p, the mass m_p, and Boltzmann's constant k_B. Variations of up to two orders of magnitude are observed on the stream-structure time scale (one to several days). Note that c_S itself strongly correlates with the stream pattern [10.109, 157] not shown here. Visual inspection of the stacked spectra in Fig. 10.23 reveals two types of concurrent patterns: spikes in e^- and $\Delta n/n$ and saw-tooth-like variations in e^+ and c_S. Spikes prevail in the e^- traces at low frequencies and saw-teeth in e^+ at high frequencies.

Another striking feature is the "self-similarity" of the $e^{\pm}$ variations, evident in the coherent "ups and downs" in the time series of $e^{\pm}(f_m)$. As the traces in this logarithmic plot are not equidistant, the existence of a permanent power law in the spectra is excluded in accord with the results of Sect. 10.5.3. This self-similarity property was quantitatively corroborated in [10.72] by means of a correlation analysis. For the restricted range between 10^{-4} and 6×10^{-3} Hz (bands 5 to 9) an "inertial range" could be established (in accord with Figs. 10.8, 10.20 and 10.22). The spectra were found to be well represented by power-law dependences on the parameters c_S and $\Delta n/n$ given by the formula

$$e(f_m) \propto c_S^{\alpha_m} (\Delta n/n)^{\beta_m} ,$$

(10.47)

with different indices α_m and β_m for each frequency band f_m. With such fits Grappin et al. [10.72] were able to reproduce the observed time series of $e^+(f_m)$ in great detail (see their Figs. 6(a,b) and 8). However, α and β were clearly

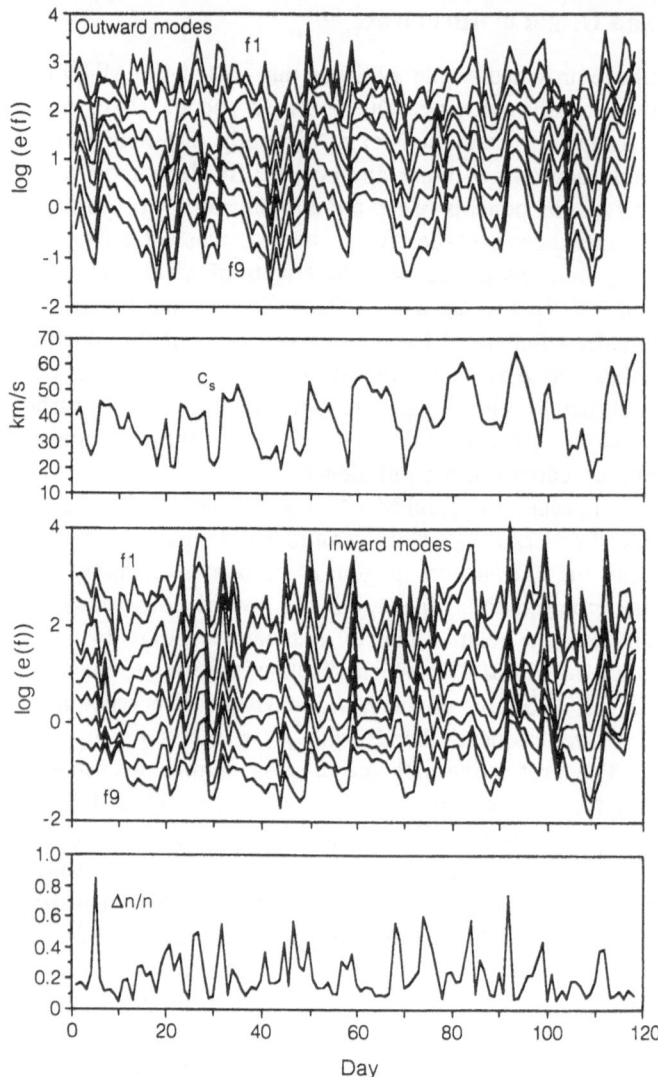

Fig. 10.23. Daily fluctuations of spectral densities and their association with the stream structure. *Top:* turbulent energy of the outward component in nine frequency bands (see text) and the proton thermal speed. *Bottom:* turbulent energies of the inward Alfvén modes and, below, the relative rms density fluctuations. The energies are normalized to $1\,\mathrm{km^2\,s^{-2}\,day} = 8.64 \times 10^4\ (\mathrm{km^2\,s^{-2}\,Hz^{-1}})$, and the abscissa is in units of days [10.72]

found to vary with frequency, thereby precluding time-invariant power laws for the entire spectra or unique global slopes.

The spectral dependence of the fit parameters in (10.47) is displayed in Fig. 10.24, showing α (left) and β (right) versus frequency normalized to $f_1 = 1.6 \times 10^{-5}\,\mathrm{Hz}$. Note that the notation $e^{\mathrm{in,out}}$ corresponds to $e^{-,+}$ for an inward-oriented mean magnetic field. Clearly $\beta \approx 1$ for e^- in the first five bands, implying that $e^- \propto \Delta n/n$. This suggests that low-frequency inward

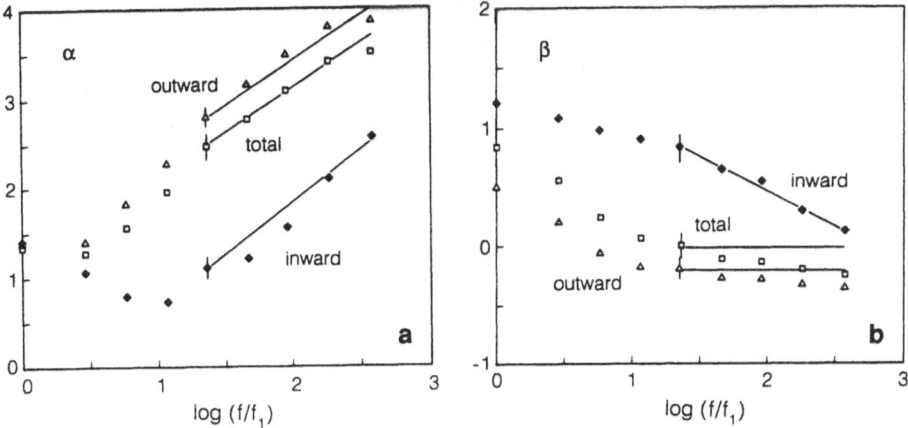

Fig. 10.24a, b. Frequency variations of the power indices for the fit (10.47) to the three energy spectra e, e^+, and e^-, obtained by logarithmically fitting the dependence of turbulent energies on $\Delta n/n$ and on c_S. For more details see [10.72]

Alfvén modes are generated in large-scale regions with enhanced relative density fluctuations. This dependence weakens towards higher frequencies. For e^+ the index β is slightly negative, and for e it is compatible with $\beta \approx 0$, thus indicating that the small-scale turbulence is generally insensitive to large-scale density variations, which is consistent with the results in the bottom panel of Fig. 10.9. Except for these two regimes β varies with frequency, and quite clearly so does α (left panel). Beyond band 5 both indices vary almost linearly with f, whereby α increases strongly, emphasizing the high sensitivity of the spectral shapes to the proton temperature.

In conclusion, while turbulent energy spectra vary strongly from day to day, in shape as well as in intensity, these variations are largely correlated with the proton thermal speed characterizing the internal energy state of the wind (see also the paper by Feldman et al. [10.57]), and with the degree of compressibility, notably for the inward mode. Consequently, the systematic variations of spectral characteristic with stream structure and radial distance are intimately linked with c_S and $\Delta n/n$, which, as may be seen in Fig. 10.24(a,b), essentially shape the high-frequency parts of the spectra. Given the radial proton temperature profiles in [10.58, 59, 109, 157, 161], one can then transform the radial dependence of c_S into a radial dependence of the spectral index by means of relation (10.47). For instance, one finds that the slope of e^+ varies from -1.8 to -1.2 as c_S varies from 16 to 63 km/s, corresponding to increasing flow speed [10.109], a result in agreement with the course of e^+ in the bottom-right frame of Fig. 10.22.

That interplanetary fluctuations, particularly the inward Alfvén mode, can originate from compression regions has also been established in individual case studies [10.15]. Figure 10.25 shows one example of a significant decline in cross-helicity σ_c below 4×10^{-5} Hz coinciding with a steep rise in the compression quantified by the ratio of P_B and P_C, i.e. of power in δB compared to the power evaluated by the trace of the correlation matrix (thin line: minimum variance

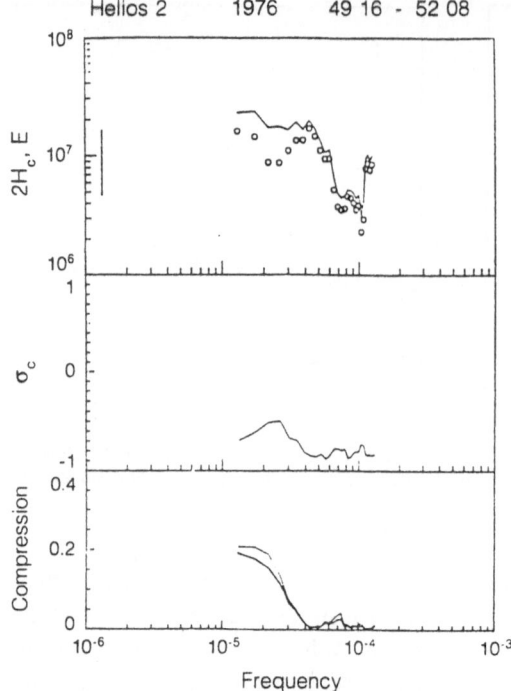

Fig. 10.25. Energy and cross-helicity, and σ_c and the compression ratio (see text) for a time interval during which the increase of compressive fluctuations below 4×10^{-5} Hz is seemingly causing a coincident decline in Alfvénicity. The *thin line* in the *bottom panel* gives results which do not account for the minimum variance component of the field fluctuations [10.15]

component excluded). Obviously, the decline in Alfvénicity can be accompanied by a sizable relative increase of the compressive turbulent energy, which confirms the notion of Fig. 10.23, that large-scale density variations seem to generate e^- and to destroy Alfvénic correlations and thereby cause a decay of σ_c. However, this does not conclusively establish compression as a cause; it could also be a simultaneous effect and dynamically secondary.

We conclude this section by addressing the spectral features of the density fluctuations themselves near solar minimum. Figure 10.26 shows the averaged spectrum of $\delta n/n$ versus reduced wave number k^* for high-speed flows at about 0.4 AU (continuous line) and 0.8 AU (long-dashed line), and for low-speed flows at about 0.4 AU (short-dashed line) and 0.9 AU (chain line). A reference line with a slope of $-\frac{5}{3}$ is drawn to guide the eye. Concerning the details of the spectral analysis see [10.115]. The k^*-values ranging between 10^{-8} and 10^{-5} km^{-1} correspond to spatial scales or wavelengths of 100 down to 0.1 million km.

In the previous sections we have analyzed frequency spectra as a function of flow speed. Since flow-speed dependence transforms into a heliomagnetic-latitude dependence (see Sect. 10.4.2), one has to go into wave-number space if he intends to study spectral features as a function of spatial structures and in absolute scales. To establish the averaged spectrum, 39 single spectra always cov-

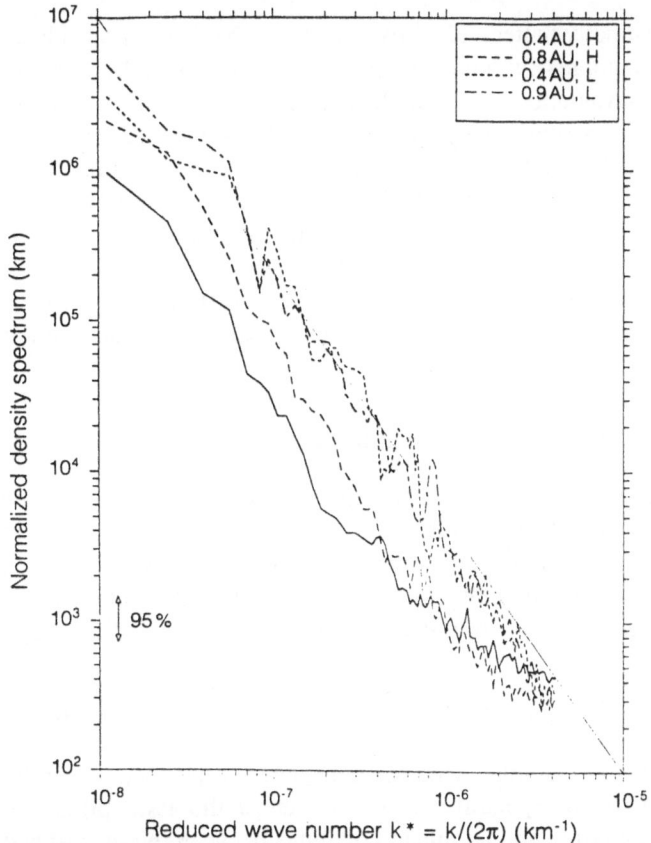

Fig. 10.26. Power spectrum of relative density fluctuations versus reduced wave number for four heliocentric distances as indicated and for high-speed (H) and low-speed (L) solar wind [10.115]

ering two days were used, which were smoothed and added binwise in k^*-space. The spectra have the dimension km, therefore the relative amplitude $\delta n/n$ at a comparable scale is obtained by multiplication with k^* and by taking the square root. Correspondingly, we obtain for $\delta n/n$ a few percent at $k^* = 10^{-6}$ km^{-1} and typically 10% at $k^* = 10^{-8}$ km^{-1}. Sizable fractional amplitude variations are thus obtained at a daily scale as shown already in Fig. 10.23.

A comparison of the high- and low-speed spectra shows striking differences in the slope and intensity of $\delta n/n$ spectra. The slow wind is more compressive at large scales, notably near 0.3–0.4 AU. Its spectrum appears to be radially invariant with an overall slope slightly above $-\frac{5}{3}$ as indicated by the reference line. However, close inspection reveals that two spectral domains are detectable (as in Figs. 10.8 and 10.22) with a break at about 3×10^{-7} km^{-1}. The small-k^* part is more aptly fitted by a -2 slope, while the large-k^* part nearly obeys a $-\frac{3}{2}$ law according to the results of Fig. 4 in [10.115]. By and large the two spectra overlap fairly well at 0.4 and 0.8 AU with no significant radial trend.

On the contrary, in fast streams the relative density fluctuations below 3×10^{-7} km^{-1} clearly increase in amplitude during the radial expansion. At small scales the spectra are distinctly flatter and partly overlap, with a flatter slope closer to the sun. As a result, compressive fluctuations in fast streams radially increase in amplitude (see also Bavassano et al. [10.10]) and reveal spectral evolution at small scales, whereas in slow wind the compressive MHD turbulence resembles fully developed fluid turbulence. The spectrum below 3×10^{-7} km^{-1} in fast wind nearly exhibits a $(k^*)^{-2}$ law, reminiscent of the spectra discussed in Fig. 10.4, and is therefore most likely associated with discontinuous structures and jumps as discussed extensively in Sect. 10.2.2. The spectra of $\delta B/B$ (see [10.115]) not shown here reveal similar radial trends, yet with significant differences in slopes, notably at large scales, where an almost invariant slope near $-\frac{3}{2}$ is found on the average.

10.5.5 Evidence for the Dissipation of Turbulence

The radial evolution of solar wind turbulence spectra and amplitudes depends on frequency, for which ample observational evidence has been provided in the previous sections. According to geometrical optics or WKB theory [10.2, 3, 76, 77, 136, 195, 196], the magnetic field amplitude of an undamped, incompressible Alfvén wave propagating away from the sun obeys the relation valid for any frequency

$$\langle (\delta B)^2 \rangle / \sqrt{\varrho} (M_{Ar} + 1)^2 = \text{const}. \tag{10.48}$$

Non-WKB Alfvén waves in the solar wind are considered in [10.74]. A similar relation can be derived for the spectral densities $e^{\pm}$ under the assumption that cross-correlations are negligible and nonlinear couplings between the inward and outward Alfvén modes can be disregarded. Under these assumptions one finds [10.113]

$$e^{\pm} \sqrt{\varrho} (M_{Ar} \mp 1)^2 = \text{const}, \tag{10.49}$$

where $M_{Ar} = V_r/V_{Ar}$ denotes the radial Alfvénic Mach-number. Of course, (10.48) is retained from (10.49) for a pure Alfvén wave with $\delta Z^+ = 0$ and an outward-oriented magnetic field. Therefore, if the waves or turbulence were evolving without dissipation, the observed fluctuation amplitude at any scale should follow the above theoretical constraints.

Several studies have been carried out to test the validity of (10.48) [10.11, 46, 48, 148, 153, 192, 193, 196] and of (10.49) [10.114] with the general conclusion that WKB seems to hold fairly well at time scales larger than a few hours but that it fails at frequences higher than 10^{-4} Hz, where the turbulence tends to decay much faster than according to (10.48, 49), thus indicating sizable interplanetary dissipation. Quite obvious is the steepening of e^+ spectra (see also Fig. 10.20) in excess of the WKB decline after (10.49) in Figs. 8 and 9 of [10.114], which demonstrates that particularly the outward "propagating" Alfvénic turbulence in high-speed streams suffers considerable interplanetary damping as well as the inward Alfvén mode, which becomes strongly eroded at higher frequencies. Remember that the fluctuations also do not remain Alfvénic but become progressively more mixed with cross-helicity radially evolving toward zero.

Villante and Vellante [10.193] performed a very detailed study on the radial evolution of magnetic field variances in comparison with theoretical models for fluctuations with periods between 5 minutes and 6 hours. Observations of the same recurrent high-speed stream by *Helios 1* and 2 at different heliocentric distances (see again Fig. 10.19 referring to the same data set) clearly indicated that the total variance of the field components decreases faster than in proportion to R^{-3}, which is to be expected from (10.48), with $M_{Ar} \propto \varrho^{-1}$ and the density declining like R^{-2}. The strongest radial decline was found at the shortest periods considered, $\Delta T = 5\,\text{min}$, where $\langle (\delta B)^2 \rangle$ scaled like R^{-4} within measurement uncertainties. Furthermore, the authors established evidence in favor of a saturated turbulence amplitude, i.e.

$$\langle (\delta B / B_0)^2 \rangle = \text{const} \tag{10.50}$$

with $B_0 = |\langle B \rangle|$ and with the constant increasing from 0.2 at 5 min to 1.0 at 360 min. Best agreement was obtained at 30 min, indicating a saturation of the fluctuation energy at about 50% of the mean magnetic field energy independent of solar distance. Remarkably, their frequency band, at which turbulence amplitudes come closest to saturation and invariance against radial distance, coincides with the domain of purest Alfvénic turbulence as indicated by the minimum in the Elsässer ratio r_E in Fig. 10.10. Whether this is purely accidental or an important dynamical feature of the turbulence remains to be clarified. For a saturated amplitude one would have [10.192]

$$\langle \delta B^2 \rangle \propto B_r^2 \left(1 + \left(\frac{\Omega R_\odot}{V_r} \right)^2 \left(\frac{R}{R_\odot} \right)^2 \right) , \tag{10.51}$$

where Ω is the angular rotation frequency of the sun (radius $R_\odot$), which gives $\Omega R_\odot = 2\,\text{km/s}$. B_r and V_r are the radial components of the average spiral field and the solar wind velocity. Of course, in the inner heliosphere (10.51) implies $\langle \delta B^2 \rangle \propto R^{-4}$ with small corrections, a result which was observationally established for the total variances of the magnetic field at scales below one hour [10.11, 43, 192, 193, 196]. At distances not too far beyond 1 AU, however, one observationally has the relation $\langle \delta B^2 \rangle \propto R^{-3}$ [10.18, 104, 131], which can hardly be discriminated from the prediction of (10.48). Therefore, in the outer heliosphere it is difficult to distinguish between undamped and saturated waves.

Recent work [10.148, 153] has again addressed the problem of turbulence-amplitude evolution by considering the comprehensive data set provided by the *Helios* and *Voyager* missions covering together the range from 0.3 AU to more than 20 AU. The radial evolution was shown to be consistent with WKB expectations in the inner heliosphere, except at scales of a few hours or less, and in the outer heliosphere, with the exception of large scales of a few days or greater. Variances of the magnetic field vector for 2.5- to 2-hour and 10- to 2-hour filters on 1-hour averaged data if normalized to the mean field showed no significant dependence upon the stream structure. However, there remain some apparent discrepancies with the results of Figs. 10.1 and 10.5 and [10.43, 45, 46, 48], which indicate a higher fluctuation level in fast than in slow streams in the high-resolution data. Similar trends are found in Table 2 of [10.153]. These differences could be related to the fact that near solar minimum the high-frequency fluctuations indeed strongly depend on the stream structures. They may also be related to the different techniques employed in filtering the data and evaluating ensemble averages that incorporate long time series including various sorts of solar wind flow. More studies are necessary to resolve these differences and to understand them.

Given the independence of $\langle |\delta B|/B_0 \rangle$ from the stream structure, Roberts et al. [10.153] could, in complementing the inner heliospheric studies discussed above, establish new features in the spatial and spectral scaling of the fluctuations, which are summarized in Table 10.3, which shows the relative magnetic fluctuation amplitude normal to the ecliptic plane for five distances (normalized to 1 AU) and for four time scales. Also, an estimate of the spectral index at large scales (first filter) and the WKB prediction are given, which tends to hold at low frequencies in the inner and higher frequencies (hourly scales) in the outer heliosphere. At scales of a few hours the WKB-scaling law (10.48) works fairly well at all distances measured between 0.3 and 20 AU. At scales of several days however, there is even a distinct excess of the fluctuation amplitudes above the WKB values, with an indication of nearly saturated behavior. The last column of Table 10.3 gives the Alfvén ratio at the 10-day scale, which shows that there is a systematic decline of r_A toward values less than unity in accord with the evolution at small scales illustrated in Fig. 10.21 and in [10.151–153]. Owing to the decay of the stream structure [10.34], the Alfvén ratio becomes less than unity with increasing heliocentric distances, thus indicating a predominance of magnetic over kinetic fluctuation energy in the outer heliosphere.

Table 10.3. $\delta B/B$ normalized to 1 AU (normal component only except for r_A)

Distance [AU]	WKB	100-2 hr	25-2 hr	10-2 hr	2.5-2 hr	α	r_A
0.4	0.75	0.71	0.74	0.84	0.86	0.75	2.9
1.0	1.00	1.00	1.00	1.00	1.00	1.0	2.8
2.0	0.90	0.94	0.89	0.84	0.83	1.1	2.5
6.0	0.54	0.91	0.82	0.68	0.57	1.4	1.4
8.0	0.48	0.97	0.74	0.53	0.43	1.7	0.8

One may consider this process as a direct large-scale dissipation of the kinetic energy of relative streaming into turbulent energy, which maintains a higher level of magnetic field fluctuations than expected from the remnant fluctuations of coronal origin. This dissipation process differs very much, of course, in nature from the one at high frequencies, which leads to an erosion of power and spectral steepening at scales below about one hour in connection with a cascade and final cyclotron damping [10.41, 54, 106, 108]. The column of Table 10.3 indicated by α gives the spectral index relevant to scales below a day for a power law $f^{-\alpha}$. Clearly the low-frequency spectrum also steepens significantly with increasing distance, changing from the typical -1 value at 1 AU. The position of this break between spectral slopes varies radially, as was discussed in Sect. 10.5.3, and moves towards lower frequencies with increasing heliocentric distance. The same observation is reported in [10.90] for the outer heliosphere. The resulting idealized spectral evolution is suggested in Fig. 3 of [10.153].

In a series of papers Tu and coworkers [10.178, 179, 181] developed a theoretical model to explain the radial evolution of the power spectrum of Alfvénic fluctuations. Based on this work a theory for the heating of solar wind protons and the associated damping of Alfvénic fluctuations was developed [10.179], which proved to be surprisingly successful in describing the observed proton temperature profiles. In the companion paper [10.106] on kinetic solar wind physics a more detailed account of this work is given. In the present context it suffices to mention that a statistical comparison of the total hourly variances of the magnetic field components and the proton temperature showed [10.180, 182] that the radial variation of the temperature between 0.3 and 5 AU in undisturbed high-speed solar wind ($V_{sw} > 550$ km/s) could be explained by heating at a rate determined by the turbulent cascade. The model failed for low-speed flows by reasons to be discussed in the subsequent sections.

It is important to remember the observed close correlation between the daily proton temperature or thermal-speed variations and e^+ as evidenced in Fig. 10.23, and of the associated determination of the spectral slopes by c_S as shown in Fig. 10.24a. This close correlation may be considered as evidence supporting Tu's cascading and heating model, which intimately links proton thermal energy with the outward turbulent energy and reveals a similar correlation to that in Table 1 of [10.179]. Of course, the model did not account for the detailed spectral features illustrated in Figs. 10.9, 10.10, 10.22 and 10.24 and assumed r_E to be frequency independent, but it seemed to incorporate essential features of the observed turbulence and emphasized the idea of a turbulent cascade heating the protons.

The dissipation of the turbulence is indicated by increasingly steeper spectra in the proton-cyclotron-frequency domain. This regime was explored in the 1970s by means of *Mariner* data by Behannon [10.17, 18], although the observations have remained sparse until the present day. In the *Helios* mission the spacecraft-spin-enhanced turbulence levels at 1 Hz spoiled the natural turbulence to the extent that this frequency domain was only poorly analyzed. Above 4.7 Hz spectra were reliably measured again with the search-coil magnetometer on-board [10.132, 133], which covered the Whistler-mode range in particular. A companion paper in this volume addresses the high-frequency turbulence and fast magnetic fluctuations [10.73].

Composite spectra extending from 10^{-4} up to 10^2 Hz are given by Denskat and Neubauer [10.47] in their Figs. 1 to 3, which summarize once again the radial and spectral trends established in the previous sections and give a spectral index of about -3 for the Whistler domain. In the proton-cyclotron regime (0.1 to 1 Hz) the spectra are therefore expected to steepen from $-\frac{5}{3}$ to about -3. A detailed early investigation [10.17] revealed that the polarization state at these frequencies is rather mixed. Left-hand and right-hand polarizations were found, and clear examples of almost sinusoidal waves were observed. Given the tendencies from Figs. 10.8, 10.10 and 10.20, that e^+ and e^- might become equipartitioned in the dissipation regime, we may thus expect to see a comparable amount of inward- and outward-propagating ion-cyclotron and fast magnetosonic waves near Ω_p (or Ω_α, the He^{2+} cyclotron frequency). The microscopic wave–particle interactions in the dissipation range are reviewed in [10.106].

10.6 Theoretical Concepts of Turbulence Generation and Evolution

10.6.1 Numerical Simulations as a Guide to Interpreting Observations

Before we dwell on theoretical concepts of MHD turbulence evolution in the solar wind let us briefly summarize the observational evidence collected in the previous sections. The wave viewpoint (see e.g. [10.3, 77]) must apparently be abandoned because the fluctuations do not propagate according to geometrical optics (WKB), their amplitudes depend on frequency, their spectral shapes evolve radially, Alfvénic correlations tend to decay, and the fluctuations tend to be saturated at high frequencies and dissipate while heating the ions. The minimum variance is along the field with sizable parallel amplitudes and increasingly compressive components at larger solar distance [10.10]. There are local sources of turbulence related to stream velocity-shear and compressive effects, leading to the generation of inward Alfvén modes in addition to fossil turbulence of coronal origin. The observed spectral and spatial evolution suggests that nonlinear effects are prevalent even in pure Alfvénic periods (radial decline of cross-helicity), reinforcing the original ideas of Coleman [10.41] of a turbulent state of the solar wind with a developed cascade. In conclusion the solar wind is evidently a dynamically evolving, turbulent magnetofluid [10.121, 125, 155].

Undoubtedly a model incorporating all these observed features will be rather complex; in fact it is at present not available. Yet several attempts have been made to simulate some of the key processes believed to influence the turbulence evolution. Of particular interest and very puzzling was the observed decay of cross-helicity, which appeared to be at odds with the expectations of dynamic alignment theory [10.53, 69, 116, 145], indicating by closure calculations and numerical simulation that MHD turbulence may evolve toward a state with ideal Alfvénic correlations. It was then even argued [10.53, 120] that Alfvénic fluctuations could result from interplanetary turbulence evolution and be generated *in situ*, and that they therefore need not reflect coronal processes. In fact the observed Alfvénic fluctuations on the large scales of a day or more (see again Figs. 10.3, 10.12, and 10.13) must originate near the sun, since the turbulence

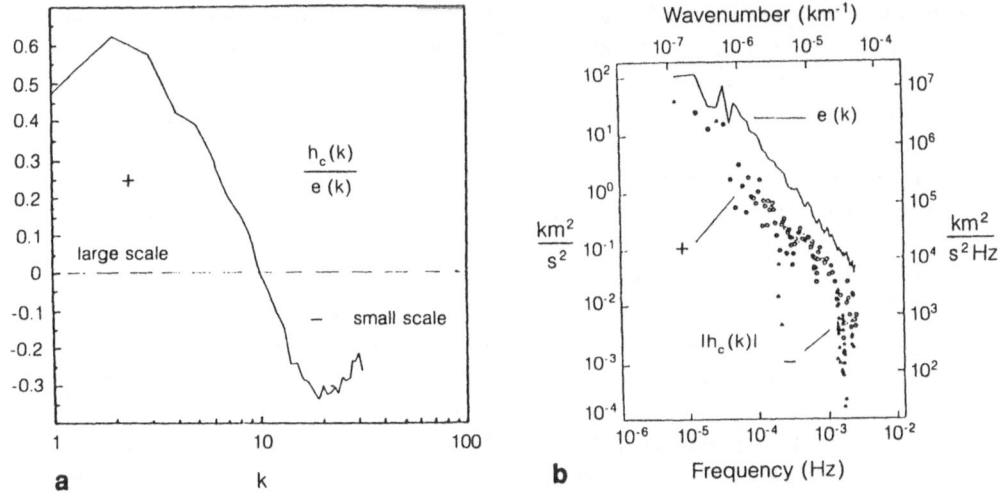

Fig. 10.27. (a) Illustration of the slow dissipation of cross-correlations in numerical simulations, showing σ_c versus k, after [10.145]. (b) Energy and cross-correlation spectra as obtained from *Voyager* data at 1 AU in solar wind at a flow speed of 352 km s^{-1}. *Triangles* give negative and *circles* positive values of σ_c. The total cross-helicity is positive, but at high k there are many negative points in qualitative agreement with the simulation results in (a) [10.119]

evolution is too slow to account for a building-up of cross-helicity by 0.3 AU or even 1 AU (see Grappin et al. [10.69]). Therefore, the above suggestions might only apply to the inertial regime beyond 10^{-4} Hz or so.

Furthermore, it should be kept in mind that the validity of simulation results is limited, because many codes are restricted in dimensions (often 2D), impose periodic boundary conditions for a homogeneous box, use perhaps unrealistically low mechanical and magnetic Reynolds numbers (in model equations such as (10.2–4) or (10.9–10)) that are inappropiate for mimicing the true solar wind dissipation process, and do not account for large-scale inhomogeneity and realistic boundary conditions. Given all these limitations, they still provide helpful guidance in interpreting dynamical processes in solar wind turbulence.

Returning to dynamic alignment, it was first found by closure calculations [10.69] and then confirmed by direct numerical simulations in many studies [10.68, 70, 120, 145] that Alfvénic correlations grew within a few nonlinear evolution times and that the resulting spectrum of cross-helicity, (10.34) or (10.38), was not positive-definite but revealed different signs at small and large scales. This effect is illustrated in Fig. 10.27, which shows a figure from Pouquet et al. [10.145] on the left-hand side (a) and from Matthaeus and Goldstein [10.119] on the right (b). The results of frame (a) were calculated by a spectral code method (with $\nu = \eta = 10^{-2}$ and a $64 \times 64 \times 64$ 3D grid) and show the plasma state after 9 time steps in units of the Alfvén transit time through a unit distance. The simulation demonstrates that correlations between velocity and magnetic field can grow out of an initially random configuration, resulting in a bi-model shape. This comes about from a slower decay of cross-helicity than of total energy, a

217

process which is permitted since h_c is not a positive-definite quantity and may avoid effective dissipation by attaining the typical shape as in frame (a).

The right-hand frame (b) of Fig. 10.27 gives *Voyager* observations of e and h_c spectra observed near 1 AU. The distribution of the signs of h_c as indicated roughly resembles that of (a) and suggests the presence of both the δZ^+ and δZ^- modes at disparate wave numbers, giving rise to a positive integrated cross-helicity. The overall course of h_c is also reminiscent of the observed high-resolution data shown before in Fig. 10.10, revealing an excess of e^- over e^+ at 10^{-5} Hz and thus the existence of long-wavelength inward Alfvén modes for an inward-oriented spiral field. These "waves" mark, according to Matthaeus et al. [10.120], the transition between the energy-containing structures driving the turbulence and the inertial-range fluctuations, which are driven (in compliance with our discussion in Sect. 10.5), at an early evolutionary stage. They also argued that as the turbulence gets older the inertial range may become equally populated by δZ^+ and δZ^-, and that interplanetary plasma being "stirred" for a longer time would show a lower degree of Alfvénicity. In this scenario the "equipartitioned" turbulence shown in Fig. 10.20, left-hand column, would be interpreted as aged and developed beyond the initial "minority species" phase. Of course, this implies considerable turbulent evolution in the near-solar region for the slow wind. However, it could also indicate that the large-scale "eddies" near the sun, containing the initial energy for turbulence generation, already possess a mixed magnetic field orientation and thus near-zero cross-helicity, in qualitative accord with the slow wind originating from closed-field coronal sources, the opening of which implies transitorily opposite field lines with subsequent reconnection.

In the model of [10.120] the outward Alfvén modes would appear as a minority species (overall $e^- \gg e^+$), which is cascaded to the dissipation range, and which, given equal cascading rates, would then die out first. Consequently, δZ^+ would rapidly populate the high-frequency domain and there dominate δZ^-, which cascades much more slowly into the inertial range. Another possibility according to Dobrowolny, Mangeney, and Veltri [10.52, 53] is that dynamic alignment only enhances a primordial asymmetry ("Alfvén waves" from the corona) in the inertial domain, which radially further develops toward higher correlations. This idea faces the problem that building-up of cross-helicity takes time, in fact many Alfvén transit times, which is not available in the super-Alfvénic flow that just transits the energy-containing scales too rapidly. The physical explanation of dynamic alignment is suggested by the form of the MHD equations (10.10) in Elsässer variables, and of the convective derivative (10.11), indicating that it is the minority species δZ^- in the sense of $|\delta Z^+| \gg |\delta Z^-|$, or vice versa, which most rapidly evolves and spreads in Fourier space. This effect is thus at the heart of the equation of motion and is unavoidable. If it does not show up clearly in the data, the conclusion can only be that it is effectively counteracted, for example by cascading of energy in the non-Alfvénic large scales into the inertial range [10.155] or by other processes destroying cross-helicity, which were not accounted for by the numerical simulations.

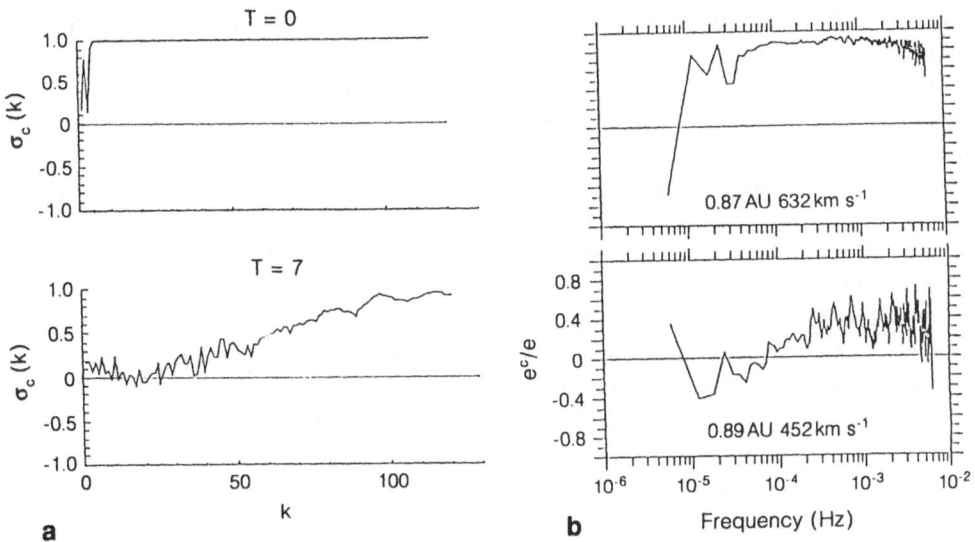

Fig. 10.28. (a) Spectra of cross-helicity in a numerical run with initially aligned magnetic and velocity fields at high wave numbers. The *top panel* shows the initial condition, the *bottom panel* the final state [10.154]. (b) Solar wind measurements of σ_c in fast (*top*) and slow (*bottom*) wind at about 0.9 AU, signifying the decay of Alfvénic correlations [10.114]

The Kelvin–Helmholtz instability [10.51, 66, 92, 154] represents such a process related to velocity shear. Motivated by results such as those shown in Figs. 10.12, 10.13, 10.15, and 10.16, the energy cascade and changes in cross-helicity associated with a velocity-shear layer were numerically examined by means of 2D MHD spectral codes simulating interplanetary dynamical processes. A model system consisting of two shear layers was set up in [10.66, 154] that was only marginally unstable (in linear theory). The nonlinear evolution produced a cascade to smaller scales, particularly in cross-helicity, that turned out to be similar in many respects to that observed in the solar wind, a result that suggests interplanetary fields to be in fact in a state of dynamically evolving turbulence. Figure 10.28 shows a comparison of numerical and observational σ_c spectra.

The right-hand panels show solar wind observations [10.114] relating to the bottom frames of Fig. 10.20, and the left-hand panels give two instances of the numerical simulation at times $T = 0$ and $T = 7$, in units of the Alfvén transit time [10.154]. If we assumed that the slow wind represented "aged" turbulence as compared to a "younger" turbulence in fast wind, then the right-hand boxes may also be regarded as giving the "initial" and "later" states of normalized cross-helicity. The overall qualitative agreement between observations and simulations is remarkable. In the model system the magnetic and flow fields were initially aligned at high k-numbers, with the exception of scales below $k = 4$, where $\sigma_c \neq 1$ and the Fourier modes were used to define the shear layer itself. By $T = 7$, only modes with the highest ks remain aligned, whereas the alignment for $k < 20$ is essentially zero. Simulation and solar wind data both show the same temporal (or spatial) decline of Alfvénicity. Not shown here is the concurrent rapid development of anisotropy in the power spectra, which arises because the cascade of magnetic energy is inhibited [10.164] in the direction parallel to the mean field.

The appearance of significant turbulent energy at high wave numbers is purely due to nonlinear processes, and unlike in the linear Kelvin–Helmholtz instability the energy does not remain confined only to scales larger than the velocity-shear scale. These numerical results indicate a possible resolution of the long-standing puzzle of the observed decay of normalized cross-helicity. They also emphasize the notion that inhomogeneity is essential for the dynamical evolution of solar wind turbulence and that shear-driven processes can rapidly affect the turbulence state. It is certainly desirable to extend the simulations to the compressible situation in order to explore the influence of compressive fluctuations upon the evolution of the shear layer. Presumably the picture will not change qualitatively, as is suggested by simulations [10.42] of dynamic alignment in compressible 2D MHD, which show growth of normalized cross-helicity with time similar to the incompressible runs.

10.6.2 Models of Turbulence for Inhomogeneous Media

The indication from the observational studies reviewed in previous sections is that the solar wind is structured and, owing to expansion, compressible on the large scales but exhibits near-incompressibility and nonlinear effects at smaller scales. Properties of a turbulent magnetofluid are prevalent in the inertial range but are intermixed with wave-like fluctuations owing their existence to the large-scale spiral magnetic field and the resulting finite Alfvén speed. Evidence has been presented that large-scale inhomogeneity, related to compression and velocity shear, can stir interplanetary turbulence. In this section we review modeling attempts to combine WKB theory with concepts of fluid turbulence [10.7, 75, 156] theory.

Whang [10.197] suggested one should consider dynamical interplanetary processes by discriminating two disparate time scales, being related to the flow (half a day and larger) and to the fluctuations (at scales below a few hours). This two-time-scale approach follows traditional techniques for modeling fluid turbulence [10.75] and allows the separation of the large-scale, background evolution from the small-scale turbulence evolution. Tu [10.177, 179, 181] and collaborators further developed this model by explicitly keeping the nonlinearities in the equations which describe the interactions between the fluctuations and which give rise to a nonlinear cascade in Fourier-space and thereby remedied failures of the simple WKB theory (for a review see again [10.3, 77]). By employing the spectral method of fluid turbulence theory [10.75] they were able to establish a transfer equation for the total-energy power spectrum, which in the stationary case is

$$\nabla \cdot \left(\left(\tfrac{3}{2} V + V_A \right) P(f, R) \right) - \tfrac{1}{2} (V \cdot \nabla) P(f, R) = -\frac{\partial}{\partial f} F(f, R) . \qquad (10.52)$$

Here $P(f, R)$ denotes the the power spectrum, which is a function of frequency and heliocentric distance R. By putting the right-hand side of (10.52) equal to zero the equation can be easily integrated to yield the WKB result (10.48). The novelty in this approach was the cascade implied by the flux function $F(f, R)$ in frequency space, which was initially derived following [10.53, 93] by means of dimensional analysis of the nonlinearity and was subsequently changed to comply with the observations better. On applying his theory to explain the damping of interplanetary fluctuations and the heating of the solar wind, Tu [10.179] then used the expression

$$F(f, R) = \alpha \alpha_1 \frac{2\pi}{\sqrt{4\pi\varrho}} \frac{1}{|V + V_A|} f^{5/2} P(f, R)^{3/2} \tag{10.53}$$

for the cascade function, which implies a spectral index of $-\frac{5}{3}$ for a developed inertial range. The constant α_1 was determined by the ratio of turbulent energy of the inward- and outward-propagating Alfvén waves and is nothing but the Elsässer ratio. Thus $\alpha_1 = r_E = e^-/e^+$ in our present nomenclature. The other constant α is related to the Kolmogorov cascade constant, which has now been determined by "calibration" with interplanetary measurements [10.183]. An empirical estimate yields $\alpha = 1.25$ in close accord with empirical hydrodynamical values [10.91, 156], under the assumption of a magnetofluid with vanishing cross-helicity. In Tu's original model [10.181] the α_1 was assumed to be independent of frequency and distance, because no empirical data on r_E existed at that time.

Clearly this is in contradiction to the recent results shown in Fig. 10.10 (and Fig. 3 of [10.114]) indicating strong frequency dispersion and radial evolution of r_E. Therefore the model was improved [10.183] so as to incorporate radial trends of α_1 in compliance with the observed r_E radially increasing toward unity. Empirically it was found for the fast streams of Fig. 10.20 that

$$\alpha_1 = 0.045 + 0.114(R - 0.65) \tag{10.54}$$

with R in AU. For the frequency regime covered by the model this ensured that $\alpha_1 \ll 1$, in accord with Fig. 10.10. The frequency dependence of α_1 was still disregarded. The results of this updated model in comparison with the magnetic power spectra of Fig. 10.19 [10.11] are displayed in Fig. 10.29, which shows (b) the spectral densities as measured and modeled at various heliocentric distances and (a) the plasma heating rate involved, if the dissipation of turbulence occurred according to the rate

$$H(R) = \frac{1}{4\pi} F(f_D, R), \tag{10.55}$$

where f_D signifies the dissipation frequency [10.178, 179], being of the order of $\Omega_p/2\pi$.

This relation establishes an important link between the particles' thermal properties and the turbulence which has been used to explain the observed proton temperature and magnetic moment profiles [10.106, 109, 160, 161] quite satisfac-

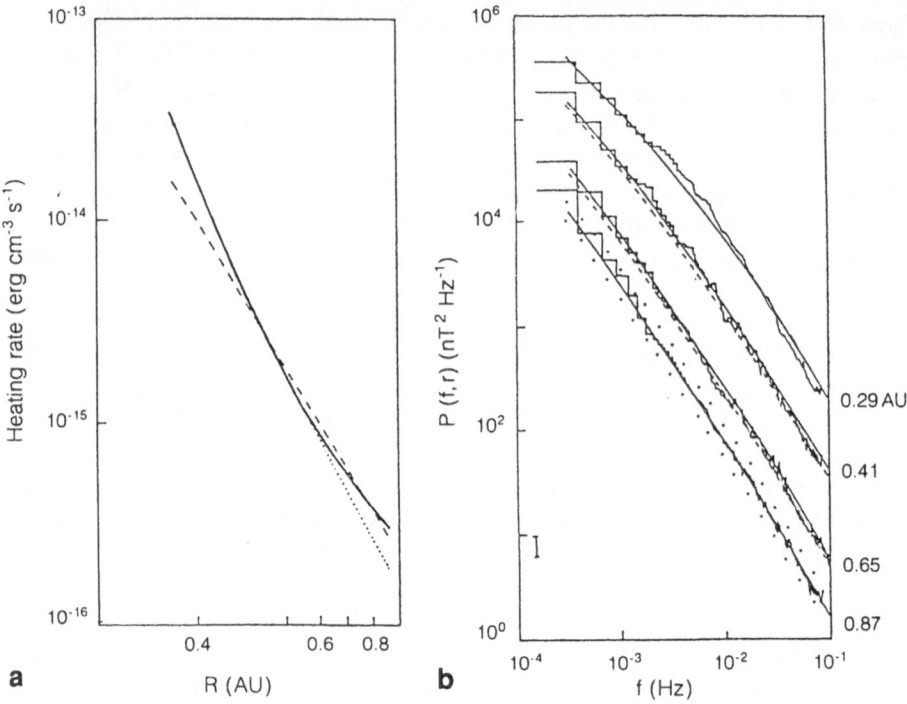

Fig. 10.29. (a) Radial variation of the cascade rate (dissipation or heating rate) given by the *thick line* calculated with $\alpha = 1.25$. Results with $\alpha \alpha_1 = 1/16$ and of the observations (*dashed line*) are also shown. (b) Power spectra (*solid lines*) calculated with (10.52) and $\alpha = 1.25$ for three distances indicated with the 0.29 AU spectrum as input. The observed spectra from Fig. 10.19 and Tu's model ([10.179], *dashed lines*) are also shown. For more details see text and [10.183]

torily. In Fig. 10.29 the power at 0.29 AU was used as an input. The three power spectra at 0.87 AU were calculated with $\alpha_1 = 1.25$ and $\alpha_1 = 0.84$ and $\alpha_1 = 1.82$ (dotted lines below and above the solid line), to indicate the variations due to a change in α_1. The heating rates in (a) equal the cascading rates by definition. For a value of $\alpha = 1.25$ and α_1 calculated with (10.54), the results are given by the thick line, whereas observations are indicated by the dashed line. These heating rates are in quite a nice agreement with those determined from the plasma data [10.160] for a particular *Helios 1* and *2* spacecraft line-up. The good agreement of the theory with observations at 0.8 AU is a consequence of the increasing flux of inward waves from (10.54) in accord with the radial trends of Fig. 10.20. A major drawback of this model remained, however: the lack of frequency variation in α_1 (or r_E) in contrast to the observations. This questionable assumption may also be the reason why the model was not able to predict the heating in slow solar wind [10.182], for which r_E is not small compared to unity but is a sizable fraction of unity. In order to remedy this situation the δZ^+ and δZ^- mode had to be treated on an equal footing, and thus separate transfer equations for e^+ and e^- needed to be developed.

To model the different spectral evolution of $e^{\pm}$ and to capture inhomogeneity effects as well as the interaction of the turbulence with large-scale structure, several theoretical attempts have recently been made, which in a two-scale (in time or length) approach start out at the general correlation functions of the type (10.20). Before we discuss them, we give the extension of the WKB theory (see e.g. Hollweg [10.77]) to Elsässer variables by Velli et al. [10.189], who obtained by a perturbation approach to the basic set of equations in Sect. 10.3.1 the result

$$\frac{\partial}{\partial t} E_k^\pm + \nabla \cdot (V_g^\pm E_k^\pm) + \frac{1}{2} E_k^\pm \nabla \cdot V = 0 , \qquad (10.56)$$

with the energy density $E_k^\pm = \frac{1}{2}\varrho(\delta Z_0^\pm)^2$ for Alfvén waves, with a spatially varying phase $\phi^\pm$ and an amplitude

$$\delta Z^\pm = \delta Z_0^\pm \exp(i\phi^\pm) , \quad \omega^\pm = -\frac{\partial \phi^\pm}{\partial t} , \quad k^\pm = \nabla \phi^\pm \qquad (10.57a)$$

and the local dispersion relation

$$\omega^\pm = k^\pm \cdot (V \mp V_A) = k^\pm \cdot V_g^\pm \qquad (10.57b)$$

defining the wave group velocity. In this picture nonlinear interactions between the two oppositely propagating Alfvén waves then give rise to modifications of the otherwise conserved wave action density $S^\pm = E^\pm/(\mp k \cdot V_A)$. Heinemann and Olbert [10.74] discuss this problem in detail. Equation (10.56) has been complemented as for (10.52) by cascading terms [10.189, 190], which will be discussed subsequently.

In a more general approach to the problem Marsch and Tu [10.113] and Zhou and Matthaeus [10.202] independently suggested starting with the correlation tensor (here we follow the nomenclature of [10.113])

$$Q^{\alpha\beta}(x, t; \xi, \tau) = \langle \delta Z_A^\alpha \, \delta Z_B^\beta \rangle \qquad (10.58)$$

with the subscripts A and B referring to two points in space and time,

$$x_{A,B} = x \mp \tfrac{1}{2}\xi , \quad t_{A,B} = t \mp \tfrac{1}{2}\tau . \qquad (10.59)$$

The task then is to calculate the dynamical evolution of the correlation tensor (10.58) of the fluctuations $\delta Z^\pm$ and to evaluate their traces, giving the desired spectral densities $e^\pm$ and e^R of Sect. 10.3.5. In order to do this the evolution equations for the background and the fluctuation were derived in various ways [10.113, 189, 202]. Starting from the MHD equations in terms of Elsässer variables (10.9–10) one finds [10.113] for incompressible fluctuations ($\delta\varrho = 0$)

$$\frac{d^\mp}{dt} \delta Z^\pm = \delta S^\pm + \delta F^\pm - \langle \delta F^\pm \rangle - \frac{1}{\varrho} \nabla \delta p_T \qquad (10.60)$$

with the nonlinear force

$$\delta F^\pm = -(\delta Z^\mp \cdot \nabla)\delta Z^\pm \pm \tfrac{1}{2}(\delta Z^+ - \delta Z^-)(\delta Z^\pm \cdot \nabla) \ln \sqrt{\varrho} \qquad (10.61)$$

and the linear force related to inhomogeneity, i.e. velocity shear and density gradients, being given by

$$\delta S^{\pm} = \pm \tfrac{1}{2} \left[(Z^+ - Z^-)(\delta Z^{\pm} \cdot \nabla) + (\delta Z^+ - \delta Z^-)(Z^{\pm} \cdot \nabla) \right]$$
$$\times \ln \sqrt{\varrho} - (\delta Z^{\mp} \cdot \nabla) Z^{\pm} . \qquad (10.62)$$

Note that $\langle \delta S^{\pm} \rangle = 0$, because of (10.19). To simplify the notation the index nought for the background quantities has been omitted. Most of the theories advanced so far neglect total pressure fluctuations, $\delta p_T = 0$. There are differences between existing models, as regards some of the terms in (10.61) and (10.62), which should be understood and eliminated in the future. Compared to the WKB results (10.56), the new effects are related to $\delta S^{\pm}$, particularly to the last term, which describes the coupling of the fluctuations to large-scale flow and Alfvén velocity shear or electric currents. Furthermore, the fluctuations are driven by the nonlinear force (10.61), which owes its existence to the convective derivative. The work done by this force on the fluctuations gives rise to the cascading effects described below.

By means of (10.60) the time evolution of the two-point correlation function can be calculated. After considerable algebra [10.113, 187] one arrives at the following set of transfer equations for the spectral densities of inhomogeneous MHD turbulence:

$$\frac{d^{\mp}}{dt} e^{\pm} + (e^{R} - e^{\pm})(Z^{\pm} \cdot \nabla) \ln \sqrt{\varrho} + S^{\pm}$$
$$= -(\xi \cdot \nabla) Z^{\mp} \cdot \frac{\partial}{\partial \xi} e^{\pm} + NL^{\pm} , \qquad (10.63)$$

where the source term $S^{\pm}$ is related to (10.62) and reads

$$S^{\pm} = \tfrac{1}{2} \left(Q^{\pm,\mp} + [Q^{\mp,\pm}]^{T} \right) : \nabla Z^{\pm}$$
$$\mp \tfrac{1}{2} V_A \cdot \left(Q^{\pm,\pm} + [Q^{\pm,\pm}]^{T} \right) \cdot \nabla \ln \sqrt{\varrho} . \qquad (10.64)$$

The superscript T indicates the transposed tensor. The symbol e^{R} denotes the residual energy or cross-correlation function

$$e^{R} = \tfrac{1}{4} \langle \delta Z_A^+ \cdot \delta Z_B^- + \delta Z_A^- \cdot \delta Z_B^+ \rangle , \qquad (10.65)$$

and the nonlinearity is defined by the ensemble average of the triple-correlations [10.113, 187], given with (10.61) as

$$NL^{\pm} = \tfrac{1}{2} \langle \delta F_A^{\pm} \cdot \delta Z_B^{\pm} + \delta F_B^{\pm} \cdot \delta Z_A^{\pm} \rangle . \qquad (10.66)$$

Several features of these equations warrant mention. First, the "source" term $S^{\pm}$ still involves the full correlation tensor (10.58) and in particular the cross-correlations defined by the nondiagonal elements of (10.58), which are dynamically independent of the trace elements and of $e^{\pm}$. The full set of equations describing the advancement of all tensor components is given in [10.113]. In principle it can describe effects related to dynamo activity (interplanetary electric field fluctuations) as well as magnetic helicity evolution [10.127] and the dynamical evolution of the Alfvén ratio (10.32). Either one accepts the complexity of the coupled set of equations (see equation (35) in [10.113]) or he needs to break the chain by approximating $S^{\pm}$ in terms of $e^{\pm}$, for example as done in [10.125, 202] in the spirit of dynamo theory [10.127]. Whatever the approximation is, the term (10.64) certainly produces qualitatively new effects compared to WKB theory in (10.56), by coupling the turbulence to the large-scale fields $Z^{\pm}$ and density ϱ and by accounting for cross-correlations between δZ^{+} and δZ^{-}.

The first term on the right-hand side of (10.63) is due to a Taylor expansion of tensor components related to the convective derivative term [10.113] and describes the distortions of the small-scale turbulence due to the inhomogeneity of the background flow and Alfvén velocity. This term does not appear in any wave-like theory and imposes, when Fourier-transformed, a spectral constraint on the fluctuations even in the absence of nonlinearities. The nonlinear term (10.66) itself is evaluated in the subsequent section.

We can finally calculate the Fourier transform of (10.63) with respect to the small-scale (spatial differences after (10.59)) variable ξ. For the sake of simplicity we consider only the case $\tau = 0$ and then obtain the spectral density function

$$e_k^{\pm}(\boldsymbol{x}, t) = \int_{-\infty}^{\infty} d^3\xi \, e^{-i\boldsymbol{k}\cdot\boldsymbol{\xi}} e^{\pm}(\boldsymbol{x}, \boldsymbol{\xi}; t, 0) \,, \qquad (10.67)$$

which obeys the transfer equation

$$\frac{d^{\mp}}{dt} e_k^{\pm} + \left(e_k^R - e_k^{\pm}\right) \left(Z^{\pm} \cdot \nabla\right) \ln \sqrt{\varrho} + S_k^{\pm}$$

$$= \frac{d}{d\boldsymbol{k}} \left(\boldsymbol{k} \, e_k^{\pm}\right) : \left[\nabla Z^{\mp}\right]^T + NL_k^{\pm} \,. \qquad (10.68)$$

One may also go from the specific spectra $e_k^{\pm}$ to the spectral densities $E_k^{\pm} = \varrho e_k^{\pm}$ by exploiting the continuity equation, after [10.112], given by

$$\frac{d^{\pm}}{dt} \ln \varrho = -\frac{1}{2} \nabla \cdot (3Z^{\pm} - Z^{\mp}) \,. \qquad (10.69)$$

The two combined equations (10.68), if supplemented by an equation for e^R [10.113, 202], allow us to calculate the radial evolution of $e_k^{\pm}$ separately and thus replace the single equation (10.52) of Tu's original model.

10.6.3 Nonlinearities and Cascading

A rigorous calculation of the nonlinear terms in (10.66) is one key problem of MHD turbulence theory. At present, only approximations have been achieved. Veltri et al. [10.191] proposed a closure relation involving complicated integral equations, which are only valid for a homogeneous system and are not of practical use in the solar wind context. Following the early work by Dobrowolny et al. [10.53], all the recent attempts to tackle the nonlinearities rely on a dimensional analysis of the triple correlations in order to come up with cascading function of the type discussed in (10.53). The subsequent calculations are concerned with reduced spectra in the variable k_1 obtained by an integration over the k_2 and k_3 component of the spectrum (10.67). Generalizations to fully three-dimensional turbulence are notoriously difficult [10.7, 75] and have not been undertaken for the solar wind. In what follows we closely follow Tu and Marsch [10.187].

The spectral transfer function may be defined after [10.138] by the relation

$$F_{k_1}^{\pm} = \int_{k_1}^{\infty} dk_1 \, NL_{k_1}^{\pm} \tag{10.70a}$$

so that one has

$$-\frac{\partial F_{k_1}^{\pm}}{\partial k_1} = NL_{k_1}^{\pm} . \tag{10.70b}$$

Inspection of (10.66) and (10.61) shows that, to leading order in l/L and by assuming proportionality of $F_{k_1}^{\pm}$ to the nonlinearity, one may have the relation

$$F_{k_1}^{\pm} \propto \delta Z^{\pm} \delta Z^{\mp} (D\delta Z^{\pm})/l , \tag{10.71}$$

where the gradient has been replaced by a "difference" $D\delta Z^{\pm}$ over a typical turbulence scale l, to be taken as $l \approx 1/k_1$. Following [10.53, 93] we may estimate the fluctuation amplitude near k_1 by assuming

$$\delta Z^{\pm} \propto \sqrt{k_1 e_{k_1}^{\pm}} . \tag{10.72}$$

To estimate $D\delta Z^{\pm}$ poses a problem that cannot be solved unambiguously. By taking simply $D\delta Z^{\mp} \approx \delta Z^{\mp}$ one finds the spectral flux function [10.187]:

$$F_{k_1}^{\pm} = \alpha^{\pm} e_{k_1}^{\pm} \sqrt{e_{k_1}^{\mp}} k_1^{5/2} . \tag{10.73}$$

For the developed inertial range this function must be constant. Spectral power laws

$$e_{k_1}^{\pm} \propto k_1^{-m^{\pm}} \tag{10.74}$$

then imply the two relations

$$m^{\pm} + \tfrac{1}{2} m^{\mp} = \tfrac{5}{2} , \tag{10.75}$$

which only admit the unique solution $m^+ = m^- = \tfrac{5}{3}$. These spectral indices are fully consistent with the observations discussed in Sect. 10.5. Observational evidence favors the particular choice of (10.73). However, we may also estimate the difference by

$$D\delta Z^{\pm} \approx k_1 \delta Z^{\pm} \delta Z^{\mp} \tau_i , \tag{10.76}$$

i.e., by the convective change of $\delta Z^{\pm}$ during the nonlinear interaction time τ_i, which of course still needs to be assessed. Possible choices [10.53] are the Alfvén time

$$\tau_A = l/V_A \qquad (10.77)$$

and the eddy lifetime

$$\tau^\pm = 1/\delta Z^\mp , \qquad (10.78)$$

the shorter of which may be considered relevant. If the fluctuating Mach number is larger than unity, i.e. if $\delta Z^\pm > V_A$, we may put $\tau_i = \tau^\pm$ and retain the Kolmogorov cascading function (10.73). Otherwise we obtain with $\tau_i = \tau_A$ the result of Kraichnan [10.93]

$$F_{k_1}^\pm = \frac{1}{V_A} \beta^\pm e_{k_1}^+ e_{k_1}^- k_1^3 \qquad (10.79)$$

with another cascading constant $\beta^\pm$. Observationally, $\delta Z^+/V_A$ is close to unity [10.190], which makes it difficult to choose the correct τ_i for the flux function $F_{k_1}^-$.

Velli et al. [10.189, 190] have recently argued that for the predominantly Alfvénic periods, with $\delta Z^+ \gg \delta Z^-$, the time τ_i would be essentially determined by the lifetime $\tau^+ = l/\delta Z^-$, the reason being that the δZ^+ and δZ^- wave packets interact coherently, since δZ^- is to lowest order described by the small refracted component of the outgoing waves owing to large-scale flow divergence and inhomogeneity. The authors estimated the amplitude by expanding to the next order with respect to the WKB result (10.57a) with $\delta Z_0^- = \mathbf{0}$. It is given by

$$\delta Z_1^- = \frac{i}{4}\{-2\delta Z_0^+ \cdot \nabla(V - V_A) + \delta Z_0^+ \nabla \cdot \left(V_A - \tfrac{1}{2}V\right)\}/(\mathbf{k} \cdot V_A) , \quad (10.80)$$

which is proportional to the outgoing wave amplitude and the small ratio l/L, where L is related to the large-scale gradient.

The energy flux may then simply be estimated as $(\delta Z^+)^2/\tau_i$. With (10.72) and $\tau_i = \tau^+$, and upon insertion of (10.80) into (10.78), we obtain the cascading function (with constant γ^+) after [10.189] as

$$-F_{k_1}^+ = \frac{1}{V_A}\gamma^+|\nabla \cdot \left(V_A - \tfrac{1}{2}V\right) - 2(\nabla Z^-)_{1,1}| \left(e_{k_1}^+ k_1\right)^{3/2} . \qquad (10.81)$$

The inertial-range solution is obtained for the flux being independent of k_1, which with (10.74) implies a spectral slope of $m^+ = 1$. This spectrum was found as "a possible outcome of nonlinear interactions in a population of waves propagating in a single direction in a medium with a large-scale gradient" [10.190]. Of course, if the flow is homogeneous, (10.81) gives no energy transfer at all, whereas in strongly divergent or sheared flows a k_1^{-1}-spectrum should result. This favors the near-sun solar wind acceleration region as the source of the observed flat perihelion spectra of e^+ shown in Sect. 10.5.

A similar kind of result may yet be linearly obtained from the first term on the right-hand side of the transfer equation (10.68), which in one dimension yields the flux function

$$-F^{\pm}_{k_1} = (\nabla Z^{\mp})_{1,1} \left(e^{\pm}_{k_1} k_1 \right) , \qquad (10.82)$$

which involves no unknown cascading constant and also implies a spectral slope of $m^{\pm} = 1$. However, there is an essential difference in that (10.82) applies equally to the inward and outward mode, whereas (10.81) relies upon the assumed asymmetry of the turbulence with predominance of δZ^+ over δZ^-. Also, (10.82) rather naturally arises from the general dynamical equation of correlation functions akin to fluid turbulence [10.75], whereas (10.81) is the result of an approximate wave-perturbation approach but is truly nonlinear in nature. Yet it seems that a turbulence mechanism for the large scales in the inner heliosphere may be ruled out because the eddy-turnover time is too large and it takes at least a few nonlinear times to develop a self-similar cascade. In this context it should be mentioned that the observed f^{-1} spectra are interpreted in [10.122] in terms of a different mechanism producing the spectra by "superposition of signals with scale-invariant distribution of correlation times".

We may finally note that a cascading function has also been proposed in [10.187] for the cross-correlation spectrum e^{R}, which depends on the spectrum of the kinetic energy (10.36) and reads

$$F^{R}_{k_1} = \delta^{R} \sqrt{e_V(k_1)} e^{R}_{k_1} k_1^{5/2} \qquad (10.83)$$

with another cascading constant δ^{R}. In the case of a Kolmogorov spectrum for $e_V(k_1)$, the cross-correlation would have an "inertial range" spectrum with index $-\frac{5}{3}$, which is consistent with the observations [10.114]. A possible cascading function for e^{R} based on numerical closure calculations was also considered by Grappin et al. [10.70], who predicted a k^{-2} spectrum, which is too steep to be relevant to the observed interplanetary turbulence. The full transfer equation for e^{R} is derived in [10.113, 187].

In conclusion, the present understanding of the processes related to the non-linearity (10.66) is far from satisfying. A more complete algebraic treatment and a decomposition of (10.66) into parts of different physical significance is given in [10.187]. Following traditional paths in modeling this term for fully developed stationary MHD turbulence, various cascading functions have been established that are tailored such as to comply with the observed spectra and their indices. However, much needs to be done to comprehend the physics of the nonlinear interactions better and, in particular, to account for the observed spectral anisotropies which invariably occur in numerical simulations and interplanetary observations in the inner heliosphere.

10.6.4 Compressibility and Pseudosound

As discussed in Sects. 10.4.3 and 10.5.4 in particular, compressive fluctuations of variable amplitude and at various characteristic scales are ubiquitous in the interplanetary medium. The observed small-scale compressive turbulence and the origin of pressure-balanced structures (PBS) is not well understood and is a matter of ongoing research. To explain the overall spectral slopes of δn and δB spectra [10.115] (see for example Fig. 10.26) let alone the spectral details in the high-frequency domain, is at present an important issue. The bearing compressive fluctuations have on the predominantly Alfvénic fluctuations remains to be explored. Theoretical models, comparable to the ones outlined in the previous section, although certainly desirable are not available for compressive solar wind turbulence.

Recently, Montgomery and coworkers [10.130, 165] proposed an explanation of the often observed $k^{-5/3}$ spectrum of δn at those times when kinetic and magnetic energy spectra of the magnetofluid themselves exhibit a Kolmogorov slope. The idea is to conceive the density fluctuations, or better their persistent part aside from transitory sound waves, as being driven nonlinearly by sources related to such V and B fluctuations that deviate from being incompressible and Alfvénic. By differentiating (10.2) with respect to time and subtracting the divergence of (10.3) they obtain with (10.5) and (10.6) an inhomogeneous sound-wave equation for the mass density fluctuation $\delta\varrho$ in the form

$$\left(\frac{\partial^2}{\partial t^2} - c_S^2(\varrho)\nabla^2\right)\delta\varrho = S' \,. \tag{10.84a}$$

Here the viscous terms have been neglected. The source term S' is given by the double divergence of the Maxwell and Reynolds stress tensor

$$S' = \nabla\nabla : \left(\varrho VV - BB\frac{1}{4\pi} + B^2\frac{1}{8\pi}1\right) \,. \tag{10.84b}$$

Clearly for ideal Alfvénic correlations and $|B|$ being strictly constant this source vanishes and a sound-wave solution is obtained from (10.84a). After the sound waves have damped out a formal solution of (10.84a) after [10.165] yields

$$\delta\varrho(\boldsymbol{x}, t) \approx c_S^{-2}\nabla^{-2}S' \,, \tag{10.85}$$

which may be Fourier-transformed to yield the power spectrum of $\delta\varrho$. According to [10.130] this density variation "rides parasitically on the back of the incompressible turbulence " and to leading order attains a value

$$\delta\varrho(\boldsymbol{k}) \approx -\frac{1}{c_S^2}\frac{1}{8\pi}(B^2)_{\boldsymbol{k}} \,, \tag{10.86}$$

which contains the Fourier transform of the magnetic pressure. Of course, this is nothing but the equation

$$\delta p_T = 0 \tag{10.87}$$

transformed into $\boldsymbol{k}$-space, with the total plasma pressure as in (10.13b). Note that in this section the system is considered to be homogeneous without a mean field and therefore B and V denote the fluctuations themselves. Formula (10.86) fits numerical simulation results [10.165] best and thus indicates a turbulent state where the total pressure gradients have relaxed away. As the authors themselves state, this "novel physical process ... remains to be elucidated at the elementary physical level". It appears interesting also to find out how this approach relates to the generation of PBS obeying (10.87) and to their evolution. How (10.86) and similar expressions explicitly yield a $-\frac{5}{3}$ density fluctuation spectrum is discussed in [10.165].

A paper by Matthaeus and Brown [10.124] further elaborates this theory of "pseudosound", in which the nonlinearly driven density fluctuations are shown to adjust quickly to the magnetic field fluctuations, thereby producing PBS. Pseudo-sound density irregularities naturally arise from a perturbation expansion at low sonic Mach numbers; they do not propagate and involve only convective variations. It remains to be substantiated whether the interesting suggestion [10.124] holds true that a closed set of dynamical equations for nearly incompressible MHD and generalized magnetoacoustics exists which comprises Alfvénic fluctuations and both magnetoacoustic waves and pseudosound density fluctuations. Recent numerical simulations [10.140] reveal a transition in compressible and viscous Navier–Stokes flows from a state with low acoustic noise level to a state dominated by longitudinal modes particularly at small scales. This transition takes place at an rms sonic Mach number of 0.3. According to [10.72, 190] the solar wind is in this critical state with sonic and Alfvénic Mach number varying between 0.2 and 1.4 and typically being about 0.3. The relevance of these results needs to be better understood.

Magnetoacoustic-type waves are inevitably obtained as the result of the parametric decay of large-amplitude waves, notably circularly polarized Alfvén waves [10.49, 63, 87, 95, 101, 158, 171, 201]. Compressive variations are also intrinsic components of localized and other large-amplitude variations such as Alfvén solitons [10.67, 83, 137, 167–169]. Such coherent structures may exist in the solar wind, but there exists no experimental verification of them at the present time. The validity of turbulence concepts and their success in describing the observations renders it unlikely that idealized large-amplitude Alfvén waves correspond to the reality of the interplanetary medium. Li and Zweibel [10.97] considered the scattering of an Alfvén wave packet by density fluctuations and concluded that the interaction with density irregularities caused an energy transfer to magnetosonic waves and attenuation of the original wave. Such a process may contribute to the observed dissipation of Alfvénic turbulence, but how the assumed localized wave envelope relates to the observed broad-band solar wind spectra remains unclear. The older literature on this topic can be found in Barnes' review [10.3].

10.7 Summary and Conclusions

10.7.1 Left-Out Topics

Before we conclude, some omitted topics will be briefly addressed. We have not looked at turbulence associated with interplanetary shocks. As regards slow shocks this topic is covered in an accompanying review by Richter [10.147] in this book. This review by Burlaga [10.34] covers MHD turbulence at interplanetary corotating shocks in the outer heliosphere. Collisionless shocks and their turbulence patterns and concurrent waves are intensively discussed in the collection of reviews in [10.176]. The power spectral signatures of interplanetary corotating and transient flows are investigated in [10.64, 98, 99, 166, 175, 194] and in the literature cited therein.

Another omitted topic is the interaction of the MHD turbulence with the solar wind background flow. As is well known [10.3], Alfvén waves exert an acceleration force on the plasma given by the gradient of their energy density. The related physical picture is well understood, see e.g. [10.85] or more recently e.g. Leer et al. [10.96]. Less well comprehended, however, is how these interactions need to be conceived in the turbulence picture. Following the papers by Whang [10.197] and Marsch [10.105] and the nomenclature of Sect. 10.3.1 one can show that the compressible magnetohydrodynamic fluctuations exert a force given by the negative divergence of the turbulent stress tensor

$$T = \varrho \left\langle \delta V \delta V + \frac{\gamma - 1}{2} c_s^2 1 \left(\frac{\delta \varrho}{\varrho}\right)^2 \right\rangle + \frac{1}{4\pi} \left\langle \frac{1}{2} \delta B^2 1 - \delta B \delta B \right\rangle, \qquad (10.88)$$

which reduces to the usual MHD wave expressions [10.50, 105], if the linear wave amplitude correlations and dispersion equations are exploited. To fully incorporate (10.88) in models with the observed fluctuation amplitudes is a task for future solar wind modeling. Hollweg [10.80] and Tu [10.178] were the first to combine the ingredients of the traditional Alfvén-wave pressure and of turbulence dissipation within one model. Tu's model quite successfully explained some basic features of high-speed-stream temperature profiles and turbulence amplitudes (based on (10.52) and (10.53)). Of course, this model needs to be improved by better accounting for the detailed observations, notably in slow wind, in which the role of turbulence is poorly understood.

An important point is the spectral features assumed at the sun, and the slope of power spectra in the corona, for which the observational evidence is reviewed by Bird and Edenhofer [10.23]. In fact, Roberts [10.148] comes to the conclusion that, given the interpanetary observational constraints and a spectral index near $-\frac{5}{3}$ for the high-frequency fluctuations at the coronal base, the Alfvén waves extrapolated to the sun are not energetic enough to accelerate the fast streams. This issue will presumably remain at the focus of solar wind research in the near future. Lack of space prohibits a further elaboration of the problem. A list of the relevant literature is found in [10.148] and the references therein.

10.7.2 Observational Summary

When Barnes [10.3] completed his review on hydromagnetic waves and turbulence in the solar wind a decade ago, the complexities and the richness in structure of the interplanetary medium revealed by the *Helios* and *Voyager* missions were neither known and nor even anticipated. The main task of this review was to collect the empirical evidence on the nature and origin of interplanetary fluctuations. These are amply illustrated by the figures of the previous sections. Observationally, this field of research has matured. Evidently, the solar wind is a dynamically evolving, inhomogeneous, anisotropic, turbulent magnetofluid. Its major features are obvious, but many facets remain to be explored and explained. Notably, the role of compressibility must be better comprehended. Its importance for the large-scale evolution of the turbulence is unquestioned. Small-scale compressive fluctuations require further studies and their coupling to the Alfvénic fluctuations deserves thorough analysis.

The origin of the turbulence has been elucidated. Quite clearly, the sun and its outer corona are the major source of interplanetary turbulence in the inner heliosphere. But local sources of turbulence generation exist, particularly in association with velocity-shear and compression regions on large and medium scales, which manifest themselves by enhanced levels of the inward Alfvén modes, or equivalently as sinks of cross-helicity. In fact, *in situ* sources of turbulence and the coupling to large-scale inhomogeneities are strong enough to counteract entirely the otherwise inescapable evolution towards dynamic alignment, which is not observed in the solar wind. All empirical power spectra steepen radially with spectral slopes apparently approaching the Kolmogorov value. The anisotropies of the fluctuations still remaining by 1 AU and the different intensities of component spectra need to be better understood.

The kinetic energy liberated by large-scale stream dissolution represents a major source for small-scale turbulence, particularly in the magnetic field at larger heliocentric distances, where it presumably gives rise to an evolution of the Alfvén ratio to values below unity. The overall strong coupling of fluctuations in fast and slow streams on large scales is suggested by the observed variations in Fig. 10.23 of the turbulence level over two orders of magnitude in close relation to the stream pattern. Undoubtedly the turbulence dissipates and heats the plasma, although the microphysics of this process demands further studies and better high-resolution measurements.

10.7.3 Conclusions and Recommendations

In concluding his review, Barnes [10.3] identified several problem areas which he considered then especially important for future investigation. Some of these problems have been solved, others are still with us. The nature and morphology of MHD fluctuations and their variations in space and time (solar cycle) are by now fairly well documented and much better comprehended in the language of turbulence rather than of waves. The "minimum-variance problem" comes in the new guise of the equally difficult anisotropy problem of the observed turbulence [10.10, 114]. The physics of MHD discontinuities was not dealt with in this article, but it also remains an active research area [10.79, 103, 135]. As mentioned in the previous section, to develop a rigorous solar wind transport theory and a theory of the dissipation of turbulent energy and momentum remains an outstanding task for future research. The associated wave–particle interactions, particularly with broad-band fluctuations compared to single large-amplitude waves, are badly understood and an appropriate kinetic theory is far from being completely developed. Progress in this area will perhaps require the theoretical efforts of many "man years".

Clear advances have been made in developing theoretical descriptions of the interplanetary turbulence. The concepts described in Sect. 10.6 provide the tools to deal successfully with the spatial and spectral evolution of MHD turbulence in the inhomogeneous solar wind plasma and to establish more realistic models. Clearly, in many areas further progress should be made. The dissipation mechanism is to be clarified. Improving the treatment of the nonlinearities is notoriously difficult if one goes beyond simple cascading functions to the full complexity of nonlinear interactions between turbulence at disparate scales. Numerical simulations will probably remain the most powerful means of gaining more insight into the physics of the nonlinearities. It appears promising to exploit the existing approximate turbulence models fully and investigate their ability to explain the *in situ* observations.

Of course, one wishes to see these observations to be extended to the near-solar regions and the corona itself. The primordial generation mechanism of the turbulence is still an enigma and the origin of Alfvénic fluctuations remains puzzling. Waves and turbulence in the solar atmosphere (see e.g. [10.80, 81, 84]) and their role in accelerating the solar wind is a research field of increasing activity and interest in view of the upcoming *SOHO* mission [10.55]. Last but not least, further progress is expected from continuing solar wind data analysis dedicated to specific questions. At present, the observations are far ahead of theory. To close this gap will be an achievement of general importance to space plasma physics.

Acknowledgements. I am very grateful to C.Y. Tu and R. Grappin for careful reading of the manuscript and many helpful comments. I owe special thanks to B. Bavassano and D.A. Roberts for their thorough review of this rather long paper and their many valuable suggestions for improvement. I also wish to thank A. Julier for transforming my handwriting into a comprehensible English typescript and M.J. Rees, D. Lynden-Bell, and the staff at the Institute of Astronomy of the University of Cambridge for their kind hospitality.

References

10.1   Barnes, A., Microscale fluctuations in the solar wind, NASA SP-308, 333, 1972.

10.2   Barnes, A., J.V. Hollweg, Large-amplitude hydromagnetic waves, J. Geophys. Res., **79**, 2302, 1974.

10.3   Barnes, A., Hydromagnetic waves and turbulence in the solar wind, in *Solar System Plasma Physics*, Vol I, ed. by E.N. Parker, C.F. Kennel, L.J. Lanzerotti, North-Holland, Amsterdam, 249, 1979.

10.4   Barnes, A., Turbulence and dissipation in the solar wind, *Solar Wind Four*, ed. by H. Rosenbauer, MPAE-Report No. W-100-81-31, 326, 1981.

10.5   Barnes, A., Interplanetary Alfvénic fluctuations: A stochastic model, J. Geophys. Res., **86**, 7498, 1981.

10.6   Barnes, A., Hydromagnetic waves, turbulence, and collisionless processes in the interplanetary medium, in *Solar–Terrestrial Physics*, ed. by R.L. Carovillano and J.M. Forbes, D. Reidel, Dordrecht, 155, 1983.

10.7   Batchelor, G.K., *Theory of Homogeneous Turbulence*, Cambridge University Press, 1970.

10.8   Bavassano, B., M. Dobrowolny, G. Moreno, Local instabilities of Alfvén waves in high-speed streams, Solar Phys., **57**, 445, 1978.

10.9   Bavassano, B., M. Dobrowolny, F. Mariani, N.F. Ness, On the polarization state of hydromagnetic fluctuations in the solar wind, J. Geophys. Res., **86**, 1271, 1981.

10.10   Bavassano, B., M. Dobrowolny, G. Fanfoni, F. Mariani, N.F. Ness, Statistical properties of MHD fluctuations associated with high-speed streams from Helios-2 observations, Solar Phys., **78**, 373, 1982.

10.11   Bavassano, B., M. Dobrowolny, F. Mariani, N.F. Ness, Radial evolution of power spectra of interplanetary Alfvénic turbulence, J. Geophys. Res., **87**, 3617, 1982.

10.12   Bavassano, B., F. Mariani, Interplanetary Alfvénic fluctuations: A statistical study of the directional variations of the magnetic field, in *Solar Wind Five*, ed. by M. Neugebauer, NASA Conference Publ. CP-2280, 99, 1983.

10.13   Bavassano, B., E.J. Smith, Radial variation of interplanetary Alfvénic fluctuations: Pioneer 10 and 11 observations between 1 and 5 AU, J. Geophys. Res., **91**, 1706, 1986.

10.14   Bavassano, B., R. Bruno, Large-scale solar wind fluctuations in the inner heliosphere at low solar activity, J. Geophys. Res., **94**, 168, 1989.

10.15   Bavassano, B., R. Bruno, Evidence of local generation of Alfvénic turbulence in the solar wind, J. Geophys. Res., **94**, 11977, 1989.

10.16   Behannon, K.W., L.F. Burlaga, Alfvén waves and Alfvénic fluctuations in the solar wind, in *Solar Wind Four*, ed. by H. Rosenbauer, MPAE-Report No. W-100-81-31, 374, 1981.

10.17   Behannon, K.W., Observations of the interplanetary magnetic field between 0.41 and 1 AU by the Mariner 10 spacecraft, Goddard Space Flight Center, GFSC Doc. 692-76-2, Greenbelt, Maryland, USA, 1976.

10.18   Behannon, K.W., Heliocentric distance dependence of the interplanetary magnetic field, Rev. Geophys. Space Phys., **16**, 125, 1978.

10.19   Belcher, J.W., L. Davis, Jr., E.J. Smith, Large-amplitude waves in the interplanetary medium: Mariner 5, J. Geophys. Res., **74**, 2303, 1969.

10.20   Belcher, J.W., L. Davis, Large-amplitude Alfvén waves in the interplanetary medium, 2, J. Geophys. Res., **76**, 3534, 1971.

10.21   Belcher, J.W., R. Burchsted, Energy densities of Alfvén waves between 0.7 and 1.6 AU, J. Geophys. Res., **79**, 4765, 1974.

10.22   Bendat, J.S., A.G. Piersol, *Random Data: Analysis and Measurement Procedures*, Wiley-Interscience, 1971.

10.23   Bird, M., P. Edenhofer, Remote sensing observations of the solar corona, in *Physics of the Inner Heliosphere*, Vol. I, ed. by R. Schwenn and E. Marsch, Springer-Verlag, Berlin, Heidelberg, New York, 1990.

10.24   Blackman, R.B., J.W. Tukey, *The Measurement of Power Spectra*, Dover, New York, 1958.

10.25   Blake, D.H., J.W. Belcher, Power spectra of the interplanetary magnetic field, 0.7-1.6 AU, J. Geophys. Res., **79**, 2891, 1974.

10.26   Bruno, R., B. Bavassano, U. Villante, Evidence for long-period Alfvén waves in the inner solar system, J. Geophys. Res., **90**, 4373, 1985.

10.27 Bruno, R., M. Dobrowolny, Spectral measurements of magnetic energy and magnetic helicity between 0.29 and 0.97 AU, Annales Geophys., **4**, A, 17, 1986.

10.28 Bruno, R., U. Villante, B. Bavassano, R. Schwenn, F. Mariani, In-situ observations of the latitudinal gradients of the solar wind parameters during 1976 and 1977, Solar Phys., **104**, 431, 1986.

10.29 Bruno, R., B. Bavassano, Latitudinal dependence of the interplanetary magnetic field spectrum, Annales Geophys., **5A**, 265, 1987.

10.30 Burlaga, L.F., Hydromagnetic waves and discontinuities in the solar wind, Space Sci. Rev., **12**, 600, 1971.

10.31 Burlaga, L.F., Micro-scale structures in the interplanetary medium, 2, J. Geophys. Res., **76**, 3534, 1971.

10.32 Burlaga, L.F., Microstructure in the interplanetary medium, NASA SP-308, 309, 1972.

10.33 Burlaga, L.F., Heliospheric magnetic fields and plasmas, Rev. Geophys. Space Phys., **21**, 363, 1983.

10.34 Burlaga, L.F., MHD processes in the outer heliosphere, Space Sci. Rev., **39**, 255, 1984.

10.35 Burlaga, L.F., K.W. Ogilvie, Magnetic and thermal pressures in the solar wind, Solar Physics, **15**, 61, 1970.

10.36 Burlaga, L.F., J.B. Turner, Microscale 'Alfvén waves' in the solar wind at 1 AU, J. Geophys. Res., **81**, 73, 1976.

10.37 Burlaga, L.F., W.H. Mish, Large-scale fluctuations in the interplanetary medium, J. Geophys. Res., **92**, 1261, 1987.

10.38 Burlaga, L.F., W.H. Mish, D.A. Roberts, Large-scale fluctuations in the solar wind at 1 AU: 1978–1982, J. Geophys. Res., **94**, 177, 1989.

10.39 Coleman, P.J., Jr., Variations in the interplanetary magnetic field: Mariner 2, J. Geophys. Res., **71**, 5509, 1966.

10.40 Coleman, P.J., Jr., Wave-like phenomena in the interplanetary plasma: Mariner 2, Planet. Space Sci., **15**, 953, 1967.

10.41 Coleman, P.J., Turbulence, viscosity, and dissipation in the solar wind plasma, Astrophys. J., **153**, 371, 1968.

10.42 Dahlburg, R.B., J.M. Picone, J.T. Karpen, Growth of correlation in compressible two-dimensional magnetofluid turbulence, J. Geophys. Res., **93**, 2527, 1988.

10.43 Denskat, K.U., Untersuchung von Alfvénischen Fluktuationen im Sonnenwind zwischen 0.29 AE und 1.0 AE, Dissertation (PhD thesis), Technische Universität Braunschweig, F.R. Germany, 1982.

10.44 Denskat, K.U., L.F. Burlaga, Multispacecraft observations of microscale fluctuations in the solar wind, J. Geophys. Res., **82**, 2693, 1977.

10.45 Denskat, K.U., F.M. Neubauer, R. Schwenn, Properties of Alfvénic fluctuations near the Sun: Helios-1 and Helios-2, in *Solar Wind Five*, ed. by H. Rosenbauer, MPAE-Report No. W-100-81-31, Max-Planck-Institut für Aeronomie, Lindau, F.R. Germany, 392, 1981.

10.46 Denskat, K.U., F.M. Neubauer, Statistical properties of low-frequency magnetic field fluctuations in the solar wind from 0.29 to 1.0 AU during solar minimum conditions: Helios 1 and Helios 2, J. Geophys. Res. **87**, 2215, 1982.

10.47 Denskat, K.U., H.J. Beinroth, F.M. Neubauer, Interplanetary magnetic field power spectra with frequencies from 2.4×10^{-5} Hz to 470 Hz from Helios observations during solar minimum conditions, J. Geophys., **54**, 60, 1983.

10.48 Denskat, K.U., F.M. Neubauer, Observations of hydromagnetic turbulence in the solar wind, *Solar Wind Five*, ed. by M. Neugebauer, NASA Conf. Publ., **CP-2280**, 81, 1983.

10.49 Derby, N.F., Jr., Modulational instability of finite-amplitude, circularly polarized Alfvén waves, Astrophys. J., **224**, 1013, 1978.

10.50 Dewar, R.L., Interaction between hydromagnetic waves and a time-dependent, inhomogeneous medium, Phys. Fluids, **13**, 2710, 1970.

10.51 Dobrowolny, M., Kelvin–Helmholtz instability in a high-β collisionless plasma, Phys. Fluids, **15**, 2263, 1972.

10.52 Dobrowolny, M., A. Mangeney, P. Veltri, Properties of magnetohydrodynamic turbulence in the solar wind, Astron. Astrophys., **83**, 26, 1980.

10.53 Dobrowolny, M., A. Mangeney, P. Veltri, Fully developed anisotropic turbulence in interplanetary space, Phys. Rev. Lett., **45**, 144, 1980.

10.54 Dobrowolny, M., G. Torricelli-Ciamponi, Alfvén wave dissipation in the solar wind, Astron. Astrophys., **142**, 404, 1985.

10.55 Domingo, V., (ed.), *The SOHO Mission, Scientific and Technical Aspects of the Instruments*, ESA SP-1104, 1988.

10.56 Elsässer, W.M., The hydromagnetic equations, Phys. Rev., **79**, 183, 1950.

10.57 Feldman, W.C., B. Abraham-Shrauner, J.R. Asbridge, S.J. Bame, The internal plasma state of the high speed solar wind at 1 AU, in *Physics of Solar Planetary Environments*, Vol. I, ed. by D.A. Williams, AGU, 413, 1976.

10.58 Freeman, J.W., Estimates of solar wind heating inside 0.3 AU, Geophys. Res. Lett., **15**, 88, 1988.

10.59 Freeman, J.W., R.E. Lopez, The cold solar wind, J. Geophys. Res., **90**, 9885, 1985.

10.60 Frisch, U., A. Pouquet, J. Léorat, A. Mazure, Possibility of an inverse cascade of magnetic helicity in magnetohydrodynamic turbulence, J. Fluid Mech., **68**, 769, 1975.

10.61 Goldstein, B.E., Analysis of fluctuations in the solar wind, Ph.D. thesis, MIT, Cambridge, USA, 1971.

10.62 Goldstein, B., G.L. Siscoe, Spectra and cross spectra of solar wind parameters from Mariner 5, in *Solar Wind Three*, ed. by C.P. Sonett, P.J. Coleman, and J.M. Wilcox, NASA Special Publ., SP-308, 506, 1972.

10.63 Goldstein, M.L., An instability of finite-amplitude, circularly polarized Alfvén waves, Astrophys. J., **219**, 700, 1978.

10.64 Goldstein, M.L., L.F. Burlaga, W.H. Matthaeus, Power spectral signatures of interplanetary corotating and transient flows, J. Geophys. Res., **89**, 3747, 1984.

10.65 Goldstein, M.L., D.A. Roberts, W.H. Matthaeus, Systematic errors in determining the propagation direction of interplanetary Alfvénic fluctuations, J. Geophys. Res., **91**, 13, 357, 1986.

10.66 Goldstein, M.L., D.A. Roberts, W.H. Matthaeus, Numerical simulation of interplanetary and magnetospheric phenomena: The Kelvin–Helmholtz instability, Proceedings of Yosemite, *Outstanding Problems in Solar System Plasma Physics: Theory and Instrumentation*, in press, 1989.

10.67 Granik, A.T., Large-amplitude waves in an anisotropic plasma, J. Geophys. Res., **86**, 5431, 1981.

10.68 Grappin, R., Onset and decay of two-dimensional magnetohydrodynamic turbulence with velocity magnetic field correlation, Phys. Fluids, **29**, 2433, 1986.

10.69 Grappin, R., U. Frisch, J. Léorat, A. Pouquet, Alfvénic fluctuations as asymptotic states of MHD turbulence, Astron. Astrophys., **105**, 6, 1982.

10.70 Grappin, R., A. Pouquet, J. Léorat, Dependence of MHD turbulence spectra on the velocity field–magnetic field correlation, Astron. Astrophys., **126**, 51, 1983.

10.71 Grappin, R., A. Mangeney, E. Marsch, On the origin of solar wind turbulence: Helios data revisited, in *Turbulence and Nonlinear Dynamics in MHD Flows*, ed. by M. Meneguzzi, A. Pouquet and P.L. Sulem, Elsevier Science Publishers B.V., North-Holland, 81, 1989.

10.72 Grappin, A. Mangeney, E. Marsch, On the origin of solar wind MHD turbulence: Helios data revisited, J. Geophys. Res., **95**, 8197, 1990.

10.73 Gurnett, D.A., Waves and Instabilities, in *Physics of the Inner Heliosphere* (this volume).

10.74 Heinemann, M., S. Olbert, Non-WKB Alfvén waves in the solar wind, J. Geophys. Res. **85**, 1311, 1980.

10.75 Hinze, J.O., *Turbulence*, Second edition, McGraw-Hill, New York, 1975.

10.76 Hollweg, J.V., Transverse Alfvén waves in the solar wind: Arbitrary k, v_o, B_o, and $|\delta B|$, J. Geophys. Res., **79**, 1539, 1974.

10.77 Hollweg, J.V., Waves and instabilities in the solar wind, Rev. Geophys. Space Phys., **13**, 263, 1975.

10.78 Hollweg, J.V., Some physical processes in the solar wind, Rev. Geophys. Space Phys., **16**, 689, 1978.

10.79 Hollweg, J.V., Surface waves on solar wind tangential discontinuities, J. Geophys. Res., **87**, 8065, 1982.

10.80 Hollweg, J.V., Transition region, corona, and solar wind in coronal holes, J. Geophys. Res., **91**, 4111, 1986.

10.81 Hollweg, J.V., Small-scale MHD wave processes in the solar atmosphere and solar wind, in Proc. of the 21st ESLAB Symposium, ed. by B. Battrick and E.J. Wolfe, ESA SP-275, Boleksjo, Norway, 161, 1987.

10.82 Hollweg, J.V., M.K. Bird, H. Voll, P. Edenhofer, C.T. Stenzelried, B.L. Seidel, Possible evidence for coronal Alfvén waves, J. Geophys. Res., **87**, 1, 1982.

10.83 Hollweg, J.V., B. Roberts, Surface solitary waves and solitons, J. Geophys. Res., **89**, 9703, 1984.

10.84 Hollweg, J.V., W. Johnson, Transition region, corona, and solar wind in the coronal holes: Some two fluid models, J. Geophys. Res., **93**, 9547, 1988.

10.85 Holzer, T.E., The solar wind and related astrophysical phenomena, in *Solar System Plasma Physics*, Vol. I, ed. by E.N. Parker, C.F. Kennel, and L.J. Lanzerotti, North-Holland, Amsterdam, 101, 1979.

10.86 Holzer, T.E., T. Flå, E. Leer, Alfvén waves in stellar winds, Astrophys. J., **275**, 808, 1983.

10.87 Inhester, B., A drift-kinetic treatment of the parametric decay of large-amplitude Alfvén waves, J. Geophys. Res., **95**, 10525, 1989.

10.88 Intriligator, D.S., J.H. Wolfe, Preliminary power spectra of the interplanetary plasma, Astrophys. J., **162**, L187, 1970.

10.89 Isenberg, P.A., J.V. Hollweg, On the preferential acceleration and heating of solar wind heavy ions, J. Geophys. Res., **88**, 3923, 1983.

10.90 Klein, L.W., Observations of turbulence and fluctuations in the solar wind, Ph.D. thesis, Catholic University of America, Washington D.C., 1987.

10.91 Kolmogorov, A.N., The local structure of turbulence in incompressible viscous fluid for very large Reynolds numbers, Compt. Rend. Acad. Sci. U.R.S.S., **30**, 301, 1941.

10.92 Korzhov, N.P., V.V. Mishin, V.M. Tomozov, On the role of plasma parameters and the Kelvin–Helmholtz instability in a viscous interaction of solar wind streams, Planet. Space Sci., **32**, 1169, 1984.

10.93 Kraichnan, R.H., Inertial-range spectrum of hydromagnetic turbulence, Phys. Fluids, **8**, 1385, 1965.

10.94 Kunow, H., G. Wibberenz, G. Green, R. Müller-Mellin, M.B. Kallenrode, Energetic particles in the inner solar system, in *Physics of the Inner Heliosphere* (this volume).

10.95 Lacombe, C., A. Mangeney, Non-linear interaction of Alfvén waves with compressive fast magnetosonic waves, Astron. Astrophys., **88**, 277, 1980.

10.96 Leer, E., T.E. Holzer, T. Flå, Acceleration of the solar wind, Space Sci. Rev., **33**, 161, 1982.

10.97 Li, H.-S., E.G. Zweibel, The scattering of Alfvén waves by density fluctuations, Astrophys. J., **322**, 248, 1987.

10.98 Luttrell, A.H., A.K. Richter, Power spectra of low-frequency MHD turbulence up- and downstream of interplanetary fast shocks within 1 AU, Annales Geophys., **4**, 439, 1986.

10.99 Luttrell, A.H., A.K. Richter, A study of MHD fluctuations upstream and downstream of quasi-parallel interplanetary shocks, J. Geophys. Res., **92**, 2243, 1987.

10.100 Luttrell, A.H., A.K. Richter, The role of Alfvénic fluctuations in MHD turbulence evolution between 0.3 and 1 AU, Proceedings of Sixth International Solar Wind Conference, Vol. I, NCAR Technical Note, 335, 1988.

10.101 Machida, S., S.R. Spangler, C.K. Goertz, Simulation of amplitude-modulated circularly polarized Alfvén waves for beta less than one, J. Geophys. Res., **92**, 7413, 1987.

10.102 Mariani, F., U. Villante, R. Bruno, B. Bavassano, N.F. Ness, An extended investigation of Helios 1 and 2 observations; The interplanetary magnetic field between 0.3 and 1 AU, Solar Phys., **63**, 411, 1979.

10.103 Mariani, F., Bavassano, B., Villante, U., A statistical study of MHD discontinuities in the inner solar system: Helios 1 and 2, Solar Phys., **83**, 349, 1983.

10.104 Mariani, F., F.M. Neubauer, Interplanetary magnetic field, in *Physics of the inner Heliosphere*, Vol. I, ed. by R. Schwenn and E. Marsch, Springer-Verlag, Heidelberg, 1990.

10.105 Marsch, E., Acceleration potential and angular momentum of undamped MHD-waves in stellar winds, Astron. Astrophys., **164**, 77, 1986.

10.106 Marsch, E., Kinetic physics of the solar wind plasma, in *Physics of the Inner Heliosphere* (this volume).

10.107 Marsch, E., K.-H. Mühlhäuser, H. Rosenbauer, R. Schwenn, K.U. Denskat, Pronounced core temperature anisotropy, ion differential speed, and simultaneous Alfvén wave activity in slow wind at 0.3 AU, J. Geophys. Res., **86**, 9199, 1981.

10.108 Marsch, E., C.K. Goertz, A.K. Richter, Wave heating and acceleration of solar wind ions by cyclotron resonance, J. Geophys. Res., **87**, 5030, 1982.

10.109 Marsch, E., K.-H. Mühlhäuser, R. Schwenn, H. Rosenbauer, W. Pilipp, F.M. Neubauer, Solar wind protons: Three-dimensional velocity distributions and derived plasma parameters measured between 0.3 and 1 AU, J. Geophys. Res., **87**, 52, 1982.

10.110 Marsch, E., A.K. Richter, Helios observational constraints on solar wind expansion, J. Geophys. Res., **89**, 6599, 1984.

10.111 Marsch, E., A.K. Richter, Distribution of solar wind angular momentum between particles and magnetic field: Inferences about the Alfvén critical point from Helios observations, J. Geophys. Res., **89**, 5386, 1984.

10.112 Marsch, E., A. Mangeney, Ideal MHD equations in terms of compressive Elsässer variables, J. Geophys. Res., **92**, 7363, 1987.

10.113 Marsch, E., C.-Y. Tu, Dynamics of correlation functions with Elsässer variables for inhomogeneous MHD turbulence, J. Plasma Phys., **41**, 479, 1989.

10.114 Marsch, E., C.-Y. Tu, On the radial evolution of MHD turbulence in the inner heliosphere, J. Geophys. Res., **95**, 8211, 1990.

10.115 Marsch, E., C.-Y. Tu, Spectral and spatial evolution of compressive turbulence in the inner solar wind, J. Geophys. Res., **95**, 11945, 1990.

10.116 Matthaeus, W.H., D. C. Montgomery, Selective decay hypothesis at high mechanical and magnetic Reynolds numbers, Ann. N. Y. Acad. Sci. **357**, 203, 1980.

10.117 Matthaeus, W.H., C. Smith, Structure of correlation tensors in homogeneous anisotropic turbulnce, Phys. Rev. A, **24**, 2135, 1981.

10.118 Matthaeus, W.H., M.L. Goldstein, Stationarity of magnetohydrodynamic fluctuations in the solar wind, J. Geophys. Res., **87**, 10347, 1982.

10.119 Matthaeus, W.H., M.L. Goldstein, Measurements of the rugged invari- ants of magnetohydrodynamic turbulence in the solar wind, J. Geophys. Res., **87**, 6011, 1982.

10.120 Matthaeus, W.H., M.L. Goldstein, D.C. Montgomery, Turbulent generation of outward traveling interplanetary Alfvénic fluctuations, Phys. Rev. Lett., **51**, 1484, 1983.

10.121 Matthaeus, W.H., M.L. Goldstein, Magnetohydrodynamic turbulence in the solar wind, *Solar Wind Five*, ed. by M. Neugebauer, NASA CP-2280, 73, 1983.

10.122 Matthaeus, W.H., M.L. Goldstein, Low frequency 1/f noise in the interplanetary magnetic field, Phys. Rev. Lett., **57**, 495, 1986.

10.123 Matthaeus, W.H., M.L. Goldstein, J. H. King, An interplanetary magnetic field ensemble at 1 AU, J. Geophys. Res. **91**, 59, 1986.

10.124 Matthaeus, W.H., M.R. Brown, Nearly incompressible magnetohydrodynamics at low Mach number, Phys. Fluids, **31**, 3634, 1988.

10.125 Matthaeus, W.H., Y. Zhou, Nearly incompressible MHD turbulence in the solar wind, in *Turbulence and Nonlinear Dynamics in MHD Flows*, ed. by M. Meneguzzi, A. Pouquet, and P.L. Sulem, Elsevier Science Publishers B.V., North-Holland, 1989.

10.126 Mavromichalaki, H., X. Moussas, J.J. Quenby, J.F. Valdes-Galicia, E.J. Smith, B.T. Thomas, Relatively stable, large-amplitude Alfvénic waves seen at 2.5 and 5.0 AU, Solar Phys., **116**, 377, 1988.

10.127 Moffatt, H.K., *Magnetic Field Generation in Electrically Conducting Fluids*, Cambridge University Press, Cambridge, 1978.

10.128 Montgomery, D.C., Theory of hydromagnetic turbulence, *Solar Wind Five*, ed. by M. Neugebauer, NASA CP-2280, 107, 1983.

10.129 Montgomery, D., Characteristic phenomena of magnetohydrodynamic turbulence, Proc. ESA Workshop on Future Missions in Solar, Heliospheric &Space Plasma Physics, ESA SP-235, 175, 1985.

10.130 Montgomery, D., M.R. Brown, W.H. Matthaeus, Density fluctuation spectra in magnetohydrodynamic turbulence, J. Geophys. Res., **92**, 282, 1987.

10.131 Musmann, G., F.M. Neubauer, E. Lammers, Radial variation of the interplanetary magnetic field between 0.3 AU and 1.0 AU; Observations by the HELIOS 1 spacecraft, J. Geophys., **42**, 591, 1977.

10.132 Neubauer, F.M., H.J. Beinroth, H. Barnstorf, G. Dehmel, Initial results from the Helios 1 search coil magnetometer experiment, J. Geophys., **42**, 599, 1977.

10.133 Neubauer, F.M., G. Musman, G. Dehmel, Fast magnetic fluctuations in the solar wind: Helios 1, J. Geophys. Res., **82**, 3201, 1977.

10.134 Neugebauer, M., C.S. Wu, J.D. Huba, Plasma fluctuations in the solar wind, J. Geophys. Res., **83**, 1027, 1978.

10.135 Neugebauer M., C.J. Alexander, R. Schwenn, A.K. Richter, Tangential discontinuities in the solar wind: correlated field and velocity changes and the Kelvin–Helmholtz instability, J. Geophys. Res., **91**, 13694, 1986.

10.136 Olbers, D.J., A.K. Richter, Wave-trains in the solar wind I, Astrophys. Space Sci., **20**, 373, 1973.

10.137 Ovenden, C.R., H.A. Shah, S.J. Schwartz, Alfvén solitons in the solar wind, J. Geophys. Res., **88**, 6095, 1983.

10.138 Panchev, S., *Random Functions and Turbulence*, Pergamon Press, Oxford, 1971.

10.139 Parker, E.N., *Interplanetary Dynamical Processes*, Interscience, New York, 1963.

10.140 Passot, T., A. Pouquet, Numerical simulation of compressible homogeneous flows in the turbulent regime, J. Fluid Mech., **181**, 441, 1987.

10.141 Pizzo, V.J., Quasi-steady solar wind dynamics, in Solar Wind Five, ed. by M. Neugebauer, NASA Conference Publication 2280, Washington, USA, 675, 1983.

10.142 Pouquet, A., Magnetohydrodynamic turbulence, in Astrophysical Fluid Dynamics, Les Houches, ed. by J.P. Zahn and J. Zinn-Justin, Elsevier Science Publishers, 1988.

10.143 Pouquet, A., U. Frisch, J. Léorat, Strong MHD helical turbulence and the nonlinear dynamo effect, J. Fluid Mech., **77**, 321, 1976.

10.144 Pouquet, A., G.S. Patterson, Numerical simulation of helical magnetohydrodynamic turbulence, J. Fluid Mech., **85**, 305, 1978.

10.145 Pouquet, A., M. Meneguzzi, U. Frisch, Growth of correlations in magnetohydrodynamic turbulence, Phys. Rev. A, **33**, 4266, 1986.

10.146 Pouquet, A., P.L. Sulem, M. Meneguzzi, Influence of velocity velocity–magnetic field correlations on decaying MHD turbulence with neutral points, Phys. Fluids, **31**, 2635, 1988.

10.147 Richter, A.K., Interplanetary slow shocks, in *Physics of the Inner Heliosphere* (this volume).

10.148 Roberts, D.A., Interplanetary observational constraints on Alfvén wave acceleration of the solar wind, J. Geophys. Res., **94**, 6899, 1989.

10.149 Roberts, D.A., Heliocentric distance and temporal dependence of the interplanetary density-magnetic field magnitude correlation, J. Geophys. Res., **95**, 1087, 1989.

10.150 Roberts, D.A., M.L. Goldstein, Spectral signatures of jumps and turbulence in interplanetary speed and magnetic field data, J. Geophys. Res., **92**, 10105, 1987.

10.151 Roberts, D.A., L.W. Klein, M.L. Goldstein, W.H. Matthaeus, The nature evolution of magnetohydrodynamic fluctuations in the solar wind: Voyager observations, J. Geophys. Res., **92**, 11021, 1987.

10.152 Roberts, D.A., M.L. Goldstein, L.W. Klein, W.H. Matthaeus, Origin and evolution of fluctuations in the solar wind: Helios observations and Helios-Voyager comparisons, J. Geophys. Res., **92**, 12023, 1987.

10.153 Roberts, D.A., M.L. Goldstein, L.W. Klein, The amplitudes of interplanetary fluctuations: Stream structure, heliocentric distance, and frequency dependence, J. Geophys. Res., **95**, 4203, 1989.

10.154 Roberts, D.A., M.L. Goldstein, Simulation of interplanetary dynamical processes, in *Proceedings of the Third International Conference on Supercomputing*, Vol. I, ed. by L.P. Kartashev and S.I. Kartashev, International Supercomputing Institute, 370, 1988.

10.155 Roberts, D.A., M.L. Goldstein, W.H. Matthaeus, L.W. Klein, Observation and simulation of MHD turbulence in the solar wind, in *Turbulence and Nonlinear Dynamics in MHD Flows*, ed. by M. Meneguzzi, A. Pouquet, and P.L. Sulem, Elsevier Science Publishers B.V., North-Holland, 87, 1989.

10.156 Rose, H.A., P.L. Sulem, Fully developed turbulence and statistical mechanics, Le Journal de Physique, **139**, 441, 1978.

10.157 Rosenbauer, H., R. Schwenn, E. Marsch, B. Meyer, H. Miggenrieder, M.D. Montgomery, K.-H. Mühlhäuser, W. Pilipp, W. Voges, S.M. Zink, A survey of initial results of the Helios plasma experiment, J. Geophys., **42**, 561, 1977.

10.158 Sakai, J.-I., B.U.Ö. Sonnerup, Modulational instability of finite amplitude dispersive Alfvén waves, J. Geophys. Res., **88**, 9069, 1983.

10.159 Sari, J.W., G.C. Valley, Interplanetary magnetic field power spectra: Mean field radial or perpendicular to radial, J. Geophys. Res., **81**, 5489, 1976.

10.160 Schwartz, S.J., E. Marsch, The radial evolution of a single solar wind plasma parcel, J. Geophys. Res. **88**, 9919, 1983.

10.161 Schwenn, R., The 'average' solar wind in the inner heliosphere: Structures and slow variations, in *Solar Wind Five*, ed. by M. Neugebauer, NASA Conf. Publ., CP-**2280**, 489, 1983.

10.162 Schwenn, R., Large-scale structure of the interplanetary medium, in *Physics of the Inner Heliosphere*, Vol. I, ed. by R. Schwenn and E. Marsch, Springer-Verlag, Berlin, Heidelberg, New York, 1990.

10.163 Schwenn, R., K.-H. Mühlhäuser, E. Marsch, H. Rosenbauer, Two states of the solar wind at the time of solar activity minimum, II. Radial gradients of plasma parameters in fast and slow streams, in *Solar Wind Four*, ed. by H. Rosenbauer, Report MPAE-W-100-81-31, MPI Aeronomie, Katlenburg-Lindau, F.R.Germany, 126, 1981.

10.164 Shebalin, J.V., W.H. Matthaeus, D. Montgomery, Anisotropy in MHD turbulence due to a mean magnetic field, J. Plasma Physics, **29**, 525, 1983.

10.165 Shebalin, J.V., D. Montgomery, Turbulent magnetohydrodynamic density fluctutations, J. Plasma Physics, **39**, 339, 1988.

10.166 Smith, C.W., M.L. Goldstein, W.H. Matthaeus, Turbulence analysis of the Jovian upstream "wave" phenomena, J. Geophys. Res., **88**, 5581, 1983.

10.167 Spangler, S.R., The evolution of nonlinear Alfvén waves subject to growth and damping, Phys. Fluids, **29**, 2535, 1986.

10.168 Spangler, S.R., Density fluctuations induced by nonlinear Alfvén waves, Phys. Fluids, **30**, 1104, 1987.

10.169 Spangler, S.R., J.P. Sheerin, Properties of Alfvén solitons in a finite-beta plasma, J. Plasma Phys., **27**, 193, 1982.

10.170 Taylor, G.I., The spectrum of turbulence, Proc. Roy. Soc. London Ser. A, **164**, 476, 1938.

10.171 Terasawa, T., M. Hoshino, J.-I. Sakai, T. Hada, Decay instability of finite-amplitude circularly polarized Alfvén waves: A numerical simulation of stimulated Brillouin scattering, J. Geophys. Res., **91**, 4171, 1986.

10.172 Thieme, K.M., Zusammenhänge zwischen raum-zeitlichen Strukturen im Sonnenwind und seinen Quellgebieten in der Korona, Dissertation (PhD thesis), Universität Göttingen, 1990.

10.173 Thieme, K.M., E. Marsch, R. Schwenn, Relationship between structures in the solar wind and their source regions in the corona, Proceedings of the Sixth International Solar Wind Conference, Vol. I, NCAR Technical Note-306, 317, 1988.

10.174 Thieme, K.M., R. Schwenn, E. Marsch, Are structures in high-speed streams signatures of coronal fine structures?, Adv. Space Res., **9**, 127, 1989.

10.175 Tsurutani, B.T., E.J. Smith, D.E. Jones, Waves observed upstream of interplanetary shocks, J. Geophys. Res., **88**, 6545, 1983.

10.176 Tsurutani, B.T., R.G. Stone (eds.), Collisionless shocks in the Heliosphere: Reviews of Current Research, Geophysical Monograph, **35**, 1985.

10.177 Tu, C.-Y., The governing equations of the variation of Alfvénic fluctuations, Chin. J. Space Sci., **3**, 269, 1983.

10.178 Tu, C.-Y., A solar wind model with the power spectrum of Alfvénic fluctuations, Solar Phys., **109**, 149, 1987.

10.179 Tu, C.-Y., The damping of interplanetary Alfvénic fluctuations and the heating of the solar wind, J. Geophys. Res., **93**, 7, 1988.

10.180 Tu, C.-Y., Explanation for the radial variation of the temperature of protons within the trailing edge of high speed streams between 1 and 5 AU, in Proceedings of the Sixth International Solar Wind Conference, Vol. II, NCAR Technical Note-306, 593, 1988.

10.181 Tu, C.-Y., Z.-Y. Pu, F.-S. Wei, The power spectrum of interplanetary Alfvénic fluctuations: Derivation of the governing equation and its solution, J. Geophys. Res., **89**, 9695, 1984.

10.182 Tu, C.-Y., J.W. Freeman, R.E. Lopez, The proton temperature and the total hourly variance of the magnetic field components in different solar wind speed regions, Solar Phys., **119**, 197, 1989.

10.183 Tu, C.-Y., D.A. Roberts, M.L. Goldstein, Spectral evolution and cascade constant of solar wind Alfvénic turbulence, J. Geophys. Res., **94**, 13575, 1989.

10.184 Tu, C.-Y., E. Marsch, K.M. Thieme, Basic properties of solar wind MHD turbulence near 0.3 AU analysed by means of Elsässer Variables, J. Geophys. Res., **95**, 11739, 1989.

10.185 Tu, C.-Y., E. Marsch, Evidence for a "background" spectrum of solar wind turbulence in the inner heliosphere, J. Geophys. Res., **95**, 4337, 1990.

10.186 Tu, C.-Y., E. Marsch, H. Rosenbauer, The dependence of MHD turbulence spectra on the inner solar wind stream structure near solar minimum, Geophys. Res. Lett., **17**, 283, 1990.

10.187 Tu, C.-Y., E. Marsch, Transfer equations for spectral densities of inhomogeneous MHD turbulence, J. Plasma Phys., **44**, 103, 1990.

10.188 Vellante, M., A.J. Lazarus, An analysis of solar wind fluctuations between 1 and 10 AU, J. Geophys. Res. **92**, 9893, 1987.

10.189 Velli, M., R. Grappin, A. Mangeney, The effect of large scale gradients on the evolution of Alfvénic turbulence in the solar wind, in *Plasma Phenomena in the Solar Atmosphere*, ed. by M.A. Dubois, Editions de Physique, Orsay, in press, 1989.

10.190 Velli, M., R. Grappin, A. Mangeney, Turbulent cascade of incompressible unidirectional Alfvén waves in the interplanetary medium, Phys. Rev. Lett., **63**, 1807, 1989.

10.191 Veltri, P., A. Mangeney, M. Dobrowolny, Cross-helicity effects in aniso- tropic MHD turbulence, Il Nuovo Cimento, **68B**, 235, 1982.

10.192 Villante, U., On the role of Alfvénic fluctuations in the inner solar system, J. Geophys. Res., **85**, 6869, 1980.

10.193 Villante, U., M. Vellante, The radial evolution of the IMF fluctuations: A comparison with theoretical models, Solar Phys., **81**, 367, 1982.

10.194 Viñas, A.F., M.L. Goldstein, M.H. Acuña, Spectral analysis of magnetohydrodynamic fluctuations near interplanetary shocks, J. Geophys. Res., **89**, 3762, 1984.

10.195 Völk, H.J., Microstructure of the solar wind, Space Sci. Rev., **17**, 255, 1975.

10.196 Whang, Y.C., Alfvén waves in spiral interplanetary field, J. Geophys. Res., **78**, 7221, 1973.

10.197 Whang, Y.C., A magnetohydrodynamic model for corotating interplanetary structures, J. Geophys. Res., **85**, 2285, 1980.

10.198 Whang, W.C., L.F. Burlaga, Evolution and interaction of interplanetary shocks, J. Geophys. Res., **90**, 10765, 1985.

10.199 Woltjer, L., A theorem on force free magnetic fields, Proc. Natl. Acad. Sci. USA, **44**, 489, 1958.

10.200 Woltjer, L., On hydrodynamic equilibrium, Proc. Natl. Acad. Sci. USA, **44**, 833, 1958.

10.201 Wong, H.K., M.L. Goldstein, Parametric instability of circularly polarized Alfvén waves including dispersion, J. Geophys. Res., **91**, 5617, 1986.

10.202 Zhou, Y., W.H. Matthaeus, Non-WKB evolution of solar wind fluctuations: A turbulence modeling approach, Geophys. Res. Lett., **16**, 755, 1989.

11. Energetic Particles in the Inner Solar System

Horst Kunow, Gerd Wibberenz, Günter Green,
Reinhold Müller-Mellin and May-Britt Kallenrode

11.1 Introduction

The distinction between the inner and the outer solar system is traditionally related to the planetary system, where the inner, earth-like planets, are distinguished from the outer planets by their different chemical composition and structure. In the particles and fields area a division into the "inner" and the "outer" solar system also makes sense, though for other reasons.

- The stream structure of the solar wind and its relation to properties of the solar corona are more clearly visible within 1 AU than beyond (see [11.292]). The interaction of fast and slow solar wind streams becomes important beyond the orbit of the earth, leading to corotating interaction regions with a forward and a reverse shock. This "stream zone" extends to about 8 AU and is followed by the pressure wave zone and wave interaction zone further out.
- The spiral angle of the average interplanetary magnetic field (IMF) varies with distance from the sun from about 30° at 0.5 AU to about 87° at 5 AU [11.172]. The approximately radial alignment of the IMF in the inner solar system and the approximately azimuthal alignment in the outer solar system have important consequences for the nature and propagation of turbulence, for the relative importance of energetic particle propagation parallel and perpendicular to the average magnetic field (Sect. 11.3), and for the acceleration of particles at corotating interplanetary shocks (Sect. 11.6).
- The largest part of the modulation of galactic cosmic rays (see Sect. 11.2) occurs way beyond 1 AU, and the additional radial variation of galactic cosmic rays amounts to only a few percent if one moves from 1 AU towards the sun.

As an important consequence, it is much easier to relate properties of the local interplanetary medium to the solar source if we have observations within 1 AU. In particular, in the study of solar cosmic rays it helps greatly to trace observations back to the magnetic connection point in the corona.

Results presented in this chapter will to a large extent be based on cosmic ray observations from *Helios 1* and 2, because the two spaceprobes covered the inner solar system between 0.29 and 1.0 AU for a long time, comprising almost one solar cycle for *Helios 1*. In addition to the advantages already mentioned, cosmic ray studies on *Helios* profited from the following.

- The close approach to the sun allows the detection of small solar events. Related to the small extent of the source and acceleration region in the solar atmosphere they display features not found by the large and dramatic eruptions.
- The close approach also allows the study of injection of energetic particles from the sun with good temporal resolution, because the effects of interplanetary diffusion are much less pronounced.
- During the missions of *Helios 1* and 2, which overlapped for about four years, azimuthal effects for different types of energetic particle populations (solar events, corotating events, energetic storm particles, super-events) could be analyzed.
- The observations are free from magnetospheric influences. In particular, interplanetary angular distributions of charged particles become distorted in the vicinity of the earth by reflections of particles at the magnetospheric bow shock.
- The *Helios* payload composition is almost ideally suited for cosmic ray studies. The information from the plasma and magnetic field instruments allows specification of the large-scale structure as well as the local and momentary state of the interplanetary medium. This information is required for connecting the point of observation magnetically back to the sun and for determining local wave–particle interactions based on the direction and turbulence of the IMF.

In dealing with energetic particles in any cosmic surrounding one has to treat properties of the sources supplying the seed population, the acceleration mechanism, and propagation processes between source and observer. When discussing energetic particles in the heliosphere it is very important to distinguish a number of populations with very different properties (for overviews see [11.95, 318, 295]). The main energetic particle components in the interplanetary medium are

(A) galactic cosmic rays,
(B) the anomalous component,
(C) solar flare produced cosmic rays (sometimes called solar cosmic rays or solar energetic particles (SEP)),
(D) low-energy cosmic ray nuclei associated with corotating interaction regions (also called corotating or recurrent events),
(E) particles accelerated by flare-initiated interplanetary shocks (also called energetic storm particles (ESP)),
(F) particles originating from planetary magnetospheres, in particular Jovian electrons.

The various components and their sources are sketched in Fig. 11.1. Their different properties can be briefly summarized as follows.

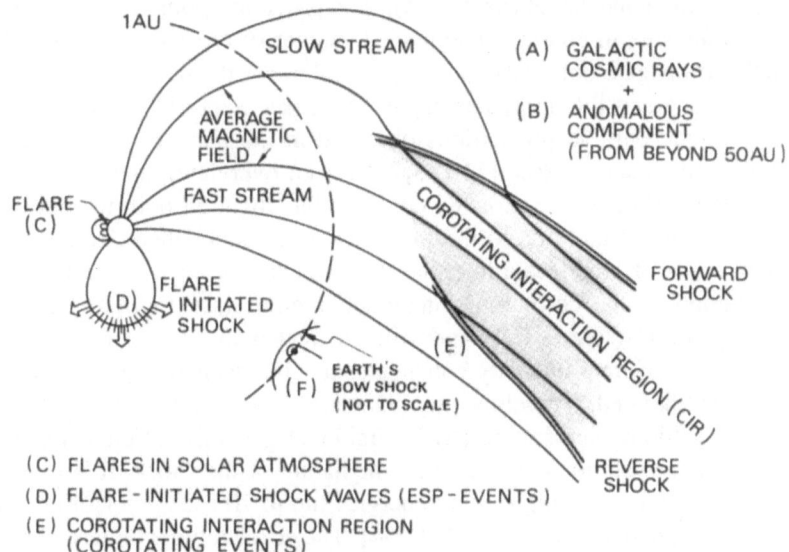

Fig. 11.1. Schematic representation of the large-scale structure of the interplanetary medium, related to the sequence of fast and slow solar wind streams. Letters (A) through (F) indicate various energetic particle components and their sources

The *galactic cosmic radiation* is incident uniformly and isotropically upon the heliosphere from the outside, extending to energies up to 10^{20} eV. For the purposes of this work we shall disregard point sources as well as a possible extragalactic origin at the highest energies. The galactic component has to find its way into the solar system with an extent of the order of 100 AU, penetrating against the combined action of the solar wind and the ordered and disordered magnetic fields. Time variations within the solar system caused by the variable structure of the interplanetary medium as a result of solar activity are called modulation and will be treated in Sect. 11.2.

The *anomalous component* takes its name from its anomalous chemical composition, charge state, spectral shape, and solar cycle variation, as compared with galactic cosmic rays (see [11.295] for an overview). The history of its discovery is as complex as the history of the individual particles in it, leading from interstellar neutrals to singly ionized energetic particles which most probably have gained their final energy at the termination shock separating the heliosphere from the interstellar medium. First observed for helium nuclei in 1972 [11.89] it soon became clear that the anomalous helium was also accompanied by anomalous nitrogen and oxygen but not carbon [11.109, 192]. A now widely accepted model for the origin of the anomalous component was proposed by Fisk and coworkers [11.81]. The suggested acceleration mechanisms, however, still need to be tested by observations [11.79].

Solar cosmic rays or *solar energetic particles* have their origin in isolated solar events. They are manifestations of energy release processes in solar active regions, e.g. during solar flares (Sect. 11.5), thus providing point sources of particles close to the sun, in contrast to extended sources of types (A) and (B). The solar particles propagate outwards and supply probes for the magnetic structures of the solar corona and the interplanetary medium (see Sects. 11.3 and 11.4).

Many energetic particle populations observed in the heliosphere can be related to collisionless shocks (see [11.285, 88] for an overview), and particle acceleration at these shocks has attracted considerable attention both theoretically and observationally. In addition to the planetary bow shocks there are basically two different types of large-scale interplanetary shock. The forward and reverse shock pair bounding the corotating interaction regions (see Fig. 11.1) lead to the *corotating events* (source (D)) and will be treated in Sect. 11.6. They are characterized by moderate intensity enhancements and steep energy spectra. No electrons are accelerated. Traveling interplanetary shocks related to flares and coronal mass ejections (see source (E) in Fig. 11.1) generate different types of energetic particle population depending on the nature of the shock. It is remarkable that acceleration to the highest ion energies (up to 70 MeV/nucleon) occurs in the outer heliosphere beyond 5 AU [11.254]. The acceleration at interplanetary shocks is generally restricted to ions; electron acceleration occurs only occasionally. Lopate [11.169] presents a few cases where electrons are observed up to relativistic energies for radial distances between 7 and 28 AU, and he summarizes necessary conditions for both electron and ion acceleration to occur at the shock at the same time. These conditions include the existence of a precursor wave in the upstream region and the strength of the shock. We shall show in Sect. 11.2 that the rare occurrences of these very strong shocks are connected to the appearance of ensembles of shocks in the inner solar system. We shall restrict our discussions to this particular case. For the large variety of studies on shock accelerated particles the reader is referred to overviews by Scholer [11.284, 285], Decker [11.55], Armstrong et al. [11.6], Lee [11.155], and references therein.

A particle species originating in a planetary magnetosphere (see (F) in Fig. 11.1) are the *Jovian electrons*. They are accelerated within Jupiter's magnetosphere and escape from the planet. Apart from electrons produced during solar flares, the interplanetary electron population is dominated by these Jovian electrons up to about 40 MeV. They can be found throughout the inner heliosphere. Near Jupiter their intensity is modulated with the 10-hour periodicity of the planet's rotation. Near the earth, peak intensities occur at times of best connection between the earth and Jupiter along the average direction of the interplanetary magnetic field every $\sim$13 months [11.33]. In their propagation away from Jupiter the electrons react to the stream structure of the interplanetary medium, and part of their motion occurs perpendicular to the average interplanetary magnetic field. The cross-field transport is affected by the degree of compression of the solar wind [11.44]. The average ratio of the perpendicular to parallel diffusion coefficient is of the order of 1–2% [11.44, 33]. Thus, the point source of electrons provided by Jupiter gives an excellent opportunity to use these particles as probes for the structure of the interplanetary medium.

This chapter does not provide a summary of all the components just sketched briefly and summarized in Fig. 11.1. The main emphasis is rather on components (C) and (D), where *Helios* made its largest contributions (see Sects. 11.3 through 11.6). Section 11.2 deals with modulation of galactic cosmic rays and serves as an introduction to solar cycle variations and the large-scale processes in the interplanetary medium.

The most familiar feature of the solar cycle is the 11-year variation of the sunspot number. It is directly related to the number of active regions on the sun

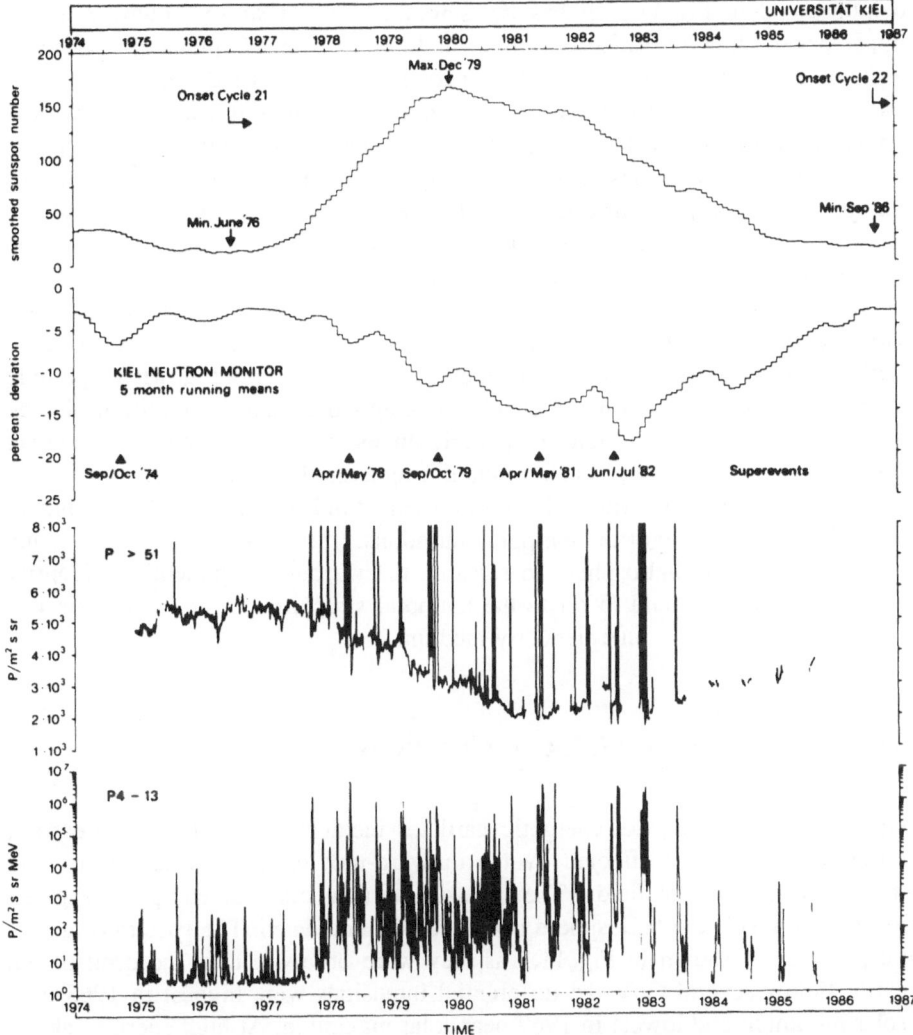

Fig. 11.2. *Upper two traces*: Solar cycle 21 given by smoothed Zürich sunspot number; 5-month running means of Kiel neutron monitor (cutoff 2.28 GV/c). *Solid triangles*: Super-events coincide with the neutron monitor minima. *Lower two traces*: *Helios 1* galactic protons above 51 MeV with flare induced increases; *Helios 1* solar protons between 4 and 13 MeV superimposed

and therefore to the number of solar flares. Sufficiently large flares generate solar energetic particles (see Sect. 11.5) and interplanetary shock waves. This means that the number of events connected with sources (C) and (D) is large during the maximum and small during the minimum of solar activity. However, there are also variations in the large-scale structure of the interplanetary medium connected with the polarity change of the general solar magnetic field and with variations of the warped current sheet [11.292, 172]. These changes modulate the intensity of particles streaming into the solar system from the outside (see Sect. 11.2).

The solar cycle variation of cosmic ray components of different energies is demonstrated in Fig. 11.2. It has the smoothed Zürich sunspot number in the upper panel, showing sunspot minima in June 1976 and September 1986, and a maximum in December 1979. The second panel shows the 5-month running mean of the Kiel neutron monitor. This monitor, located at sea level, responds typically to primary cosmic ray protons of galactic origin with energies of the order of 5 GeV. It varies in time roughly in anticorrelation with the sunspot number. The third panel in Fig. 11.2 is the integral counting rate of nuclei above 51 MeV/nucleon (essentially protons) as measured on *Helios*. The lower envelope of this curve shows a variation similar to the 5 GeV protons, with an onset of the decrease in September 1977. However, superimposed on this smooth decrease are a number of sharp increases which are due to solar particle events produced during individual solar flares (see Sect. 11.5). Going further down in particle energies we see in the lowest panel the temporal structure of protons in the energy range 4–13 MeV. Here the particle fluxes may increase by more than six orders of magnitude above background during a single solar event.

The three selected particle channels presented in Fig. 11.2 give a brief indication of the complexity of the temporal variations. It is one of the tasks of cosmic ray research in the heliosphere to separate the various components, to identify the particle sources, and to relate the temporal structures to the large-scale and local properties of the interplanetary medium.

11.2 Modulation of Galactic Cosmic Radiation

Galactic cosmic rays, i.e. energetic particles incident on the heliosphere from interstellar space, are affected by the outward-moving solar wind plasma and its frozen-in magnetic fields. The intensity as a function of energy at a given location r_0 inside the heliosphere is reduced or modulated. The spectra of galactic protons are shown in Fig. 11.3 as they were observed near the orbit of the earth. Below several GeV the differential intensities were highest in 1965 near solar minimum and lowest in 1969 near solar maximum. At high energies above 10 GeV no perceptible variation with time is observed. The modulation with the 11-year cycle of solar activity was shown in Fig. 11.2 for protons of more than 51 MeV from *Helios 1* and protons of more than 1.4 GeV from the Kiel neutron monitor.

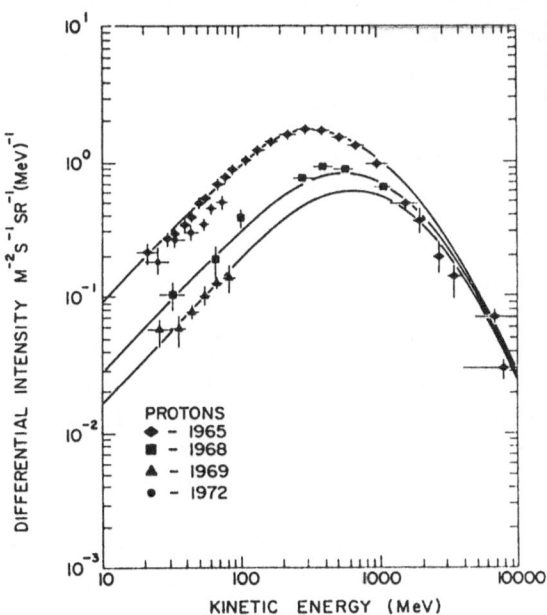

Fig. 11.3. Proton energy spectra observed at 1 AU in 1965 (solar minimum), 1968, 1969 (solar maximum), and 1972 (from [11.78])

Modulation research is intimately related to the exploration of the heliosphere and can make important contributions towards a unified view of cosmic ray propagation which furthers the understanding not only of the structure and electrodynamical processes near the sun but also of other astrophysical objects throughout our galaxy. The physical processes causing the modulation are still a matter of considerable debate, and the various models include the effects of diffusion, convection, adiabatic deceleration, large-scale drifts, and the influence of traveling interplanetary shocks.

In what follows we shall first briefly introduce these models followed by *in situ* and ground-based observations. The *Helios* spacecraft is well suited to carrying out modulation studies since it covers a whole solar cycle from the decaying phase of cycle 20 to the decaying phase of cycle 21 at a vantage-point outside the disturbing magnetosphere of the earth and close to the source of the modulation. Although – as we will show below – most of the modulation will take place far out in the heliosphere, it is still the changes in the condition of the sun which trigger off changes in conditions in the heliosphere, which in turn cause cosmic ray intensity variations.

11.2.1 Model Concepts

The solar modulation of galactic cosmic rays in the interplanetary medium can be discussed with the aid of a model based on Parker's concept [11.242], in which particles undergo outward convection with the solar wind and inward diffusion driven by the radial gradient and adiabatic deceleration resulting from the expansion of the solar wind. The cosmic ray number density U per unit interval of kinetic energy T satisfies a Fokker–Planck equation,

$$\frac{\partial U}{\partial t} = \nabla \cdot (K \cdot \nabla U) - \nabla \cdot (V U) + \frac{\nabla \cdot V}{3} \frac{\partial}{\partial T} (\alpha T U) - V_D \cdot \nabla U , \quad (11.1)$$

where K is the diffusion tensor, V the solar wind velocity, V_D the mean particle drift velocity, and $\alpha = (T + 2 T_0) / (T + T_0)$, with T_0 denoting the particle rest energy.

In the early model the balance of the three physical processes diffusion, convection, and energy change – i.e. the first three terms on the right-hand side of (11.1) – determines the spectra of galactic cosmic rays [11.242, 93, 125], and the model was quite successful in reproducing the observations from the beginning of modulation research in 1954, when Forbush discovered the anticorrelation between the level of solar activity and the cosmic ray intensity [11.84], until the launch of *Helios 1* and 2 some twenty years later and further on.

It is not possible to solve (11.1) analytically for realistic forms of the diffusion coefficient. As the scattering process is not isotropic, gradients along the magnetic field are more easily relieved than gradients across the field, and the spatial diffusion coefficient K is a tensor. Its antisymmetric elements describe the guiding center drift of charged particles which is caused by the large-scale gradient and curvature of the interplanetary field [11.125]. In early discussions this drift term was neglected, and the relative importance of it is still a debatable issue.

As conditions in the solar wind change on time scales long compared to the dwell time of >100 MeV particles in the solar cavity, only steady-state solutions of (11.1) were considered, i.e. $\partial U / \partial t = 0$, and the long-term modulation was visualized as a sequence of quasi-equilibrium conditions.

A further simplification to the transport equation led to the force-field approximation of Gleeson and Axford [11.93, 94]. In the quasi-steady, spherically symmetric model of the interplanetary medium they considered the case where the net radial streaming is small, which holds for relatively high energies (>100 MeV/N). Neglecting the streaming modulation theory, one arrives at a convenient definition of the radial gradient:

$$G = \frac{1}{U} \frac{\partial U}{\partial r} = \frac{C V}{D_r} , \quad (11.2)$$

where D_r is the radial diffusion coefficient and $C = 1 - (U/3)\partial(\alpha T U)/\partial T$ is the so-called Compton–Getting factor which incorporates the effects of the motion of the observer relative to the moving solar wind and can be determined directly from the observed energy spectrum $U(T)$. G is usually quoted in percent per astronomical unit and provides a direct measure of the local interplanetary diffusion coefficient D_r once C and V are known. For a constant Compton–Getting factor and solar wind speed, (11.2) can easily be integrated between the point of observation, say $r = r_E = 1$ AU, and the outer boundary r_M of the modulation region, and we obtain for the modulation factor M_0, i.e. the ratio of the observed density to the density in interstellar space,

$$M_0 = \frac{U(r)}{U_\infty} = e^{-\text{const} \cdot \Phi} \quad \text{with} \quad \Phi = \frac{3\,V\,C}{v} \int_{r_E}^{r_M} \frac{dr'}{\lambda(r')}\,. \tag{11.3}$$

Here Φ is the modulation parameter, v is the particle speed, and the integral corresponds to the number of mean free paths between r_E and r_M. The parameter Φ can be interpreted in terms of an effective potential of a conservative force field, varying from about 350 MeV at solar minimum to about 750 MeV at solar maximum, i.e. even at solar minimum there is considerable residual modulation.

The virtue of this simplified approach is that the modulation can be described by a single function where only three parameters have to be fitted to arrive at a given modulation level: the solar wind speed V, the extent r_M of the modulation region, and the cosmic ray scattering mean free path λ.

Although we can obtain solutions of (11.1) either numerically [11.74, 75] or by approximations such as the force-field solution, we have to characterize the modulation by certain functions which change in time (e.g. the solar wind speed or the mean free path). However, none of the observed interplanetary plasma parameters show temporal variations necessary to describe the observed cosmic ray variations [11.291, 185].

Various attempts have been made to overcome this problem. In particular, the simple radially symmetric, stationary solutions of the diffusion–convection equation (11.1) have been supplemented in the following way:

- The interplanetary medium is not spherically symmetric, particle propagation occurs in three dimensions, and off-ecliptic variations of the propagation medium which we have not been able to observe so far may well explain the solar cycle modulation.
- The 11-year modulation results from pile-up of individual intensity decreases caused by shocks and other disturbances in the solar wind [11.167], i.e. the modulation is a fundamentally dynamic, nonequilibrium process, which cannot be described by steady-state solutions of the transport equation [11.245, 168, 86].
- Gradient and curvature drifts play a significant role in modulation and may not be neglected. These effects are due to the large-scale magnetic field configuration and should give rise to observable changes when the sun's polarity is reversed every 11 years during solar maximum [11.123, 159, 141, 251].

The first effect was treated first by Fisk [11.75] and is contained in many modulation models. The second effect can in principle be modeled by systems of shocks propagating outward. They generate regions of increased turbulence, thus modifying the simple diffusion–convection model [11.245, 86]. The third effect is incorporated into the modulation equation through the fourth term on the right-hand side of (11.1): the drift term $V_D \cdot \nabla U$ where V_D depends on the antisymmetric part of the diffusion tensor. For high-energy particles (>100 MeV/N) the particle drift speed V_D is comparable to the solar wind speed or larger [11.123]. The direction of the drift velocity is, however, opposite for positive and negative particles. Thus, the drift is the only term in the transport equation which introduces a charge dependence when particles of charge q are moving in Parker's Archimedean spiral magnetic field

$$B(r, \theta) = \frac{A}{r^2} \left(e_r - \frac{r \omega_\theta \sin \theta}{V_{sw}} e_\varphi \right) ,$$

where A is a constant, ω_θ is the angular rotation rate of the sun, and V_{sw} is the constant solar wind velocity. From 1970 to 1980, when the north pole of the sun was magnetically positive (i.e. outward-pointing field, $A > 0$), protons would drift from the poles to lower latitudes and outward along the equatorial current sheet ($qA > 0$). Electrons would enter the heliosphere along the tightly wound magnetic field lines in the equatorial plane and drift polewards ($qA < 0$). After the sun reversed its polarity in 1980 [11.317], the drift paths for positive and negative particles would be exchanged. If cosmic ray propagation were dominated by drifts, marked differences in intensities, gradients, spectra, and nucleon to electron ratios ought to be expected after such a field reversal, as particles probe entirely different regimes of the heliosphere.

11.2.2 Observations

1. Radial Gradients. Determination of the radial gradient is a fundamental measurement for modulation theory since the size of the gradient (at least for high-energy particles) is related to the scale size of the modulation region.

Using *Helios* data we have determined the integral radial gradient of protons > 51 MeV for the solar minimum period in 1975–76 between 0.3 and 1 AU [11.220]. The observed value of $4 \pm 3\%$/AU over this radial distance is in agreement with the values measured in the outer solar system up to > 35 AU (*Pioneer 10/11*) of about 2–3%/AU as shown in Fig. 11.4(B) for protons of more than 70 MeV [11.200]. The latter values are measured with much higher accuracy than *Helios* was able to do because statistical errors are magnified when the radial range interval is small. At energies below 100 MeV/N the gradient value is somewhat higher than for relativistic particles during solar minimum, however, it was a surprise to find that at maximum modulation in 1981 the gradient at these energies nearly disappeared as shown in Fig. 11.4 (C and D), i.e. all the modulation in this energy range takes place outside 35 AU.

In 1968/1969, at the time when the *Helios* mission was planned, the measurements of *Mariner 4* and *Pioneer 8* and *9* had suggested large radial gradients of the order of 10–70%/AU [11.237, 236], and at the time of launch of *Pioneer 10* in 1972 this spacecraft was expected to reach interstellar space within a few years. It was not at all expected to find the above-quoted low gradient values, which could only be reconciled with the diffusion–convection model when assuming a large solar cavity and a large diffusion coefficient of about 2×10^{22} cm^2 s^{-1} [11.203] corresponding to a mean free path of 0.1–0.2 AU. This in turn produced a new problem, because quasi-linear theory which derives the mean free path from observations of the power spectra in magnetic field fluctuations arrived at a value

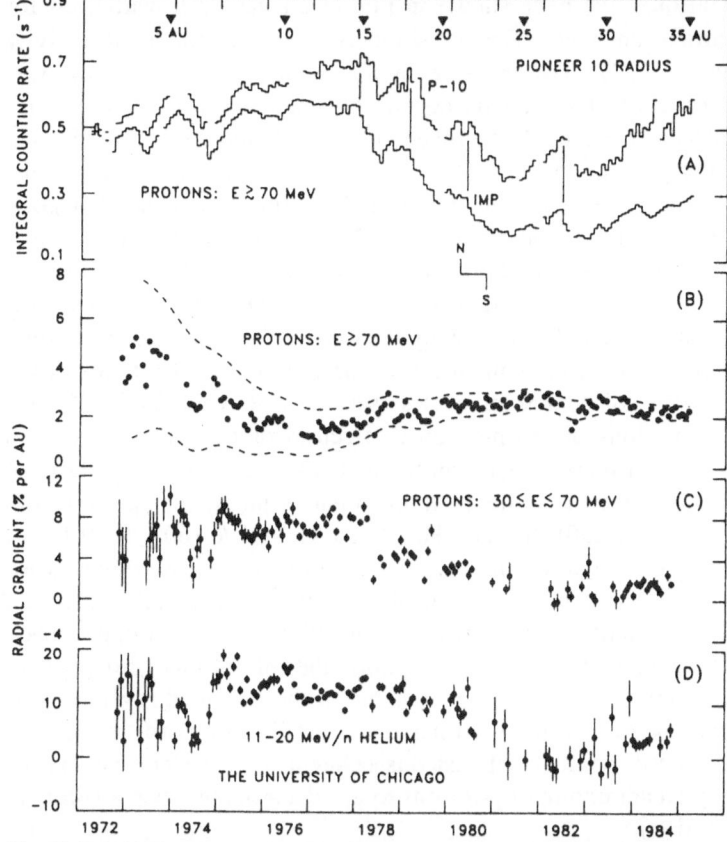

Fig. 11.4. (A) 27-day averages of quiet-selected integral cosmic ray intensity ($E > 70$ MeV) at *Pioneer 10* and at 1 AU. N, S indicate solar magnetic field reversal. (B) 27-day average radial gradients for the integral flux. *Dashed lines* show probable systematic uncertainty. (C) and (D) same as (B) for low-energy protons and helium (from McKibben [11.200])

for the mean free path lower by a factor of 10. Meanwhile it has been shown that the mean free paths obtained from the study of solar particle events are of the same order of magnitude as those from galactic cosmic ray gradients. These observations and the discrepancy of about a factor of 10 with early theoretical estimates will be discussed extensively in Sect. 11.3.

The low gradient of relativistic protons is now confirmed to beyond 35 AU [11.202], implying a radius of the heliosphere of at least 50 AU. The most striking feature of the gradient is its relative constancy in time, which has led not only to a determination of the scale size of the modulation region but also to the suggestion of its change during a solar cycle: $\Delta r_\mathrm{M} = 50$ AU [11.319]. The radial intensity gradients remain even constant at the time of the sun's magnetic field reversal in 1980 in contrast to predictions of the drift model. In view of the large distance to the modulation boundary the *Helios* spacecraft can be used as a 1 AU baseline for the *Pioneer* and *Voyager* missions to the outer planets [11.190].

2. Intensity Variations. We have shown in Fig. 11.2 solar cycle variations for three selected particle channels. The transition from low to moderate activity started in September 1977. The increase in the number of solar events was followed by a decrease in the galactic cosmic ray intensity a short time later. Various attempts have been made to relate the modulation to observed solar parameters, e.g. flare counts [11.2], solar activity index [11.24], or size of the polar coronal holes [11.116]. Though these studies supply in general a good correlation between the index under study and the cosmic ray intensity, a clear mechanism for the cause–effect relation, e.g. via the release of turbulence into the interplanetary medium, is still missing. Meanwhile it has become clear that a 22-year cycle is more appropriate for describing the long-term modulation of cosmic rays than an 11-year cycle. This is manifest in different temporal profiles of the modulated intensity in adjacent solar cycles (see the discussion by McKibben [11.200]) and in variations of the nucleon to electron ratio from one cycle to the next [11.90]. Both of these effects can be related to the changing polarity of the sun from one 11-year cycle to the next; quantitative predictions are given in [11.140]. According to [11.90] the ratio He(70–95 MeV/N)/e (600–1000 MeV) within the same magnetic rigidity interval 600–1000 MV/c increased by a factor of 2 around the time of the solar field reversal in 1970 and stayed fairly constant at the increased level until the next reversal in 1980, when it dropped back to its lower level. Whereas this behavior supports the role of charge-dependent drifts the intensity time profile of electrons between the two field reversals was rather broad in contrast to a pronounced maximum predicted by drift models for $qA < 0$. It was recently pointed out that this behavior of the electrons can be partly explained by an admixture of positrons to the electron component (Moraal, private communication).

It was an important observation that the modulation is not simply characterized by a smooth decrease and increase of intensities, but that decays and recoveries on shorter time scales are superimposed on the overall profile (see second curve from top in Fig. 11.2). In particular, during increasing solar activity, the changes in the level of modulation take place in a series of steps related to individual large interplanetary shocks causing Forbush decreases. Lockwood and Webber [11.168] have modeled the cumulative effects of Forbush decreases and were able to reproduce the 11-year modulation cycle as seen by the Mount Washington neutron monitor from 1955 to 1965. A quantitative explanation how a single shock actually influences cosmic ray propagation was given in [11.245]. Here the shock-related structures are characterized by a higher degree of turbulence leading to a smaller mean free path. The number of shocks between the observer and the outer boundary of the modulation region directly gives a measure for the total amount of modulation, and the outward motion of a single large shock naturally explains that changes in the level of modulation also move outward, as found in [11.190]. It is interesting to note that even the recovery from maximum modulation propagates outwards [11.73, 202].

In the model [11.245, 87] it is obvious that ensembles of shocks following each other in close sequence are particularly effective in modulating the galactic cosmic ray intensity. By using this model we were able to reproduce the phase and the amplitude of the short-term changes in the solar cycle variation ("mini-cycles") [11.274]. We took Forbush decreases observed with the Kiel neutron monitor as an indicator of shocks passing by the earth; each shock was assumed to reach the outer boundary of the modulation region within two months and to produce a 1% modulation.

3. Mini-Cycles, Shock Ensembles, and Super-Events. We want to present now a phenomenon which at first sight has little to do with modulation. Increases in the intensity of low-energy cosmic rays which could not be associated with individual solar flares have been a well-known phenomenon for several solar cycles (e.g. increases associated with long-lived activity centers [11.71], quiet time increases of electrons [11.38], and corotating events [11.189]). A new phenomenon, called super-events because of the long duration at high intensities [11.219], differs distinctly from these intensity increases in the following characteristics: its onset and decay is much longer than for flare-associated solar particle events, the intensity may remain above background for a period of the order of 40 days, its intensity variation with heliolongitude is small, its radial gradient is small, there is no sign of anisotropy, the proton to helium ratio is constant, and ions and electrons are enhanced simultaneously.

These long-lasting events are rare: four super-events were observed on *Helios* between 1975 and 1983. It is interesting that in coincidence with the super-events in the inner heliosphere one finds large increases in the >10 MeV proton intensities in the outer heliosphere out to 20 AU and beyond [11.254]. Figure 11.5 shows a super-event in 1978 which was observed on board *Helios 1* and *2*, *IMP 8*, *Pioneer 11* and *10* between 0.3 AU and 16 AU. Inspection of the intensity maxima shows that they seem to be roughly constant out to about 10 AU and decline further out. In [11.219] an interplanetary origin of the super-events is suggested consistent with the creation of an expanding "shell" [11.28] when coronal mass ejections, flare-associated shocks, and corotating interaction regions coalesce to form a compound stream. This propagating barrier can accelerate locally energetic particles [11.254, 157], confine these particles in the inner solar system, and simultaneously block incoming galactic cosmic rays effectively [11.40]. The latter observation is of particular relevance. We have found that the super-events occur at time periods where the long-term variation of galactic cosmic rays shows relative intensity minima. This is depicted in Fig. 11.2, where 5-month running means of the Kiel neutron monitor counting rates are shown along with the occurrence of super-events (solid triangles). The underlying common cause for both effects are the ensembles of shocks mentioned earlier. They are discussed in [11.219] for the 1978 period, in [11.40] for the 1982 period, and can also be identified for the 1979 period (Cliver, private communication). The introduction of the super-events as a new class of particle events was partly based on their occurrence with shock ensembles. This interpretation gets further support by the acceleration of electrons to relativistic energies in the outer heliosphere between 7 and 28 AU [11.169]. In all cases where we find super-events in the inner so-

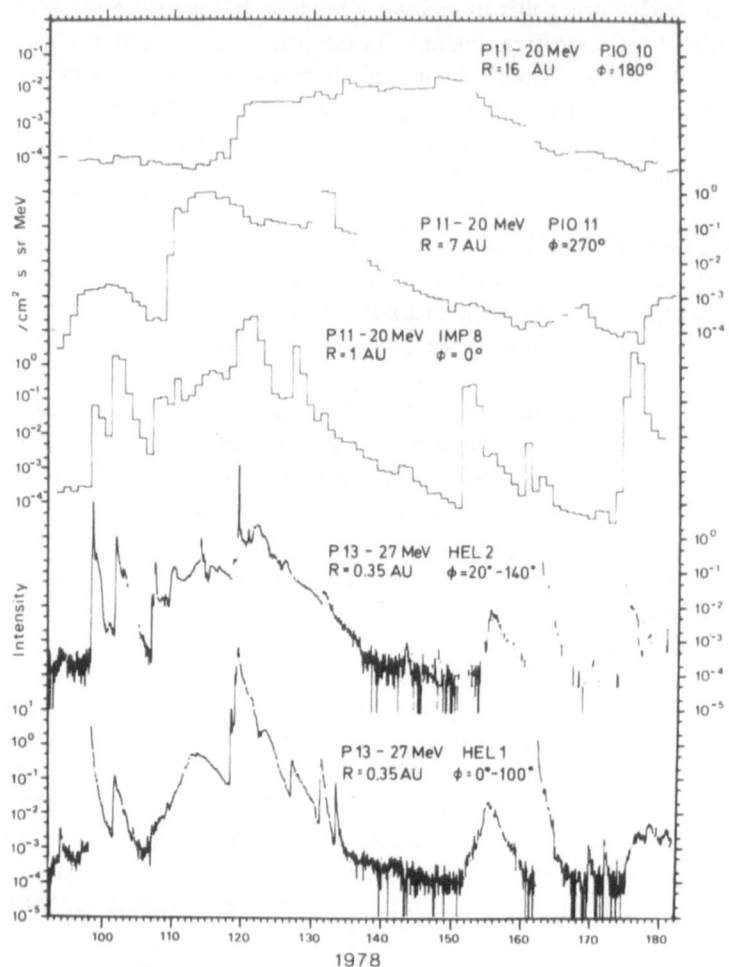

Fig. 11.5. Intensity–time profiles of energetic protons during super-event April/May 1978 at various heliocentric longitudes ϕ and radial distances R

lar system, Lopate [11.169] sees accelerated electrons in the outer heliosphere. He can associate them with *individual* shocks with large compression ratios and a precursor wave in the upstream region. We interpret the set of observations tentatively by a merging of several shocks out of the original ensemble to form particularly strong shocks at large distances from the sun.

We wanted to present this as an interesting example of the coupling of various phenomena in the heliosphere and for the necessity of observations of various physical quantities (solar wind plasma, magnetic fields, and energetic particles) at widely separated locations.

4. Conclusions. Understanding modulation is still based on the standard model with diffusion, convection, and adiabatic deceleration effects, where the path of individual particles through the heliosphere is determined by drift processes. This leads to characteristic differences between adjacent solar cycles owing to the different polarity of the solar and large-scale interplanetary magnetic fields. Medium-scale structures in the heliosphere caused in particular by propagating interplanetary shocks are effective as additional modulating agents and seem to be responsible for temporal structures ("mini-cycles") superimposed on the overall modulation. These structures may also disturb the regular drift motion, leading to a complex overall picture which is far from being understood in all details. This also refers to the extent of the heliosphere and processes at the termination shock, the boundary between the heliosphere and the interstellar medium. Recent reviews on modulation and the structure of the heliosphere can be found in [11.174, 201, 85, 300]. They all agree on the necessity to model the three-dimensional heliosphere and to extend the observations towards high heliographic latitudes. The *Ulysses* spacecraft launched in 1990 will gather data above the solar poles – eagerly awaited by the scientific community.

11.3 Interplanetary Propagation of Solar Energetic Particles

Energetic charged particles can travel through the heliosphere with a negligible chance of hitting each other or even hitting one of the much more numerous particles of thermal energy which constitute the solar wind, yet they do not travel in straight lines. The forces which shape the trajectories of charged particles are provided by the interplanetary electromagnetic fields. Even though solar flare particles are much more energetic than the solar wind particles their energy density is much too low to modify the fields through which they pass. This makes energetic particles ideal probes for studying the structure of these fields.

Unfortunately there is no way to observe the complete trajectory of an individual energetic charged particle from its source to the point of observation. What is measurable is the intensity of charged particles of a given type as a function of time, of energy, and of direction of incidence – preferably relative to the local direction of the magnetic field, i.e. as a function of pitch angle.

In the inner heliosphere there are two main fields of interest in which these data can be utilized:

- *Interplanetary phenomena*: What are the large- and small-scale structures of the interplanetary magnetic field which determine the particle trajectories? Which propagation properties of the medium can we derive from particle measurements? On the other hand, can we understand particle propagation from observed features of the interplanetary magnetic field as presented in Vol. 1 of this book [11.292].
- *Solar phenomena*: What are the sources of energetic particles (the acceleration process, propagation in the vicinity of the sun, release of particles into the propagation medium)? Only after interplanetary propagation has been

understood sufficiently can conclusions about the source be drawn from particle measurements in interplanetary space. Particularly favorable conditions are, however, offered by the two *Helios* missions because of their periodic close approaches to the sun during substantial portions of solar cycle 21.

We shall first treat the theoretical concepts (propagation models and parameters) and then turn to experimental results.

11.3.1 Transport Models

In the first, pre-space-age interpretations of energetic solar flare particle time intensity profiles measured by neutron monitors at ground level Meyer et al. [11.208] model a spherically symmetrical interplanetary medium in the following way. An inner nearly field-free ($B < 0.1$ nT) sphere extends beyond the earth to about 1.4 AU, followed by a finite region of disordered magnetic fields with $B \sim 1$ nT which partly scatters particles back into the field-free cavity and partly releases them into the galaxy.

In the early 1960s, the diffusive time profiles of solar flare particle intensities gave rise to a description of interplanetary propagation by diffusion equations, the characteristic quantity being a diffusion coefficient which depends on radial distance from the sun. The medium was still considered spherically symmetric initially. However, direct measurements of the interplanetary magnetic field near the earth and satellite observations of energetic solar particles made clear that the interplanetary medium is strongly anisotropic. This has formed a picture in which energetic charged particles move primarily along magnetic "flux tubes", each consisting of a bundle of magnetic field lines originating on the sun, corotating with the sun, and extending throughout the heliosphere.

The reason that charged particles follow the corotation is the $V \times B$ electric field caused by the solar wind speed V. If the solar wind speed is known then any position in interplanetary space can be mapped back onto the coronal footpoint of its magnetic flux tube [11.234].

For the solar energetic particles under discussion here, motion perpendicular to the average field by curvature and gradient drift can be neglected because of their small Larmor radii compared to the scale lengths of the field variations. The role of perpendicular diffusion had been under debate for some time. One underlying physical process is the random walk of field lines [11.124] which is assumed to originate in the supergranular motion of the photospheric magnetic field. In addition, in strongly turbulent interplanetary fields the guiding centers of particles may be displaced across the field in single scattering processes by as much as a Larmor radius [11.121]. Perpendicular diffusion plays an important role in the outer solar system, because even a small amount of transport *across* the nearly azimuthal average field leads to an efficient radial transport competing with the long paths *along* the tightly wound magnetic spiral.

The situation is different inside 1 AU from the sun because of the nearly radial alignment of the magnetic field. Observations demonstrating that cross-field diffusion is indeed small there are summarized in [11.269]. The arguments are based in particular on persistent field-aligned anisotropies and azimuthal density gradients [11.142].

This picture may only break down in regions of strong turbulence in corotating interaction regions occurring beyond the orbit of the earth (see Fig. 11.1). We shall therefore neglect perpendicular diffusion inside 1 AU and use only one spatial coordinate, the distance z along a flux tube. The magnetic field within such a flux tube can be characterized as a smooth average field, represented by an Archimedean spiral, with superimposed irregularities.

Energetic particle propagation can be considered to consist of two components: adiabatic motion along the smooth field and pitch-angle scattering caused by the irregularities. A quantitative model of the evolution of the gyro-averaged particle intensity $j(z, \mu, t)$ in space z, time t, and pitch-angle cosine μ is given by the partial differential equation as developed by Jokipii [11.119] and by Hasselmann and Wibberenz [11.104] for homogeneous guiding fields and by Roelof [11.270] in its current form, which also covers inhomogeneous guiding fields,

$$\frac{\partial j}{\partial t} + \mu v \frac{\partial j}{\partial z} + \frac{1 - \mu^2}{2 L} v \frac{\partial j}{\partial \mu} - \frac{\partial}{\partial \mu} \left(\kappa \frac{\partial j}{\partial \mu} \right) = Q(t) . \tag{11.4}$$

The particle velocity v remains constant in this model, in which the magnetic field is assumed to be static and the electric field to be zero. The systematic forces are characterized by a focusing length $L(z) = B(z)/(-\partial B/\partial z)$, while the stochastic forces are described by a pitch-angle diffusion coefficient $\kappa(z, \mu)$. In the case of solar flare particles the source $Q(t)$ is assumed in the high corona. Effects of convection and of adiabatic deceleration in the expanding solar wind are neglected in this equation, which means that it should only be applied to particles with diffusion velocities much higher than the solar wind speed.

IDEALIZED PROPAGATION MODELS

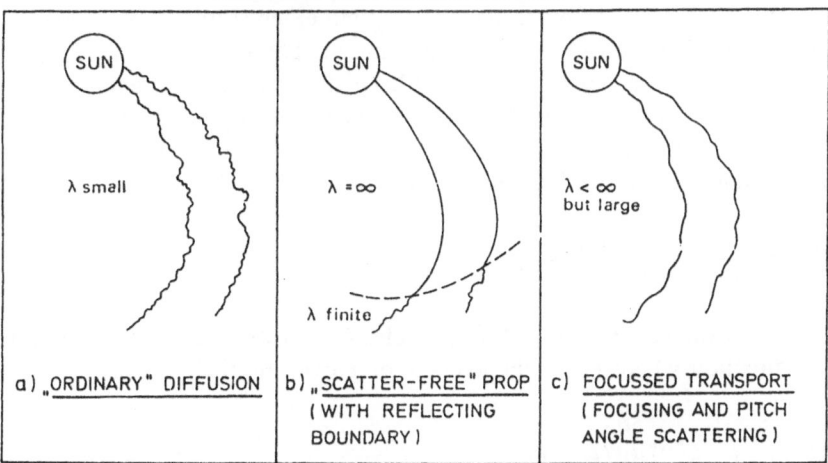

Fig. 11.6a–c. Propagation schemes for various degrees of turbulence superimposed on the average Archimedean spiral

The nature of particle propagation varies considerably with the relative strength of focusing and scattering forces. Several idealized situations are sketched in Fig. 11.6.

(a) If pitch-angle scattering is strong compared to focusing, then propagation can be adequately described by spatial diffusion with a small mean free path.
(b) Scatter-free propagation occurs in a sufficiently smooth magnetic field up to some outer boundary which may have reflecting properties.
(c) In an intermediate range both focusing and pitch-angle scattering must be considered, a situation which has been termed "focused transport" by Earl [11.65].

1. Scatter-Free Transport. In the case of scatter-free propagation Liouville's equation states that the phase-space density along the trajectory of an individual particle remains constant. Particle motion is then governed by the constancy of the first adiabatic invariant or, equivalently,

$$\frac{\sin^2 \alpha}{B} = \text{const.} \tag{11.5}$$

This means that the pitch angle α decreases when a particle moves in a weaker magnetic field, i.e. the particle motion becomes more focused in the field direction. In the idealized case of a monopolar magnetic field, approximately given in the inner heliosphere, the field strength $B(r)$ varies as r^{-2}. Charged particles injected isotropically at $z_0 = 0.05$ AU then appear to an observer at 0.5 AU to be coming from the solar direction in a narrow cone only 6.4° wide. In this case an interplanetary particle intensity time profile is nearly identical to the injection time profile at the sun except for the travel time. This situation is only rarely observed, but an example coming close to it was found when *Helios 1* was in an extremely quiet solar wind stream about 0.5 AU from the sun [11.226]. In the case depicted in Fig. 11.6b the scatter-free regime extends to a certain distance from the sun, and is followed by increased scattering beyond that distance. Based on ideas of Roelof and Nolte [11.233, 235], Green [11.100] has given analytical solutions of the transport problem for the situation that pitch angle scattering is concentrated at a certain solar distance in an otherwise scatter-free medium.

We mention for later applications that the anisotropy

$$\xi(t) = 3 \frac{S}{vU} = 3 \frac{\langle \mu\, j(t, \mu)\rangle}{\langle j(t, \mu)\rangle} \tag{11.6}$$

is a measure of the net streaming S of particles of density U and velocity v into a certain direction where $\langle x \rangle$ is the average of x over all pitch angles:

$$\langle x(\mu)\rangle = \int_{-1}^{1} x(\mu)\, d\mu/2 \,.$$

2. Focused Transport. If there is pitch-angle scattering then particles diffuse in pitch-angle space, and solar particles arrive in interplanetary space in a wider cone of pitch angles than in a scatter-free medium. Unfortunately, exact analytical solutions of (11.4) are not known in general. Therefore, approximate analytical solutions have been developed by Earl [11.65, 66] based on an eigenfunction expansion and by Kunstmann [11.152] in a perturbation approach for the case that scattering is too weak to justify a diffusion approximation (see below). The solution for an impulsive solar injection is a particle pulse which moves along the magnetic field with the "coherent" velocity V_{coh} and widens at a rate given by a diffusion coefficient D_{coh}. Both parameters are determined by the focusing to scattering ratio. The coherent pulse is followed by a "diffusive" wake.

One of the first attempts to compare theoretical predictions of (11.4) with observed solar particle data was performed by Bieber et al. [11.17]. Their approach was limited by the fact that scattering and focusing had to be independent of solar distance and that only specific forms of the pitch-angle diffusion coefficient $\kappa(\mu)$ could be used. A spatially constant focusing length means that any magnetic flux-tube area increases – rather unrealistically – exponentially with distance from the sun. These restrictions could be overcome by numerical solutions of (11.4) [11.232]. Such solutions for 9 MeV protons and a fairly large mean free path of 0.5 AU are shown in Fig. 11.7. The two omnidirectional intensities at the

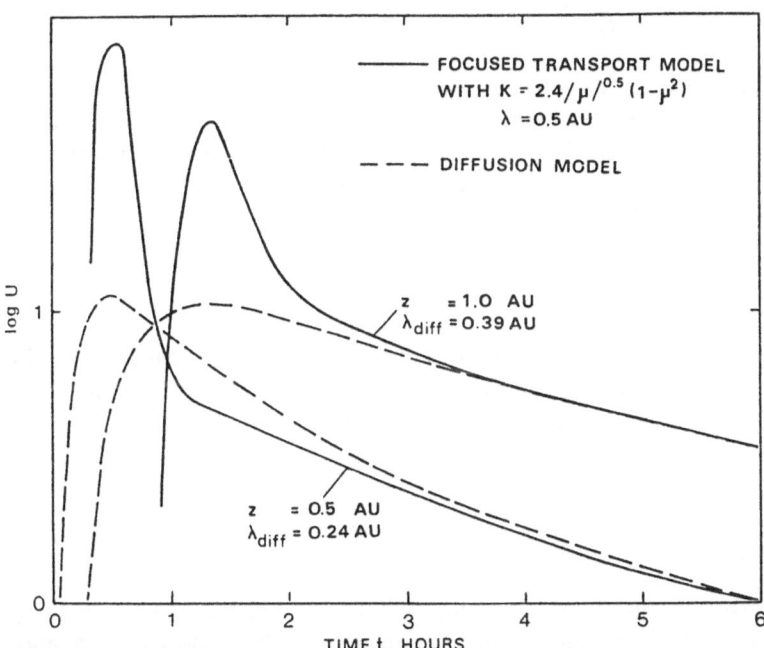

Fig. 11.7. Numerical solution of the transport equation (11.4) (*solid line*) for a mean free path $\lambda = 0.5$ AU and an observer at 0.5 AU (*left*) and at 1 AU (*right*). For comparison the solution of a diffusion approximation (*dashed line*) with a mean free path λ_{diffus} adjusted to give the same time to maximum (from Wong [11.328])

right (left) correspond to an observer at 1 AU (0.5 AU). The strong "coherent peak" mentioned above followed by a "diffusive wake" is clearly visible. For comparison, solutions of an approximation by a diffusion equation (see below) are given as dashed lines. They have been adjusted to give the same time-to-maximum as the full numerical solutions (solid lines). This example shows the limitations of the diffusion approximation in the case of large mean free paths.

The terminology describing the focused transport was introduced by Earl [11.65] and is a characteristic feature of the scatter-free electron events [11.161] for electrons in the $> 40 \, \mathrm{keV}$ range. However, a coherent peak is observed only for a sufficiently short solar injection. For longer solar injections interplanetary intensity time profiles are indistinguishable from those in diffusive transport. Careful studies of particle angular distributions are required in this case to obtain correct propagation parameters.

3. Diffusive Transport.

In a strongly scattering medium, pitch angles may easily be changed by more than 90°, which reverses the velocity parallel to the field and leads to a stochastic motion back and forth along the average field. The transport equation (11.4) can then be reduced to the simpler diffusion equation for the omnidirectional intensity $\langle j \rangle$ or, equivalently, the spatial particle density U [11.100, 14]:

$$\frac{\partial U}{\partial t} = \frac{1}{A} \frac{\partial}{\partial z} \left(A \, D_\| \frac{\partial U}{\partial z} \right) \tag{11.7}$$

for the spatial particle density $U(z)$. $A(z)$ is the cross-sectional area of a flux tube, related to the focusing length by $L(z) = A(z)/(\partial A/\partial z)$. In this approximation the spatial diffusion coefficient $D_\|(z)$ is related to the pitch-angle diffusion coefficient $\kappa(z, \mu)$ by

$$D_\|(z) = v \, L \frac{\langle \mu e^G \rangle}{\langle e^G \rangle}. \tag{11.8}$$

The early merely phenomenological descriptions of particle propagation by diffusion equations are thus explained by the physical process of pitch-angle scattering. The function $G(z, \mu)$, which also plays an important role in the discussion of angular distributions (see below), is defined by

$$G(z, \mu) = \frac{v}{2L} \int_0^\mu \frac{\left(1 - \nu^2\right)^2}{\kappa(z, \nu)} \, d\nu \, . \tag{11.9}$$

For a nearly homogeneous magnetic field ($L \to \infty$) we can use a linear approximation of the exponential functions in (11.8). This reduces the relation between the pitch-angle scattering coefficient and the coefficient for spatial diffusion parallel to the average field to

$$D_{\parallel}(z) = \left(\frac{v}{2}\right)^2 \left\langle \frac{(1-\mu^2)^2}{\kappa(z,\mu)} \right\rangle , \qquad (11.10)$$

as given by Jokipii [11.119] and Hasselmann and Wibberenz [11.104, 105]. A careful discussion of the situations in which spatial diffusion is an adequate description is given by Earl [11.64]. The parallel mean free path $\lambda_{\parallel}$ is related to $D_{\parallel}$ by

$$\lambda_{\parallel} = 3 \frac{D_{\parallel}}{v} . \qquad (11.11)$$

In the approximation that the field irregularities are stationary in the solar wind frame the mean free path $\lambda_{\parallel}$ is the same for particles of different velocities (or energies) but the same rigidity

$$P = \frac{p}{q} , \qquad (11.12)$$

since these particles have the same trajectories. This fact allows one to order experimentally observed mean free paths as a function of a single particle parameter only and will be used below. Because of its figurative meaning, the mean free path as given by (11.11) may also be used to characterize the strength of scattering in cases where a diffusion approximation is not applicable.

The diffusion coefficient $D_{\parallel}$ given in (11.8) and (11.10) relates to particle motion parallel to the average field. The corresponding *radial* diffusion from the sun is described by

$$\frac{\partial U}{\partial t} = \frac{1}{r^2} \frac{\partial}{r} \left(r^2 D_r \frac{\partial U}{\partial r} \right) \qquad (11.13)$$

with a diffusion coefficient

$$D_r = D_{\parallel} \cos^2 \psi , \qquad (11.14)$$

where ψ is the spiral angle between the radial and the magnetic field direction. A somewhat different derivation of (11.13) has been given by Ng and Gleeson [11.228]. For typical solar wind conditions we have $D_{\parallel} \sim \left(1 + (r/\mathrm{AU})^2\right) D_r$. Near 1 AU this gives $D_{\parallel} \sim 2D_r$.

The smaller the mean free path becomes the more important convection and adiabatic deceleration become (see e.g. [11.80]). These additional effects shift the intensity maximum to earlier times and may steepen the decay of the event considerably and must be taken into account in fits of theoretical models to observed solar particle events. They are of particular importance at distances beyond 1 AU from the sun (see [11.102]).

After having discussed how spatial diffusion can be explained by the elementary process of pitch-angle scattering let us now see how pitch-angle scattering can be derived from the properties of the interplanetary magnetic field.

11.3.2 Magnetic Field Properties and Pitch-Angle Scattering

Pitch-angle scattering is caused by irregularities of the interplanetary magnetic field which violate the first adiabatic invariant. The nature of irregularities can differ so largely that a generally applicable theoretical derivation of the scattering properties from the field structure is impossible. An important special case which has been studied extensively is the quasilinear theory (QLT) of pitch-angle scattering and various modifications.

1. Quasilinear Theory. When the irregularities superimposed on the average magnetic field are sufficiently small such that in a perturbation approach terms higher than the linear ones can be neglected, then a pitch-angle diffusion coefficient can be calculated in closed form [11.119, 120, 104]; for an introduction to the theoretical framework the reader is referred to [11.78]. Small relative irregularities mean that the changes in pitch angle during a single gyration are small, and several gyrations are required to accumulate a certain amount of scattering. This implies that particles are scattered essentially by irregularities which are in spatial resonance with the particle gyration. The relationship between the magnetic field irregularities, expressed by the power spectral tensor, and the pitch-angle diffusion coefficient is rather complicated. A particularly simple case is given when the fluctuating part of the field only contains axially symmetric transverse components with wave vectors parallel to the average field. This case is generally referred to as the "slab model" and is used in many applications to experimental results. We shall refer to this combination of QLT with the slab model as the "standard model" of particle scattering.

Let us assume that the power spectrum of the fluctuating field can be expressed by a power law,

$$f(k_\parallel) = C \, k_\parallel^{-q} , \tag{11.15}$$

where $k_\parallel$ is the wavenumber parallel to the magnetic field. A particle experiences resonant influences from irregularities with wavenumber $k_\parallel$ when it travels a distance $v_\parallel T_g = 2\pi/k_\parallel$ along the field during one gyration time T_g. The standard model then gives a pitch-angle diffusion coefficient of the form

$$\kappa(\mu) = A \left(1 - \mu^2\right) |\mu|^{q-1} , \tag{11.16}$$

where the rigidity-dependent factor A is proportional to the level C of field fluctuations. In the diffusive limit, the integral (11.8) or (11.10) yields a finite mean free path (11.11)

$$\lambda_\parallel \sim P^{2-q} \tag{11.17}$$

as long as $q < 2$.

The relationship between the functional forms of the power spectrum $f(k_\parallel)$, the pitch-angle diffusion coefficient $\kappa(\mu)$, and the mean free path $\lambda_\parallel(P)$ for the standard model is sketched in Fig. 11.8. We have inserted spectral slopes q quoted

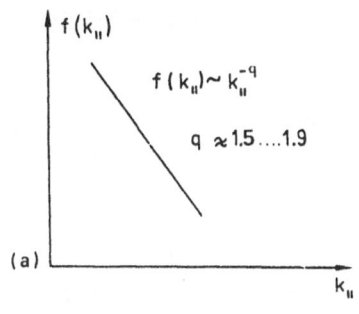

(a)

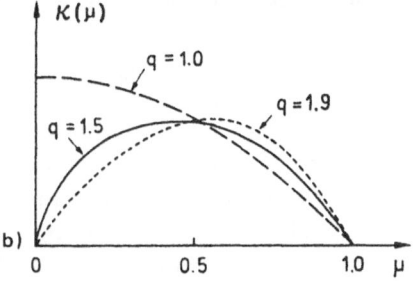

(b)

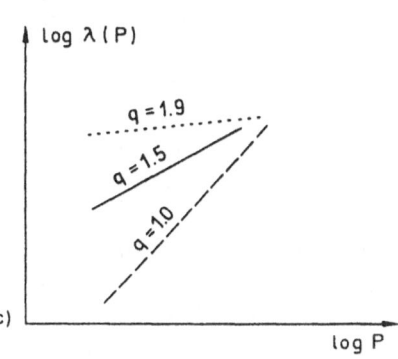

(c)

log P

Fig. 11.8. (a) Predictions of the standard model for a power-law spectrum. (b) Shape of the pitch-angle diffusion coefficient $\kappa(\mu)$ from (11.16). (c) Mean free path λ as a function of magnetic rigidity P (11.17)

in the literature (see below). The case $q = 1$ corresponds to isotropic scattering, $\kappa(\mu) \sim \left(1 - \mu^2\right)$.

For $q \geq 2$ the gap near $\mu = 0$ becomes so large that particles are not scattered at all through a pitch angle of 90°. This leads to totally coherent propagation where particle transport in the two pitch-angle hemispheres is decoupled. It is treated theoretically by Earl [11.63] and Kunstmann and Alpers [11.150]. The sign of the parallel component of the particle velocity remains unchanged, but its magnitude varies between 0 and v. Wibberenz et al. [11.325] have found a situation where particles seem to travel coherently between 0.3 and 1.0 AU (see the discussion in [11.76]).

Measured magnetic field spectra give spectral slopes in the range $q = 1.3 \ldots 1.9$ [11.42, 277, 106, 59] with an often quoted average value of $q = 1.63$ [11.106]. Measuring the spatial structure of the magnetic field fluctuations along a flux tube is not trivial. The data are time series of vector components carried by the solar wind across an observer. Depending on the changing direction of the guiding field they may belong to different fluxtubes. Typically, the power of the relative fluctuations in the field *strength* is considerably smaller than those in the field *direction*. This has generally been taken as evidence that the slab model is a good representation of the magnetic field structure.

Unfortunately, predictions of the standard model are not in agreement with particle observations. It was noted by Wibberenz et al. [11.323] that mean free paths deduced from observed interplanetary solar particle intensities are larger by about an order of magnitude than those predicted by the standard model (see also [11.240, 255]). This situation together with general doubts in QLT has led to a large number of corrections and improvements which will now be briefly sketched.

2. Modifications of the Slab Model and Nonlinear Corrections. Several nonlinear corrections to the quasilinear theory of pitch-angle scattering have been developed (see [11.313, 78] for an introduction into the problem). They enhance scattering for μ-values close to zero, particularly for a large relative strength of fluctuations (see [11.98, 126]). According to (11.8–11) these corrections *reduce* the theoretical mean free path and enhance the discrepancy with observations. This has led to suggestions that the assumptions about the nature of the magnetic field fluctuations should be modified (see [11.321]).

Lee and Völk [11.158] and Morfill [11.213] consider scattering by Alfvén waves oblique to the magnetic field which widen the scattering gap around $\mu = 0$. The resulting mean free path is considerably larger than in the standard model even though the gap at small pitch cosines is partly filled by non-resonant interactions, e.g. reflections by compressional waves [11.213, 214]. In a different approach, Fisk et al. [11.81] start from the assumption of isotropic magnetic field fluctuations. This adds a factor $|\mu|$ to $\kappa(\mu)$ which also widens the scattering gap and has the interesting consequence that pitch-angle diffusion from one hemisphere into the other now becomes impossible for spectral slopes with $q \geq 1$ compared to $q \geq 2$ in case of the slab model (see also [11.78]). The question of the correct form of the pitch-angle diffusion coefficient has recently also been discussed by Bieber et al. [11.22]. They use an isotropic model that includes dissipation, i.e. at high wave numbers the omnidirectional power spectrum falls off faster than k^{-3}. Consideration of dissipation leads to a transition from the "slab" form of the pitch-angle diffusion coefficient at low energies to the isotropic form at intermediate energies, where the particle Larmor radius is of the order of the dissipation scale. The authors conclude that a correct higher-order treatment is required to describe the actual behavior of $\kappa(\mu)$ and the absolute value of the mean free path.

Pitch-angle diffusion coefficients $\kappa(\mu)$ have been calculated numerically in the quasilinear approximation by Kunstmann [11.151] for different types of magnetic field fluctuation. Figure 11.9 gives a comparison for a power-law spectrum with $q = 1.5$: (a) slab model, (b) isotropic fluctuations, and (c) Alfvén waves with an isotropic distribution of wave vectors.

Figure 11.10 shows an example of nonlinear corrections, based on the "partially averaged field approach" by Jones et al. [11.126] with a pitch-angle diffusion coefficient $\kappa(\mu)$ calculated for a spectral slope of $q = 2$ and the "oblique slab model", i.e. Alfvén waves moving into a fixed direction in space inclined by 30° with respect to the magnetic field direction. In contrast to the QLT prediction the gap at $\mu = 0$ is filled up by nonlinear effects (curve P.A.F.). The amount of correction depends on the relative strength of magnetic field fluctuations, so that for sufficiently strong fluctuations one ought to reach a situation similar to the case of isotropic scattering (see curve $q = 1$ in Fig. 11.8b).

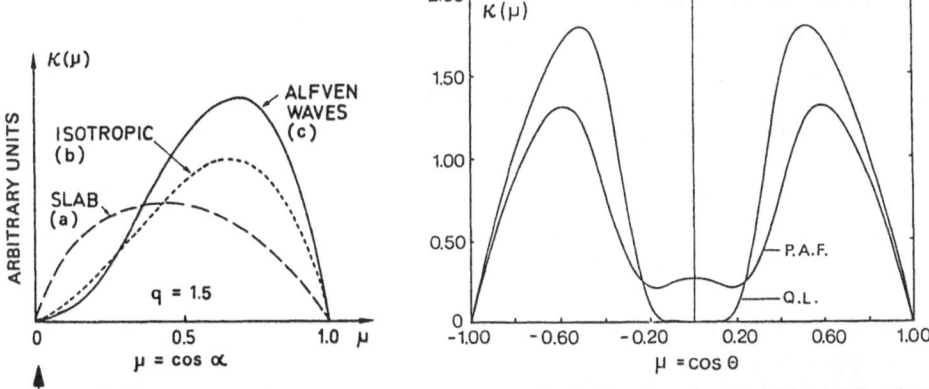

Fig. 11.9. Examples of the influence of different magnetic field structures on the shape of the pitch diffusion coefficient $\kappa(\mu)$. A spectral exponent $q = 1.5$ is assumed in all three cases. Case (a) corresponds to the slab model. The cases of isotropic fluctuations (b) and Alfvén waves with an isotropic distribution of wave vectors (c) are taken from Kunstmann [11.151]

Fig. 11.10. Pitch diffusion coefficient $\kappa(\mu)$ for Alfvén waves moving inclined by 30° with respect to the average field in the quasilinear approximation (Q.L.) and in the partically averaged field approach (P.A.F.). The relative strength of fluctuations amounts to 10% (from Jones et al. [11.126])

A qualitatively similar result is obtained by the resonance broadening concept [11.312]. Here the value of $\kappa(\mu)$ is replaced by $\kappa(\mu_c)$ for $|\mu|$ below a critical value μ_c. This approach is combined by Morfill [11.213] with particle scattering due to oblique Alfvén waves. The critical value μ_c increases with the amount of magnetic field fluctuations, so that we obtain the same effect as above: the resonance gap is filled up more and more with increasing turbulence in the field.

Pitch-angle diffusion coefficients as in Fig. 11.10 should be taken as representative examples. The correct treatment of the pitch-angle scattering theory is not yet totally clear. Nonlinear corrections are not the only way to bridge the resonance gap. An important effect for particles near 90° pitch-angle is mirroring due e.g. to magnetosonic waves. For average conditions this leads to a constant mean free path of the order of 0.3 AU for rigidities below a few GV/c [11.99].

Davila and Scott [11.48] reduce scattering near a pitch angle of 90° by collisionless damping of waves with high wavenumbers. This has the interesting consequence that the width of the resonance gap (see Fig. 11.10) varies with particle rigidity, and that below a limiting rigidity we would observe the transition to scatter-free propagation. They also add mirroring due to field compressions and obtain a typical mean free path of 0.04 AU for protons between 5 MeV and 2 GeV, roughly independent of energy.

A different aspect is taken into consideration by Schlickeiser [11.283]. He modifies the slab model by not having the magnetic field fluctuations frozen in the solar wind, but allowing for Alfvén waves to move parallel and anti parallel to the magnetic field. This increases the small values of $\kappa(\mu)$ near $\mu = 0$, and yields finite values of the mean free path even for magnetic field spectra with $q \geq 2$.

One important consequence of the modifications is that the relationship between the functional shapes of $f(k_\parallel)$, $\kappa(\mu)$, and $\lambda(P)$ as derived from the standard model (see Fig. 11.8) no longer exists. In particular, we might consider a situation where the shape of the magnetic field spectrum remains constant, but the overall level fluctuates. If the field turbulence increases, not only a higher overall level of scattering, but also a partly filled resonance gap could be expected. For weaker field fluctuations we would have a pronounced dip in $\kappa(\mu)$ near $\mu = 0$, whereas for a high level we might end up with a situation which is close to isotropic scattering (see curve $q = 1$ in Fig. 11.8b).

3. Particle Orbit Simulation. Instead of computing pitch-angle scattering in the QLT approach and the various modifications, particle trajectories can be simulated in a model based on the power spectral representation of the field (e.g. [11.129]) or on actual satellite magnetic field and plasma data (e.g. [11.216]). Such simulations allow us to study the diffusion process independently of any theoretical assumptions in QLT or its modifications. In the particular case considered by Kaiser et al. [11.129] the agreement with the nonlinear corrections considered by Jones et al. [11.126] is rather good. We shall come back to this point below when we discuss new results by Valdes-Galicia et al. [11.307].

11.3.3 Observations of the Mean Free Path

Solutions of the transport equation (11.4) show that the time profiles of interplanetary omnidirectional particle intensities $\langle j(\mu, t) \rangle$ and of anisotropy $\xi(t)$ and the instantaneous pitch-angle distribution $j(\mu, t)$ reflect the influences of the injection time profile, of the focusing length $L(z)$, and of the pitch-angle diffusion coefficient $\kappa(z, \mu)$. Observed intensity and anisotropy time profiles as well as pitch-angle distributions can therefore be utilized to determine these properties of the source and of the medium. If purely diffusive transport is anticipated, then the diffusion coefficient $D(z)$ from (11.8) or, equivalently, a scattering mean free path $\lambda(z)$ may replace $\kappa(z, \mu)$. The following difficulties must be considered in such an analysis:

- During a solar event a spacecraft does not stay in the same magnetic flux tube because the flux tubes corotate across it. Therefore, what we see are samples taken from different flux tubes in which particle distributions may have evolved independently and differently. The analysis of solar events in a stream-structured solar wind is discussed by Scholer et al. [11.288]. According to a study by Morfill et al. [11.212] the mean free path can vary by as much as a factor of five from the leading to the trailing edge of a solar wind stream. A transition to a solar wind regime with different propagation characteristics can often be detected by a significant change in the shape of pitch-angle distributions [11.101].
- Simultaneously with the ejection of MeV particles a flare region often generates a shock wave which travels through the solar wind with speeds up to

the order of 1000 km/s. This is considerably slower than the energetic solar particles, but the shock may accelerate particles either from thermal solar wind energies or from the background of solar particles which arrived earlier. Thus, a shock can appear as a moving source of energetic particles which are superimposed on the intensities of particles of prompt solar origin. At first glance a compound time profile of this kind could give the impression of double or multiple injections on the sun. A careful analysis of the related magnetic field and plasma properties is required to diagnose the arrival of the shock, and the shock-accelerated particles have to be separated from the prompt particles of solar origin by studying the spectral characteristics, chemical composition, and angular distributions.

- Scattering strengths (or mean free paths) deduced from time profiles of intensity and anisotropy can differ from those obtained from pitch-angle distributions. Solutions of the transport equation (11.4) show that the latter give *local* scattering values while scattering strengths at different distances along a flux tube shape the time profiles and can thus – to a certain extent – be deduced from the profiles as functions of distance. In other words, pitch-angle distributions reflect the local properties and time profiles the large-scale structure of the medium.

An extensive summary of near-earth-scattering mean free paths $\lambda_\parallel$ is given by Palmer [11.240]. He has carefully selected solar events where effects of the solar injection have been taken into account. Values for the mean free path versus rigidity P are shown in Fig. 11.11. The data suggest that there is no systematic dependence on rigidity. A "consensus range" for $\lambda_\parallel$ exists between 0.08 and 0.3 AU over a large range of rigidities. Relativistic electrons and tens of MeV protons have very similar mean free paths even though their rigidities differ by a factor of 100. This small variation of the mean free path with rigidity which is consistent with $\lambda_\parallel$ = const has inspired a number of theoretical predictions which are also sketched in Fig. 11.11. In curve G, based on [11.99], the mean free path is determined by a small amount of compressive fluctuations, whereas large-scale Alfvén waves do not produce particle scattering through a pitch angle of 90°. Under average conditions this leads to a constant mean free path of about 0.3 AU. Scholer and Morfill [11.287] consider the superposition of small-scale fluctuations on medium-scale variations. The resulting scattering can lead to a constant mean free path as shown in curve SM. The letter J in Fig. 11.11 represents the standard model discussed in Sect. 11.3.2 and is based on [11.121]. Near 100 MV/c the discrepancy between this "standard model" and observations amounts to about the factor 10 as discussed in the preceding section. For some events one even finds a rigidity range with a negative slope of $\lambda_\parallel(P)$. This slope would also be suggested by comparison of the bulk of data around 200 MV/c with the so-called scatter-free protons (> 0.3 MeV) postulated by Roelof and Krimigis [11.273] (see [11.322]).

In general, the variation of the mean free path with particle rigidity provides important insights into the correct theoretical description. We have just mentioned

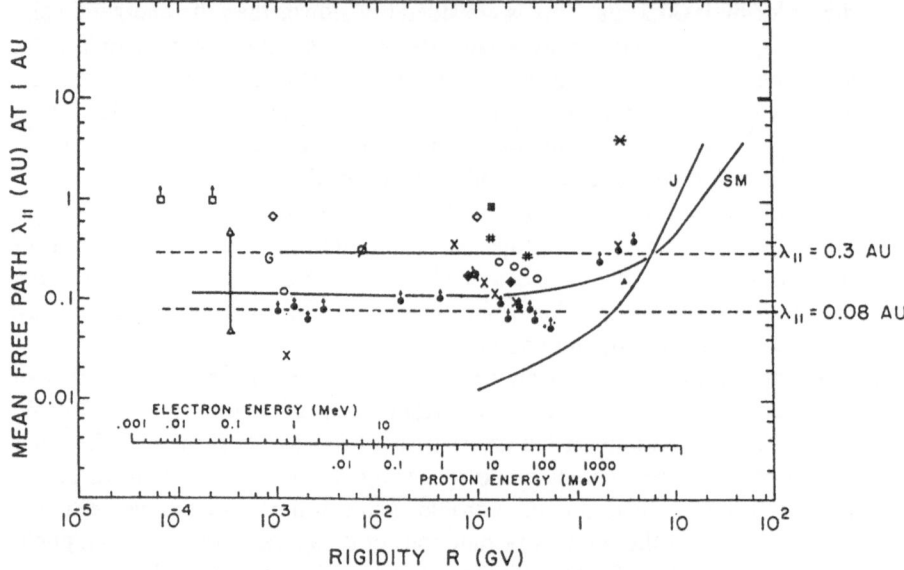

Fig. 11.11. Parallel mean free path as a function of rigidity for electrons and protons. Events have been selected [11.240] where finite injection at the sun is taken into account. The "consensus range" is indicated by the two *dashed lines* at 0.08 AU and 0.3 AU. Theoretical predictions are curves J [11.122], SM [11.287], G [11.99] (from Palmer [11.240])

the occasional negative slope, $\partial\lambda/\partial P < 0$, over a limited range of rigidities below about 200 MV/c. The resulting relative minimum in λ is difficult to explain in the context of most existing theories. However, Scholer and Morfill [11.287] consider the superposition of small-scale fluctuations – which are in resonance with the particle motion – on medium-scale variations. This could lead to a constant mean free path (see curve labeled SM in Fig. 11.11), but also to a negative slope.

Beeck et al. [11.13] discuss the rigidity dependence of λ based on measurements by various instruments on different spaceprobes covering a large range of particle species. They show that the negative slope found earlier for the event of 22 November 1977 for certain rigidity ranges [11.327, 180] is not confirmed. The intensity profiles for this event have to be explained by different solar injection profiles for different particles species. As a result, the P-dependence is obtained as $\lambda(P) \propto P^{0.45}$ for rigidities between 30 and 600 MV/c. A positive slope for $\lambda(P)$ is also found for the event of 27 December 1977, from two totally different types of studies. The authors use data from three spaceprobes at different radial distances from the sun and find $\lambda(P) \propto P^{0.2}$ for the rigidity range between 30 and 250 MV/c [11.13]. Wibberenz et al. [11.324] use data from three spaceprobes at different heliolongitudes to decouple the influence of coronal propagation from the observed intensity profiles. Their mean free path of 13–27 MeV protons agrees well with the nucleon mean free paths in [11.13]. They also analyzed $\sim$0.5 MeV electrons, finding electron mean free paths slightly smaller than those

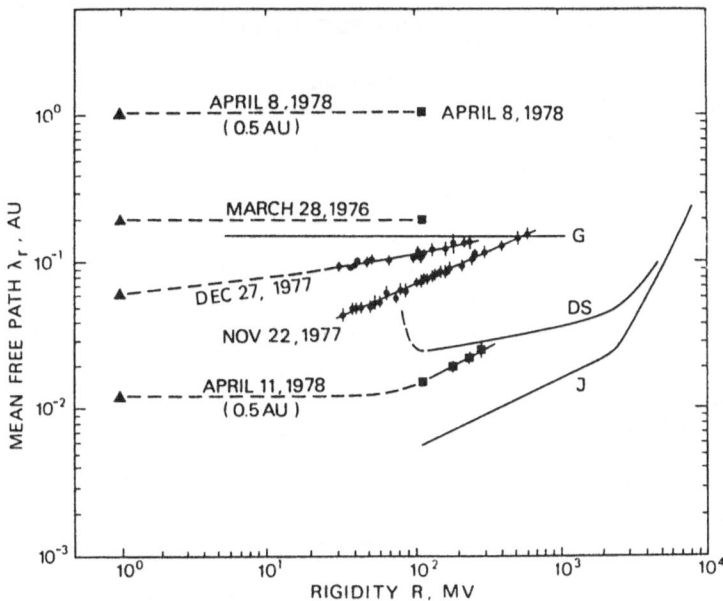

Fig. 11.12. The radial mean free path as a function of rigidity during several solar particle events

of nucleons, giving a monotonic continuation of the $\lambda(P)$ results from higher rigidities.

Figure 11.12 summarizes the rigidity dependence of the mean free path corresponding to (11.14). The results were collected at different distances from the sun, but here we want to discuss the variation with rigidity only. Figure 11.12 contains a number of remarkable results:

- It demonstrates the high variability in the scattering efficiency of the interplanetary medium. The limiting values differ by about two orders of magnitude, but these two solar events occurred only three days apart from each other in April 1978.
- It demonstrates the high correlation in the behavior of relativistic electrons and tens of MeV protons which differ in rigidity by a factor of 100. The mean free path is either high or low simultaneously for both particle groups.
- For nucleons above 30 MV/c there is a definite monotonic increase of the mean free path with rigidity for the three events of 22 November 1977, 27 December 1977, and 11 April 1978. For events which show the stronger variation of λ with P the mean free path is generally smaller. This may or may not be accidental. It would be quite interesting to see if the case $\lambda(P) = $ const is approached when the general level of scattering is further reduced, i.e. the mean free path is increased.

We have inserted in Fig. 11.12 three theoretical curves based on average conditions of the interplanetary medium. Curve J represents the standard model

271

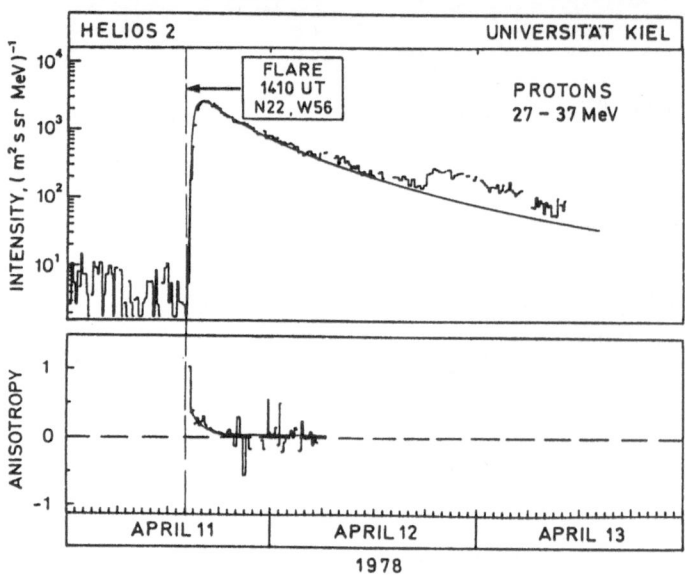

Fig. 11.13. A diffusion–convection model fit to a solar particle event (from Valdes-Galicia et al. [11.306])

discussed above [11.121]. Curve G is from Goldstein [11.99], where the constancy of the mean free path results from considering mirroring effects only. Curve DS is based on Davila and Scott [11.48] with a certain mixture of resonant and nonresonant interactions. Their suggestion that, owing to damping of waves at high wave numbers, a turnup of the mean free path should occur at low rigidities marking a transition to scatter-free propagation is in contradiction to observations. It is obvious from the large variations in the mean free path from one event to the other that a comparison with average interplanetary conditions is not meaningful. We need an analysis where during the same event the mean free path is obtained over a large rigidity range and where the magnetic field structure at the time of the event is used as an input for the theory of particle transport.

The radial dependence of λ is not yet well determined. On the basis of fits of intensity time profiles of individual events it has often been described by a power law $\lambda_r(r) \sim r^\beta$ with values of $\beta \leq 1$ in several cases. Observing solar event particle intensities on spaceprobes at different radial distances, e.g. *Pioneer 10* and *11* and near-earth instruments yielded values between $\beta = 0.0 \pm 0.2$ [11.333] and $\beta = 0.4 \pm 0.2$ [11.102].

Including anisotropy time profiles in the fits adds information and makes more exact and more detailed conclusions possible about the radial dependence of λ and simultaneously about the solar injection time profile. As an example Fig. 11.13 shows an event on 11 April 1978. In contrast to the events mentioned above, a negative value of $\beta = -1$ is required here to obtain a good fit to the intensity profile observed when *Helios 2* was at a position of 0.49 AU from the sun [11.307]. The fast decay of the anisotropy indicates a short injection, so that the width of the intensity profile must be essentially due to interplanetary diffusion. The profile implies unusually low values of the mean free path.

An example for considerably weaker scattering is the event of 28 March 1976. This event has been studied very carefully by Ng et al. [11.231, 230] by fitting numerical solutions of the transport equation (11.4) to intensity and anisotropy time profiles and to pitch distributions throughout the event. A detailed structure of the scattering distribution along the magnetic flux tube was obtained in which scattering was concentrated in a shell around 1 AU for 4–13 MeV protons. The spatial distribution of scattering differed slightly for ~0.5 MeV electrons but showed the same tendency. The important point here is that for this event one finds a small degree of scattering within about 0.5 AU. There are other cases where events with weak scattering are observed on *Helios* close to 0.5 AU (3 March 1975 at 0.3 AU [11.325], 8 April 1978 at 0.52 AU [11.14], 7 June 1980 [11.226], 8 June 1980 [11.130]). This indicates that an increase of the mean free path towards the sun is likely.

A qualitatively similar radial variation is predicted by Morfill et al. [11.212]. The radial development of the amplitudes and wave vectors of Alfvén waves produced at the sun leads to a certain radial dependence of the mean free path which depends on the phase in a solar wind stream. At a fixed distance the mean free path may vary up to a factor of 5; this would explain part of the variations shown in Fig. 11.12. The small degree of scattering closer to the sun is in qualitative agreement with our suggestion based on several events studied with *Helios*.

11.3.4 The μ-dependence of the Scattering Coefficient

Interplanetary pitch-angle scattering is phenomenologically described by the μ-dependence of the scattering coefficient in the transport equation (11.4). As shown above, this coefficient is of critical importance for theories which relate particle propagation to magnetic field structures.

Solutions of the transport equation (11.4) show how $\kappa(z, \mu)$ together with the focusing length $L(z)$ shapes the μ-dependence of interplanetary particle intensities. Special cases have been treated analytically (diffusive approximation with no focusing [11.104] and a steady-state solution for a radially constant ratio of scattering to focusing forces with "exponential anisotropy" $\exp(G(z, \mu)$ with $G(z, \mu)$ from (11.9) [11.66]). The stationary solution has been applied by Bieber et al. [11.19] to solar energetic particles, and by Bieber and Pomerantz [11.20, 21] to angular distributions of galactic cosmic rays.

Whereas time profiles of omnidirectional intensity reflect the large-scale distribution of scattering, angular distributions are essentially determined by the *local* propagation properties of the interplanetary medium, i.e. by $L(z)$ and $\kappa(z, \mu)$ via $G(z, \mu)$. It has been shown by Green and Schlüter [11.101] and Beeck and Wibberenz [11.14] that after proper normalization the shape of the anisotropic part $j(\mu, t) - \langle j(\mu, t) \rangle$ of the distribution function typically remains constant

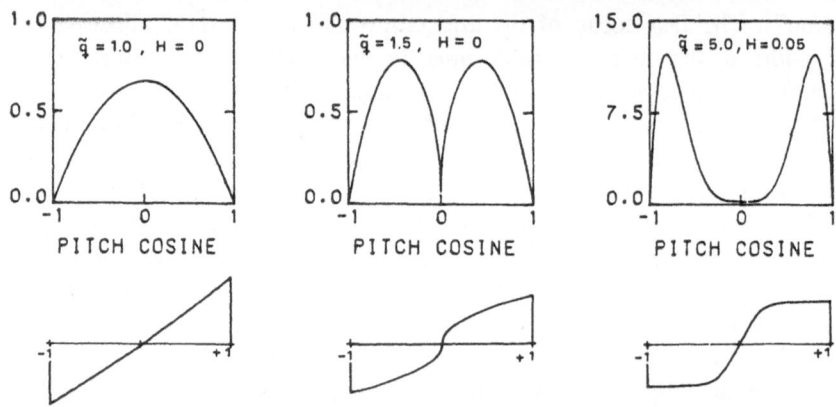

Fig. 11.14. Relation between the shape of the pitch-angle diffusion coefficient $\kappa(\mu)$ and the normalized angular distribution. A more prominent dip in $\kappa(\mu)$ is modeled by increasing values of $\tilde{q}$. The value of H simulates mirroring or nonlinear effects (after Green and Schlüter [11.101])

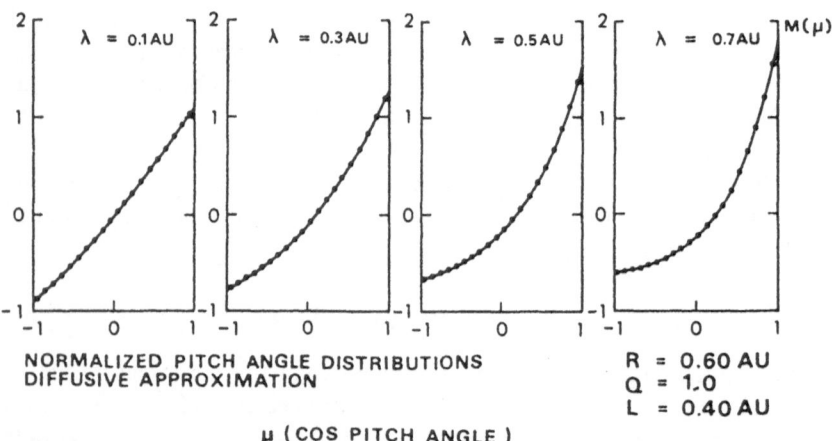

Fig. 11.15. Normalized angular distributions in the case of isotropic scattering ($\tilde{q} = 1$) for different scattering to focusing ratios. The focusing length $L = 0.4$ AU is kept constant; the mean free path λ increases from left to right (after Beeck and Wibberenz [11.14])

throughout a solar particle event provided that quasi-stationary conditions are given either by a sufficiently long-lasting injection or by sufficiently strong scattering.

The relationship between the scattering coefficient and particle pitch-angle distributions is illustrated in Figs. 11.14 and 11.15. In these examples the pitch-angle diffusion coefficient is parametrized (see [11.64]) in the form

$$\kappa(\mu) = A\left(1 - \mu^2\right)\left(|\mu|^{\tilde{q}-1} + H\right). \tag{11.18}$$

The exponent $\tilde{q}$ is used to characterize the width of the dip around $\mu = 0$. The

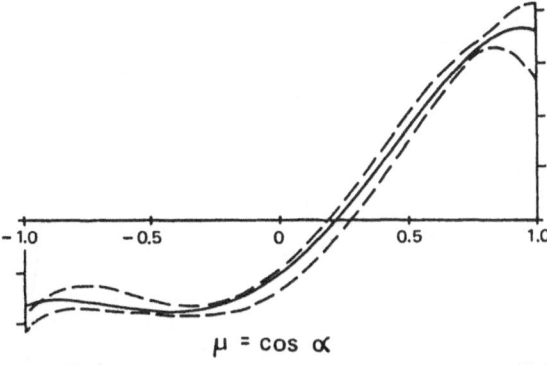

$$\mu = \cos \alpha$$

Fig. 11.16. Normalized angular distributions of 4–13 MeV protons during the event of 28 March 1976. The two *dashed lines* represent the 1-σ tolerances of a sequence of 18 pitch distributions. The *solid line* is the fit corresponding to the scattering coefficient given by the solid line in Fig. 11.17

additive term H determines its bottom height. Note that the standard model discussed in Sect. 11.3.2 corresponds to $H = 0$, $\tilde{q} = q$.

In Fig. 11.14 negligible focusing ($L \rightarrow \infty$) has been assumed. The upper panel shows the shape of $\kappa(\mu)$ for various combinations of $\tilde{q}$ and H. The corresponding normalized distribution functions are given in the lower panel, showing a linear curve in case of $\tilde{q} = 1$, whereas in the case of an increasing dip the distribution function approaches an S-shape.

For the case $\tilde{q} = 1$ we see the combined effects of scattering and focusing in Fig. 11.15. Calculations are for a position at $r = 0.6$ AU from the sun, where for an Archimedean spiral the focusing length is $L = 0.4$ AU. The mean free path λ increases from left to right corresponding to a growing role of focusing. This is reflected in the steepening of the angular distributions. If angular distributions are written as Legendre polynomial series, then the second-order coefficient may serve as a approximate measure for the ratio of scattering to focusing forces, whereas the third-order coefficient reflects the size of the resonance gap (see [11.14]).

An example of this kind of determination of the pitch-angle diffusion coefficient is the analysis of interplanetary directional particle intensities of the 28 March 1976 solar flare event measured on *Helios 2* about 0.5 AU from the sun by Ng et al. [11.230]. For the focusing length $L(z)$ the mean interplanetary magnetic field could be assumed to be an Archimedean spiral corresponding to the quiet solar wind conditions observed during that day. The pitch-angle diffusion coefficient was written as the product $\kappa(z, \mu) = \kappa_1(z)\,\kappa_2(\mu)$ of a distance-dependent factor $\kappa_1(z)$ and a normalized μ-dependent factor $\kappa_2(\mu)$. Figure 11.16 shows normalized pitch-angle distributions of 4–13 MeV protons obtained during this event. The angular resolution of the detector system leaves some uncertainty in the reconstructed $\kappa_2(\mu)$, which was assumed to be an even function of μ. Three possible shapes equally well fitting the observations are presented in Fig. 11.17. They have in common, however, that pitch-angle scattering is significantly reduced around $\mu = 0$, i.e. for mirroring particles. As discussed by Ng

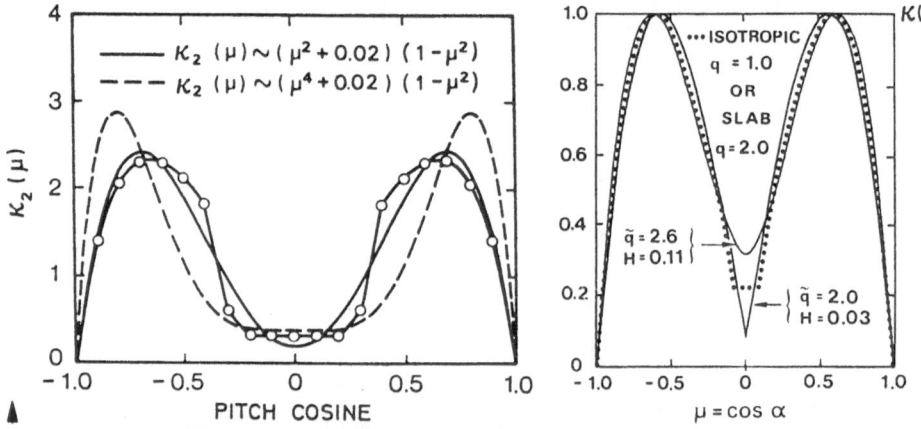

Fig. 11.17. The μ-dependent part of the pitch diffusion coefficient. All three curves give equally good fits to the angular distributions in Fig. 11.16. The *open circles* connected by *straight lines* are taken from Ng et al. [11.231]. The two other curves are analytical curves as given by (11.18), with $\bar{q} = 3$, $H = 0.02$ (*solid curve*), $\bar{q} = 5$, $H = 0.02$ (*dashed curve*) (from Ng et al. [11.230])

Fig. 11.18. The pitch diffusion coefficient $\kappa(\mu)$ determined from directional particle intensities observed during the solar event of 8 April 1978 (from Beeck and Wibberenz [11.14])

et al. [11.230], the valley around $\mu = 0$ is not consistent with the slab model of magnetic field fluctuations. They suggest that isotropic or oblique Alfvénic fluctuations are responsible for the valley. One possibility that explains the finite degree of scattering near $\mu = 0$ is the existence of mirroring effects. This interpretation is supported by the behavior of electrons. For the same event the shape of $\kappa_2(\mu)$ for ~ 0.5 MeV electrons has been determined as well. It is found that within observational uncertainties it is identical to that for protons. This is a very interesting result because it suggests quantitatively similar mechanisms of pitch-angle scattering for these two particle types with Larmor radii differing by more than two orders of magnitude.

Similar conclusions for another solar event (8 April 1978) with weak interplanetary scattering have been drawn by Beeck and Wibberenz [11.14]. They discuss the angular distributions in terms of the Legendre polynomial expansion mentioned above. The resulting $\kappa(\mu)$ is shown in Fig. 11.18. The two full lines give the theoretical fits to the angular distributions within statistical uncertainty when the analytical form (11.18) is used for $\kappa(\mu)$. We see that the dip around $\mu = 0$ is not quite as pronounced as for the event of 28 March 1976 (Fig. 11.17). The dotted line in Fig. 11.18 is a theoretical prediction from QLT, supplemented by a finite degree of pitch-angle scattering around $\mu = 0$. This curve can be obtained by two different forms of the field fluctuations. It either requires a slope of $q = 1$ for isotropic fluctuations or of $q = 2$ for the slab model. Beeck and Wibberenz [11.14] argue that a mixture of isotropic and slab-type fluctuations would provide the same result if one uses a spectral slope $q = 1.5$ which is closer to the average conditions for observed magnetic field spectra.

Again the angular distributions are similar for electrons and protons. This is only a particular case out of a general survey of *Helios* pitch-angle distributions: the shape of the angular distributions varies greatly from one event to the other, but is always very similar for electrons and protons. This observation puts rather severe limits to any theory of pitch-angle scattering.

As mentioned above, the shape of pitch-angle distributions reflects the *local* scattering and focusing influences while time profiles depend on the *large-scale* distribution of these forces. This would suggest a systematic investigation where the mean free paths determined from the two independent types of information are compared with each other. A first attempt in this direction has been undertaken by Beeck and Wibberenz [11.15]. They have determined the focusing to scattering ratios λ/L from pitch-angle distributions and found that they tend to be larger than those from intensity and anisotropy time profiles. In a subsequent paper Beeck and Wibberenz [11.16] find that an asymmetry of the pitch-angle scattering coefficient $\kappa(\mu)$ distorts angular distributions similar to the way focusing does. However, careful analysis shows that the asymmetry σ, the ratio λ/L, and the parameter q can be determined separately from the angular distributions. In the analyzis one assumes that the focusing length L is given by the nominal Parker spiral. It was found that for nine solar events the asymmetry σ is in the range 0.2–0.6, and the resulting mean free paths of the order of 0.15 AU are in rather good agreement with the consensus range discussed above in relation to Fig. 11.11. One way to explain asymmetric pitch-angle scattering coefficients is the existence of helicity, which measures the departure from mirror symmetry of the field fluctuations (see [11.186]). In the case of the slab model, helicity corresponds to the polarization of waves propagating parallel to the field [11.104]. If interpreted in this way, the helicity decreases with distance from the sun, which is in qualitative agreement with recent results by Marsch and Tu [11.173], who analyzed the magnetic field turbulence between 0.3 and 1.0 AU. Their new method using Elsässer variables [11.305] seems to give more definite results than the earlier analysis by Bieber et al. [11.18]. The importance of helicity effects for the propagation of energetic charged particles in the interplanetary medium is just beginning to emerge.

It is obvious that one of the reasons we cannot decide which model for the field fluctuations and which theoretical treatment is correct is the lack of knowledge about the actual magnetic field structure at the time of a solar event. A recent attempt in this direction has been made by Valdes-Galicia et al. [11.307]. In this work the computer simulation of particle orbits in a magnetic field model derived from spacecraft data [11.215] is used and applied to the solar event of 11 April 1978. The shape of $\kappa(\mu)$ derived from interplanetary magnetic field data does not show the "90° dip" which was found for the two events presented in Figs. 11.17 and 11.18. The result from the particle trajectory simulation agrees within a factor of 2 with direct particle observations. This remaining difference could be due to tangential discontinuities which are seen by the magnetometer, but not experienced by charged particles traveling along the magnetic field.

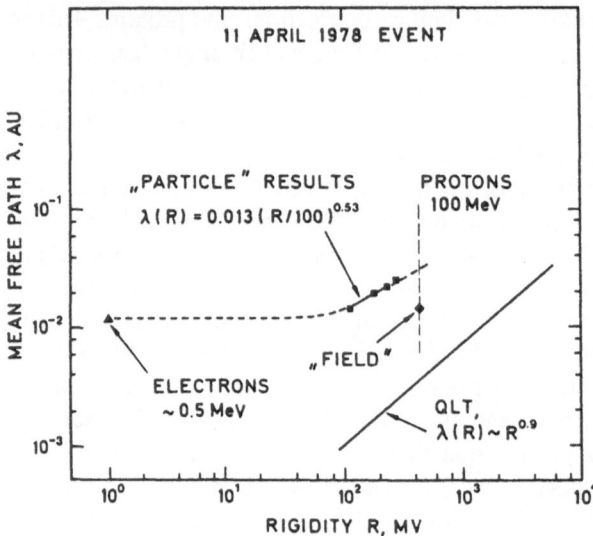

Fig. 11.19. Mean free path during the solar particle event on 11 April 1978, derived by three different methods

The mean free path resulting from different methods for this event is shown in Fig. 11.19. The "field" results represent the particle orbit simulation. The "particle" results are derived from fits to time profiles of the intensity. One example was shown in Fig. 11.13. The "QLT" curve is based on the measured magnetic field power spectrum and the standard model (QLT + slab). It shows roughly the same discrepancy (see Sect. 11.3.3) of about an order of magnitude with the "particle" results. However, the result from the standard model is also about a factor of 4 smaller than the "field" result obtained from the particle orbit simulation. Valdes-Galicia et al. [11.307] suggest a breakdown of the quasilinear treatment as a reason for this discrepancy. QLT requires small fluctuations, which are not guaranteed during this event. Note the large amount of scattering during this event which can be seen from the lowest curve in Fig. 11.12.

The large degree of scattering for this event is one possible explanation for the nearly isotropic form of the pitch-angle scattering mentioned above. The mean free path for the event of 11 April 1978 is of the order of 0.03 AU and is almost two orders of magnitude smaller than in the two cases of 8 April 1978 and 28 March 1976. In other words, the turbulence superimposed on the average magnetic field is much larger for the 11 April event. Jones et al. [11.126] predict that for a given structure of the magnetic field the "90° dip" is filled up more and more with an increased relative strength of turbulence $\eta = \sqrt{\langle \delta B \rangle}/\langle B \rangle$. This means we find a strong dip in the 90° scattering for the events with small turbulence (large mean free path), whereas we approach the isotropic scattering in case of large turbulence (small mean free path). We draw the preliminary conclusion that the shape of the pitch-angle diffusion coefficient κ is not universal, but seems to vary with the degree of turbulence. This result will have to be confirmed by a systematic study of a much larger number of solar particle events.

11.3.5 Summary, Open Questions, and Suggestions for Further Studies

Mathematical tools are available to solve the transport equations for both weak and strong scattering. Effects of long-lasting solar injection and the a priori unknown radial variation of the mean free path require particular care to determine the transport parameters from the solutions. Modifications of the "standard model" (consisting of QLT combined with the slab model) are required in various directions: one has to assume structures of the interplanetary magnetic field different from the slab model, e.g. isotropic or Alfvénic fluctuations, as well as nonlinear corrections and nonresonant interactions, e.g. by mirroring effects. The method of particle orbit simulations can help to distinguish between effects due to the field structure and effects due to inadequate theoretical treatment. In spite of the "consensus range" for the mean free path between 0.08 AU and 0.3 AU the variations from event to event are considerable. For "diffusive" events careful determination of $\lambda(P)$ shows that the mean free path is not constant but increases monotonically with rigidity; the variation is flatter than predicted by the standard model. The radial variation of the mean free path (along a fixed field-line bundle) is very difficult to determine. There are preliminary indications for a shell of increased scattering around 1 AU, with a marked increase of the mean free path toward the sun.

Angular distributions can be used to determine the variation of pitch-angle scattering with pitch angle and to determine locally the ratio between focusing and scattering forces. In two cases of weak scattering one finds a marked dip in scattering near a pitch angle of 90°. The interpretation is in terms of field fluctuations different from the slab model and a small amount of nonresonant interactions. In one case of strong scattering one finds that the dip near a pitch angle of 90° has been filled up. A tentative interpretation consists in a systematic filling up of the resonance gap when the level of field fluctuations increases. A systematic study of angular distributions for a number of solar events shows that pitch-angle scattering is asymmetric. This asymmetry is tentatively interpreted in terms of helicity effects. If this interpretation is correct then the effect should be most pronounced close to the sun [11.173]. Further studies in this area are required.

Observationally, one should also attempt to determine the mean free path over a large range of rigidities by studying various particle species, including in particular electrons, and use multispaceprobe observations to separate longitudinal and radial effects. The spectral tensor of the magnetic field fluctuations should be determined at the time of solar energetic particle observations, with the aim of determining the nature of the magnetic field structure directly. For this field structure, particle orbit simulations should be continued, and the pitch-angle scattering coefficient be computed theoretically, based on a mixture of resonant and nonresonant interactions. Reames [11.263] points out that energetic protons from large solar flares can generate Alfvén waves which increase scattering and impede further flow of particles. This makes the study of magnetic field properties during solar particle events even more important.

11.4 Coronal Propagation

In [11.292] we have seen to what extent the large-scale structure of the interplanetary medium is determined by the magnetic field topology of the sun, in particular by regions of open and closed magnetic field lines. In this section we shall demonstrate the relevance of the magnetic field properties to the transport of solar energetic particles in the solar corona and their escape into the interplanetary medium. It is well known from the adiabatic approximation of charged-particle orbits that particles of sufficiently small Larmor radii are trapped on "closed" dipole-like magnetic field lines and that they can freely escape from a region along "open" field lines. The distinction open/closed holds for a static magnetic field and breaks down in a dynamic situation where field-line reconnection leads to changes of the field topology. The "flux transfer events" are examples from the earth's magnetosphere where reconnection allows the transfer of particles from closed regions inside the magnetopause into the magnetosheath (for a recent summary see [11.67]). Probing the solar atmosphere by *in situ* observations is not possible. But there is sufficient indirect evidence that reconnection processes play an important role in the solar photosphere and corona [11.249, 303, 253]. The flare process which generates high-energy particles is the most prominent example. But the very existence of accelerated particles in interplanetary space is probably an indication of dynamic processes involving magnetic fields. Particles seen in space must have traveled out from the sun on open magnetic field lines. But the complex magnetic field configuration in active regions responsible for flares makes it very likely that large numbers of particles are accelerated within closed configurations. How do fast particles get out of these loops? Which part remains trapped and which part escapes? Are there always tiny "threads" of fields leading out of these regions as suggested by radio observations of type III bursts (see [11.23])? Are the closed field configurations broken up by strong flares thus giving the particles free access to interplanetary field lines? Are particles transferred from closed to open fields via drift or reconnection processes? How are particles transported in the corona over large azimuthal distances in very short time? Can we resolve these puzzles by postulating that the particles we see in space have been accelerated directly on open field lines by extended coronal shocks? Let us keep such questions in mind when we proceed now to observational results and models which have been designed to explain the observations.

11.4.1 Brief Overview of Early Observations

First observations on the relation between solar particles, mainly protons, observed at the earth and the solar longitude of the associated H_α-flare showed that the number of events detected at the earth decreases considerably with increasing longitude of the parent flare on the eastern hemisphere of the sun, and, for those events which are detected, the delay between the optical flare and the arrival of particles increases (for a summary see [11.27, 320]). This east–west effect is due to the spiral structure of the interplanetary magnetic field (IMF). A bundle of IMF lines observed near the earth connects back to a point on the sun which is close to 60°W. We mentioned in Sect. 11.3 that particle motion perpendicular to the average magnetic field direction in space can be neglected. This means that perpendicular transport of particles close to the sun is required to explain the azimuthal propagation of energetic flare particles (see [11.244] for a detailed discussion).

This transport process has been named "coronal propagation" and has been studied in numerous papers [11.294, 153, 271, 272, 269, 310, 47, 184, 229]. Let us briefly summarize a few of the early results based on these studies (see also [11.244]):

1. Protons may be transported within an hour to coronal distances up to a certain distance from the flare site. This region with nearly uninhibited access of particles to the interplanetary magnetic field is called the fast propagation region (FPR) [11.269] or the region of preferred connection [11.310]. The extension of the FPR is about 50–60°; its actual extension seems to vary with the actual solar magnetic field topology and can be as small as about 25°.

2. Outside this region, the temporal delay increases considerably with increasing azimuthal distance, leading to delay times for the arrival of the bulk of particles amounting to tens of hours in the several MeV proton energy range. As shown by Ng and Gleeson [11.229], this behavior can in principle be described by a model of coronal diffusion [11.268]. For coronal diffusion as the main transport mechanism, one would need different diffusion coefficents for the two regions [11.320].

3. The transition from one region to the other is connected with the large-scale structure of the solar corona, in particular with certain types of sector boundary [11.271, 272]. It should be noted that there is no abrupt cessation of particle transport outside the FPR, but rather a drop in the efficiency of transport.

4. Several authors have concluded that the azimuthal transport process is rigidity independent, covering rigidities from nearly relativistic electrons to roughly 100 MeV protons (see [11.244] for discussion). This feature has played an important role in some of the models developed subsequently. It is probably correct if one considers protons in the energy range 5–100 MeV. As we shall see below it breaks down totally if electrons and protons are compared.

These features have been used to construct a number of theoretical models. We should keep in mind, however, that some of the models described in the next section are essentially based on observations of protons. As we shall see below, electrons behave differently. In general they are transported more efficiently over a given azimuthal distance. Some of the models are mainly mathematical descriptions (e.g. the diffusion model by Reid [11.268]) while other models relate directly to physical mechanisms in the corona. Because of the lack of *in situ* measurements the possible validity of a model can only be examined by fitting the parameters contained in that particular model and trying to derive these parameters from independent observations or theoretical estimates. As we shall see, no consistent set of physical processes is available so far to take all observations into account.

11.4.2 Models

We shall select and briefly describe some models as characteristic examples. They are based on the magnetic field structure of the solar corona and, in addition, on plasma processes related with the flare itself. Depending on the relative importance of different processes, models can be grouped into three different types:

1. Particles are accelerated instantaneously and close to the flare. They propagate in the coronal magnetic field structure which existed before the occurrence of the flare.
2. Particles are accelerated instantaneously and close to the flare. Part of their lateral transport occurs by processes related to the flare, e.g. an expanding magnetic bottle. This type of model provides an easy distinction between the fast and slow transport regions.
3. Particles are partly accelerated by a shock wave traveling through the solar corona; they are thus directly generated at large angular distances from the flare. If this occurs on open field lines, they have direct access to the interplanetary medium.

One of the key questions is whether a fundamental process exists throughout the solar atmosphere or whether the transport is restricted to the latitude belts where magnetic loops span large distances on the solar surface. We note in passing that this question will probably only be answered uniquely by the *Ulysses* mission, which will cover positions in space which are connected with high solar latitudes.

1. Coronal Propagation in Magnetic Field Structures Undisturbed by the Flare. In this section we will describe three kinds of model:

- The particles can escape along open magnetic field lines connecting from the flare site to interplanetary space (open cone of propagation).
- The particles can diffuse through the coronal magnetic field.
- The particle transport is controlled by the large-scale structure of the coronal magnetic field ("bird cage" model).

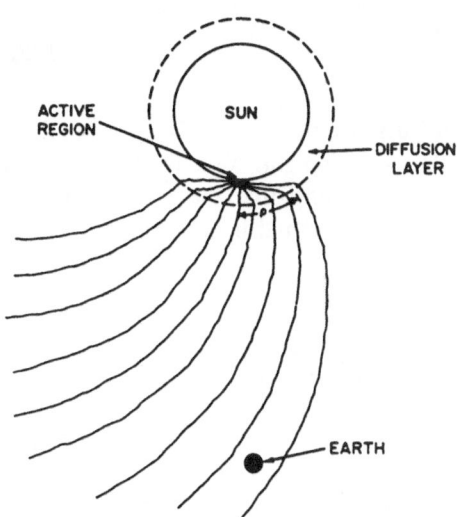

Fig. 11.20. Open cone of propagation: up to a certain distance from the flare site electrons are released promptly after acceleration. Within this cone open magnetic field lines are connected with the flare site (after Wang [11.316])

One of the basic ideas was the existence of an "open cone of propagation" i.e. an extended region of open magnetic field lines directly connected to the active region, as postulated for electrons by Lin [11.160] and Wang [11.316], for protons by Fan et al. [11.71], and McKibben [11.199]. A sketch of this configuration by Wang [11.316] is reproduced in Fig. 11.20. This model is certainly oversimplified. Recent studies [11.23] showed from the observations of type III radio bursts that at least electron escape also occurs in regions where the white-light coronal images show closed magnetic loops. This is an interesting observation. Though it is not clear how the field lines extend back to the inner corona, it shows that the existence of open field lines leading out from the sun is not limited to extended unipolar regions providing the origin of coronal holes (see [11.292]).

In contrast to this direct access a diffusion process was suggested to account for the observations at larger distances from the flare. The first quantitative description of coronal diffusion was given by Reid [11.268] in a more mathematical than physical model. Nevertheless, this model is very popular and because of its simple mathematical handling is often used in connection with models describing the interplanetary transport, e.g. as injection function $Q(t)$ in (11.4). Particles are assumed to diffuse in a shell at distance r_c from the center of the sun, and their injection rate is proportional to the (two-dimensional) particle density U. This leads to the transport equation

$$\frac{\partial U}{\partial t} = \kappa_c \, \Delta U - k \, U \, , \tag{11.19}$$

where κ_c is the diffusion coefficient, Δ the Laplace operator, and k the loss rate. The number of particles released from the sun at a given solar longitude φ at time t, the injection or Reid profile, is then given by

$$I(\varphi, t) = \frac{N_0}{4\pi t_L \kappa_c t} \exp\left[-\frac{\varphi^2 r_c^2}{4\kappa_c t} - \frac{t}{t_L} \right] \tag{11.20}$$

with the escape time $t_L = 1/k$ and the diffusion time $t_c = r_c^2/D_c$. The angle φ gives the distance between observer and flare location and N_0 is the total number of particles accelerated in the flare. It is interesting that the combination of diffusion and losses leads to a time of maximum coronal injection:

$$t_m = \sqrt{\frac{t_L}{4}\left(t_L + \varphi^2 t_c\right)} - \frac{t_L}{2} . \tag{11.21}$$

For large angular distances we get the approximation

$$t_m = \frac{\varphi}{2}\sqrt{t_L t_c} . \tag{11.22}$$

This means that far from the flare the time to maximum for coronal injection increases linearly with angular distance φ, in contrast to what one would expect from simple diffusive propagation. This linear relationship was found experimentally by Reinhard and Wibberenz [11.269], MaSung et al. [11.184], McGuire at al. [11.193], and Schellert [11.281]. Insertion of (11.22) into (11.20) leads to an exponential variation of the intensity maximum with coronal distance,

$$I_{\max}(\varphi) \propto \frac{1}{\varphi} e^{-\sqrt{t_c/t_L}\,\varphi} . \tag{11.23}$$

This feature can be used to determine the ratio of the two coronal time constants from the observed intensity variation with coronal distance [11.324].

The combination of coronal propagation and interplanetary anisotropic transport along the magnetic field has been treated by Axford [11.7]. Numerical solutions of the combined transport equation of coronal and interplanetary transport were given by Englade [11.68] and Schulze et al. [11.290]. Ng and Gleeson [11.229] extended the plane geometry used by Reid [11.268] to a spherical geometry and folded this with the solution of the interplanetary propagation. By fitting their model to existing observations Ng and Gleeson [11.229] found parameters in the range 10–15 hr for the escape time and 50–100 hr for the diffusion time at distances $\varphi > 50°$.

A physical explanation for the parameters of coronal diffusion was attempted by Fisk and Schatten [11.83]. They assume that particles undergo drifts along current sheets separating discontinuous field structures in the corona. A diffusion-like process arises from a randomly oriented series of current sheets related to the filamentary field structure, e.g. from extension of the granulation and super-granulation. For the supergranulation with a scale size of the order of 3×10^4 km one would estimate a coronal diffusion time of the order of 1 hr for 10 MeV protons. This is almost two orders of magnitude too small compared with the value estimated by Ng and Gleeson [11.229], but it is roughly of the right order of magnitude to account for the transport of protons in the fast propagation region;

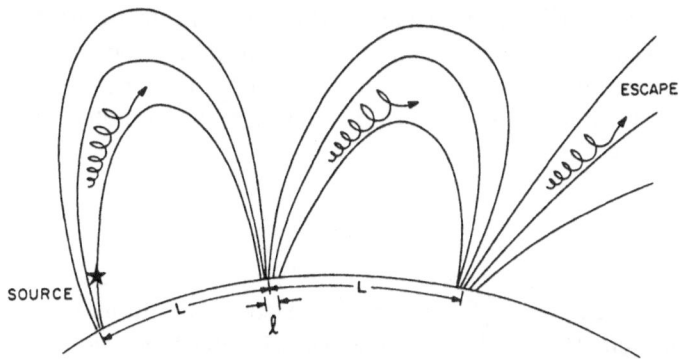

Fig. 11.21. Transport of energetic particles along magnetic loops as described by the "bird cage" model. The particles are eventually released from open magnetic field structures (after Newkirk and Wentzel [11.227])

but if current-sheet diffusion is a fundamental process acting in the solar corona, we would have fast transport everywhere, and not a transition to a much slower process beyond a certain angular distance from the flare. Moreover, the fixed magnetic field structure determines the mean free path λ, roughly independent of particle properties, whereas the active motion of particles along the sheets is proportional to the particle speed v, giving rise to a diffusion coefficient $\kappa \propto v$. Such a strong variation with particle energy is in contrast to observations. The process is totally inadequate for the lateral motion of electrons, because their Larmor radius is small compared to the estimated current-sheet thickness, so that the efficient transport along the current sheet breaks down. In summary, it seems unlikely that the current-sheet diffusion provides an adequate model for coronal propagation.

After this discussion of the diffusion model and its extensions we will now describe a transport model based on the large-scale structure of the coronal magnetic field. It has always appeared an attractive idea that particles can span wide gaps in the corona by traveling along large magnetic field loops. Such loops were observed in white light and calculated from photospheric field images and span typical distances of about 30°. However, apart from particles in the GV/c rigidity range which may leave these loops by gradient and curvature drifts (see [11.218]) particles will remain essentially trapped in the loops, have no access to open field lines, and will not be able to travel by more than the extension of a loop of about 30°–35° in angle. A mechanism to overcome this difficulty was proposed by Newkirk and Wentzel [11.227] in their "bird cage" model. Here particles are transferred between adjacent flux tubes, loops or bundles of magnetic field lines connecting to interplanetary space, by field-line reconnection produced by the rearrangement of the field in the supergranular network. In this way particles traverse large distance along magnetic loops, and they migrate from loop to loop, and occasionally to open field lines. A simplified sketch is shown in Fig. 11.21. The crucial parameter is the time T_{reset} which is needed for the small-scale magnetic field structure to reorganize significantly. Because of the dependence of the diffusion coefficient on the reconnection time as the parameter controlling the transport, the coronal propagation is independent of velocity and rigidity, except for the few high-energy particles escaping from the loops owing to the drift mechanism. For randomly oriented arcs of length L one obtains an effective diffusion coefficient

$$\kappa_c = \frac{L^2}{2\,T_{reset}} \ .$$

A typical loop spans 26° to 36° ($L = (3.2 - 4.0) \times 10^{10}$ cm). With the estimated reconnection time $T_{reset} = 10^4$ s [11.227] the diffusion coefficient turns out to be $\kappa_c = (5 - 8) \times 10^{16}$ cm^2/s. This leads to coronal diffusion times of 40 to 60 hr if the migration of particles occurs at a height of $r_c = 1.5$ solar radii, in good agreement with results for the slow propagation outside the fast propagation region.

2. Coronal Propagation Influenced by the Flare Process. One of the essential features of coronal transport is the two-step process consisting of a fast initial transport over a limited range of longitudes and a subsequent less efficient process acting further away from the flare. Schatten and Mullan [11.279] assumed that the fast particle propagation results from the influence of the flare on the closed magnetic loops over the active region (see Fig. 11.22(a)). The hot plasma forces the loop to expand in radial and azimuthal direction. Energetic particles remain trapped in this expanding magnetic bottle (Fig. 11.22(b)). The expanding structure develops a Rayleigh–Taylor instability, the structure becomes unstable, the bottle opens, and the particles are released into interplanetary space (see Fig. 11.22(c)). At this moment, which is estimated to be 40–50 minutes after the flare, the loop may have expanded over a region of ±60° around the flare location as estimated from the observation of type II radio bursts. The particle injection time does not depend on particle parameters, but only on the time scale σ_0 of the opening of the bottle. Thus, this expanding bottle concept has the attractive features that it

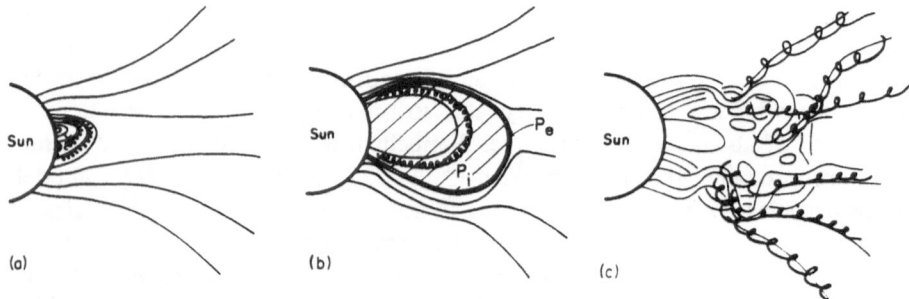

(a)   (b)   (c)

Fig. 11.22a–c. Particle release from an expanding magnetic bottle. (a) Particles accelerated in the flare are trapped in closed magnetic loops above the flare site. (b) The hot plasma expands the bottle. (c) The bottle opens and releases the particles into interplanetary space (after Schatten and Mullan [11.279])

explains the lateral extent of the fast propagation region and the filling-up of this extended region within an hour, independent of particle properties.

Particle propagation after the opening of the bottle is assumed to occur in magnetic structures undisturbed by the flare, owing to drift and cross-field diffusion (see [11.218] for details). The propagation time σ then is given by $\sigma = \sigma_0 + \tau(\beta, P)$, where τ is the time between the opening of the bottle and the arrival of the particles at the observer. τ depends on the particle velocity β and the rigidity P. As an alternative, one could assume that loop reconnection correctly models the second (slow) coronal transport process [11.227]. The essential point is that field lines in the vicinity of the flare are profoundly modified by the flare itself, so that the preflare coronal field structures within 30–60° of the flare site are of little relevance for the particle transport. Based on this general idea, Perez-Peraza [11.244] develops a global model for coronal transport over small and large distances.

3. The Role of Coronal Shocks for Acceleration and Transport. Here the idea from the preceding paragraph is carried one step further. The flare is not only responsible for the lateral transport via an expanding bottle, but the particles are directly accelerated at the leading edge of a coronal shock which is driven by the expanding material. The idea goes back to Palmer and Smerd [11.241] and was discussed by Reinhard and Wibberenz [11.269] as one possible origin for the fast propagation region. Coronal shocks may move over distances of the order of 60° within an hour, and many authors use this feature to explain fast coronal transport (e.g. [11.41, 181] and references therein). Mason et al. [11.181] develop in detail the picture of the LSSA (large-scale shock acceleration) which gains support in particular from unsystematic variations of the chemical abundance ratios. The authors interpret these changes as being due to relative abundance fluctuations of particles at different locations in the corona which are successively sampled as the field-line connection point of the observer moves across the corona. The particles leak out into the interplanetary space very near the sites where they have acquired their final energies. Weakening of the shock as it moves away from the flare site results in lower particle-flux levels.

In this model, coronal propagation is limited to the range which is reached by the expanding shock. Changes occurring at solar sector boundaries [11.271] might in this case be related to modifications in the shock properties when it hits neutral sheets in the corona (see e.g. Steinolfson and Mullan [11.299]). Mullan [11.217] discusses the weakening of a shock during filament crossing. Particles in the slow mode of propagation, beyond the range covered by the shock, can reach the earth only by corotation. To give an example: if the outer border of the shock-accelerated particle population is located 40° east of the magnetic connection point of the observer, it reaches the earth by corotation within about three days, thus explaining the long delay times if the flare occurs far east. This idea, that shock processes are not responsible for the long delays to maximum intensity for eastern events, is also stressed by Cane et al. [11.30]. They present a detailed study of how the intensity profiles during solar particles events vary with the location of the flare on the sun. They suggest that the major controlling agent for the appearance of the profiles is the existence of an interplanetary shock. We shall come back to this point in the next section where we discuss more recent observations and try to single out the valid model(s) for coronal transport.

11.4.3 Recent Observations

Many observations summarized in Sect. 11.4.1 were based on statistical studies from a fixed point in interplanetary space (the earth), ordered with the solar longitude of the flare. Different types of information could be gained from multi-spaceprobe observations with several spaceprobes located at different heliolongitudes. Early results from the *Pioneer* spaceprobes are summarized by McCracken and Rao [11.187], McCracken et al. [11.188], and McKibben [11.198, 199]. The improved instrumentation on the *Helios* and *Voyager* spaceprobes, covering a large range of energies and different particles, allowed us to discover principally new features of coronal transport (see e.g. [11.146, 193, 324]).

Let us start the discussion with a characteristic example showing the advantage of two spaceprobes at different heliolongitudes, demonstrating also two

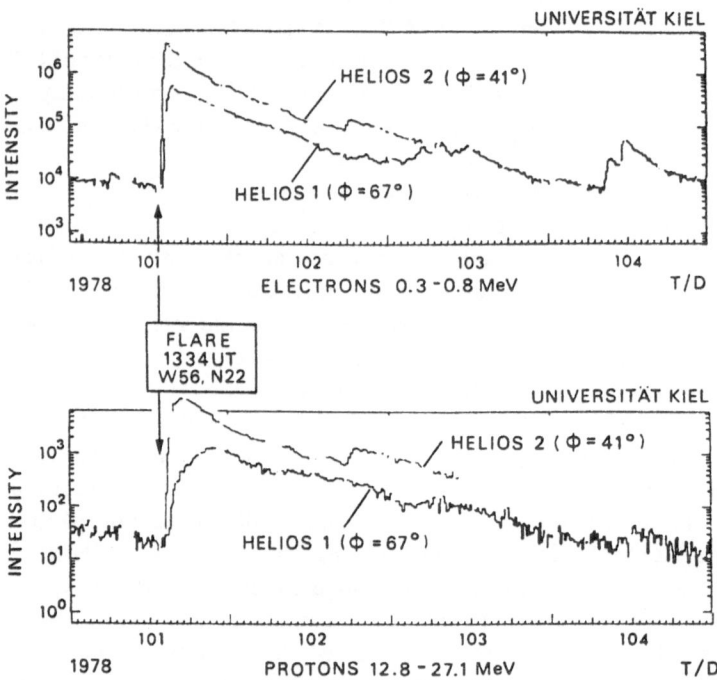

Fig. 11.23. Intensity–time profiles of ~0.5 MeV electrons and 4–13 MeV protons measured on *Helios 1* und 2 during the 11 April 1978 flare event. The magnetic connection points of the two spacecraft and the flare site are shown in the insert. The larger distance in heliolongitude of *Helios 1* from the flare causes a later time of maximum intensity and a lower maximum intensity than on *Helios 2*

new features of coronal transport. The 11 April 1978 event observed by *Helios 1* and 2 is a good example of the influence of coronal propagation. Both spacecraft were at a radial distance of about 0.5 AU from the sun, but the angular distance to the flare site was much larger for *Helios 1* than for *Helios 2*. The constellation of the spaceprobes and the flare is shown in the insert of Fig. 11.23, where we present intensity–time profiles for $\sim$0.5 MeV electrons and 4–13 MeV protons. All profiles show a common slow decay of the intensity following the maximum, a result of particle storage in the interplanetary magnetic field. This decay phase is not much different for both particle species and both spacecraft, but the particle intensities are markedly lower at the location of *Helios 1* for both particle types. The rise time for the protons is much larger at *Helios 1* than at *Helios 2*. This is in agreement with the general trend for the "slow" coronal transport at distances beyond the FPR. The flare is located about 90° west of *Helios 1*. This poses an immediate problem for the concept mentioned above where the particle population observed at *Helios 2* should have reached *Helios 1* by corotation only. For the electrons we find the interesting result that they are reduced in intensity, but appear with practically no temporal delay with respect to *Helios 2*. We shall see below that electrons are generally transported in the corona more efficiently than protons. This is an important constraint for any propagation model and contrasts with the "rigidity-independent transport" which dominated the discussion for a while.

1. Coronal Transport of Protons. The coupling between the temporal delay and the intensity decrease as shown in Fig. 11.23 is one of the remarkable results of the multispaceprobe studies. McGuire et al. [11.193] summarized solar proton data from events observed simultaneously on *Helios 1/2* and *IMP 7/8* and tried to generate a clean sample by taking multiple injections and/or contamination by particles owing to interplanetary shocks into account. The results are shown in the next two figures. Figure 11.24 summarizes for 11–60 MeV protons the time of maximum flux (relative to the flare H_α-onset) vs. the difference between flare location and spacecraft connection longitude (negative for flares east of the observer). Points observed for the same event on different spacecraft are connected by straight lines. The curve confirms the fast propagation region around the flare site [11.269, 310]. For points near 0° connection longitude the time to maximum flux is due to the interplanetary propagation. Its value of 5±3 hr would correspond to an interplanetary mean free path in the range 0.07 AU to 0.3 AU, in good agreement with the results presented in Sect. 11.7.3. The average slope for increasing longitude differences corresponds to $(\Delta T_{max})/(\Delta \varphi) = (38 \pm 19)$ hr/rad for longitudes between $-60°$ and $-120°$. This is consistent with the value 27 ± 10 hr/rad for the slope found by van Hollebeke et al. [11.310].

The variation of peak intensities with connection longitude for 11–60 MeV protons is shown in Fig. 11.25. Since the total intensities of each event are different, individual event peak fluxes have been renormalized to give a minimum overall spread of the points. Data points obtained at different radial positions have been corrected with an empirical radial correction factor. The overall distri-

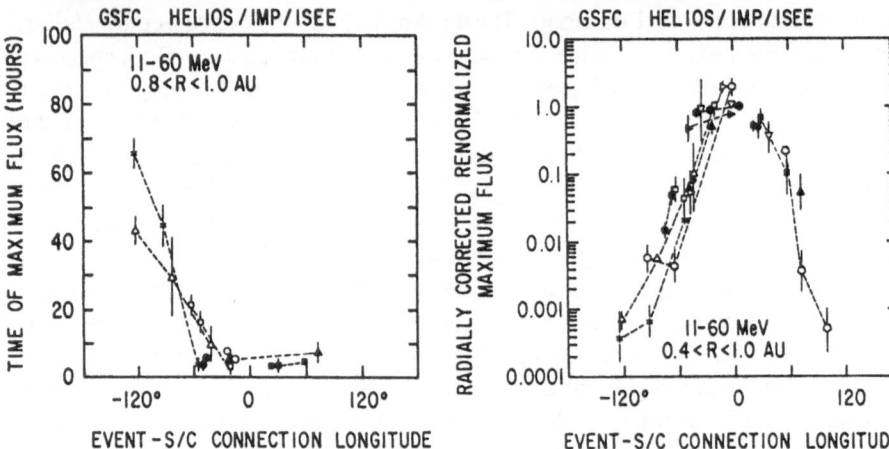

Fig. 11.24. Time of maximum of 11–60 MeV protons plotted versus coronal distance. Data points from the same event observed on different spacecraft are connected by *dashed lines* (after McGuire et al. [11.193])

Fig. 11.25. Dependence of interplanetary maximum intensities of protons on coronal distance between flare and magnetic footpoint of observer (after McGuire et al. [11.193])

bution is nearly symmetric with respect to longitudes east and west of the flare. A relatively flat distribution close to the flare is followed by a steep variation, with a characteristic intensity drop of two orders of magnitude for a 60° increase in longitudinal distance. This corresponds to an e-folding angle of 13°. McGuire et al. [11.194] discuss their results in terms of the coronal diffusion model of Reid [11.268], with the two characteristic time constants for escape t_L and for coronal diffusion t_c; see (11.20) above. As discussed above (see also [11.324]) these two time constants control the variation of the time to maximum intensity and of the peak intensity with longitude. McGuire et al. [11.194] also allow for an initial finite width φ_0 of the injection at time zero. Careful inspection shows that a simultaneous fit of the results in Figs. 11.24 and 11.25 by the three parameters t_L, t_c, and φ_0 is not possible. With some widening of the degree of uncertainty for the observed parameters, the authors claim that values of $t_c = 91$ hr, $t_L = 9$ hr, and $\varphi_0 = 10°$ simultaneously match the observations. These results are in qualitative agreement with one set of parameters obtained by Ng and Gleeson [11.229], namely $t_L = 15$ hr and $t_c = 100$ hr, for a totally different set of observations. We believe that the coronal diffusion time by McGuire et al. [11.194] is a slight underestimate. Their parameters would give an e-folding angle of $\sqrt{t_L/t_c} = 18°$, in contrast to the value of 13° which represents the data in Fig. 11.25 better.

However, we should take the value of $t_c \approx 100$ hr as a good representative value for the coronal diffusion time if transport over large distances is involved. If we model the transport to occur in a spherical shell of 1.5 r_0 (r_0 = solar radius), we obtain a diffusion coefficient $\kappa_c = 3 \times 10^{16}$ cm^{2}s^{-1}. As discussed above this is the correct order of magnitude obtained by the "bird cage" model (reconnection of large coronal magnetic field loops, see Fig. 11.21).

2. Coronal Transport of Electrons. We discussed above (see Fig. 11.23) that electrons are transported through the corona more efficiently than protons. This is confirmed by a study of a number of electron events on *Helios 1/2* by Schellert [11.280] and Schellert et al. [11.281]. Figure 11.26 shows the variation of the time to maximum intensity as a function of angular distance from the flare for electrons with an average energy of 0.5 MeV. Only events where the footpoint of *Helios* was west of the flare are shown. A fit to the data was performed by assuming that the constant delay within a certain distance φ_0 is due to interplanetary propagation and that t_m increases linearly beyond φ_0; see (11.22). A best fit was obtained for $\varphi_0 = 26°$. Interplanetary propagation would correspond to a mean free path of the order of 0.12 AU or below, in good agreement with the results presented in Sect. 11.7.3. The linear increase beyond 26° is fitted by a straight line corresponding to $\Delta T_{max}/\Delta \varphi = 3.9$ hr/rad. This delay rate is roughly an order of magnitude smaller than for 11–60 MeV protons (see above). This is a clear indication that the concept of rigidity-independent transport cannot be extended to electrons, in contrast to the findings by MaSung et al. [11.184]. A similar conclusion can be drawn when we consider the variation of the maximum intensities as shown in Fig. 11.27. The maximum intensities of ∼0.5 MeV electrons from 12 solar events seen simultaneously on *Helios 1* (open squares) and *Helios 2* (open triangles) are connected by straight lines and arbitrarily shifted in absolute magnitude. We see some characteristic differences

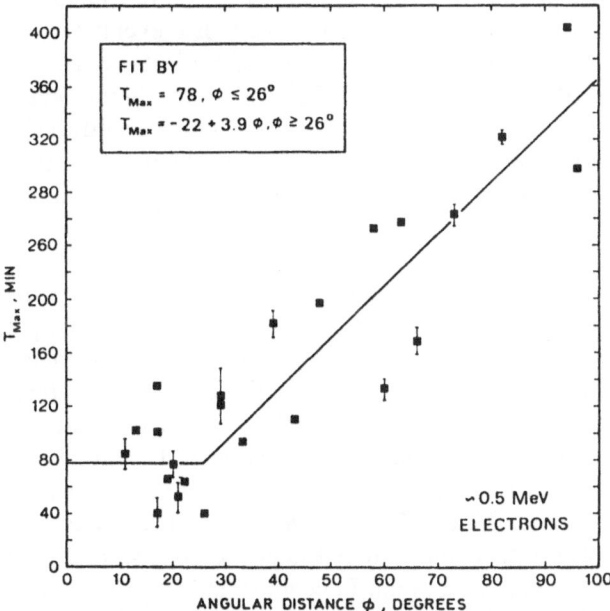

Fig. 11.26. Time of maximum of ∼0.5 MeV electrons versus coronal distance. Separate least-squares fits were performed inside and outside the FPR with an extension of 26° (after Schellert et al. [11.281])

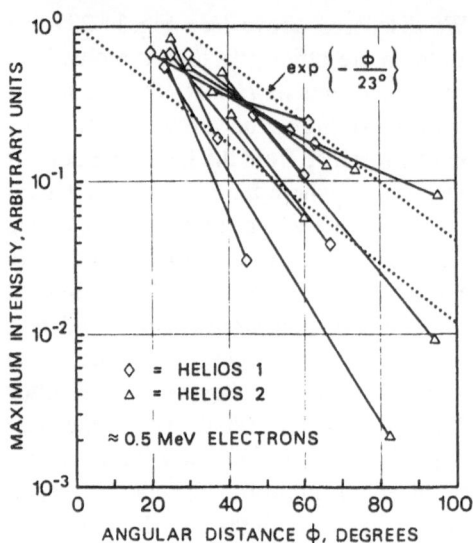

from the behavior of protons (Fig. 11.25). The data do not seem to be so regularly organized as in case of the protons. We may find a rather moderate variation with longitude even at large distances from the flare. The general trend seems to be well represented by an e-folding angle of 23°, see the dotted lines in Fig. 11.27. This value is markedly larger than the corresponding one for protons of about 13°. This general trend of a markedly stronger decrease of proton intensities with longitude is confirmed by a careful inspection of individual events seen on both *Helios 1* and 2. Two events are extensively discussed by Wibberenz et al. [11.324]. Let us again describe the variations in the time to maximum and in the maximum intensity by the two time constants t_c and t_L for coronal diffusion and escape, respectively. Applying (11.22) and (11.23) we obtain $t_c = 20$ hr, $t_L = 3$ hr, in contrast to the corresponding values of the order of 100 and 9 hr for protons.

3. Comparison of Protons and Electrons in Individual Events. Before discussing the implications for the various transport models, let us close this part with two characteristic examples demonstrating the considerably more efficient coronal transport of electrons. It is an interesting consequence of the difference in e-folding angles that one may still find a considerable number of relativistic electrons far away in longitude from the flare when proton intensities are already hidden in the background. An instructive example is seen in Fig. 11.28. During the event of 28 March 1976 *Helios 2* was 67° and *Helios 1* was 147° away from the flare site. The event is clearly seen in electrons and protons on *Helios 2*, with a fast rise indicative of a well-connected event. In contrast, the event is not seen at all in the proton channel on *Helios 1* but definitely appears in the electron channel, though with a markedly reduced intensity compared to *Helios 2*. There are quite a few examples found during the double mission *Helios 1/2* where

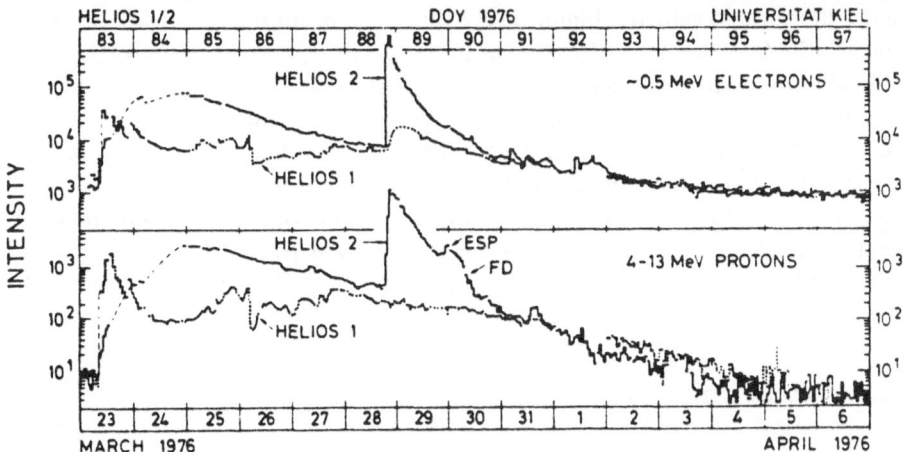

Fig. 11.28. Intensity as a function of time for an electron and proton energy channel during a 15-day period in March/April 1976. During the 28 March event *Helios 2* was at a distance of 67° and *Helios 1* at a distance of 147° west of the flare site (from Kunow et al. [11.148])

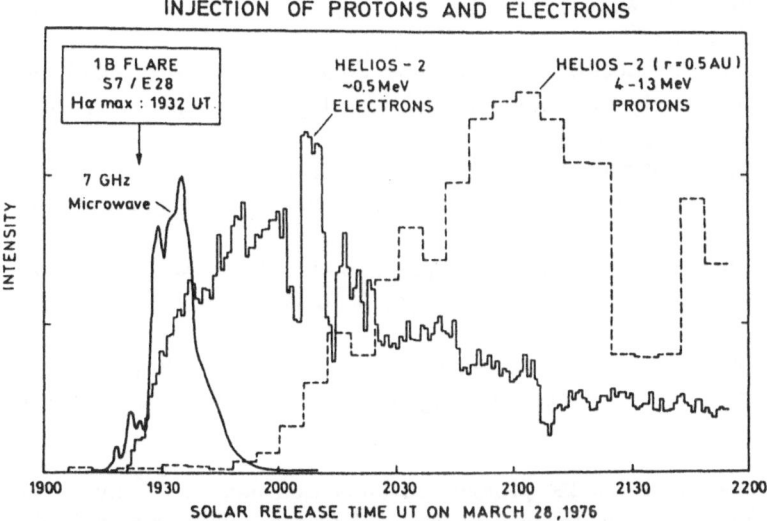

Fig. 11.29. Injection profiles for one electron and one proton channel for the event of 28 March 1976. The nearly relativistic electrons are released almost simultaneously with the appearance of the 7 GHz microwave burst, whereas the protons are delayed by about 40 minutes (after Neustock [11.225])

small, but definite, electron increases with slow rise times are found more than 150° away from the flare. It is interesting that the two particle types also show differences within the FPR. Bieber et al. [11.17] and Ng et al. [11.230] analyzed the same event in great detail. At the position of *Helios 2* at a distance of 0.5 AU from the sun it is characterized by conditions of weak interplanetary scattering. This means that the earliest part of the intensity profile can be used to obtain the solar injection by simply shifting the observations in time by an amount $\Delta t = s/v$ to earlier times, where s is the distance along the Archimedean spiral

293

and v the particle velocity. Figure 11.29 shows the result for ~0.5 MeV electrons and 4–13 MeV protons. Also shown is the 7 GHz microwave emission as an indication of particle acceleration on the sun. It is noteworthy that the injection of electrons starts immediately after acceleration for a situation when *Helios 2* was connected with a coronal distance 67° away from the flare. In contrast, protons are injected with a delay of about 40 minutes. According to our earlier idea, we are still within the "fast" propagation region (delays with less than one hour). A similar difference between electrons and protons was found by Wibberenz et al. [11.324] for the 27 December 1977 event with *Helios 1* at 62° distance from the flare.

11.4.4 Discussion

We can now summarize the essential observed features of coronal transport as follows:

- A clear separation between fast and slow transport processes is confirmed. The separation between the corresponding regions may vary with the general coronal properties.
- Within the FPR, electrons seem to have direct access to open field lines, whereas protons appear with some delay, at least if the observer is of the order of 60° away from the flare.
- Within the FPR, particle intensities are diminished the further away from the flare we go. This process is more pronounced for protons than for electrons [11.324].
- Outside the FPR, both particle types appear delayed with delay rates $\Delta t_m / \Delta \varphi$ which are roughly a factor of 5–10 larger for protons than for electrons.
- Outside the FPR, the maximum intensities decrease markedly with increasing longitudes. The average e-folding angle is smaller for protons than for electrons.

So in general we find that the protons are influenced much more by coronal transport than the electrons. Because of the small Larmor radii of electrons, drift processes (as in [11.244, 227]) are negligible, and on the basis of drift processes one would expect a *less* efficient lateral transport of electrons, in contrast to observations. We also reject all models based on a rigidity-independent transport, such as e.g. the "bird cage" model in its original form [11.227] or the model of the opening of a magnetic bottle [11.279], because they do not consider the different behavior of electrons and protons.

For the fast access over limited longitudinal range it seems obvious that we need some process which is connected with the structure of the flare-producing active region or with the flare itself. Considering the different temporal behavior of electrons and protons Wibberenz et al. [11.324] suggest two fundamentally different processes. Electrons should have direct access to open field lines extending out from the complex active region as suggested by Lin [11.160] or Wang [11.316]; see also Bougeret et al. [11.23], Dulk et al. [11.62]. It is interesting that protons do not seem to have the same fast access to such open field lines as seen in events where despite large coronal distances the electrons arrive fast while the protons are delayed (e.g. in the 28 March 1976 event). This may be connected with their considerably larger Larmor radius, envisaging "holes" of some kind in the system of nested loops or reconnection mechanisms which allow only electrons to get out. For protons direct acceleration on open field lines by coronal shocks seems to be a possibility.

For the slow transport over larger distances Wibberenz et al. [11.324] reject the possibility of direct acceleration on open field lines by a propagating shock, based on the very slow delay rates and on the differences between electrons and protons. They suggest a modification of the "bird cage" model by Newkirk and Wentzel [11.227] with electrons having the chance to move from one cage to the other during the reconnection process more efficiently than protons.

While the processes discussed above are connected with the structure of the active region we will now consider the possibility that a fundamental transport process is acting throughout the solar atmosphere. For protons the current-sheet diffusion by Fisk and Schatten [11.83] would supply a fast lateral transport. Some boundaries distributed more or less at random over the solar surface, connected with regions of different magnetic polarity, inhibit the lateral transport [11.271], thus generating the slow transport over large distances. As we can see, a number of alternatives to explain coronal transport still does exist. A distinction between them does not seem possible at present and may have to await the off-ecliptic results of the *Ulysses* mission.

11.5 Particle Acceleration in Solar Flares

Solar flare events are the most energetic phenomena in the solar system. They can release energies up to 10^{25} Ws [11.111, 252]. Flares are impulsive energy transformations of complex active magnetic field configurations and result in acceleration of ions and electrons from the ambient solar atmospheric matter. Some of the accelerated particles can reach energies of several GeV for ions and more than 100 MeV for electrons.

The complexity of various flare models and their relation to different magnetic topologies is discussed extensively by Sturrock [11.301, 302]. A simplified model based on a group of sunspots with different polarity is sketched in Fig. 11.30. Sketches of this type have been widely used to relate acceleration, escape, and storage of energetic particles to different electromagnetic phenomena. Field lines of opposite direction approach each other (1), caused by motion of the inner spots (S, N) in the underlying solar material. Merging of magnetic field lines (reconnection) accelerates particles in the compression region (see [11.253] for a discussion of the physical mechanism). Particles moving towards the denser

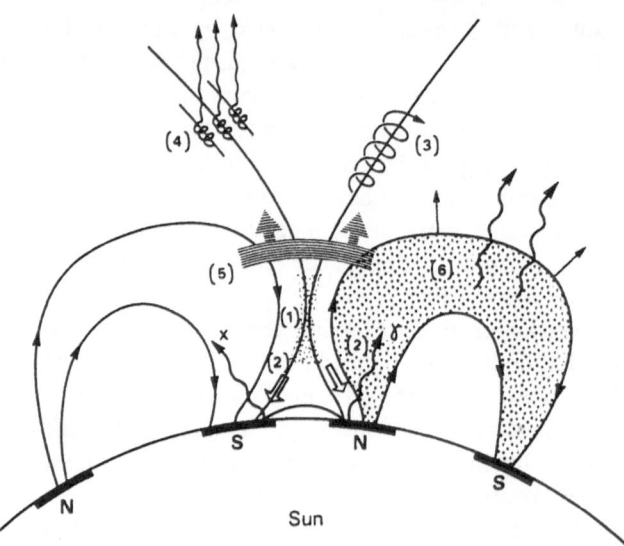

Fig. 11.30. Phenomena related to a solar flare: (*1*) reconnection region, (*2*) particles moving downward after acceleration to produce X- and γ-rays, (*3*, *4*) escaping particles, producing type III radio bursts in the case of electrons (*4*), (*5*) shock wave, and (*6*) trapped particles generating type IV radio bursts

regions of the solar atmosphere (2) lose their energy by heating and ionization of atmospheric matter and generate X-rays (X) in the case of electrons and γ-rays (γ) in the case of protons and ions. Particles moving in the opposite direction escape (3, 4) if they reach open field lines and are observed as energetic particles in interplanetary space. During their escape fast electrons generate type III radio bursts (4). The abrupt heating can generate a shock wave (5), which can be observed in interplanetary space a few days later. A part of the accelerated particles will be stored in closed magnetic field loops (gray dotted region). This is a reservoir for energetic particles which can still leave the solar corona after several hours. Electrons within this magnetic bottle will generate type IV radio emission (6).

The electromagnetic radiation (X-ray, γ-ray, and radio emission) depends critically on the conditions in the ambient solar atmosphere through which the particles travel (e.g. density, magnetic field, and temperature) and provides information on the acceleration mechanism and time sequence. The solar cosmic rays observed in the interplanetary medium allow one to derive information on the processes of acceleration, coronal storage and propagation, injection into the interplanetary medium, and interplanetary propagation. Some of these processes were discussed in the two preceding sections. A brief overview of acceleration mechanisms will be given in Sect. 11.5.1. While the particles observed in space are influenced by propagation effects the electromagnetic radiation arrives nearly undisturbed at the observer. Recent progress in our knowledge of the flare process has only been possible by correlating observations of particles in interplanetary space with observations of the electromagnetic radiation (see Sect. 11.5.2).

11.5.1 Acceleration Mechanisms

For reviews of particle-acceleration theories and their relation to observations we refer to Spicer [11.298], de Jager [11.50,51], Forman et al. [11.87], Ramaty [11.256], Lin [11.163], Sakai and Ohsawa [11.275], and Scholer [11.285]. Today it is widely believed that the source of energy in solar flares is the annihilation of magnetic fields [11.303]. This process is also called reconnection or merging of fields. The principle of this process is described e.g. in [11.249], [11.250], and [11.243]. Strong electric fields accompanying the annihilation of magnetic fields are one possible mechanism for direct particle acceleration. However, the rapid energy deposition in the flare process is also the source of hydromagnetic turbulence and shocks which may both lead to stochastic particle acceleration. Reconnection can occur on different time scales (steady, fast, and explosive; for a summary see [11.275]), which seem to play an important role in the different phases of flare development (e.g. in the model by Sturrock [11.302]).

Particle acceleration by shock waves can be due to three mechanisms: resonant acceleration in the turbulence generated by shock waves [11.275], diffusive shock-wave acceleration [11.8], the shock-drift mechanism [11.297, 54, 6], or a combination of the two latter effects [11.57]. Švestka and Fritzova [11.314] found a strong correlation between type II radio bursts interpreted as a coronal shock and the occurrence of solar energetic particles in interplanetary space. This is only one example of the close relation between particles accelerated in a flare and a shock. Nevertheless, it is not clear whether a type II burst is indicative of shock acceleration of energetic particles observed in interplanetary space or whether another agent accelerates these particles (see e.g. [11.315]). The role of coronal shocks in the acceleration of particles at some coronal distance from the flare site is still open as was discussed in Sect. 11.4.

The Fermi process or stochastic acceleration was first suggested by Fermi [11.72] to explain the acceleration of galactic cosmic rays. As an individual process Fermi acceleration can be described as being due to the collision of a charged particle with a moving magnetic irregularity connected e.g. to a shock front or waves moving both parallel and antiparallel to the average magnetic field. Waves leading to stochastic particle acceleration include Alfvén waves and magnetosonic waves [11.143, 204, 1, 52], as well as Langmuir waves and electrostatic waves with plasma velocities of the order of the particle speed [11.204]. If the random energy increment is small compared to the particle energy, stochastic acceleration can be described as a diffusion in momentum space (e.g. [11.282]). The spectra of stochastic acceleration can then be obtained from the momentum diffusion coefficient by solving the transport equation [11.61, 87].

A terminology widely accepted for a long time was the existence of two "phases" of the acceleration process, based on radio, X-ray, and energetic particle observations [11.326, 49]. During the first phase (impulsive or flash phase) electrons are accelerated up to energies of about 100 keV. For small flares the energy transferred to the electrons may be a substantial fraction of the total flare energy. Therefore the primary energy release mechanism during flares could be the initial conversion of stored magnetic energy into energetic electrons. Most of the flare phenomena observed in the first or impulsive phase can be explained by interactions of these electrons with the solar atmosphere. The second-phase acceleration was believed to occur only occasionally during large solar flares, producing both ions and electrons of energies above several MeV. This second phase was thought to be delayed by about ten minutes with respect to the first phase (see e.g. [11.11, 165, 49]). A close association of the second acceleration phase with coronal shock waves manifested by type II radio bursts is observed. This sequence of events was also underlying the sketch in Fig. 11.30. However, the delay of several minutes between the two phases could no longer be postulated when γ-ray observations from the *Solar Maximum Mission* showed that the separation between electron and ion acceleration can be as short as one second (see [11.37]). In these cases no clear distinction of different acceleration "phases" is possible; rather two or more "steps" are assumed consisting of different physical mechanisms and occurring in close temporal association with each other (see e.g. [11.9, 205]) or by one acceleration mechanism [11.37] only.

Furthermore, it turned out that an important distinction between two classes of flare is the topology of the underlying coronal structures and the height in the atmosphere where the main energy release occurs (Pallavicini et al. [11.239]). The distinction is related to the duration of the soft X-ray emission, and they call flares with a soft X-ray emission shorter (longer) than one hour impulsive (gradual). It is interesting that this distinction is not related to the *size* of the flare in soft X-ray emission. Similar classifications are made on the basis of other frequency ranges (e.g. hard X-rays [11.238] and microwaves [11.46]). On a statistical basis, the different definitions of impulsive and gradual events seem to lead to a similar classification, but it is not yet totally clear how far these two classes indicate the occurrence of different physical processes (see the discussion in [11.58]). A series of subsequent observations showed that the classification into impulsive and gradual flares also leads to differences in other phenomena. In particular, Cane et al. [11.29] found that gradual events are associated with coronal mass ejections and with interplanetary shocks and show in general a higher proton to electron ratio than impulsive events. Nevertheless, the helium to electron ratio seems to be the same in both classes of flare [11.133]. So the protons but not the ions in general seem to behave differently in impulsive and gradual events. Additionally, events with emission of γ-rays [11.39] and ^{3}He-rich events [11.182] are mostly classified as impulsive.

It is not totally clear whether this categorization by the duration of the soft X-ray emission allows a strict separation into two distinct classes or if one should rather introduce a continuous transition (see e.g. [11.39]). A more physical distinction may be related to the appearance of a coronal mass ejection. These are in general connected with interplanetary shocks. Though it is well known that these shocks are able to accelerate particles, the particles accelerated by interplanetary shocks appear as a separate feature superimposed on the decay phase of a solar event. Many authors believe that the same shock when passing through the outer solar corona is also responsible for a large part of the solar energetic particles (see [11.181, 30]), but a direct proof is still missing. If it should turn out that shocks are indeed responsible for a large part of the SEP population, it is still unknown which of the three physical mechanisms mentioned above in connection with shocks is mainly responsible.

In this brief description we saw that flares can be very variable. A huge body of observations exists, and so far there is no model explaining them in a consistent manner. Different models have been developed that agree with each other in some aspects and differ in many other aspects, describing only parts of the observations.

In the next section we shall discuss a selection of observations of the chemical composition, the shape of energy spectra, the relation between interplanetary particles and γ-ray emission, and the particle injection. The energy spectrum gives hints as to the acceleration mechanism, the relation between γ-emission and interplanetary particles refers to injection and storage mechanisms as well as different acceleration processes, and the analysis of the particle injection relates to the temporal sequence of particle acceleration. A separate section will be devoted to the unusual ^{3}He enrichments and to preferential heating of particles in the source region. ^{3}He-rich solar particle events represent a particle injection and/or acceleration process which differs from those for normal solar events. These usually small events often show ^{3}He/^{4}He ratios near unity, an extremely large number compared to the ratios in the normal solar atmosphere or in the solar wind of a few times 10^{-4}. Because of the small radial distance, *Helios* was in a more favorable position than earth-orbiting instruments to study these low-intensity events in detail and with good time resolution. A major part of this section will focus on ^{3}He-rich events.

11.5.2 Selected Observations

1. Chemical Composition. During the last three decades it has turned out that studies of the chemical composition of solar energetic particles (SEP) are well suited to probing the solar corona (see e.g. [11.207, 25]). Though it was found that SEP abundance varies significantly from flare to flare, the cause of the flare-to-flare variation seems to have meanwhile been understood. Fractionation of the abundances of different species for the same velocity (or, equivalently, the same energy per nucleon) in individual flares can be represented well by a variation with $(Q/M)^\alpha$, where Q is the charge state and M the mass of the nucleus. The charge state of SEP was analyzed by Luhn et al. [11.170, 171] and found to be similar to that of coronal material. It can be seen from the discussion in Sect. 11.3 that the ratio Q/M is proportional to magnetic rigidity for particles of given velocity. Variation of the fractionation with $(Q/M)^\alpha$ could then be related to a rigidity-dependent acceleration process in diffusive shock acceleration (see [11.300] for details).

The average SEP abundance exhibits little energy dependence in the 1 to 10 MeV/nucleon interval. If this SEP-derived coronal abundance is compared with spectroscopically determined photospheric abundances, one finds a step-function-like fractionation with first ionization potential (FIP). For FIP < 11 eV both values agree reasonably well, whereas a depletion in the corona of a factor of about four is found above 11 eV. The origin of this FIP-dependent fractionation is not yet understood. It is unlikely to occur in the corona, because owing to the high temperatures all of the elements are highly ionized. In contrast, elements with high FIP are essentially neutral at photospheric temperatures. It is suggested that the fractionation takes place during the transition from the photosphere to the corona.

Inspection of the accompanying electromagnetic radiation shows that the chemical composition of SEP discussed so far, which is very similar to solar abundances, is related to the gradual events [11.163, 177]. Totally different types of chemical (and isotopic) composition are observed in a class of relatively small solar events, related to effects of preacceleration or "preferential heating". It turns out that many of these events are of the impulse type. They will be discussed in connection with the enrichments of ^{3}He in Sect. 11.5.3.

2. Energy Spectra. Energy spectra of particles in interplanetary space reflect the combined effects of the acceleration at the sun and of coronal and interplanetary propagation. The influence of propagation effects can be minimized if only flares occurring in the fast propagation region are chosen and the time of maximum intensity is used to derive the energy spectra [11.310].

Lin et al. [11.166] found that interplanetary electron energy spectra from solar events show a double power law with a smooth transition between 100 and 200 keV. This can be interpreted as a result of a two-step particle acceleration. The spectra are similar to those inferred from hard X-ray and microwave observations, which suggest that the observed electron spectra are representative of the spectra of the electrons accelerated at the sun. Evenson et al. [11.69] found that the energy spectrum for relativistic electrons also depends on whether the accompanying soft X-ray events were impulsive or gradual. Spectra of gradual events can be fitted if a first step is assumed that accelerates particles up to about 100 keV and a second step that accelerates electrons by a shock up to 100 MeV. In impulsive flares the steep spectrum above 1 MeV shows that the acceleration is dominated by the first step. In impulsive flares the electron spectra in the range 0.3–3 MeV are similar for interplanetary particles and the γ-ray continuum, while in gradual events the electron spectrum seems to be harder than the γ-spectrum [11.130]. This not only is evidence for different acceleration mechanisms in the two classes of events, but provides information on the origin of the particle population escaping into the interplanetary medium.

McGuire and von Rosenvinge [11.195] analyzed the energy spectra of protons and α-particles over a broad energy range (1–400 MeV). They found that the spectra could be fitted best by exponentials in rigidity and by Bessel functions. Decker et al. [11.56] showed that spectral forms close to exponentials in rigidity could result from the passage of a flare-induced shock wave through the corona. Bessel-function spectra are expected from stochastic acceleration. Proton spectra can also be derived from the observation of γ-rays and neutrons

[11.222, 221]. They lead to Bessel functions similar to those obtained from the protons observed in interplanetary space. Like the electrons, the proton spectra in impulsive events are similar for particles escaping and particles interacting with the solar atmosphere, while in gradual events the spectra of these two proton species are different (see [11.258] and references therein).

3. Gamma-Ray Flares. Solar γ-rays and neutrons are produced by the interaction of accelerated protons and ions with deeper layers of the solar atmosphere. Production mechanisms are summarized by Kocharov [11.137] and Murphy and Ramaty [11.222]. The study of these high-energy neutrals has become a comprehensive topic in itself. For many of the details relevant to the acceleration process the reader is referred to Chupp [11.35, 36], Bai and Dennis [11.10], and Ramaty and Murphy [11.261]. We already mentioned the short time delay down to the order of one second between the first indication of electron acceleration and the appearance of γ-rays indicative of proton acceleration. In the simplest case one population of accelerated protons would partly escape and partly interact with the solar atmosphere. It is surprising that the correlation between γ-ray line emission and interplanetary protons is rather broad (see e.g. [11.246, 39], and references therein). Observations on *Helios* [11.131] confirmed the poor correlation but also showed a correlation between interplanetary electrons and γ-continuum emission, produced by electrons interacting with the solar atmosphere. It is interesting that the interplanetary electron to proton ratio fluctuates considerably more than the ratio of the γ-ray continuum and the γ-ray line emission. The differences in electromagnetic and interplanetary particle observations may have several reasons:

- γ-rays are indicative of particles moving toward the solar atmosphere, whereas particles observed in space have escaped from the sun. These two populations may be very different, depending on details of acceleration, storage, and subsequent motion of charged particles. Zaitsev and Stepanov [11.329] discuss a particular model with different populations of interacting and escaping protons.
- The distinction between gradual and impulse events discussed above (proton-rich vs. proton-poor) suggests that in gradual events an additional proton component is accelerated by a traveling shock with little probability of moving toward the solar atmosphere owing to the large distance from the surface and the magnetic topology. So in gradual events the number of escaping protons compared to the number of interacting protons is expected to be larger than in impulsive events.

Recently reported observations (for a summary see [11.258]) seem to support this latter point.

4. Particle Injection. If the interplanetary scattering is weak, then the injection of particles close to the sun can be determined from the time intensity profile measured in interplanetary space by a simple temporal shift according to the travel time along the Archimedean spiral. Examples of weak or negligible scattering are the events on 28 March 1976 [11.17, 230], 7 and 8 June 1980 [11.132]. and a series of four homologous flares on 28 May 1980 [11.132]. The injection profiles show two interesting features. In all these events the electrons leave the sun immediately after their acceleration, as determined from the microwave and hard X-ray radiation. This is remarkable, since in the 28 March 1976 event *Helios* was connected via magnetic field lines with regions on the sun which are about 60° apart from the flare location (see also Sect. 11.4). In the other events the coronal distance is much smaller. Unlike the electrons the protons show different behavior: proton release simultaneous with the electron injection is observed as well as a delay of the proton injection up to 40 minutes.

Simultaneous injection of protons and electrons is observed in one of the events on 7 June 1980, and in the four events on 28 May 1980. The latter four events all are impulsive and show no indication of shock acceleration. Nevertheless, acceleration of nuclei to energies of more than 50 MeV/nucleon is observed. Acceleration up to these energies is suggested in a model by de Jager [11.51]. He relates the impulsive phase of a flare to a sudden energy release due to reconnection followed immediately by an intermediate phase, accelerating particles up to tens of MeV by ion-acoustic turbulences in the current sheet. This stochastic acceleration occurs on time scales of less than a second. The turbulence will also lead to differential heating and to the corresponding enrichment in ^{3}He as observed during the 28 May events. Acceleration and injection of relativistic protons and electrons simultaneous with the hard X-ray and microwave burst was also demonstrated by Debrunner et al. [11.53].

In other cases protons appear delayed. As an example we will discuss briefly the injection on 8 June 1980, as shown in Fig. 11.31. The intensity time profiles of electrons, protons, hard X-ray, 8800 MHz microwave, and 1980 kHz radio emission are plotted versus solar release time (SRT means that the time profiles are shifted by the mean travel time of the particles along the nominal Archimedean field line). The protons were injected about twelve minutes after the electrons, but simultaneously with a second increase in the electromagnetic emission. The time and frequency profile of the radio emission shows evidence of a shock-associated event in the sense introduced by Cane et al. [11.32]. At the same instance the electron intensities show a well-defined increase associated with a hardening of the electron spectrum. These observations can be interpreted in terms of the original model of two phases of acceleration [11.326]. Contrary to the 28 March 1976 event as shown in Fig. 11.29 the delay of the proton injection cannot be understood in terms of coronal propagation because of the small coronal distance of about 10° and the close association of the proton injection with the second electron injection and the second increase in electromagnetic emission. Despite the different injection history, compared to the 28 May events, this event can also be reconciled with the model of de Jager [11.51]. The primary energy release in the impulsive phase leads to acceleration and injection of electrons with energies up to 1 MeV observed as the first injection, while the second injection is a result of shock acceleration in the gradual phase. The

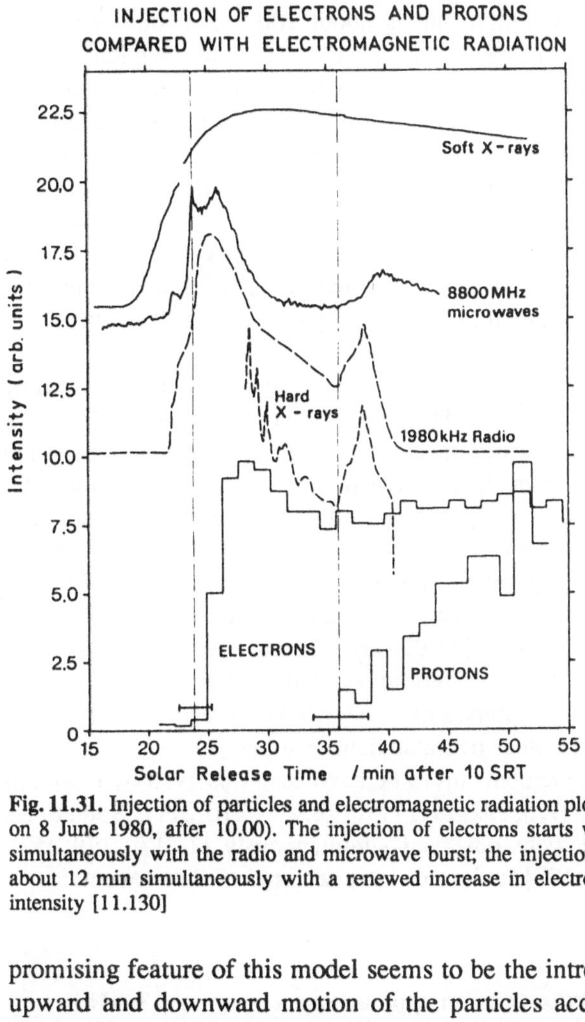

INJECTION OF ELECTRONS AND PROTONS
COMPARED WITH ELECTROMAGNETIC RADIATION

Fig. 11.31. Injection of particles and electromagnetic radiation plotted versus solar release time (SRT, on 8 June 1980, after 10.00). The injection of electrons starts within the time resolution of 1 min simultaneously with the radio and microwave burst; the injection of protons occurs with a delay of about 12 min simultaneously with a renewed increase in electromagnetic radiation and in electron intensity [11.130]

promising feature of this model seems to be the introduction of a highly variable upward and downward motion of the particles accelerated in the intermediate phase. This would make possible an explanation of γ-events. γ-events tend to be impulsive ([11.39] and references therein), just like ^{3}He-rich events ([11.182] and references therein). Impulsive γ-flares could then be explained by a suitable mixture of intermediate-phase particles moving upward and downward. Nevertheless, gradual γ-events, which do exist, could then be explained only if both an intermediate and a gradual phase are working after the primary energy release.

It is not possible to base far-reaching conclusions on these few examples. Nevertheless, it seems interesting that the analysis of solar injection profiles can be related to existing flare models.

11.5.3 ^{3}He-Rich Solar Particle Events

Solar particle events rich in ^{3}He present a striking example of an enrichment mechanism operating in the corona. These small particle events often have

303

^{3}He/^{4}He ratios of the order of unity, that is about 1000 to 10 000 times the abundance observed in the photosphere or in the solar wind. Originally they were considered rare since they were observed only by instruments with relatively high energy thresholds, above a few MeV/nucleon (see e.g. the review by Kocharov and Kocharov [11.138]). Recent studies with lower-threshold instrumentation have shown that they may occur more often than once a month and are associated with impulsive kilovolt electron events and with type III radio bursts [11.267, 265]. Since on the average more than one keV-electron event is observed per day, it is reasonable to anticipate that advanced instruments with still lower energy thresholds and increased sensitivity may detect ^{3}He-rich events with a similar frequency.

Since the first observation of a ^{3}He overabundance in a structural element of *Discoverer 17* with a ratio of 20% ^{3}He/^{4}He by Schaeffer and Zähringer [11.278] and the first direct measurement during a solar flare by Hsieh and Simpson [11.110], every effort has been made to solve the "mystery" of the unusually high ^{3}He fluxes [11.60, 91, 4, 293, 107]. A huge amount of information has been collected in the meantime. The problem, however, is still not completely solved.

At a first glance it is very surprising to observe ^{3}He overabundances of two or three orders of magnitude above the known abundances in the solar atmosphere or the solar wind of 5×10^{-4} (e.g. [11.92, 45]). Strongly selective processes are required either to collect ^{3}He in certain areas of the solar atmosphere from which the unusual composition is afterwards accelerated or to accelerate by some selective mechanisms ^{3}He and other overabundant isotopes [11.117].

Large overabundances of elements up to Fe have been observed by more advanced instruments [11.108, 97, 5, 332, 197, 179, 178, 266, 210, 211]. Sometimes strong depletions of carbon were observed simultaneously [11.179, 178, 257]. Kocharov and Kocharov [11.138] reviewed many features of these events.

1. Review of Observations

Helios Observations. The *Helios* mission approaching the sun to 0.3 AU is very well suited to observing ^{3}He events owing to the higher fluxes observed close to the sun. In the energy range above about 5 MeV/N most ^{3}He-rich events cannot be identified from earth orbit because ^{3}He events are generally low in intensity. Compared to those in earth orbit the observable total number of particles is about an order of magnitude higher and considerably compressed in time at 0.3 AU. The first ^{3}He event on *Helios* was observed on 19 March 1975 at a distance of 0.35 AU from the sun (Fig. 11.32). Three different energetic particle events can be distinguished during the period of 19 and 20 March 1975. A detailed discussion is given by Kunow et al. [11.147]. It is remarkable that the three injections have very different relative abundances of electrons, protons, and helium nuclei. Emission of relativistic electrons is restricted to the first event. The peculiar feature of event no. 2 is the absence of a new proton emission, whereas event no. 3 is definitely accompanied by a new proton injection which can also be confirmed by the anisotropy. All three events have remarkably short

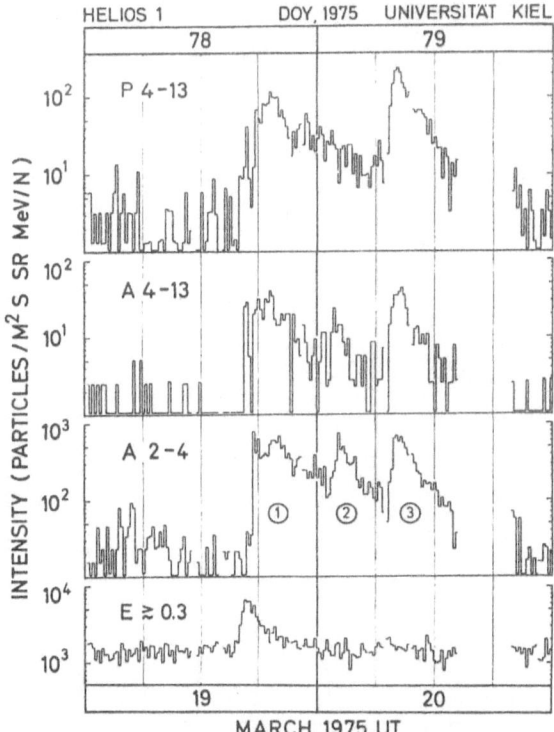

Fig. 11.32. Intensity variations in selected channels during the ^{3}He-rich events on 19 and 20 March 1975 as observed with the University of Kiel Cosmic Ray Experiment on board *Helios 1*. The time intensity profiles of three ^{3}He-rich events can most clearly be observed in the A2–4 channel measuring helium nuclei between 2 and 4 MeV/N. Electrons above 0.3 MeV are observed only during event no. 1 [11.144]

decay constants of about two hours. No optical flare has been identified on the sun as associated with either of these ^{3}He-rich events.

The relative contribution of ^{3}He in the energy range 4–13 MeV/N is shown in Fig. 11.33. In the second panel the ^{4}He-part is marked light and the ^{3}He part dark. This figure also shows another small ^{3}He-rich event on 22 March. No distinction between ^{3}He and ^{4}He can be made in channel A 2 (energy range 2–4 MeV/N).

The mass histogram in the He range for the three events on 19 and 20 March is shown in Fig. 11.34. The ratios ^{3}He/^{4}He for these events are:

Event-No.	1	2	3
^{3}He/^{4}He-ratio	2.4	2.8	0.55

The ratio of ^{3}He to ^{4}He is very high in the first two peaks (2.4 and 2.8) and is reduced to 0.55 in the third peak. Observations of several other solar events confirm the tendency, that for several accelerations which follow each other closely in time within the same active region the ^{3}He content decreases. Theoretical models discussed in more detail below suggest preferential heating of ^{3}He due

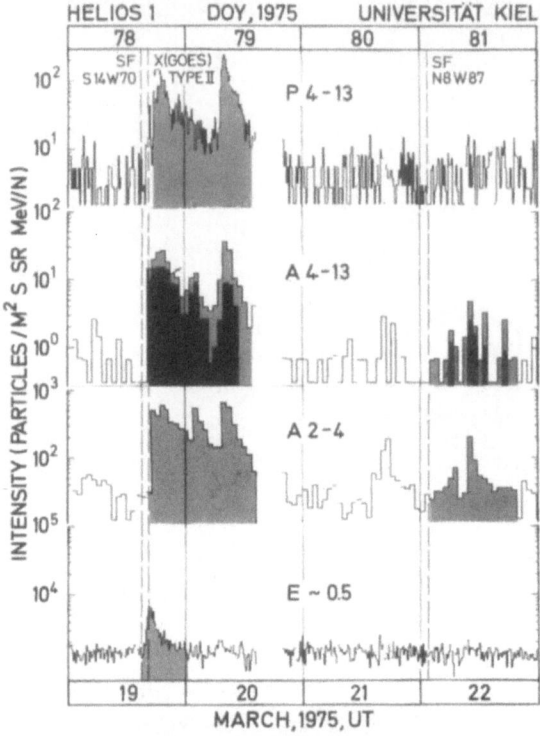

Fig. 11.33. The sequence of ^{3}He-rich particle events from Fig. 11.32 on a compressed time scale. The two species of helium are marked in the second panel showing helium nuclei in the energy range 4–13 MeV/N. Here the ^{3}He-content is marked dark, the ^{4}He-part is marked light. No distinction between ^{3}He and ^{4}He can be made in channel A2–4. An additional small ^{3}He-rich event was observed on 22 March (from [11.145])

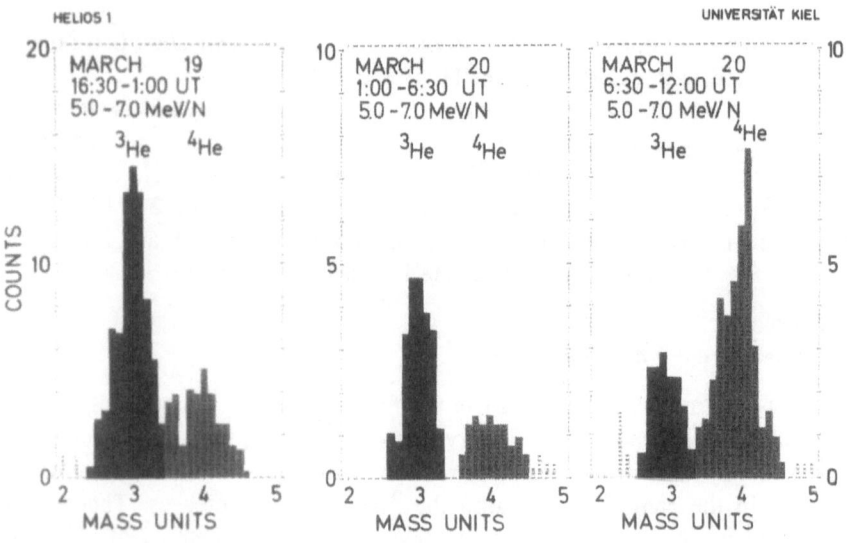

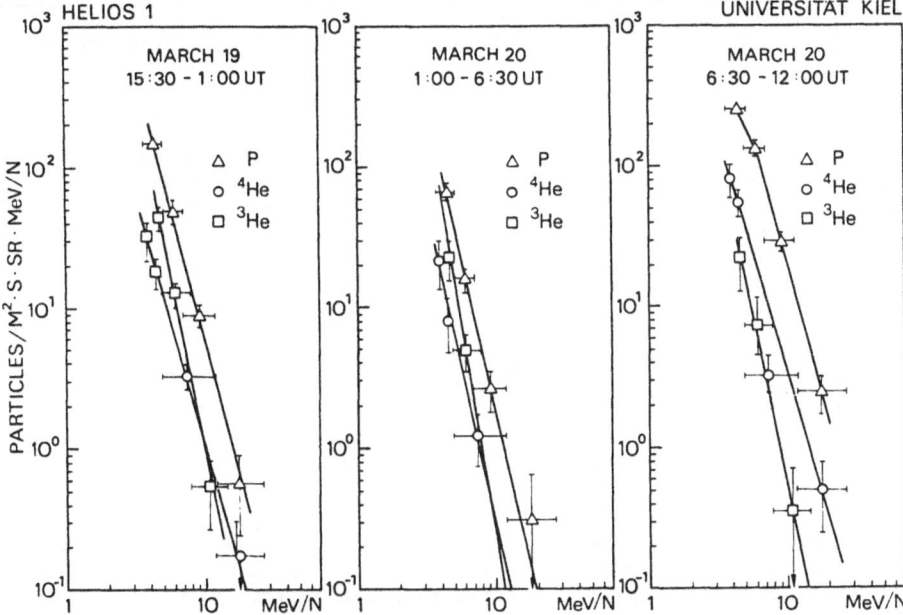

Fig. 11.35. Energy spectra for hydrogen, ³He, and ⁴He as measured by the University of Kiel Cosmic Ray Experiment on board *Helios 1* for the three ³He-rich events on 19 and 20 March 1975. Owing to the small total number of analyzed particles it was only possible to construct one average spectrum over the whole time period of enhanced intensity for each of the three events

to suitable resonance conditions in certain areas of the chromosphere. This heating could possibly last for several days prior to the flare. If a flare occurs in this part of the chromosphere, the conditions for accelerating ³He nuclei are very favorable because of their high initial velocity. If several events occur in the same region, a ³He rarefaction in the preheated matter might occur, leading to smaller ³He abundances in later events.

The energy spectra during the three events over a very restricted energy range are shown in Fig. 11.35. They follow a power-law dependence in energy for protons, ³He, and ⁴He. The three spectra are very similar to each other, with a tendency of the ³He-spectra to be slightly steeper. The similarity of the spectra would be consistent with the idea that preheated ³He is accelerated by the same mechanisms as ⁴He and hydrogen.

A summary of ³He-observations during the first years of the *Helios* mission was presented by Hempe et al. [11.107]. Results combined with earlier measurements are shown in Fig. 11.36. Here the ratio ³He/⁴He is plotted as a function of event size expressed by the maximum proton intensity at 10 MeV. It is obvious that the occurrence of ³He-rich events is restricted to small events. So far we have discussed the ratio of the two helium components. The question arises

Fig. 11.34. Mass histogram for the three peaks of 19 and 20 March 1975. The ³He contribution is very high in the first two peaks (2.4 and 2.8) and is reduced to 0.55, related to ⁴He in the third peak

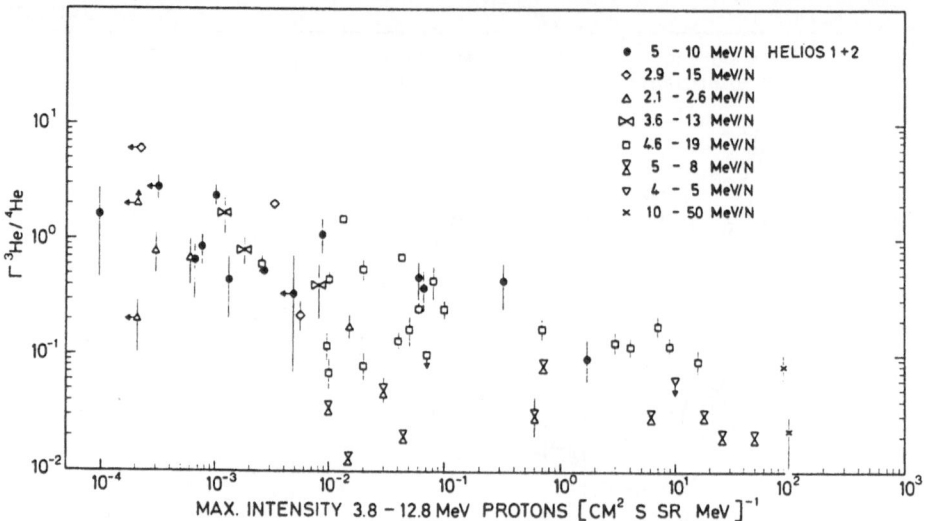

Fig. 11.36. The ratio ^{3}He/^{4}He as a function of the maximum proton intensity at 10 MeV as reported by several authors. Measurements from the University of Kiel Cosmic Ray Experiment on board *Helios 1* and 2 are marked by *black dots*. The other values are taken from Ramaty et al. [11.257]

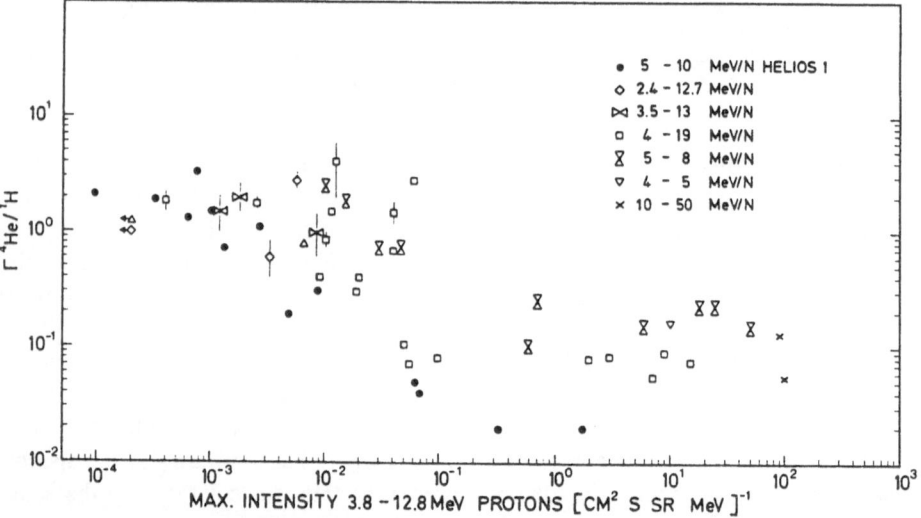

Fig. 11.37. The ratio of ^{4}He/^{1}H as a function of the maximum proton intensity at 10 MeV as reported by several authors. Measurements of the University of Kiel Cosmic Ray Experiment on board *Helios 1* and 2 are marked by *black dots*. The other values are taken from Ramaty et al. [11.257]

whether the increase in ratio is due to a genuine increase in ^{3}He or to a depletion in ^{4}He. The answer can be found in Fig. 11.37. Here the ratio ^{4}He/^{1}H is plotted as a function of event size. The ratio shows a clear tendency to be larger for small event sizes. Closer inspection shows that the ^{3}He-rich events are typically connected with α/p ratios of the order of 10%. In contrast, the α/p ratios for large solar events without ^{3}He enrichment are close to 1%.

Electron and X-ray Observations Correlated with ^{3}*He-Rich Events.* Reames et al. [11.267], Reames and Lin [11.264], and Reames and Stone [11.265] found that small nonrelativistic electron events that generate type III radio bursts [11.162] are frequently accompanied by ^{3}He-rich events. After it was shown that ^{3}He-rich events have a clear impulsive nature, Reames et al. [11.267] used the association between ^{3}He and electrons and type III metric radiation to determine their source. The associated X-ray profiles are extremely impulsive, which is also reflected in the electron and ^{3}He profiles.

Kahler et al. [11.127] found nearly no correlation of ^{3}He-rich events with type II emission and coronal mass ejections except for a few events with high proton abundances, which might belong to the gradual (long duration) solar particle events. It should be mentioned here that large gradual solar particle events show ^{3}He/^{4}He $< 2.6 \times 10^{-3}$ [11.206], an upper limit which is consistent with the solar wind value of $(4.9 \pm 0.5) \times 10^{-4}$ [11.45].

Starting from 153 electron events in a 9-month period, Reames and Lin [11.264] found that more than half of them had detectable levels of ^{3}He, and for the larger events with measurable intensities of 20 keV electrons (86 events), $\frac{2}{3}$ of them were accompanied by ^{3}He. This might lead to the suggestion that the availability of more sensitive ^{3}He detectors would reveal a one-to-one correlation of nonrelativistic electron events and ^{3}He-rich events. The similarity of the acceleration process is further supported by the spectral indices of electrons and ^{3}He for the event of 17 May 1979 observed by instruments aboard *ISEE 3* as shown in Fig. 11.38 [11.267].

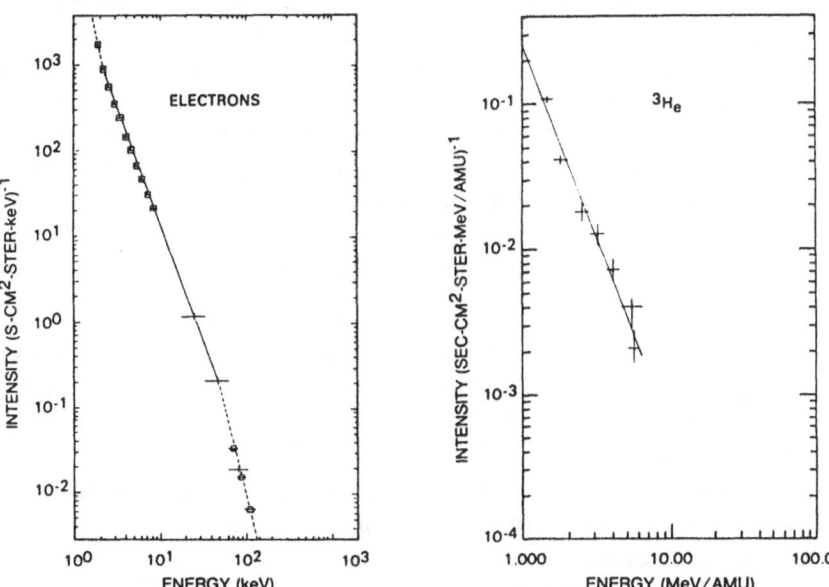

Fig. 11.38. Event-averaged energy spectra for electrons and ^{3}He in the 17 May 1979 event. *Lines* through the data are least-squares fit lines to power law spectra with resulting spectral indices of 2.7 ± 0.1 for electrons and of 2.7 ± 0.3 for ^{3}He. The electron fit was confined to the 2.5–60 keV region (from [11.267])

Kahler et al. [11.127] investigated the coronal source regions for ^{3}He/electron events as identified by Reames et al. [11.267]. They found a good correlation with kilometric type III radio bursts which are excited by solar electrons streaming outward through interplanetary space. The association of the ^{3}He-rich particle events with kilovolt electrons and type III bursts has made it possible often to identify the source regions, which are in active regions in the western hemisphere [11.332, 267, 265]. Like the small impulsive electron events, in general [11.31] these events seem to originate relatively high in the corona. This hypothesis is also confirmed by the shape of the electron spectrum, which typically extends down to ~ 2 keV without any obvious bending over (see Fig. 11.38). The correlation with phenomena deeper in the solar atmosphere, e.g. H_α brightenings, is poor. Most of the keV electron events are not associated with a reported H_α flare. It is apparent that the acceleration mechanism in the ^{3}He-rich flares is different from that operating in large solar particle events (e.g. [11.128, 127, 163]), and in fact appears to be the same mechanism which impulsively accelerates kilovolt electrons.

Relation to Heavy Ion Observations. The ^{3}He enrichments are most likely due to a resonant plasma heating mechanism [11.77, 311] (see also below) which preferentially heats the ^{3}He before some additional mechanism energizes the particles to the MeV energies observed in interplanetary space. It may be that this plasma resonance mechanism occurs in coronal sites where ^{4}He and heavy-ion abundances are enhanced [11.77]. This is consistent with the fact that these events are also associated with substantial heavy-nuclei enrichments of 10–20 times normal SEP composition, as shown in Fig. 11.39 [11.183]. The heavy-ion observations in correlation with ^{3}He-rich events have been studied by several investigators (see [11.177] for references). It is interesting that the heavy-nuclei enrichment pattern does not correlate well with the first ionization potential (FIP) for these events: for example, Ne, a high-FIP element, is enhanced about as much as Mg, a low-FIP element. Ionization-state measurements show that the heavy ions in these events come from hotter coronal regions than ions accelerated in large flares [11.135]. Mason et al. [11.183] studied 66 ^{3}He-rich events in great detail and summarized the heavy-ion observations as follows:

- There exists a characteristic abundance pattern (within a factor of 2) wherein heavy ions C, O, Ne, Mg, Si, Fe are enriched compared to large solar flare abundances; e.g. Fe/O is enriched by a factor of ~ 10 and Fe/^{4}He is enriched by a factor of ~ 20 (see Fig. 11.39).
- The heavy-ion enrichment
  - increases with atomic mass A and/or atomic charge Z (Fig. 11.39), but
  - does not correlate well with the first ionization potential,
  - is energy independent,
  - is independent of the spectral exponent, and
  - is independent of the ^{3}He/^{4}He ratio.

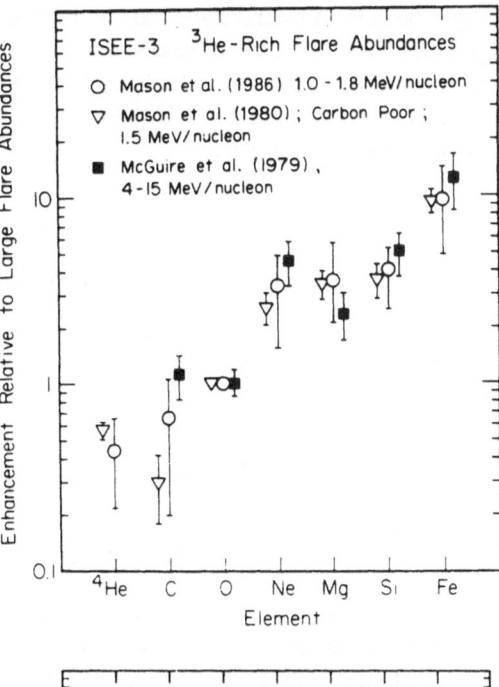

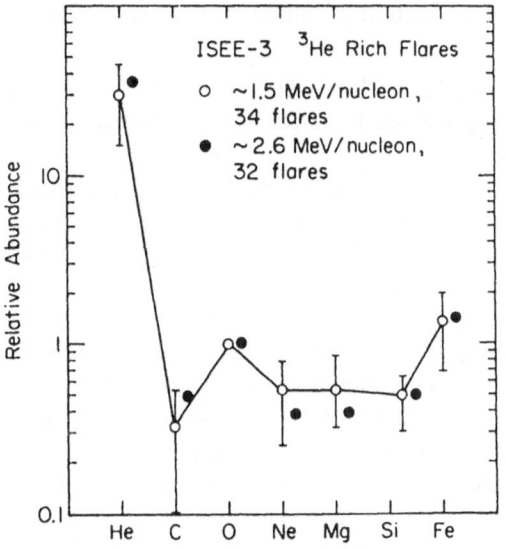

Fig. 11.39. Enhancement of element abundances in ^{3}He-rich events compared to large solar particle event abundances [11.183]

Fig. 11.40. Averaged element abundances relative to oxygen in ^{3}He-rich events for two different energies. (For helium the abundance is given for ^{4}He only.) *Error bars* are given for *open circles* only and represent the range of values observed for two-thirds of the events considered in the study of Mason et al. [11.183]

- Heavy-ion abundances correlate well with each other. They show a smaller range of variations than the ^{3}He/^{4}He ratio.
- ^{3}He is much more overabundant than heavy ions are.
- Deviations from the average ^{3}He-rich-event heavy-ion pattern are rare (Fig. 11.40). Two observed cases show apparent mixing of normal event and ^{3}He-rich-event abundance patterns.

While the heavy nuclei enrichments might be caused by the same plasma heating mechanism which affects the ^{3}He, it could also be that these heavy ion enrichments are due to thermal diffusion which could be important at high temperature sites in the corona. It should be noted, however, that our knowledge to date is based only on the abundances of a few major heavy ion species.

2. Theoretical Models

Several possibilities have been considered to explain the unusual abundance observed in ^{3}He-rich events:

- ^{3}He ascending from the solar interior,
- nuclear reactions in the solar atmosphere,
- preferential heating before or during the acceleration process.

Models based upon these possibilities will be discussed in the following sections.

Ascending ^{3}He. Ibragimov and Kocharov [11.118] suggested a local ^{3}He enrichment in the solar atmosphere by ^{3}He ascending from the interior of the sun. An increased ^{3}He abundance in active regions cannot be excluded; however, the amount of ^{3}He produced is too small to explain the observed enrichment of ^{3}He in certain energetic particle events. A contribution cannot be totally ruled out.

Nuclear Reactions and Thermonuclear Effects. It is well known that ^{3}He can be produced in nuclear reactions. This process has therefore been considered as the source of the observed ^{3}He abundance.

During solar flares, protons and α-particles might be accelerated in a relatively dense environment. Depending on the amount of matter traversed, p and α nuclear reactions occur with environmental material which produce the rare isotopes ^{3}He, ^{2}H, and ^{3}H. In normal flares a maximum of 2 g/cm^2 of traversed matter is estimated leading to a small abundance of ^{2}H, ^{3}H, and ^{3}He. To account for ^{3}He-rich flares, at least 10 g/cm^2 would be necessary; however, in this case one would expect large amounts of ^{2}H and ^{3}H, which are not observed. Therefore, special mechanisms have to be included to absorb the hydrogen isotopes after their generation. All further attempts to explain ^{3}He events with nuclear reactions focused on the problem of low abundances of ^{2}H and ^{3}H [11.259, 260, 43]. The models demanded very special conditions and yet were still unable to explain the observed ratio of He and H isotopes together with the absence of major γ-emissions during many ^{3}He-rich flares.

Preferential Heating. Spallation and thermonuclear reactions have been ruled out as sources for ^{3}He-rich events. Different effects are necessary which act highly selectively on different isotopes depending on A and Z. In two areas such selective processes can be expected:

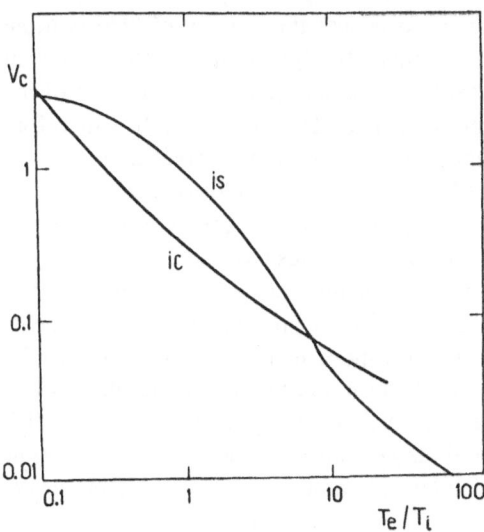

Fig. 11.41. Dependence of the minimum critical velocity V_c in a hydrogen plasma on the electron to ion temperature ratio T_e/T_i for ion acoustic (is) and ion cyclotron (ic) instabilities. Units of V_c are $\sqrt{2T_e/m_e}$ [11.134])

- If a threshold velocity exists below which no acceleration takes place, but above which acceleration is very effective, particle species with different velocity distributions could be separated effectively.
- The acceleration rate above a threshold could be different for different species (e.g. the Z^4/A^2 dependence) and could work selectively if the acceleration time is too short to accelerate all particles to maximum velocity.

Preferential Heating by Electrostatic Ion Cyclotron Waves. One possible mechanism for preferential heating of certain isotopes has been proposed by Fisk [11.77]. The source of the particles accelerated in ^{3}He-rich flares is believed to be in the lower corona. The thermal energy in the acceleration region is proposed to be not too large, so in this region a low-β plasma is assumed with a thermal to magnetic pressure ratio $\beta \equiv p_{th}/p_m < 10^{-3}$. This assumption has the consequence that the model requires only weak currents and operates best in small flares or perhaps even under nonflare conditions. Under normal coronal circumstances electrostatic ion cyclotron waves are likely to be excited. They are generated if electrons move parallel to the magnetic field relative to the ions at speeds which are small compared to the electron thermal speed but higher than the ion thermal speed.

In the normal proton–electron plasma, electrostatic ion cyclotron waves have frequencies just above the proton cyclotron frequency (Fig. 11.41). If heavier elements are also present, e.g. ^{4}He^{2+}, the range between the proton cyclotron frequency and the helium cyclotron frequency can also be excited and will be excited more rapidly. A necessary condition is that the density of ^{4}He is more than 20% of the proton density, which is unusually high compared to the normal 7%. Processes exist which tend to concentrate heavier elements at the base of the corona, as observed e.g. in plasma ejecta which are driving shocks. It is therefore possible to find regions which fulfill the high-helium-content condition.

The instability will be partly damped by ^{4}He^{2+} and p cyclotron resonance and thus will heat these elements. The heating process by the waves is particularly effective for those ions which have cyclotron frequencies close to the frequency of the excited waves, e.g. between the proton and helium electrostatic ion cyclotron waves. Therefore, if a sufficient amount of ^{4}He is present, this process heats very effectively ^{3}He^{2+} and in addition neutron-poor isotopes in the corona with $Z/A > 0.5$, while heating ^{4}He and protons to a smaller but not negligible amount. The ions must be totally stripped since any partially stripped ion has $Z/A < 0.5$.

Mechanisms which enhance ^{4}He will also enhance heavier elements, in some cases more dramatically than ^{4}He, which leads to a plasma which is rich in heavy elements and ^{4}He.

It should be noted that heavier ions can only be heated by higher harmonics of their cyclotron frequency which lie between the proton and helium electrostatic ion cyclotron frequency. Only certain isotopes and charge states have higher harmonics in the respective frequency range: e.g. isotopes of C, N, and O: ^{12}C^{4+}, ^{16}O^{5+}, and ^{56}Fe^{17+}, ^{56}Fe^{16+}, and ^{56}Fe^{11+}.

Preferential Heating of ^{3}He *Owing to Induced Ion-Acoustic Turbulence.*
Kocharov and Kocharov [11.138] describe a model where ^{3}He is preferentially heated by induced scattering by ion-acoustic turbulence. They showed that for typical parameters for the ^{3}He preheating region the turbulence cannot exist simultaneously throughout the active region. The generation of the turbulence takes place at the front of the wave which propagates through the region.

The physical model of the acceleration is described in Fig. 11.42. During the first stage of acceleration, electrons are accelerated up to energies above 10 keV. They enter the cold chromospheric plasma and create oppositely directed currents

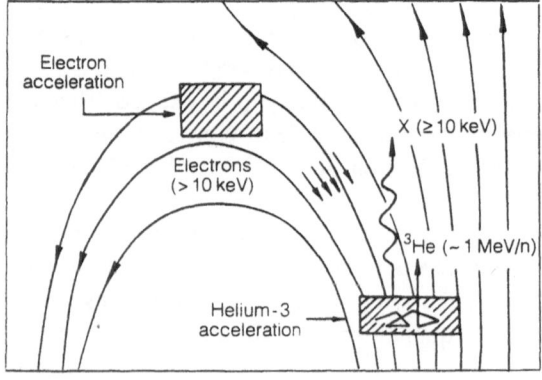

Fig. 11.42. A possible schematic representation of a ^{3}He-rich flare [11.138]

which compensate the electron current. The onset of the ion-acoustic turbulence starts if the electron density n_e is sufficiently high.

A nonisothermal plasma layer develops directly ahead of the shock. Inside the shock ^{3}He is heated preferentially by nonlinear scattering at the ion-acoustic turbulence. Neither carbon nor heavier ions have enough time to become ionized up to the equilibrium charge state. This results in a heating rate lower than for ^{3}He.

A consequence of the model is the relation of ^{3}He-events with hard X-rays and microwave bursts from the suprathermal electrons. The heated electrons produce soft X-rays.

The theory is remarkably well elaborated. For example, an expression has been derived theoretically for the total number of electrons necessary to generate the observed ^{3}He/^{4}He ratios and absolute intensities.

3. Tests of Models and Summary

Two of the models discussed seem to predict enhancement factors for ^{3}He and certain heavier isotopes which can be compared with observations. Both mechanisms provide preferential heating of ^{3}He with respect to both ^{4}He and protons; however, the degree of heating of ^{4}He is different: ion-acoustic turbulence heats ^{4}He much more strongly (approximately by a factor of 4) than electrostatic ion cyclotron resonance.

Ion-acoustic turbulence needs a velocity threshold V_s in the main acceleration process, which depends on Z/A to arrive at the observed values of ^{3}He/p. Both models would in principle be able to explain the observed ^{3}He and heavy-ion enrichments. The ion cyclotron model operates in the lower corona from where an escape is easier.

A crucial test for the models was given by Mason et al. [11.183], showing that there seems to be no strong correlation between heavy-ion enrichment and ^{3}He enrichment. Kahler et al. [11.127] go even further and question whether there is an association at all, stating that there are ^{3}He events without iron-rich events and vice versa. Klecker et al. [11.136] reported ^{56}Fe$^{+19.5\pm2}$, an ionization state higher than the prediction from the Fisk model. However, if ^{3}He enrichment and heavy-ion enrichment are not related, this would not count against the Fisk model.

The heavy-ion abundances reflect the source abundance in the lower corona for which e.g. Nakada [11.223, 224] and others had proposed a model of heavy-ion enrichment due to transport processes from the photosphere including thermal and pressure gradient effects (Fig. 11.43).

We can now briefly summarize the scenario for events rich in ^{3}He and heavy ions as follows:

1. Enrichment of heavy elements in the lower corona (still under discussion).
2. Preacceleration (preferential heating) of ^{3}He and heavy ions due to induced ion-acoustic turbulence or electrostatic ion cyclotron waves.

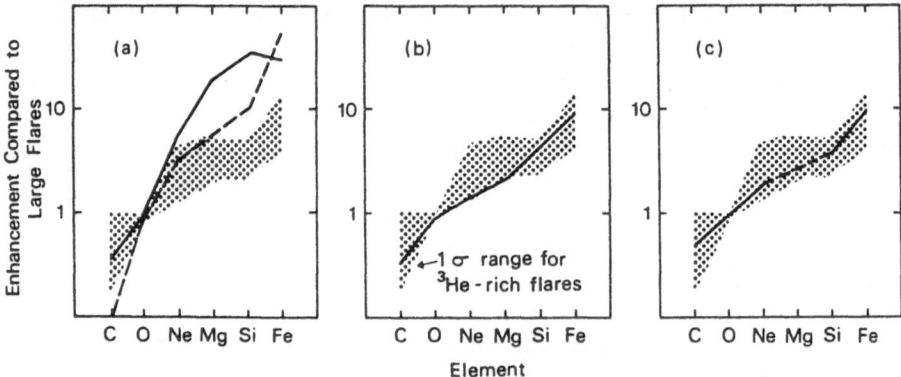

Fig. 11.43a–c. The ^{3}He-rich-event enrichment pattern (*shaded area*) is compared to (a) the model of Kocharov and Orishchenko [11.139] for a short period of heating (*solid line*) and a long period of heating (*dashed line*). (b) The plasma-heating model of Fisk [11.77] and (c) the model of Nakada [11.223] on the lower corona heavy-ion abundance enhancements [11.183]

3. First-phase (impulsive) acceleration of electrons up to 100 keV and of the preheated ions to the order of 1 MeV/N.

It is still open whether process 1 is required or not and whether one finds chemical compositions in the corona which vary with location and time. In addition, it is not clear whether processes 2 and 3 have to be considered separate processes or whether they occur naturally as part of impulsive flares.

A summary of a large number of observations has recently lead Reames [11.262] to suggest that *all* solar energetic particles from impulsive flares have an unusual composition in the sense that the abundance ratios of ^{3}He/^{4}He and Fe/O are enhanced. We introduced the distinction between impulsive and gradual flares in Sect. 11.5.1 and related it to the different heights in the solar atmosphere and different topologies of the active region. Reames [11.263] proposes that the particles are accelerated directly on open field lines near the flare site, because it seems difficult to remove low-energy electrons and ions from flare loops on a fast time scale. At the end of Sect. 11.4 we pointed out the fast access of electrons to finite coronal distances and suggested direct access to open field lines after they have been accelerated within the active region. It is still open whether it is correct to postulate that the energetic particles, in particular electrons, are first accelerated within the active region and then transferred to open field lines, or whether they are directly accelerated on open field lines. Hempe et al. [11.107] found that the ^{3}He-rich events on *Helios* were restricted to slow solar wind streams, i.e. those regions in interplanetary space which are connected to looped structures in the corona including the complex magnetic field topologies close to active regions. This is an interesting observation. It suggests a study where regions in interplanetary space inside *fast* solar wind streams are searched for the appearance of solar energetic particles. If it is correct that during impulsive events particles are accelerated directly on open field lines close to the flare, they would not be seen within fast solar wind streams which are connected to large unipolar region on the sun, remote from complex active regions.

So it turns out that the ^{3}He-rich events which were considered rare and exotic for some time after their discovery finally may provide one of the key solutions to understanding the flare phenomenon. The picture which is beginning to emerge relates the impulsive flares occurring low in the corona to unusual chemical compositions (enrichment in ^{3}He and heavy ions) and to electron acceleration up to moderate energies. They are not associated with coronal mass ejections or interplanetary shocks. In contrast, the gradual or long-duration flares, occurring higher in the corona, are believed to be related with extended shocks which accelerate the ambient (unprocessed) coronal material and lead to elemental abundances similar to those in the corona and the solar wind. Many details in this oversimplified picture are still lacking, and it must be expected that a superposition of the two "pure" cases could produce mixtures of energetic particles in space without clear signatures.

11.6 Particle Acceleration in Corotating Interplanetary Regions

Moderate intensity enhancements of MeV ions which differ markedly from prompt flare events were observed in the early years of space exploration. They were named corotating events because of their strong tendency to reappear after 27 days, the rotation period of the sun close to the solar equator [11.26, 70]. These observations were not correlated with major flares nor was an increase in radio emission observed. While prompt events typically show a fast increase and a slow decay in their intensity profile, corotating events have a symmetric profile with typical durations between 5 and 12 days. The variation of intensity with time is practically the same for particles with different velocities, whereas in prompt events of solar flare origin marked velocity dispersion is observed. Corotating events show in general moderate intensity increases. The energy spectrum is relatively steep, no intensity increase is observed above 20 MeV/N [11.191], and no electrons are observed [11.330].

The question whether these particles are accelerated at the sun or in interplanetary space has been discussed by various authors [11.164, 71, 3, 187, 189, 142, 273]. The first models try to explain the observations by continuous acceleration of particles near the sun. The particles should be accelerated and stored in active magnetic field regions which corotate with the sun and from which they leak out continuously [11.189]. This model requires a stable magnetic field configuration over several months and would result in a decrease of the intensity with increasing solar distance.

The discovery of a positive radial gradient by *Pioneer 10* and *11* between 1 and 4 AU led McDonald et al. [11.191] to propose interplanetary acceleration as the most plausible explanation for the formation of these streams and to suggest the suprathermal distribution of the solar wind as a possible source of these particles. Detailed studies using a large network of cosmic ray experiments on board spaceprobes located between 0.3 AU and 10 AU confirm the earlier explanations. A positive gradient of the intensity of about 300%/AU between 0.3 and 1 AU was observed by Kunow et al. [11.146]. Corotating events of highest intensities were observed at a radial distance between 4 and 5 AU from the sun. At larger distances the maximum intensity decreases again [11.308].

Barnes and Simpson [11.12] showed that the intensity enhancements correlate with the forward and reverse shocks from "corotating interaction regions" (CIR); see Fig. 11.1. These regions develop at a distance of about 1.5 to 2 AU from the sun when corotating fast solar wind streams collide with slower plasma. The phenomena associated with colliding solar wind streams are theoretically discussed by Hundhausen [11.113, 112]. Measurements on *Pioneer 10* and *11* are compared with theoretical models by Smith and Wolfe [11.296] and Hundhausen and Gosling [11.115].

Fast solar wind streams develop in coronal holes on the sun (see [11.292]). Although mainly located in the solar polar region, around solar minimum large and stable coronal holes frequently extend down to the solar equator [11.114]. This is the reason corotating events are preferentially observed around solar minimum. They do show, however, a considerable time dependence even at conditions close to solar minimum, as observed by *Helios 1* and 2 between October 1975 and June 1976 (see Fig. 11.2 and Fig. 11.44). Reasons for these intensity variations could be:

- a varying acceleration process,
- variation of the source population for the acceleration,
- varying propagation conditions between acceleration region and observer.

Before discussing these variations and their possible causes in Sect. 11.6.2 we shall describe properties of single events as observed in the inner solar system, their relation to solar wind phenomena, dependence on radial distance, and effects of interplanetary propagation. An acceleration model will be briefly discussed at the end of this section. For further details, the reader is referred to summaries by Lee [11.155] and Mihalov [11.209] and references therein.

11.6.1 Features of Corotating Events and Their Relation to the Solar Wind

Corotating events in the inner solar system have a close relation to solar wind features and interplanetary field structures. Figure 11.44 shows the correlation between the proton intensity in the energy range 3.7–12.6 MeV, the solar wind speed, and the polarity of the interplanetary magnetic field. The shaded areas in the upper part of the panel represent the observed corotating events. They coincide with fast solar wind streams (second panel) in those sectors where the interplanetary magnetic field is pointing away from the sun (third panel). Prompt solar events identified by their much harder spectra, appearance of velocity dis-

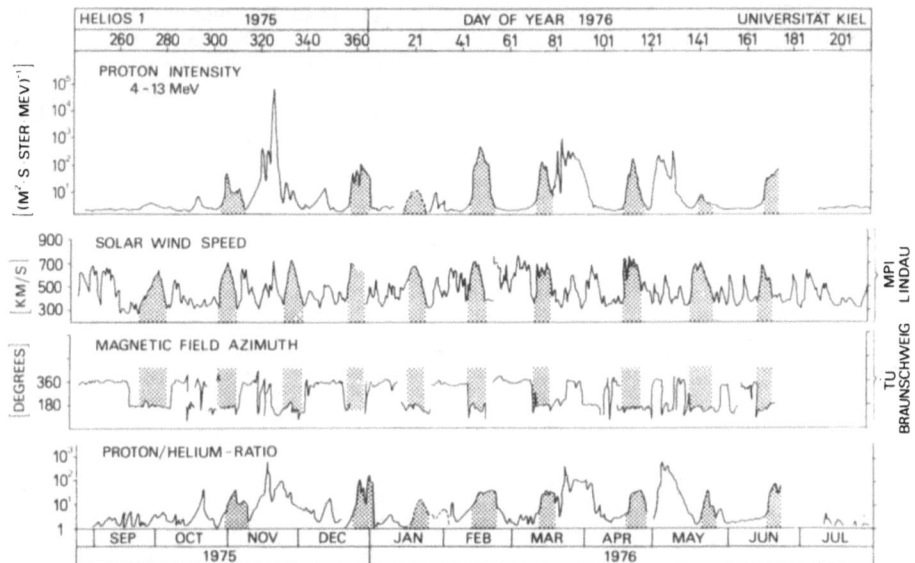

Fig. 11.44. Corotating events observed from September 1975 through August 1976 by the University of Kiel Cosmic Ray Experiment on board *Helios 1*. *Upper panel*: Intensity of 4–13 MeV protons. *Second panel*: Solar wind speed as measured by the Max-Planck-Institut Lindau solar wind experiment on *Helios 1*. *Third panel*: Magnetic field azimuth as observed by the TU Braunschweig magnetometer experiment on *Helios 1*. *Lower panel*: The proton to helium ratio for particles between 4–13 MeV/N. Identified corotating events are marked in the *upper panel*, identical time periods are shaded in the *other panels* as well, indicating that during this time period corotating events are always correlated with high-speed solar wind streams and magnetic sectors pointing away from the sun. The proton to helium ratio is always between 10 and 20

person, and the association with solar flares are left unshaded in the upper panel. The increases of the corotating events start simultaneously with increases in the solar wind speed above approximately 300 km/s; the particle intensity disappears in the background when the solar wind speed drops below 300 km/s. The situation on *Helios 2* is qualitatively very similar to Fig. 11.44. A characteristic difference is that the event in April 1976 which was clearly seen on *Helios 1* disappears on *Helios 2* (see the discussion in Sect. 11.6.2).

A typical corotating event as observed in the inner heliosphere is shown in Fig. 11.45. The upper panel of this figure shows the corotating event of 17 February 1976 as observed by the University of Kiel Cosmic Ray Experiment on board *Helios 2* [11.330]. This event happens to be the largest one in this sequence, the first of the single fast-stream-structure events. Obvious is the very slow increase lasting for four days and the even longer decay phase of five days. Note also the double peak on 17 and 19 February, a feature which is typical for most corotating events observed further out in the solar system. However, it is not yet clear whether this small effect is the remnant of a very pronounced intensity dip observed further out.

From the second panel of Fig. 11.45, which shows the magnetic field azimuth as measured by the Technical University of Braunschweig Instrument on board

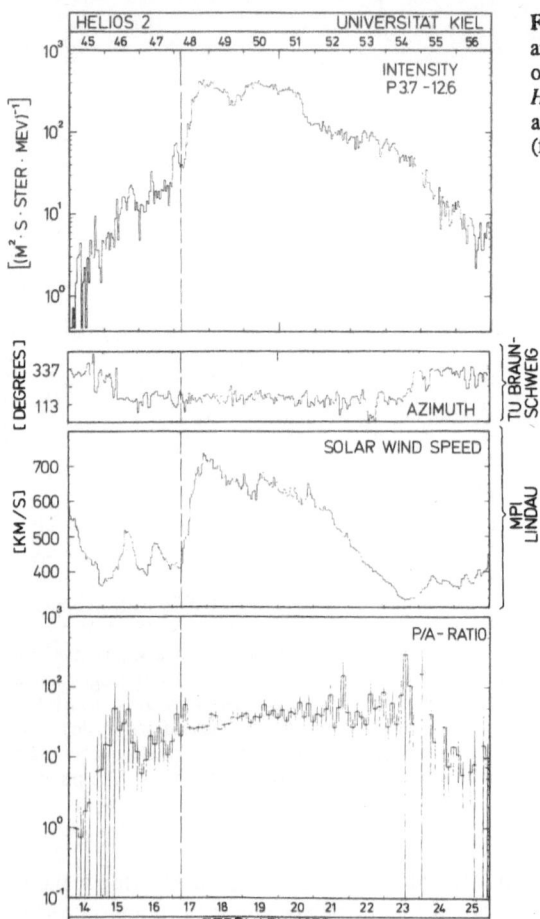

Fig. 11.45. The corotating event of February 1976 as observed by the University of Kiel Cosmic Ray Experiment on board *Helios 2*. The panels show hourly averages of the same data as in Fig. 11.44 (from [11.330])

the same spacecraft, the magnetic field direction is measured to be directed away from the sun during almost the whole event time. The third panel shows the relation with the fast-speed solar wind stream as measured by the Max-Planck-Institut Lindau Experiment. Note that the fast solar wind stream is of shorter duration than the corotating event. The stream starts from about 400 km/s on 17 February with a fast rise to more than 700 km/s within half a day, followed by a moderate decay and a second increase on 19 February. The major increases coincide well with relatively fast increases of the 3.7–12.6 MeV proton intensity which shows its maximum intensities at the time of the highest solar wind speeds. During the three days prior to 17 February, however, the proton intensity already increases slowly prior to the major increase of the solar wind speed. This portion will be called a precursor and is observed in connection with many of the corotating events. Similarly, the proton intensity decays lasts even longer than the solar wind speed decay.

1. The Proton/Alpha Ratio. Figure 11.45 shows in the lower panel the variation of the proton to alpha ratio over the entire corotating event of February 1976. Despite the large variation of intensity it is remarkable that the proton to alpha ratio remains relatively constant over the entire event, including the precursor. The helium content of about 5 to 10% is relatively high compared to solar-flare-associated events and is of the order of the solar wind helium content. Similar values are observed for all other corotating events at various energies. This again supports the hypothesis that the solar wind plasma provides the seed particles for the acceleration process of corotating particle events. This is supported by additional measurements of the chemical composition of corotating events which also agrees with the solar wind composition [11.96, 196, 103, 286].

2. Multispaceprobe Observations and Radial Dependence. During the *Helios*-mission we had for the first time the possibility of observing the same corotating event in the inner solar system simultaneously by means of three different spaceprobes which are located nearly on the same magnetic field line [11.149]. In the middle of March 1976 *Helios 1* and *Helios 2* and *IMP 7* and *8* were located within about 15° on the same interplanetary magnetic field line, but at different distances from the sun. Fig. 11.46 shows the observation of this event at three different locations in the inner heliosphere. It is remarkable that the event shows similar fine structures at the three positions, some of which are marked by numbers 1 to 7 in the different panels. The features are most clearly observed on *IMP 7* and *8*. The arrows attached to the numbers are chosen for identical Carrington longitude of the magnetic field lines connecting the respective observers to the sun. Features 1 and 2 occur during the precursor phase, which is more pronounced closer to the sun. For all other features, including the different peaks which are best resolved at 1 AU, the intensities are higher at larger distances from the sun.

The fact that features at the three positions correspond to the same Carrington longitude (Fig. 11.46), i.e. to the same corotating field line, is a very clear indication of a quasistationary, corotating structure. The intensities measured on identical magnetic field lines by *Helios 1* and *Helios 2* and *IMP 7/8* (see marks 1 through 7 in Fig. 11.46) are presented vs. radial distance from the sun in Fig. 11.47. Marks 1 and 2 belong to a precursor to the corotating event shortly prior to an abrupt increase of the intensity which is most pronounced at *IMP 7/8* between marks 2 and 3. In the precursor the intensity is independent of solar distance or decreases slightly with increasing distance from the sun. We cannot exclude a solar origin of the particles in the precursor, but we might also have propagation conditions different from the rest of the event.

The intensity increases remarkably with increasing solar distance beyond mark 2, with a radial gradient of 330%/AU. This is clear evidence for an outer source of the energetic particles. Kunow et al. [11.149] have interpreted the radial variation in terms of a diffusion–convection model with a radially constant mean free path. Application of the force-field solution leads to a mean free path of $\lambda_r = 0.06$ AU, in excellent agreement with the average behavior of

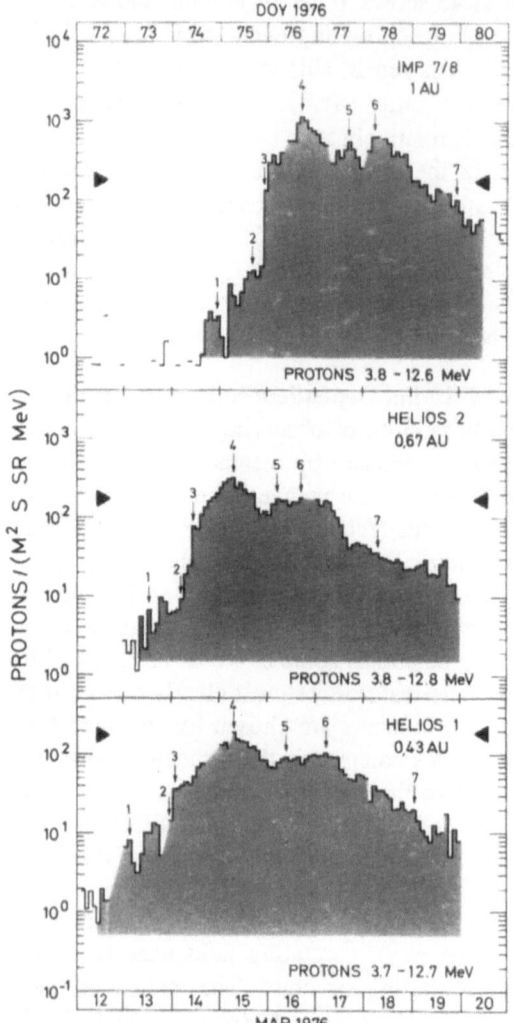

Fig. 11.46. Hourly averages of intensities during the corotating event of March 1976 observed at 0.43 AU (*Helios 1*), 0.673 AU (*Helios 2*), and 1 AU (*IMP 7/8*). The *arrows* 1 through 7 mark measurements which were taken on the same interplanetary magnetic field line. Multiple peak structures which corotate past the respective observer are clearly seen. The intensity increases with increasing distance from the sun are made more evident by *triangular marks* at the maximum intensity at *Helios 1* (from [11.149])

the cosmic ray mean free path derived from prompt solar particle events (see Sect. 11.3). Marshall and Stone [11.175] observed during corotating events at 1 AU an inward-directed anisotropy (in the frame moving with the solar wind), which confirms the hypothesis of a particle source beyond 1 AU. Their analysis leads to a local mean free path of 0.05 AU. Values of the mean free path ranging from 0.03 to 0.11 AU have been found by Christon [11.34] who used sunward-directed anisotropies and radial gradients. The positive intensity gradients were

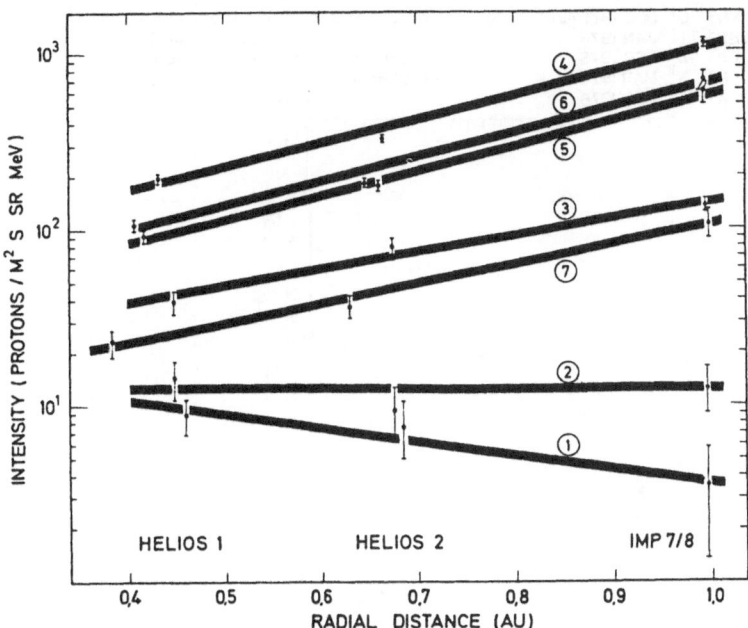

Fig. 11.47. Intensities of 4–13 MeV protons for various phases during the March 1976 corotating event as described in Fig. 11.46 plotted versus radial distance from the sun (from [11.149])

also confirmed by van Hollebeke et al. [11.308], who included more events from spacecraft on closely neighboring field lines and in addition data from *Pioneer 10* and *11* between 2.6 and 10 AU. The results are shown in Fig. 11.48. The radial gradient is calculated to be 350%/AU between 0.3 and 1 AU, 100%/AU between 1 and 3 to 5 AU, and negative between 4 and 9 AU with a variability between −40 and −100%/AU. Pesses et al. [11.247] also found a sunward flow at larger distances close to the corotating interaction regions. Considering these results, Scholer et al. [11.289] conclude that acceleration of part of the thermal solar wind plasma at the corotating shocks can explain many of the observed features of corotating events. In particular, they have correlated the observed absolute proton intensity near the shock with a plasma temperature measured on the same field line at 1 AU.

3. Spectra. The shape and the radial dependence of the energy spectrum of energized particles is of special importance to obtaining an adequate model for the acceleration process. Van Hollebeke et al. [11.309] found that an exponential spectrum in momentum of the form $dJ/dP \sim \exp(P/P_0)$ gives a good fit to the data for both protons and α-particles, as shown in Fig. 11.49. P_0 ranges typically from 9 to 16 MV/N for most of the events for both protons and helium nuclei and shows little variation with radial distance from 0.3 to 4 AU. Harder spectra, however, have been observed upstream of the reverse shock compared to measurements upstream of the forward shock.

323

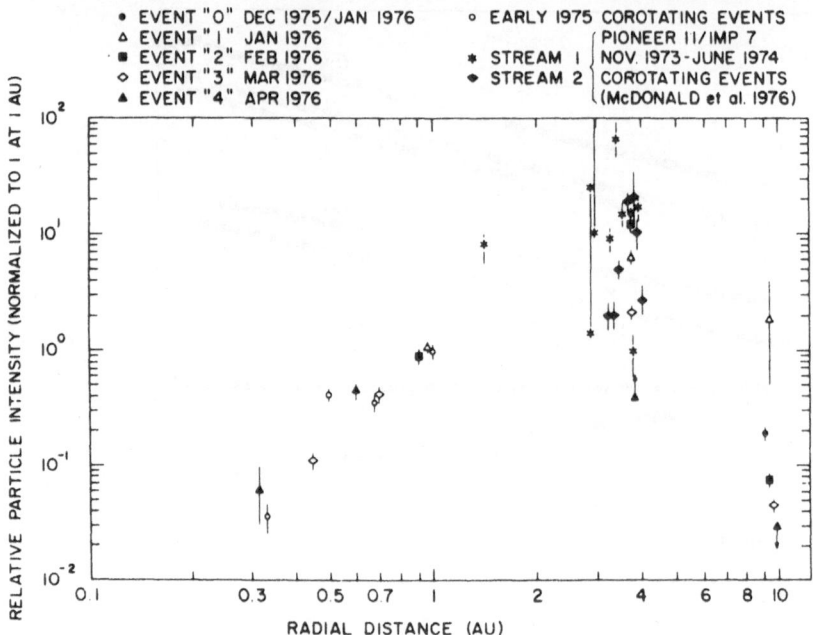

Fig. 11.48. Relative intensity of the 0.96–2.2 MeV protons versus radial distance with respect to *IMP 7* at 1 AU. Data between 0.3 and 1 AU are from *Helios 1* and 2 and data outside 1 AU from *Pioneer 10* and *11* (from [11.308])

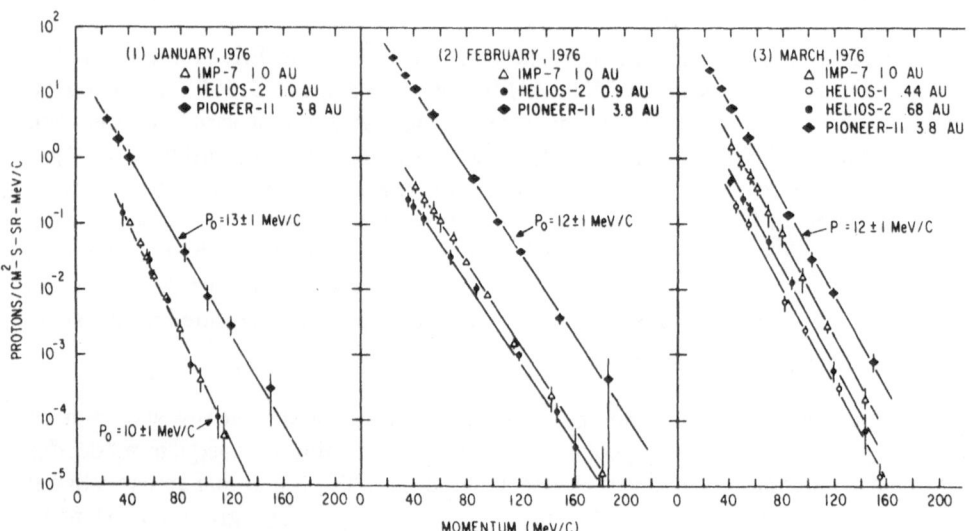

Fig. 11.49. Energy spectra during corotating energetic particle streams observed for three consecutive solar rotations in 1976. Data between 0.3 and 1 AU are from *Helios 1* and 2 and data outside 1 AU from *Pioneer 10* and *11* (from [11.309])

11.6.2 Corotating Events Throughout the Solar System and Solar Cycle Dependence

As already discussed, corotating events have their highest intensity in the 3–5 AU region. Here the intensity of MeV ions is closely correlated with the passage of the corotating interaction region, and the particle intensities peak near the passage of the forward and reverse shock and substantially decrease within the CIR [11.12, 56, 248]. A study of long-lasting cosmic ray interaction regions revealed that 50–60% of all identified shocks were accompanied by proton enhancements. At small distances (2–3 AU) the intensities at the forward and reverse shocks tend to be about equal, whereas at larger distances the reverse shock dominates over the forward shock enhancement.

Forward and reverse shocks at the edges of CIRs start to build-up generally beyond 1 AU, on average at about 2 AU. Depending on the interplanetary structure and the differences in solar wind velocity they extend out to 5–10 AU. It is assumed that the acceleration of corotating particles takes place in the forward and reverse shocks and that the accelerated particles are propagating along the magnetic field lines with some cross-field-line diffusion to the observer (see [11.176]). This can explain the observed intensity decreases in the inner solar system and beyond 5 AU.

Corotating events in the inner solar system are mostly observed within fast solar wind streams. The fast streams are connected to the reverse shock region which suggests that the particles observed in corotating events in the inner solar system are accelerated in the reverse shock. This is schematically demonstrated in Fig. 11.50. In the slow solar wind which is magnetically connected to the forward shock no corotating events are observed in the inner solar system (see

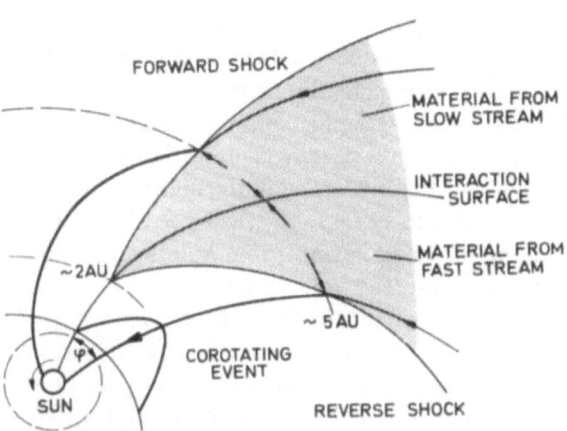

Fig. 11.50. Schematic view of the interplanetary magnetic field lines and their connection to the corotating interaction region. Slow solar wind regimes are connected to forward shock regions whereas field lines in fast solar wind streams are connected to the reverse shock of the CIR. The azimuthal intensity variation for a corotating event is sketched schematically above the 1 AU circle (from [11.330])

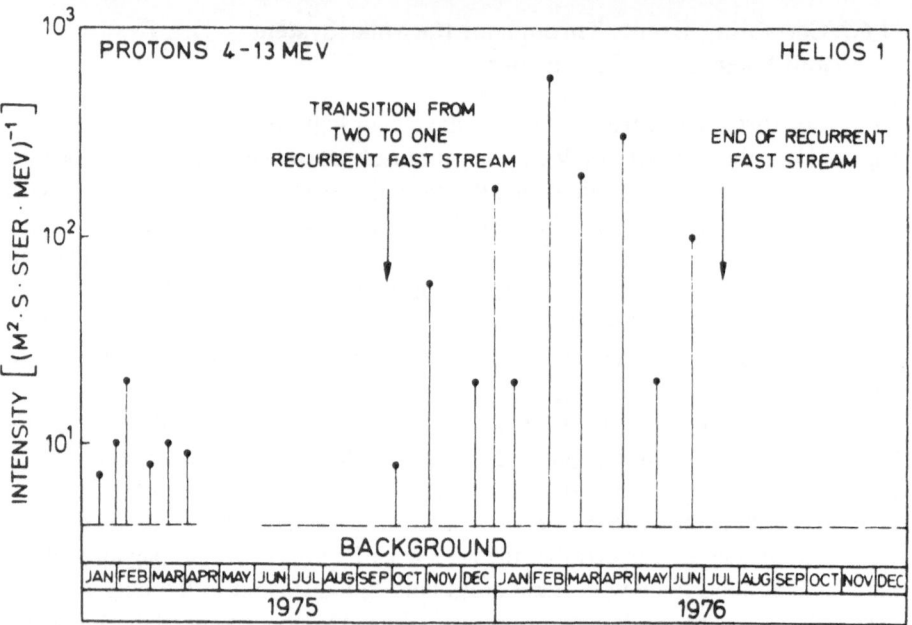

Fig. 11.51. Intensity maxima correlated with recurrent fast solar wind streams as observed by the University of Kiel Experiment on board *Helios 1* for the years 1975 and 1976 (Zöllich et al. [11.331])

also [11.34]). As the intensities around 3 AU from the forward and reverse shocks are about equal, the absence of an intensity enhancement related to the forward shock portion in the inner solar system might be due to different propagation conditions in the fast and slow speed solar wind regime. It should be recalled that a precursor outside the fast stream was observed and that in this portion of the corotating event the gradient was zero or slightly negative (see Fig. 11.47). Models of corotating events so far have not been able to explain these differences.

Let us discuss in a bit more detail the conditions for the appearance of corotating events in the inner solar system. A schematic presentation of maximum recurrent event intensities over extended time periods from January 1975 to December 1977 is shown in Fig. 11.51. For the period of the highest intensities from October 1975 to June 1976 we refer to Fig. 11.44, where we also discussed the correlation with the recurrent high speed streams. From January until April 1975 a number of small increases was observed followed by a period without such events from June until October. In the January through March 1975 period the interplanetary space was dominated by two recurrent fast solar wind streams per solar rotation. Afterwards a restructuring of the interplanetary space close to the ecliptic occurred. From October onwards only one recurrent high-speed stream was present, and until June 1976 we observed a number of events about an order of magnitude larger than previously (see Fig. 11.44). From then until the end of 1976 no recurrent fast solar wind stream and no corotating event could be observed on board *Helios*, although this period was very quiet on the sun and showed typical solar minimum conditions. The reason is the very symmetric nature of the general solar magnetic field. The current sheet is located very close to the equator and does not show large warps; no equatorial coronal hole is seen in the white light coronograph pictures.

Thus, the disappearance of the corotating events after June 1976 is obviously due to the lack of high-speed solar wind streams in the plane of the ecliptic. The difference between the size of the events in early and late 1975 and their total disappearance in mid-1975 is more striking. Early in 1975 recurrent high-speed streams did exist [11.330], but there were two streams per solar rotation. One possibility is that, owing to the interference between the two streams, with two pairs of shocks they do not extend so far out into interplanetary space and result in only moderate efficiency for acceleration. The transition from two to one recurrent fast solar wind stream was connected with a major reconfiguration of the coronal magnetic field and, as a consequence, the interplanetary magnetic field, leading to strong transient disturbances in the interplanetary medium. We conclude that this is the reason for disappearance of corotating events. From October 1975 until June 1976 one recurrent high-speed solar wind stream dominated the interplanetary magnetic field structure, and this configuration was extremely stable. This allowed the development of very intense corotating events as shown in Fig. 11.51.

We conclude that for the development of corotating events a stable high-speed solar wind stream extending out into interplanetary space without major disturbance to at least 3 AU seems to be necessary. This picture of a very stable, undisturbed near-equatorial recurrent fast stream as a necessary condition for acceleration is confirmed by the action of various flare-initiated interplanetary disturbances which occurred near the end of March 1976. The intensities of 4–13 MeV protons from March to May 1976 are shown in Fig. 11.52. The times of connection to Carrington longitudes 288 and 291 which were characteristic for the rising phases of recurrent events in preceding rotations are indicated, together with the magnetic polarities as observed by the TU Braunschweig magnetic field instruments. The missing recurrent event on *Helios 2* in April 1976, which ought to have appeared in the middle of the outward-pointing magnetic sector, is clearly seen. The position of *Helios 1* and *2* relative to various solar disturbances is discussed by Zöllich [11.330]. He argues that the most likely interpretation for the missing energetic particles lies in the source strength: a decrease of the plasma temperature in the driver gas following flare-initiated shocks reduces the tail in the velocity distribution of the ambient plasma, and thus the suprathermal seed particles for the diffusive shock acceleration are reduced [11.82].

It is interesting that with the onset of the new solar cycle in 1977 with more complex solar magnetic field configurations, which ought to lead to warped current sheets again, no corotating events are observed. This is in agreement with the perception that we do not need only fast solar wind streams, but that they have to persist for a sufficiently long time. This does not occur in the rising phase of the solar cycle. Finally, at solar maximum, coronal holes cover a much smaller percentage of the solar surface, and only few extensions reach the solar equator.

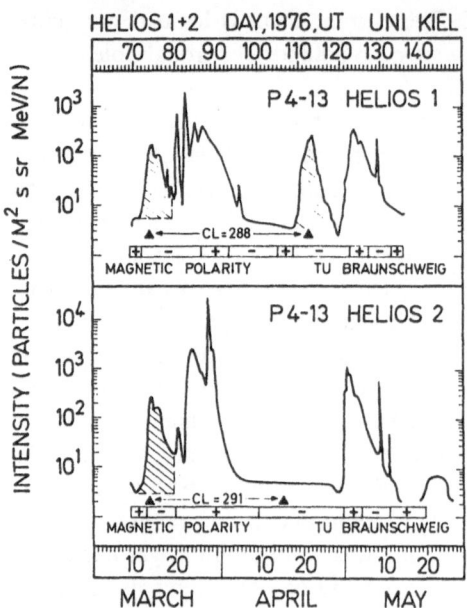

HELIOS 1+2   DAY,1976,UT   UNI KIEL

Fig. 11.52. Comparison of 4–13 MeV proton intensities on *Helios 1* and 2 during the period March – May 1976. The missing recurrent event on *Helios 2* during a period of an outward-directed field (see magnetic polarity line below the particle data) should have appeared around the *triangle* marked CL = 291

11.6.3 Models of Corotating Event Generation and Transport

The above-described observations support the ideas of several authors that the corotating particles are accelerated at the forward and reverse shocks bounding the corotating interaction region. Accelerated particles diffuse towards the sun against the solar wind flow into the inner solar system predominantly in the fast solar wind stream. The spectral shape suggests a stationary acceleration process requiring a very long shock lifetime. A theory describing the corotating events presented by Fisk and Lee [11.82] is based upon diffusive shock acceleration of ions at the forward and reverse shock. An injection energy of the ions > 5 keV is required; thus it is assumed that seed particles are particles from the suprathermal tail of "shock heated" downstream solar wind within the corotating interaction region which leak back into the upstream shock regimes. The model also describes the diffusive interplanetary transport, taking adiabatic deceleration into account. The following observational features of the spatial and energy dependence can be explained:

- exponential spectra near earth,
- spectral changes with radial distance from the sun,
- radial intensity gradients,
- different intensities and spectra generated at the forward and reverse shocks of the corotating interaction region.

Tsurutani et al. [11.304], using *Pioneer 10*, observed magnetosonic waves upstream of the trailing edge of the corotating interaction region. These waves satisfy cyclotron resonance conditions for ~100 keV protons. This is strong evidence that ions in the vicinity of corotating interaction regions can excite waves and thus modify the interplanetary diffusion coefficient. This modification of the undisturbed parallel mean free path seems to be absolutely necessary. Estimates of the acceleration time for diffusive shock acceleration based on the average mean free path lead to acceleration times for MeV protons of the order of 100 days. This time scale seems to be inconsistent with the appearance of accelerated particles in the corotating events not more than 10 days after the development of a sufficiently stable configuration (see [11.330] for discussion). The existence of upstream waves as an important means for efficient acceleration in the diffusive shock acceleration process has been clearly demonstrated for two other types of shock in the solar system: the flare-initiated interplanetary shocks and the earth's bow shock. This point is discussed e.g. by Lee [11.154, 155, 156], Sanderson [11.276], and Scholer [11.284]. In addition to diffusive shock acceleration, the process of shock drift acceleration plays an important role (see Armstrong et al. [11.6]). The drift acceleration in the presence of electromagnetic waves on both sides of the shock is discussed by Decker [11.55]. There are still quite a number of open problems in shock-acceleration physics. Of particular interest are shocks in the outer solar system where the shock-associated energetic particles in the MeV range provide the most intense source of energetic particles. The super-events discussed in Sect. 11.2 are probably a manifestation of these shock-accelerated particles in the inner solar system.

Remark. The acceleration process which seems to take place very efficiently in the interplanetary medium could also exist in other star systems where the stellar wind may interact even more strongly to form shocks which could accelerate the suprathermal stellar wind to even higher energies than those observed near our sun. Such a mechanism may possibly form some of the low-energy particle components of the galactic cosmic radiation, but at present this is speculation and both indirect observations and model calculations would be required.

Acknowledgements. Major parts of the material presented in this chapter are based on the results of the University of Kiel cosmic ray instrument on board the solar probes *Helios 1* and 2. A large group of people has contributed to the success of this mission. We would like to express our thanks in particular to Drs. H.-G. Hasler, M. Witte, and H. Hempe and to J. Fuckner for their contributions in the early stages of experiment development and data evaluation, and to U. Mende for his assistance in various steps of the data-handling procedures. We also thank J. Beeck, C. K. Ng, and W. Schlüter for permission to use previously unpublished results. We owe thanks to the referee for carefully reading the manuscript and pointing out a number of mistakes and inconsistencies. This work was supported by the Bundesministerium für Forschung und Technologie under the *Helios* program.

References

11.1   Achterberg, On the nature of small amplitude Fermi-acceleration, Astron. Astrophys., **97**, 259–264, 1981.

11.2   Akasofu, S.I., C. Olmstedt, L.A. Lockwood, Solar activity and modulation of cosmic ray intensity, J. Geophys. Res., **90**, 4439–4447, 1985.

11.3   Anderson, K.A., Electrons and protons in long-lived streams of energetic particles, Solar Phys., **6**, 111, 1969.

11.4   Anglin, J.D., The relative abundances and energy spectra of solar-flare-accelerated Deuterium, Tritium, and Helium-3, Astrophys. J., **198**, 733–753, 1975.

11.5   Anglin, J.D., W.F. Dietrich, J.A. Simpson, Super enrichments of Fe-group nuclei in solar flares and their association with large ^{3}He enrichments, in *Proc. 15th Internat. Cosmic Ray Conf.*, Vol. 5, Plovdiv, Bulgaria, 43, 1977.

11.6   Armstrong, T.P., M.E. Pesses, R.B. Decker, Shock drift acceleration, in *Collisionless Shocks in the Heliosphere, Reviews of Current Research*, ed. by B.T. Tsurutani and R.G. Stone, Geophysical Monograph 35, American Geophysical Union, Washington, 271–285, 1985.

11.7   Axford, W.I., Anisotropic diffusion of solar cosmic rays, Planet. Space Sci., **13**, 1301, 1965.

11.8   Axford, W.I., E. Leer, G. Skadron, The acceleration of cosmic rays by shock waves, in *Proc. 15th Int. Cosmic Ray Conf.*, Vol. 11, Plovdiv, Bulgaria, 132, 1977.

11.9   Bai, T., Two classes of gamma-ray/proton flares: impulsive and gradual, Astrophys. J., **308**, 912, 1986.

11.10   Bai, T., B. Dennis, Characteristics of gamma-ray line flares, Astrophys. J., **292**, 699, 1985.

11.11   Bai, T., R. Ramaty, Gamma-ray and microwave evidence for two phases of acceleration in solar flares, Solar Phys., **49**, 343, 1976.

11.12   Barnes, C.W., J.A. Simpson, Evidence for interplanetary acceleration of nucleons in corotating interaction regions, Astrophys. J., **207**, 977, 1976.

11.13   Beeck, J., G.M. Mason, D.C. Hamilton, G. Wibberenz, H. Kunov, D. Hovestadt, B. Klecker, A multispacecraft study of the injection and transport of solar energetic particles, Astrophys. J., **322**, 1052–1072, 1987.

11.14   Beeck, J., G. Wibberenz, Pitch angle distribution of solar energetic particles and the local scattering properties of the interplanetary medium, Astrophys. J., **311**, 437, 1986.

11.15   Beeck, J., G. Wibberenz, Pitch angle distributions of energetic particles during solar events, in *Proc. 20th Int. Cosmic Ray Conf.*, Vol. 3, ed. by V.A. Kozyarivsky et al., Moscow, USSR, 147–150, 1987.

11.16   Beeck, J., G. Wibberenz, Asymmetries in pitch angle scattering of solar energetic particles, in preparation, 1989.

11.17   Bieber, J.W., J.A. Earl, G. Green, H. Kunow, R. Müller-Mellin, G. Wibberenz, Interplanetary pitch angle scattering and coronal transport of solar energetic particles: New information from HELIOS. J. Geophys. Res., **85**, 2313–2323, 1980.

11.18   Bieber, J.W., P. Evenson, W.H. Matthaeus, Magnetic helicity of the IMF and the solar modulation of cosmic rays, Geophys. Res. Lett., **14**, 864–867, 1987.

11.19   Bieber, J.W., P. Evenson, M.A. Pomerantz, Focusing anisotropy of solar cosmic rays, J. Geophys. Res., **91**, 8313–8724, 1986.

11.20   Bieber, J.W., M.A. Pomerantz, A unified theory of cosmic ray diurnal variations, Geophys. Res. Lett., **10**, 920, 1983.

11.21   Bieber, J.W., M.A. Pomerantz, Magnetic fluctuations and cosmic ray diurnal variations, in *Proc. 19th Int. Cosmic Ray Conf.*, Vol. 5, La Jolla, USA, 159, 1985.

11.22   Bieber, J.W., C.W. Smith, W.M. Matthaeus, Cosmic-ray pitch angle scattering in isotropic turbulence, Astrophys. J., **334**, 470–475, 1988.

11.23   Bougeret, J.L., S. Hoang, J.L. Steinberg, 3-D coronal and heliospheric structure from radio observations, in *The Sun and the Heliosphere in three Dimensions*, ed. by R.G. Marsden, D. Reidel, Dordrecht, 213–222, 1986.

11.24   Bowe, G.A., C.J. Hatton, A Study of the modulating effect of solar flares on the cosmic ray intensity using time series analysis, Solar Phys., **89**, 351–359, 1982.

11.25   Breneman, H.H., E.C. Stone, Solar coronal and photospheric abundances form solar energetic particle measurements, Astrophys. J., (Letters), **299**, L57, 1985.

11.26   Bryant, D.A., T.L. Cline, U.D. Desai, F.B. McDonald, Continual acceleration of solar protons in the MeV range, Phys. Rev. Lett., **14**, 481, 1965.

11.27 Burlaga, L.F., Anisotropic diffusion of solar cosmic rays, J. Geophys. Res., **72**, 4449–4466, 1967.

11.28 Burlaga, L.F., F.B. McDonald, N.F. Ness, R. Schwenn, A.J. Lazarus, F. Mariani, Interplanetary flow systems associated with cosmic ray modulation in 1977–1980, J. Geophys. Res., **89**, 6579–6587, 1984.

11.29 Cane, H.V., S.W. Kahler, N.R. Sheeley, Jr., Interplanetary shocks preceeded by solar filament eruptions, J. Geophys. Res., **91**, 13321, 1986.

11.30 Cane, H.V., D.V. Reames, T.T. von Rosenvinge, The role of interplanetary shocks in longitude distribution of solar energetic particles, J. Geophys. Res., **93**, 9555-9567, 1988.

11.31 Cane, H.V., R.G. Stone, Type II solar radio bursts, interplanetary shocks and energetic particle events, Astrophys. J., **282**, 339–344, 1984.

11.32 Cane, H.V., R.G. Stone, J. Fainberg, Radio evidence for shock acceleration of electrons in the solar corona, Geophys. Res. Let., **8**, 1285–1288, 1981.

11.33 Chenette, D.L., T.F. Conlon, K.R. Pyle, J.A. Simpson, Observations of jovian electrons at 1 AU throughout the 13 month jovian synodic year, Astrophys. J., **215**, L95–L99, 1977.

11.34 Christon, S.P., On the origin of the MeV energy nucleon flux associated with CIRs, J. Geophys. Res., **86**, 8852, 1981.

11.35 Chupp, E.L., High-energy particle acceleration in solar flares – observational evidence, Solar Phys., **86**, 383–393, 1983.

11.36 Chupp, E.L., High-energy neutral radiations from the sun, Ann. Rev. Astron. Astrophys., **22**, 359, 1984.

11.37 Chupp, E.L., D.J. Forrest, J.M. Ryan, J. Heslin, C. Reppin, K. Pinkau, G. Kanbach, E. Rieger, G.H. Share, A direct observation of solar neutrons following the 0118 UT flare in 1980 June 21, Astrophys. J., **263**, 95, 1982.

11.38 Cline, T.L., G.H. Ludwig, F.B. McDonald, Detection of interplanetary 3- to 12-MeV electrons, Phys. Rev. Lett., **13**, 786–789, 1964.

11.39 Cliver, E.W., D.J. Forrest, H.V. Cane, D.V. Reames, R.E. McGuire, T.T. von Rosenvinge, S.R. Kane, R.J. MacDowall, Solar flare nuclear gamma rays and interplanetary proton events, Astrophys. J., **343**, 953–970, 1989.

11.40 Cliver, E.W., D.J. Forrest, R.E. McGuire, T.T. von Rosenvinge, D.V. Reames, H.V. Cane, S.R. Kane, Solar flare nuclear gamma-rays and interplanetary proton events, in *Proc. 20th ICRC*, Vol.3, ed. by V.A. Kozyarivsky et al., Moscow, USSR, 61–64, 1987.

11.41 Cliver, E.W., S.W. Kahler, M.A. Shea, D.F. Smart, Injection onsets of $\sim$ 2 GeV protons, 1 MeV electrons and $\sim$ 100 keV electrons in solar cosmic ray flares, Astrophys. J., **260**, 362, 1982.

11.42 Coleman, P.J., Jr., Variations in the interplanetary magnetic field: Mariner 2, 1, observed properties, J. Geophys. Res., **71**, 5509, 1966.

11.43 Colgate, S.A., J. Audouze, Possible interpretation of the isotopic composition of hydrogen and helium in solar cosmic rays, Astrophys. J., **213**, 849–860, 1977.

11.44 Conlon, T.F., The interplanetary modulation and transport of jovian electrons, J. Geophys. Res., **83**, 541–552, 1978.

11.45 Coplan, M.A., K.W. Ogilvie, P. Bochsler, J. Geiss, Interpretation of ^{3}He abundance variations in the solar wind, Solar Phys., **93**, 415–434, 1984.

11.46 Daiborg, E.I., V.G. Kurt, Y.I. Logachev, V.G. Stolpovskii, V.F. Melnikov, T.S. Podstrigach, Solar cosmic ray events with low and high e/p-ratios: comparison with X-ray and radio emission data, in *Proc. 20th Int. Cosmic Ray Conf.*, Vol.3, ed. by V.A. Kozyarivsky et al., Moscow, USSR, 45–48, 1987.

11.47 Datlowe, D., Relativistic electrons in solar particle events, Solar Phys., **17**, 436–458, 1971.

11.48 Davila, J.M., J.S. Scott, The interplanetary scattering mean free path: collisionless wave-damping effects, Astrophys. J., **285**, 400–410, 1984.

11.49 de Jager, C., Solar flares; properties and problems, in *Solar Flares and Space Research, Proceedings of the Symposium Held on the Occasion of the 11th Plenary Meeting of the Committee of Space Research, COSPAR, 1969*, ed. by C. de Jager and Z. Svetska, North-Holland, 1–15, 1969.

11.50 de Jager, C., Solar flares and particle acceleration, Space Sci. Rev., **40**, 43, 1986.

11.51 de Jager, C., Energetic phenomena in impulsive solar flares, in *Proc. 20th Int. Cosmic Ray Conf.*, Vol.7, ed. by V.A. Kozyarivsky et al., Moscow, USSR, 66–76, 1988.

11.52 de la Beaujardiere, J.-F., E.G. Zweibel, Magnetohydrodynamic waves and particle acceleration in a coronal loop, Astrophys. J., **336**, 1059–1072, 1989.

11.53 Debrunner, H., E. Flückiger, H. Gräadel, J.A. Lockwood, R.E. McGuire, Observations related to the acceleration, injection and interplanetary propagation of energetic protons during the solar cosmic ray event on February 16, 1984, J. Geophys. Res., **93**, 7206–7216, 1988.

11.54 Decker, R.B., Formation of shock-spike events at quasi-perpendicular shocks, J. Geophys. Res., **88**, 9959, 1983.

11.55 Decker, R.B., Computer modeling of test particle acceleration at oblique shocks, Space Sci. Rev., **48**, 195–262, 1988.

11.56 Decker, R.B., M.E. Pesses, T.P. Armstrong, On the acceleration of thermal coronal ions by flare induced shock waves, in *Proc. 17th Int Cosmic Ray Conf.*, Vol. 3, Paris, France, 406, 1981.

11.57 Decker, R.B., L. Vlahos, Modeling of ion acceleration through drift and diffusion at interplanetary shocks, J. Geophys. Res., **91**, 13349, 1986.

11.58 Dennis, B.R., Solar hard X-ray bursts, Solar Phys., **100**, 465–490, 1985.

11.59 Denskat, K.U., F.M. Neubauer, Statistical properties of low-frequency magnetic field fluctuations in the solar wind from 0.29 to 1.0 AU during solar minimum conditions: HELIOS 1 and HELIOS 2, J. Geophys. Res., **87**, 2215–2223, 1982.

11.60 Dietrich, W.F., The differential energy spectra of solar flare ^{1}H, ^{3}He, and ^{4}He, Astrophys. J., **180**, 955–973, 1973.

11.61 Dröge, W., R. Schlickeiser, Particle acceleration in solar flares, Astrophys. J., **305**, 909, 1986.

11.62 Dulk, C.A., J.L. Steinberg, S. Hoang, A. Lecacheux, Latitude distributions of interplanetary magnetic field lines rooted in active regions, in *The Sun and the Heliosphere in Three Dimensions*, ed. by R.G. Marsden, D. Reidel Publishing Co., Dordrecht, 229–233, 1986.

11.63 Earl, J.A., Coherent propagation of charged-particle bunches in random magnetic fields, Astrophys. J., **188**, 379–397, 1974.

11.64 Earl, J.A., The diffusive idealization of charged particle transport in random magnetic fields, Astrophys. J., **193**, 231–242, 1974.

11.65 Earl, J.A., The effect of adiabatic focussing upon charged-particle propagation in random magnetic fields, Astrophys. J., **205**, 900–919, 1976.

11.66 Earl, J.A., Analytical description of charged particle transport along arbitrary guiding field configurations, Astrophys. J., **251**, 739–755, 1981.

11.67 Elphic, R.C., The bowshock and magnetopause, Rev. Geophys., **25**, 510–522, 1987.

11.68 Englade, R.G., A computational model for solar flare particle propagation, J. Geophys. Res., **76**, 768, 1971.

11.69 Evenson, P., D. Hovestadt, P. Meyer, D. Moses, The energy spectra of solar flare electrons, in *Proc. 19th Int. Cosmic Ray Conf.*, Vol. 4, La Jolla, USA, 74–77, 1985.

11.70 Fan, C.Y., G. Gloeckler, J.A. Simpson, Protons and helium nuclei within interplanetary magnetic regions which co-rotate with the sun, in *Proc. 9th Int. Cosmic Ray Conf.*, Vol. 1, London, 109, 1966.

11.71 Fan, C.Y., M. Pick, R. Pyle, J.A. Simpson, D.R. Smith, Protons associated with centers of solar activity and their propagation in interplanetary magnetic field regions corotating with the sun, J. Geophys. Res., **73**, 1555–1582, 1968.

11.72 Fermi, E., On the origin of cosmic radiation, Phys. Rev., **75**, 1169, 1949.

11.73 Fillius, W., I. Axford, Large scale solar modulation of >500 MeV/nucleon galactic cosmic rays from 1 to 30 AU, J. Geophys. Res., **90**, 517–520, 1985.

11.74 Fisk, L.A., Solar modulation of galactic cosmic rays 2, J. Geophys. Res., **76**, 221–226, 1971.

11.75 Fisk, L.A., Solar modulation of galactic cosmic rays. 4. Latitude-dependent modulation, J. Geophys. Res., **81**, 4646–4650, 1976.

11.76 Fisk, L.A., Coronal propagation, interplanetary propagation, and interplanetary acceleration, in *Rapp. Palper 15th Int. Cosmic Ray Conf.*, Vol. 10, Plovdiv, Bulgaria, 324–343, 1977.

11.77 Fisk, L.A., ^{3}He-rich flares and some related problems of solar cosmic ray composition, in *Proc. of X Leningrad Seminar on Space Phys.*, Leningrad, USSR, 21, 1978.

11.78 Fisk, L.A., The interaction of energetic particles with the solar wind, in *Solar System Plasma Physics*, Vol. I, ed. by C.F. Kennel, L.J. Lanzerotti, and E.N. Parker, North-Holland Publ. Co., Amsterdam, 177–274, 1979.

11.79 Fisk, L.A., The anomalous component, its variation with latitude and related aspects of modulation, in *The Sun and the Heliosphere in Three Dimensions*, ed. by R.G. Marsden, D. Reidel, Dordrecht, 401–411, 1986.

11.80 Fisk, L.A., W.I. Axford, Effect of energy changes on solar cosmic rays, J. Geophys. Res., **73**, 4396–4399, 1968.

11.81   Fisk, L.A., B. Koslovsky, R. Ramaty, An interpretation of the observed oxygen and nitrogen enhancements in low-energy cosmic rays, Astrophys. J. (Letters), **190**, L35, 1974.

11.82   Fisk. L.A., M.A. Lee, Shock acceleration of energetic particles in corotating interaction regions in the solar wind, Astrophys. J., **237**, 260, 1980.

11.83   Fisk, L.A., K.H. Schatten, Transport of cosmic rays in the solar corona, Solar Phys., **23**, 204–210, 1972.

11.84   Forbush, S.E., World-wide cosmic ray variations 1937–52, J. Geophys. Res., **59**, 525–542, 1954.

11.85   Forman, M.A., Solar modulation of galactic cosmic rays, in *Proc. 20th Int. Cosmic Ray. Conf., rapporteur talk*, Vol. 8, ed. by V.A. Kozyarivsky et al., Moscow, 165–169, 1988.

11.86   Forman, M.A., F.C. Jones, J.S. Perko, Phase propagation of the solar modulation of galactic cosmic rays, J. Geophys. Res., **91**, 2914–2916, 1986.

11.87   Forman, M.A., R. Ramaty, E.G. Zweibel, The acceleraton and propagation of solar flare energetic particles, in *Physics of the Sun*, Vol. I, ed. by P.A. Sturrock, D. Reidel, Dordrecht, 1986.

11.88   Forman, M.A., G.M. Webb, Acceleration of energetic particles, in *Collisionless Shocks in the Heliosphere: A Tutorial Review*, ed. by R.G. Stone and B.T. Tsurutani, Geophysical Monograph **34**, American Geophysical Union, Washington D.C., 91–114, 1985.

11.89   Garcia-Munoz, M., G.M. Mason, J.A. Simpson, A new test for solar modulation theory: the 1972 May–July low-energy galactic cosmic ray proton and helium spectra, Astrophys. J. (Letters), **182**, L81, 1973.

11.90   Garcia-Munoz, M., P. Meyer, K.R. Pyle, J.A. Simpson, P. Evenson, The dependence of solar modulation on the sign of the cosmic ray particle charge, J. Geophys. Res., **91**, 2858–2866, 1986.

11.91   Garrard, T.L., E.C. Stone, R.E. Vogt, The isotopes of H and He in solar cosmic rays, in *Proc. Symposium on High Energy Phenomena on the Sun*, ed. by R. Ramaty and R.G. Stone, SP-342, NASA, 341, 1973.

11.92   Geiss, J., H. Reeves, Cosmic and solar system abundances of Deuterium and Helium-3, Astr. Ap., **18**, 126, 1972

11.93   Gleeson, L.J., W.I. Axford, Cosmic rays in the interplanetary medium, Astrophys. J. (Letters), **149**, L115–L118, 1967.

11.94   Gleeson, L.J., W.I. Axford, Solar modulation of galactic cosmic rays, Astrophys. J., **154**, 1011–1026, 1968.

11.95   Gloeckler, G., Characteristics of solar and heliospheric ion populations observed near earth, Adv. Space Res., **4**, 127–137, 1984.

11.96   Gloeckler, G., D. Hovestadt, L.A. Fisk, Observed distribution functions of H, He, C, O, and Fe in corotating energetic particle streams: implications for interplanetary acceleration and propagation. Astrophys. J., **230**, L191, 1979.

11.97   Gloeckler, G., D. Hovestadt, O. Vollmer, C.Y. Fan, Unusual emission of iron nuclei from the sun, Astrophys. J. (Letters), **200**, L45–L48, 1975.

11.98   Goldstein, M.L., A nonlinear theory of cosmic-ray pitch-angle diffusion in homogeneous magnetostatic turbulence, Astrophys. J., **204**, 900–919, 1976.

11.99   Goldstein, M.L., The mean free path of low-rigidity cosmic rays, J. Geophys. Res., **85**, 3033–3036, 1980.

11.100 Green, G., Pitchwinkelverteilungen energiereicher geladener Teilchen: ihre Rekonstruktion aus sektorierten Intensitätsmessungen und ihre Bedeutung für die Untersuchung interplanetarer Ausbreitungsmechanismen, Habilitationsschrift, 1984.

11.101 Green, G., W. Schlüter, The local characteristic function of interplanetary particle propagation, submitted to Astrophys. J., 1989.

11.102 Hamilton, D.C., The radial transport of energetic solar flare particles from 1 to 6 AU, J. Geophys. Res., **82**, 2157, 1977.

11.103 Hamilton, D.C., G. Gloeckler, T.P. Armstrong, W.I. Axford, C.O. Bostrom, C.Y. Fan, S.M. Krimigis, L.J. Lanzerotti, Recurrent energetic particle events associated with forward/reverse shock pairs near 4 AU in 1978, in *Proc. 16th Int. Cosmic Ray Conf.*, Vol. 5, Kyoto, Japan, 363, 1979.

11.104 Hasselmann, K., G. Wibberenz, Scattering of charged particles by random electromagnetic fields, Z. Geophys., **34**, 353–388, 1968.

11.105 Hasselmann, K., G. Wibberenz, A note on the parallel diffusion coefficient, Astrophys. J., **162**, 1049–1051, 1970.

11.106 Hedgecock, P.C., Measurements of the interplanetary magnetic field in relation to the modulation of cosmic rays, Solar Phys., **42**, 497, 1975.

11.107 Hempe, H., R. Müller-Mellin, H. Kunow, G. Wibberenz, Measurement of ^{3}He-rich flares on board HELIOS-1 and -2, in *Proc. 16th Int. Cosmic Ray Conf.*, Vol. 5, Kyoto, Japan, 95, 1979.

11.108 Hovestadt, D., B. Klecker, O. Vollmer, G. Gloeckler, C.Y. Fan, Heavy particle emission of unusual composition from the sun, in *Proc. 14th Int. Cosmic Ray Conf.*, Vol. 5, Munich, Germany, 1613, 1975.

11.109 Hovestadt, D., O. Vollmer, G. Gloeckler, C.Y. Fan, Differential energy spectra of low-energy (<8.5 MeV per nucleon) heavy cosmic rays during solar quiet times, Phys. Rev. Lett., **31**, 650, 1973.

11.110 Hsieh, K.C., J.A. Simpson, The relative abundances and energy spectra of ^{3}He and ^{4}He from solar flares, Astrophys. J. (Letters), **162**, L191–L196, 1970.

11.111 Hundhausen, A.J., *Coronal Expansion and Solar Wind*. Springer-Verlag, Berlin, Heidelberg, 1972.

11.112 Hundhausen, A.J., Evolution of large-scale solar wind structures beyond 1 AU, J. Geophys. Res., **98**, 2035, 1973.

11.113 Hundhausen, A.J., Nonlinear model of high speed solar wind streams, J. Geophys. Res., **78**, 1528, 1973.

11.114 Hundhausen, A.J., An interplanetary view of coronal holes, in *Coronal holes and high speed wind streams*, ed. by B. Zirker, Colorado Associated Press, Boulder, 225, 1977.

11.115 Hundhausen, A.J., J.T. Gosling, Solar wind structure at large heliocentric distances: An interpretation of PIONEER 10 observations, J. Geophys. Res., **81**, 1436, 1976.

11.116 Hundhausen, A.J., D.G. Sime, R.T. Hansen, S.F. Hansen, Polar coronal holes and cosmic-ray modulation, Science, **207**, 761–763, 1980.

11.117 Hurford, G.J., R.A. Mewaldt, E.C. Stone, R.E. Vogt, Enrichment of heavy nuclei in ^{3}He-rich flares, Astrophys. J. (Letters), **201**, L95–L97, 1975.

11.118 Ibragimov, I.A., G.E. Kocharov, Izv. Akad. Nauk. SSSR, Ser. Fiz., **39**, 287, 1975.

11.119 Jokipii, J.R., Cosmic ray propagation, I. Charged particles in a random magnetic field, Astrophys. J., **146**, 480, 1966.

11.120 Jokipii, R.R., Addendum and erratum to 'Cosmic ray propagation. I.' Astrophys. J., **152**, 671–672, 1968.

11.121 Jokipii, J.R., Cosmic ray propagation in the solar wind, Rev. Geophys. Space Phys., **9**, 27, 1971.

11.122 Jokipii, J.R., P.J. Coleman, Jr., Cosmic ray diffusion tensor and its variation observed with MARINER 4, J. Geophys. Res., **73**, 5495, 1968.

11.123 Jokipii, J.R., E.H. Levy, W.B. Hubbard, Effects of particle drift on cosmic ray transport. 1. General properties, applications to solar modulation, Astrophys. J., **213**, 861–868, 1977.

11.124 Jokipii, J.R., E.N. Parker, Random walk of magnetic lines of force in astrophysics, Phys. Rev. Lett., **21**, 44–47, 1968.

11.125 Jokipii, J.R., E.N. Parker, On the convection, diffusion, and adiabatic deceleration of cosmic rays in the solar wind, Astrophys. J., **160**, 735–744, 1970.

11.126 Jones, F.C., T.J. Birmingham, T.B. Kaiser, The partially averaged field approach to cosmic ray diffusion, Phys. Fluids, **21**, 347–360, 1978.

11.127 Kahler, S.W., R.P. Lin, D.V. Reames, R.G. Stone, M. Liggett, Solar source regions of the ^{3}He-rich solar particle events, in *Proc. 19th Int. Cosmic Ray Conf.*, Vol. 4, 269, 1985.

11.128 Kahler, S.W., D.V. Reames, N.R. Sheeley, Jr., R.A. Howard, M.J. Koomen, D.H. Michels, A comparison of solar ^{3}He-rich events with type II bursts and coronal mass ejections, Astrophys. J., **290**, 742–747, 1985.

11.129 Kaiser, T.B., T.J. Birmingham, F.C. Jones, Computer simulation of the velocity diffusion of cosmic rays, Phys. Fluids, **21**, 370, 1978.

11.130 Kallenrode, M.-B., Helios-Messungen solarer energetischer Teilchen bei Gammaflares unter Berücksichtigung ausgewählter Teilcheninjektionen. Master's thesis, Christian-Albrechts-Universität, Kiel, 1987.

11.131 Kallenrode, M.-B., E. Rieger, G. Wibberenz, D.J. Forrest, Energetic charged particles resulting from solar flares with gamma-ray emission, in *Proc. 20th ICRC*, Vol. 3, ed. by V.A. Kozyarivsky et al., 70–73, 1987.

11.132 Kallenrode, M.-B., G. Wibberenz, Particle injection in events with weak interplanetary scattering, in *Proc. 21st ICRC*, Vol. 5, ed. by R.J. Protheroe, 112–115, 1987.

11.133 Kallenrode, M.-B., G. Wibberenz, E. Cliver, Particle ratios in impulsive and gradual flares, in *Proc. 21st ICRC*, Vol. 5, ed. by R.J. Protheroe, 104–107, 1990.

11.134 Kindel, J.M., C.F. Kennel, Topside current instabilities, J. Geophys. Res., **76**, 3055–3078, 1971.

11.135 Klecker, B., D. Hovestadt, G. Gloeckler, F.M. Ipavich, M. Scholer, C.Y. Fan, L.A. Fisk, Direct determination of the ionic charge distribution of helium and iron in ^{3}He-rich solar energetic particle events, Astrophys. J., **281**, 458, 1984.

11.136 Klecker, B., D. Hovestadt, G. Gloeckler, E. Moebius, F.M. Ipavich, M. Scholer, Direct determination of the ionic charge distribution of heavy ions in Fe-rich solar energetic particle events, in *Proc. 18th Int. Cosmic Ray Conf.*, Vol. 10, Bangalore, India, 330, 1983.

11.137 Kocharov, G.E., Plasma and nuclear processes in solar matter, proceedings of a course and workshop in plasma astrophysics, in ESA SP-207, Varenna, Italy, 1984.

11.138 Kocharov, L.G., G.E. Kocharov, ^{3}He-rich solar flares, Space Sci. Rev., **38**, 89, 1984.

11.139 Kocharov, L.G., A.V. Orishchenko, On the mechanism of solar cosmic ray enrichment by heavy ions, in *Proc. 18th Int. Cosmic Ray Conf.*, Vol. 4, Bangalore, India, 37, 1983.

11.140 Kota, J., J.R. Jokipii, Effects of drift on the transport of cosmic rays. VI. A three-dimensional model including diffusion, Astrophys. J., **265**, 573–581, 1983.

11.141 Kota, J., J.R. Jokipii, Effects of a wavy neurtral sheet on cosmic ray anisotropies, in *Proc. 19th Int Cosmic Ray Conf.*, Vol. 4, La Jolla, USA, 453–456, 1985.

11.142 Krimigis, S.M., E.C. Roelof, T.P. Armstrong, J.A. van Allen, Low-energy (>0.3 MeV) solar particle observations at widely separated points (>0.1 AU) during 1967, J. Geophys. Res., **76**, 5921, 1971.

11.143 Kulsrud, R.M., A. Ferrari, The relativistic quasilinear theory of particle acceleration by hydrodynamic turbulence, Astrophys. Space. Sci., **12**, 302, 1981.

11.144 Kunow, H., Selected results from the University of Kiel cosmic ray experiments on board the solar probes HELIOS 1 and HELIOS 2, in *Proc. X Leningrad Seminar on Space Phys.*, 73, 1978.

11.145 Kunow, H., ^{3}He-rich solar particle events observed aboard Helios-1 and -2, heavy ion composition results, and source effects, in *Laboratory and Space Plasmas*, ed. by H. Kikuchi, Springer-Verlag New York, Berlin, Heidelberg, 1989.

11.146 Kunow, H., R. Müller-Mellin, B. Iwers, M. Witte, H. Hempe, G. Wibberenz, G. Green, J. Fuckner, MeV protons, alpha particles and electrons as observed aboard HELIOS-1 and -2 during STIP interval II, Report UAG, 61, Technical report, World Data Center A, 1977.

11.147 Kunow, H., M. Witte, G. Wibberenz, H. Hempe, R. Müller-Mellin, G. Green, B. Iwers, J. Fuckner, Cosmic Ray Measurements on board Helios 1 from December 1974 to September 1975: Quiet time spectra, radial gradients, and solar events, J. Geophys., **42**, 615–631, 1977.

11.148 Kunow, H., G. Wibberenz, G. Green, R. Müller-Mellin, M. Witte, H. Hempe, Cosmic ray experiment (E6), in *HELIOS Solar Probes Science Summaries*, ed. by J.H. Trainor, TM 82005, NASA-GSFC, Greenbelt, 36–51, 1980.

11.149 Kunow, H., G. Wibberenz, G. Green, R. Müller-Mellin, M. Witte, H. Hempe, R.A. Mewaldt, E.C. Stone, R.E. Vogt, Simultaneous observations of cosmic ray particles in a corotating interplanetary structure at different solar distances between 0.3 and 1 AU from HELIOS 1 and 2 and IMP 7 and 8, in *Proc. 15th Int. Cosmic Ray Conf.*, Vol. 3, Plovdiv, Bulgaria, 227–233, 1977.

11.150 Kunstmann, J., W. Alpers, A perturbation approach to coherent propagation of energetic charged particles propagating in random magnetic fields, Astrophys. J., **211**, 587, 1977.

11.151 Kunstmann, J.E., Zur Ausbreitung energetischer geladener Teilchen in den schwach fluktuierenden Magnetfeldern des interplanetaren Raumes, PhD thesis, University of Hamburg, 1977.

11.152 Kunstmann, J.E., A new transport mode for energetic charged particles in magnetic fluctuations superposed on a diverging mean field, Astrophys. J., **229**, 812–820, 1979.

11.153 Lanzerotti, J.L., Coronal propagation of low energy solar protons, J. Geophys. Res., **78**, 3942, 1973.

11.154 Lee, M.A., Coupled hydromagnetic wave excitation and ion acceleration upstream of the earth's bow shock, J. Geophys. Res., **87**, 5093, 1982.

11.155 Lee, M.A., The association of energetic particles and shocks in the heliosphere, Rev. Geophys. Space Phys., **21**, 324, 1983.

11.156 Lee, M.A., Coupled hydromagnetic wave excitation and ion acceleration at interplanetary traveling shocks, J. Geophys. Res., **88**, 6109, 1983.

335

11.157 Lee, M.A., Acceleration of energetic particles at solar wind shocks, in *The Sun and the Heliosphere in Three Dimensions*, ed. by R.G. Marsden, D. Reidel, Dordrecht, 305–318, 1986.

11.158 Lee, M.A., H.J. Völk, Hydromagnetic waves and cosmic ray diffusion theory, Astrophys. J., **198**, 485–492, 1975.

11.159 Levy, E.H., Theory of the solar magnetic cycle wave in the diurnal variation of energetic cosmic rays: physical basis of the anisotropy, J. Geophys. Res., **81**, 2082–2088, 1976.

11.160 Lin, R.P., The emission and propagation of ∼40 keV solar flare electrons. II. The electron emission structure of large active regions. Solar Phys., **15**, 453–478, 1970.

11.161 Lin, R.P., Observations of scatter-free propagation of 40 keV solar electrons in the interplanetary medium. J. Geophys. Res., **75**, 2583, 1970.

11.162 Lin, R.P., Non-relativistic solar electrons, Space Sci. Rev., **16**, 189–256, 1974.

11.163 Lin, R.P., Solar particle acceleration and propagation, Rev. Geophys. and Space Phys., **25**, 1987.

11.164 Lin, R.P., K.A. Anderson, Electrons >40 keV and protons >500 keV of solar origin, Solar Phys., **1**, 446–464, 1967.

11.165 Lin, R.P., H.S. Hudson, Non-thermal processes in large solar flares, Solar Phys., **50**, 153, 1976.

11.166 Lin, R.P., R.A. Mewaldt, M.A.I. van Hollebeke, The energy spectrum of 20 keV–20 MeV electrons accelerated in large solar flares, Astrophys. J., **253**, 949–962, 1982.

11.167 Lockwood, J.A., On the long-term variation in the cosmic radiation, J. Geophys. Res., **56**, 19–25, 1960.

11.168 Lockwood, J.A., W.R. Webber, Observations of the dynamics of the cosmic ray modulation, J. Geophys. Res., **89**, 17–25, 1984.

11.169 Lopate, C., Electron acceleration to relativistic energies by traveling interplanetary shocks, J. Geophys. Res., **94**, 9995–10010, 1989.

11.170 Luhn, A., D. Hovestadt, B. Klecker, M. Scholer, G. Gloeckler, F.M. Ipavich, A.B. Galvin, C.Y. Fan, L.A. Fisk, The mean ionic charges of N, Ne, Mg, Si, and S in solar magnetic particle events, in *Proc. 19th Int. Cosmic Ray Conf.*, Vol. 4, La Jolla USA, 241, 1985.

11.171 Luhn, A., B. Klecker, D. Hovestadt, Mean charge of silicon in ^{3}He-rich solar flares, in *Proc. 19th Int. Cosmic Ray Conf.*, Vol. 4, La Jolla, USA, 285, 1985.

11.172 Mariani, F., F.M. Neubauer, The interplanetary magnetic field, in *Physics of the Inner Heliosphere*, Vol. I, ed. by R. Schwenn and E. Marsch, Springer-Verlag, Berlin, Heidelberg, New York, 1990.

11.173 Marsch, E., C.-Y. Tu, On the radial evolution of MHD turbulence in the inner heliosphere, submitted for publication, 1989.

11.174 Marsden, R.G., *The Sun and the Heliosphere in Three Dimensions*, D. Reidel Publishing Co., ASSL No. 123, Dordrecht, 1986.

11.175 Marshall, F.E., E.C. Stone, Persistent sunward flow of 1.6 MeV protons at 1 AU, Geophys. Res. Lett., **4**, 57, 1977.

11.176 Marshall, F.E., E.C. Stone, Characteristics of sunward flowing proton and α-particle fluxes of moderate intensity, J. Geophys. Res., **83**, 3289, 1978.

11.177 Mason, G.M., The composition of galactic cosmic rays and solar energetic particles, Rev. Geophys. and Space Phys., **25**, 685, 1987.

11.178 Mason, G.M., L.A. Fisk, D. Hovestadt, G. Gloeckler, A survey of ∼1 MeV/nucleon solar flare particle abundances, $1 \leq Z \leq 26$, during the 1973–1977 solar minimum period, Astrophys. J., **239**, 1070–1088, 1980.

11.179 Mason, G.M., G. Gloeckler, D. Hovestadt, Composition anomalies in solar flares, in *Proc. 16th Int. Cosmic Ray Conf.*, Vol. 5, Kyoto, Japan, 128, 1979.

11.180 Mason, G.M., G. Gloeckler, D. Hovestadt, Temporal variations of nucleonic abundances in solar flare energetic particle events. I. Well-connected events, Astrophys. J., **267**, 844, 1983.

11.181 Mason, G.M., G. Gloeckler, D. Hovestadt, Temporal variations of nucleonic abundances in solar flare energetic particle events. II. Evidence for large scale shock acceleration, Astrophys. J., **280**, 902, 1984.

11.182 Mason, G.M., C.K. Ng, B. Klecker, G. Green, Impulsive acceleration and scatter free transport of ∼1 MeV per nucleon ions in ^{3}He-rich solar particle events, Astrophys. J., 529–544, 1989.

11.183 Mason, G.M., D.V. Reames, B. Klecker, D. Hovestadt, T.T. von Rosenvinge, The heavy ion compositional signature in ^{3}He-rich solar particle events, Astrophys. J., **303**, 849, 1986.

11.184 Ma Sung, L.S., M.A.I. van Hollebeke, F.B. McDonald, Propagation characteristics of solar flare particles, in *Proc. 14th Int. Cosmic Ray Conf.*, Vol. 5, Munich, Germany, 1767–1772, 1975.

11.185 Mathews, T., J. Quenby, J. Sear, Mechanism for cosmic ray modulation, Nature, **229**, 246–247, 1971.

11.186 Matthaeus, W.H., M.L. Goldstein, Measurements of the rugged invariants of magnetohydrodynamic turbulence in the solar wind, J. Geophys. Res., **87**, 6011, 1982.

11.187 McCracken, K.G., K.R. Rao, Solar cosmic ray phenomena, Space Sci. Rev., **11**, 155, 1970.

11.188 McCracken, K.G., U.R. Rao, R.P. Bukata, E.P. Keath, The decay phase of solar flare events, Solar Phys., **18**, 100–132, 1971.

11.189 McDonald, F.B., U.D. Desai, Recurrent solar cosmic ray events and solar M regions, J. Geophys. Res., **76**, 808–827, 1971.

11.190 McDonald, F.B., N. Lal, J.H. Trainor, M.A.I. van Hollebeke, W.R. Webber, The solar modulation of galactic cosmic rays in the outer heliosphere, Astrophys. J. (Letters), **249**, L71–L75, 1981.

11.191 McDonald, F.B., B.J. Teegarden, J.H. Trainor, T.T. von Rosenvinge, The interplanetary acceleration of energetic nucleons, Astrophys. J., **203**, L149, 1976.

11.192 McDonald, F.B., B.J. Teegarden, J.H. Trainor, W.R. Webber, The anomalous abundance of cosmic-ray nitrogen and oxygen nuclei at low energies, Astrophys. J., **187**, L105–L108, 1974.

11.193 McGuire, R.E., M.A.I. van Hollebeke, N. Lal, A multi-spacecraft study of the coronal and interplanetary transport of solar cosmic rays. I. Introduction and observations, in *Proc. 18th Int. Cosmic Ray Conf.*, Vol. 10, Bangalore, India, 353–356, 1983.

11.194 McGuire, R.E., M.A.I. van Hollebeke, N. Lal, A multi-spacecraft study of the coronal and interplanetary transport of solar cosmic rays. II. Model and fitting, in *Proc. 18th Int. Cosmic Ray Conf.*, Vol. 10, Bangalore, India, 357–360, 1983.

11.195 McGuire, R.E., T.T. von Rosenvinge, The energy spectra of solar energetic particles, Adv. Space Res., **4(2–3)**, 117, 1984.

11.196 McGuire, R.E., T.T. von Rosenvinge, F.B. McDonald, The composition of corotating energetic particle streams, Astrophys. J., **224**, L87, 1978.

11.197 McGuire, R.E., T.T. von Rosenvinge, F.B. McDonald, A survey of ^{3}He enriched events between 1974 and 1978, in *Proc. 16th Int. Cosmic Ray Conf.*, Vol. 5, Kyoto, Japan, 90, 1979.

11.198 McKibben, R.B., Azimuthal propagation of low-energy solar-flare protons as observed from spacecraft very widely separated in solar azimuth, J. Geophys. Res., **77**, 3957, 1972.

11.199 McKibben, R.B., Azimuthal propagation of low-energy solar flare protons: Interpretation of observations, J. Geophys. Res., **78**, 7184–7204, 1973.

11.200 McKibben, R.B., Modulation of galactic cosmic rays in the heliosphere, in *The Sun and the Heliosphere in Three Dimensions*, ed. by R.G. Marsden, D. Reidel, Dordrecht, 305–318, 1986.

11.201 McKibben, R.B., Galactic cosmic rays and anomalous components in the heliosphere, Reviews of Geophysics, **25(3)**, 711–722, 1987.

11.202 McKibben, R.B., K.R. Pyle, J.A. Simpson, The recovery of the cosmic ray flux from maximum solar modulation at IMP 8 (1 AU) and at Pioneer 10 ($R > 30$ AU), in *Proc. 19th Int. Cosmic Ray Conf.*, Vol. 5, La Jolla, 206–209, 1985.

11.203 McKibben, R.B., K.R. Pyle, J.A. Simpson, A.J. Tuzzolino, J.J. O'Gallagher, Cosmic ray radial intensity gradients measured by Pioneer 10 and Pioneer 11, in *Proc. 14th Int. Cosmic Ray Conf.*, Vol. 4, Munich, 1512–1517, 1975.

11.204 Melrose, D.B., *Plasma Astrophysics*, Vol. II, Gordon and Breach, New York, 1980.

11.205 Melrose, D.B., Prompt acceleration of >30 MeV per nucleon ions in solar flares, Solar Phys., **89**, 149, 1983.

11.206 Mewaldt, R.A., J.D. Spalding, E.C. Stone, A high-resolution study of the isotopes of solar flare nuclei, Astrophys. J., **280**, 892, 1984.

11.207 Meyer, J.-P., The baseline composition of solar energetic particles, Astrophys. J., Suppl. **57**, 151, 1985.

11.208 Meyer, P., E.N. Parker, J.A. Simpson, Solar cosmic rays of February 1946 and their propagation through interplanetary space. Phys. Rev., **104**, 768, 1956.

11.209 Mihalov, J.D., Heliospheric shocks (excluding planetary bow shocks), Revs. Geophys., **25**, 697–710, 1987.

11.210 Möbius, E., D. Hovestadt, B. Klecker, G. Gloeckler, Energy dependence and temporal evolution of the ^{3}He/^{4}He ratios in heavy-ion-rich energetic particle events, Astrophys. J., **238**, 768–779, 1980.

11.211 Möbius, E., M. Scholer, D. Hovestadt, B. Klecker, G. Gloeckler, Comparison of helium and heavy ion spectra in ^{3}He-rich solar flares with model calculations based on stochastic fermi acceleration in Alfvén turbulence, Astrophys. J., **259**, 397–410, 1982.

11.212 Morfill, G., A.K. Richter, M. Scholer, Average properties of cosmic ray diffusion in solar wind streams, J. Geophys. Res., **84**, 1505–1513, 1979.

11.213 Morfill, G.E., A two-component description of energetic particle scattering in a turbulent magnetic plasma, J. Geophys. Res., **80**, 1783–1794, 1975.

11.214 Morfill, G.E., H.J. Völk, M.A. Lee, On the effect of directional medium-scale interplanetary variations on the diffusion of galactic cosmic rays and their solar cycle variations, J. Geophys. Res., **81**, 5841, 1976.

11.215 Moussas, X., J.J. Quenby, The evaluation of the interplanetary diffusion coefficient for energetic particles employing real magnetic field data, Astrophys. Space Sci., **56**, 483–502, 1987.

11.216 Moussas, X., J.J. Quenby, J.F. Valdes-Galicia, Interplanetary acceleration of energetic particles at 1 and 5 AU, Astrophys. Space Sci., **85**, 99–120, 1982.

11.217 Mullan, D.J., Possible evidence for attenuation of an MHD shock by a magnetic neutral sheet in the solar corona, in *Proc. 17th Int. Cosmic Ray Conf.*, Vol. 3, Paris, France, 51, 1981.

11.218 Mullan, D.J., K.H. Schatten, Motion of solar cosmic rays in the coronal magnetic field, Solar Phys., **62**, 153–177, 1979.

11.219 Müller-Mellin, R., K. Röhrs, G. Wibberenz, Super-events in the inner solar system and their relation to the solar cycle, in *The Sun and the Heliosphere in Three Dimensions*, ed. by R.G. Marsden, D. Reidel Publishing Co., ASSL No. 123, Dordrecht, 349–354, 1986.

11.220 Müller-Mellin, R., G. Wibberenz, The inner heliosphere from solar minimum to solar maximum: short- and long-term variations in the energetic particle population, Adv. Space Res., **4**, **7**, 353–356, 1984.

11.221 Murphy, R.J., D.J. Forrest, R. Ramaty, B. Kozlovsky, Solar flare gamma-ray line spectroscopy, in *Proc. 19th Int. Cosmic Ray Conf.*, SH 2.1-14, La Jolla, USA, 1985.

11.222 Murphy, R.J., R. Ramaty, Solar flare neutrons and gamma-rays, Adv. Space Res., **4**, 127, 1984.

11.223 Nakada, M.P., A study of the composition of the lower solar corona, Solar Phys., **7**, 302–320, 1969.

11.224 Nakada, M.P., A study of the composition of the solar corona and solar wind, Solar Phys., **14**, 457–479, 1970.

11.225 Neustock, H.-H., HELIOS Messungen der Injektion energetischer Elektronen und Nukleonen bei solaren Flares mit Gammastrahlungsemission. Master's thesis, Christian-Albrechts-Universität, Kiel, 1984.

11.226 Neustock, H.-H., G. Wibberenz, B. Iwers, Injection of energetic particles following the gamma-ray flares on June 7, 1980, in *Proc. 19th Int. Cosmic Ray Conf.*, Vol. 4, La Jolla, USA, 102–105, 1985.

11.227 Newkirk, Jr., G., D.G. Wentzel, Rigidity-independent propagation of cosmic rays in the solar corona, J. Geophys. Res., **83**, 2009–2015, 1978.

11.228 Ng, C.K., L.J. Gleeson, Propagation of solar flare cosmic rays along corotating interplanetary flux-tubes, Solar Phys., **43**, 475, 1975

11.229 Ng, C.K., L.J. Gleeson, A complete model of the propagation of solar-flare cosmic rays, Solar Phys., **46**, 347–375, 1976.

11.230 Ng, C.K., G. Green, W. Schlüter, G. Wibberenz, H. Kunow, Interplanetary transport conditions for the 1976 March 28 solar particle event, submitted to Astrophys. J., 1989.

11.231 Ng, C.K., G. Wibberenz, G. Green, H. Kunow, Absolute value and functional dependence of interplanetary pitch angle scattering derived from HELIOS observations at 0.5 AU, in *Proc. 18th Int. Cosmic Ray Conf.*, Vol. 10. Bangalore, India, 381, 1983.

11.232 Ng, C.K., Kok-Yong Wong, Solar particle propagation under the influence of pitch-angle diffusion and collimation in the interplanetary magnetic field, in *Proc. 16th Int. Cosmic Ray Conf.*, Vol. 5, 252, 1979.

11.233 Nolte, J.T., Interrelationship of energetic particles, plasma and magnetic fields in the inner heliosphere, PhD thesis, University of New Hampshire, 1974.

11.234 Nolte, J.T., E.C. Roelof, Large-scale structure of the interplanetary medium. High coronal source longitude of the quiet-time solar wind, Solar Phys., **33**, 241, 1973.

11.235 Nolte, J.T., E.C. Roelof, Mathematical fomulation of scatter free propagation of solar cosmic rays, in *Proc. 14th Int. Cosmic Ray Conf.*, Vol. 5 Munich, Germany, 1722–1726, 1975.

11.236 O'Gallagher, J.J., Observations of the radial gradient of galactic cosmic radiation over a solar cycle, Rev. Geophysics and Space Physics, **10(3)**, 821–835, 1972.

11.237 O'Gallagher, J.J., J.A. Simpson, The heliocentric intensity gradients of cosmic-ray protons and helium during minimum solar modulation, Astrophys. J., **147**, 819–827, 1967.

11.238 Ohki, K., T. Takakura, B. Tsumata, N. Nitta, General aspects of hard X-ray flares observed by Hinotori: gradual events and impulsive bursts, Solar Phys., **86**, 301–312, 1983.

11.239 Pallavicini, R., S. Serio, G.S. Vaiana, A survey of soft X-ray limb flare images: The relation between their structure and other physical parameters, Astrophys. J., **216**, 108, 1977.

11.240 Palmer, I.D., Transport coefficients of low-energy cosmic rays in interplanetary space, Rev. Geophys. Space Phys., **20**, 335, 1982.

11.241 Palmer, I.D., S.F. Smerd, Evidence for a two-component injection of cosmic rays from the solar flare of 1969, March 30, Solar Phys., **26**, 460, 1972.

11.242 Parker, E.N., The passage of energetic charged particles through interplanetary space, Planet. Space Sci., **13**, 9–49, 1965.

11.243 Parker, E.N., *Cosmical Magnetic Fields – Their Origin and Their Activity*. Clarendon Press, Oxford, 1979.

11.244 Perez-Peraza, J., Coronal transport of solar flare particles, Space Sci. Rev., **44**, 91–138, 1986.

11.245 Perko, J.S., L.A. Fisk, Solar modulation of galactic cosmic rays. 5. Time dependent modulation, J. Geophys. Res., **88**, 9033–9036, 1983.

11.246 Pesses, M.E., B. Klecker, G. Gloeckler, D. Hovestadt, Observations of interplanetary energetic charged particles from gamma-ray line solar flares, in *Proc. 17th Int. Cosmic Ray Conf.*, Vol. 3, Paris, France, 36–39, 1981.

11.247 Pesses, M.E., J.A. van Allen, C.K. Goertz, Energetic protons associated with interplanetary active regions 1–5 AU from the sun, J. Geophys. Res., **83**, 553, 1978.

11.248 Pesses, M.E., J.A. van Allen, B.T. Tsurutani, E.J. Smith, On the acceleration of ions by interplanetary shock waves. III. High time resolution observations of CIR proton events, submitted to J. Geophys. Res., 1982.

11.249 Petchek, H.E., Magnetic field annihilation, in *Symp. on Physics of Solar Flares*, 425, NASA SP-50, Goddard Space Flight Center, 1964.

11.250 Piddington, J.H., *Cosmic Electrodynamics*, Wiley and Sons, New York, 1969.

11.251 Potgieter, M.S., H. Moraal, A drift model for the modulation of galactic cosmic rays, Astrophys. J., **294**, 425–440, 1985.

11.252 Priest, E.R., *Solar Flare Magnetohydrodynamics*, Gordon and Breach Sci. Publ., New York, Chap. 1, 1981.

11.253 Priest, E.R., Magnetic reconnection at the sun, in *Magnetic Reconnection in Space and Laboratory Plasmas*, ed. by E.W. Hones, Geophysical Monograph 30, American Geophysical Union, Washington D.C., 1984.

11.254 Pyle, K.R., J.A. Simpson, A. Barnes, J.D. Mihalov, Shock acceleration of nuclei and electrons in the heliosphere beyond 24 AU, Astrophys. J., **282**, L107–L111, 1984.

11.255 Quenby, J.J., Theoretical studies of interplanetary propagation and acceleration, Space Sci. Rev., **34**, 137–153, 1983.

11.256 Ramaty, R., Nuclear processes in solar flares, in *Physics of the Sun*, Vol. II, ed. by P.A. Sturrock, Reidel, Dordrecht, 291, 1986.

11.257 Ramaty, R., S.A. Colgate, G.A. Dulk, P. Hoyng, J.W. Knight, R.P. Lin, D.B. Melrose, F. Orrall, C. Paizis, P.R. Skapio, D.F. Smith, M.A.I. van Hollebeke, Energetic particles in solar flares, in *Solar Flares*, ed. by P.A. Sturrock, Colorado Ass. University Press, Boulder, 117, 1980.

11.258 Ramaty, R., B.R. Dennis, A.G. Emslie, Gamma-ray, neutron, and hard X-ray studies and requirements for a high energy solar physics facility, Solar Phys., **118**, 17–48, 1988.

11.259 Ramaty, R., B. Kozlovski, in *Proc. of VI Leningrad Seminar on* Space Phys., 58, 1974.

11.260 Ramaty, R., B. Kozlovski, Deuterium, tritium, and helium-3 production in solar flares, Astrophys. J., **193**, 729–740, 1974.

11.261 Ramaty, R., R.J. Murphy, Nuclear processes and accelerated particles in solar flares, Space Sci. Rev., **45**, 213–268, 1987.

11.262 Reames, D.V., Energetic particles from impulsive solar flares, Ap. J. Suppl., **73**, 235–251, 1990.

11.263 Reames, D.V., Wave generation in the transport of particles from large solar flares, Astrophys. J. (Letters), **342**, L51, 1989.

11.264 Reames, D.V., R.P. Lin, ^{3}He in solar non-relativistic electron events, in *Proc. 19th Int. Cosmic Ray Conf.*, Vol. 4, La Jolla, USA, 265, 1985.

11.265 Reames, D.V., R.G. Stone, The identification of solar ^{3}He-rich events and the study of particle acceleration at the sun, Astrophys. J., **308**, 902, 1986.

11.266 Reames, D.V., T.T. von Rosenvinge, Heavy-element abundances in ^{3}He-rich events, in *Proc. 17th Int. Cosmic Ray Conf.*, Vol. 3, Paris, France, 162, 1981.

11.267 Reames, D.V., T.T. von Rosenvinge, R.P. Lin, Solar ^{3}He-rich events and non-relativistic electron events: A new association, Astrophys. J., **292**, 716, 1985.

11.268 Reid, G.C., A diffusive model for the initial phase of a solar proton event, J. Geophys. Res., **69**, 2659, 1964.

11.269 Reinhard, R., G. Wibberenz, Propagation of flare protons in the solar atmosphere, Solar Phys., **36**, 473–494, 1974.

11.270 Roelof, E.C., Propagation of solar cosmic rays in the interplanetary magnetic field, in *Lectures in High Energy Astrophysics*, ed. by H. Ögelmann and J.R. Wayland, NASA SP-199, 1969.

11.271 Roelof, E.C., Coronal magnetic fields and the structure of low-energy solar charged particle events, in *Symposium on High Energy Phenomena on the Sun*, NASA Publ. X-693-73-193, 486–502, 1973.

11.272 Roelof, E.C., Coronal structure and the solar wind, in *Solar Wind Three, Proc. Asilomar Conf.*, ed. by C.T. Russell, Institute of Geophysics and Planetary Physics, University of California, Los Angeles, 1974.

11.273 Roelof, E.C., S.M. Krimigis, Analysis and synthesis of coronal and interplanetary energetic particle, plasma and magnetic field observations over three solar rotations, J. Geophys. Res., **78**, 5378–5410, 1973.

11.274 Röhrs, K., R. Müller-Mellin, Relations between long-term modulation of cosmic rays and superimposed Forbush decreases, in *Proc. 20th Int. Cosmic Ray Conf.*, Vol. 3, ed. by V.A. Kozyarivsky et al., Moscow, USSR, 356, 1987.

11.275 Sakai, J.-I., Y. Ohsawa, Particle acceleration by magnetic reconnection and shocks during current loop coalescence in solar flares, Space Sci. Rev., **46(1–2)**, 113–198, 1987.

11.276 Sanderson, T.R., ISEE-3 observations of energetic protons associated with interplanetary shocks, Adv. Space Res., **4(2–3)**, 305–313, 1984.

11.277 Sari, J.W., N.F. Ness, Power spectra of the interplanetary magnetic field, Solar Phys., **8**, 155–165, 1969.

11.278 Schäffer, O.A., J. Zähringer, Solar flare helium in satellite materials, Phys. Rev. Lett., **8**, 389–390, 1962.

11.279 Schatten, K.H., D.J. Mullan, Fast azimuthal transport of solar cosmic rays via a coronal magnetic bottle, J. Geophys. Res., **82**, 5609–5620, 1977.

11.280 Schellert, G., Koronale und interplanetare Ausbreitung Flare-erzeugter energetischer Teilchen gemessen auf HELIOS 1 und 2. PhD thesis, Inst. f. Reine und Angewandte Kernphysik, Christian-Albrechts-Universität, Kiel, 1985.

11.281 Schellert, G., G. Wibberenz, H. Kunow, Coronal propagation of flare associated electrons and protons, in *Proc. 19th Int. Cosmic Ray Conf.*, Vol. 4, La Jolla, USA, 305, 1985.

11.282 Schlickeiser, R., Stochastic particle acceleration in cosmic objects, in *Cosmic Radiation in Contemporary Astrophysics, Proc. 4th Int. Summer School on Cosmic Ray Astrophysics*, ed. by M.M. Shapiro, D. Reidel, Dordrecht, 1984.

11.283 Schlickeiser, R., On the interplanetary transport of cosmic rays, J. Geophys. Res., **93**, 2725–2729, 1988.

11.284 Scholer, M., Diffusive acceleration, in *Collisionless Shocks in the Heliosphere: Reviews of Current Research*, ed. by B.T. Tsurutani and R.G. Stone, Geophysical Monograph 35, American Geophysical Union, Washington, D.C., 287–301, 1985.

11.285 Scholer, M., Acceleration of energetic particles in solar flares, in *Activity in Cool Star Envelopes*, ed. by Havnes et al., Kluwer, 195–210, 1988.

11.286 Scholer, M., D. Hovestadt, B. Klecker, G. Gloeckler, The composition of energetic particles in corotating events, Astrophys. J., **227**, 323, 1979.

11.287 Scholer, M., G. Morfill, On the rigidity dependence of the interplanetary scattering mean free path, unpublished manuscript, 1980.

11.288 Scholer, M., G. Morfill, A.K. Richter, energetic solar particle events in a stream-structured solar wind, Solar Phys., **64**, 391–401, 1979.

11.289 Scholer, M., G. Morfill, M.A.I. van Hollebeke, On the origin of corotating energetic particle events, J. Geophys. Res., **85**, 1743, 1980.

11.290 Schulze, B.M., A.K. Richter, G. Wibberenz, Influence of finite injections and of interplanetary propagation on time-intensity and time-anisotropy profiles of solar cosmic rays, Solar Phys., **54**, 207–228, 1977.

11.291 Schwenn, R., The 'average' solar wind in the inner heliosphere: structures and slow variations, in *Solar Wind Five*, NASA Conf. Publication 2280, Woodstock, 489–507, 1982.

11.292 Schwenn, R., Large-scale structure of the interplanetary medium, in *Physics of the Inner Heliosphere*, Vol. 1, ed. by R. Schwenn and E. Marsch, Springer-Verlag, Berlin, Heidelberg, New York, 1990.

11.293 Serlemitsos, A.T., V.K. Balasubrahmanyan, Solar particle events with anomalously large relative abundance of ^{3}He, Astrophys. J., **198**, 195–204, 1975.

11.294 Simnett, G.M., The release of energetic particles from the sun, Solar Phys., **20**, 448–461, 1971.

11.295 Simpson, J.A., Evolution of our knowledge of the heliosphere, Adv. Space Res., **9(4)**, 5–20, 1989.

11.296 Smith, E.J., J.H. Wolfe, Observations of interaction regions and corotating shocks between one and five AU: Pioneers 10 and 11, Geophys. Res. Lett., **3**, 137, 1976.

11.297 Sonnerup, B.V.O., Magnetic field reconnection and particle acceleration, in *High Energy Phenomena on the Sun*, ed. by R. Ramaty and R.G. Stone, NASA-X-693-73-193, Washington, D.C., 1973.

11.298 Spicer, D.S., Magnetic energy storage and conversion in the solar atmosphere, Space Sci. Rev., **31**, 351, 1982.

11.299 Steinolfson, R.S., D.J. Mullan, Magnetohydrodynamic shock propagation in the vicinity of a magnetic neutral sheet, Astrophys. J., **241**, 1186–1194, 1980.

11.300 Stone, E.C., Cosmic ray studies out of the ecliptic, in *Proc. 20th Int. Cosmic Ray Conf.*, Vol. 7, ed. by V.A. Kozyarivsky et al., Moscow, 105–114, 1988.

11.301 Sturrock, P.A., Flare models, in *Solar Flares, Skylab Solar Workshop II*, ed. by P.A. Sturrock, Colorado Ass. University Press, Boulder, 411–449, 1980.

11.302 Sturrock, P.A., Solar flares and magnetic topology, Solar Phys., **113**, 13–30, 1987.

11.303 Syrovatskii, S.I., Pinch sheets and reconnection in astrophysics, Ann. Rev. Astron. Astrophys., **19**, 163, 1981.

11.304 Tsurutani, B.T., E.J. Smith, K.R. Pyle, J.A. Simpson, Energetic protons accelerated at corotating shock: Pioneer 10 and 11 observations from 1 to 6 AU, J. Geophys. Res., **87**, 7389, 1982.

11.305 Tu, C.-Y., E. Marsch, K.M. Thieme, Basic properties of solar wind MHD turbulence near 0.3 AU analyzed by means of Elsässer variables, J. Geophys. Res., **94**, 11739, 1989.

11.306 Valdes-Galicia, J.F., G. Wibberenz, J.J. Quenby, X. Moussas, G. Green, F.M. Neubauer, Comparative studies of pitch angle scattering of solar particles for the 11 April 1978 event, in *Proc. 20th Int Cosmic Ray Conf.*, Vol. 3, ed. by V.A. Kozyarivsky et al., Moscow, USSR, 143–146, 1987.

11.307 Valdes-Galicia, J.F., G. Wibberenz, J.J. Quenby, X. Moussas, G. Green, F.M. Neubauer, Pitch angle scattering of solar particles: comparison of 'particle' and 'field' approach. I. Strong scattering. Solar Phys., **117**, 135–156, 1988.

11.308 van Hollebeke, M.A.I., F.B. McDonald, J.H. Trainor, T.T. von Rosenvinge, The radial variation of corotating energetic particle streams in the inner and outer solar system, J. Geophys. Res., **83**, 4723, 1978.

11.309 van Hollebeke, M.A.I., F.B. McDonald, J.H. Trainor, T.T. von Rosenvinge, Corotating energetic particles and fast plasma streams in the inner and outer solar system in radial dependence and energy spectra, in *Solar Wind Four*, ed. by H. Rosenbauer, Katlenburg-Lindau and Garching, Max-Planck-Institut für Aeronomie and Max-Planck-Institut für Extraterrestrische Physik, 1979.

11.310 van Hollebeke, M.A.I., L.S. Ma Sung, F.B. McDonald, The variation of solar proton energy spectra and size distribution with heliolongitude, Solar Phys., **41**, 189–223, 1975.

11.311 Varvoglis, H., K. Papadopoulos, Selective nonresonant acceleration of ^{3}He^{++} and heavy ions by H$^+$ cyclotron waves, Astrophys. J., **270**, L95, 1983.

11.312 Völk, H.J., Nonlinear perturbation theory for cosmic ray propagation in random magnetic fields, Astrophys. Space Sci., **25**, 471–490, 1973.

11.313 Völk, H.J., Cosmic ray propagation in interplanetary space, Rev. Geophys. Space Phys., **13**, 547–566, 1975.

11.314 Švestka, Z., L. Fritzova, Type II radio bursts and particle acceleration, Solar Phys., **36**, 417–431, 1974.

11.315 Wagner, W.J., SERF studies of mass motions arising in flares, Adv. Space Res., **2(11)**, 1983.

11.316 Wang, J.R., The coronal transport of the flare-associated scatter-free electrons, NASA-X-661-72-76. Technical report, Goddard Space Flight Center, 1972.

11.317 Webb, D.F., J.M. Davis, P.S. McIntosh, Observation of the reappearance of polar coronal holes after reversal of the polar magnetic field, Solar Phys., **92**, 109–132, 1984.

11.318 Webber, W.R., Cosmic rays in the heliosphere, in *Essays in Space Science*, ed. by R. Ramaty, T.L. Cline, and J.F. Ormes, NASA Goddard Space Flight Center, Greenbelt, Md., USA, 125–154, 1987.

11.319 Webber, W.R., A.C. Cummings, E.C. Stone, Radial and latitudinal gradients of anomalous oxygen during 1977–1985, in *Proc. 19th Int. Cosmic Ray Conf.*, Vol. 5, La Jolla, 172–175, 1985.

11.320 Wibberenz, G., Coronal propagation: variation with solar longitude and latitude, in *Proc. Symp. on the Study of the Sun in the Interplanetary Medium in Three Dimensions*, Goddard Space Flight Center, 1976.

11.321 Wibberenz, G., Energetic particles throughout solar system, in *Physics of Solar Planetary Environments*, Vol. 1, ed. by D.J. Williams, American Geophysical Union, 346–365, 1976.

11.322 Wibberenz, G., Signatures of cosmic ray events and their relations to propagation and acceleration processes, in *Study of Travelling Interplanetary Phenomena*, ed. by M.A. Shea, D.F. Smart, and T.S. Wu, D. Reidel, Dordrecht, 323–342, 1977.

11.323 Wibberenz, G., K. Hasselmann, D. Hasselmann, Comparison of particle–field interaction theory with solar proton diffusion coefficients, Acta Phys. Acad. Sci. Hungaricae, Suppl., **29(2)**, 37, 1970.

11.324 Wibberenz, G., K. Kecskeméty, H. Kunow, A. Somogyi, B. Iwers, Y.I. Logachev, V.I. Stolpovskii, Coronal and interplanetary transport of solar energetic protons and electrons, Solar Phys., **124**, 353–392, 1989.

11.325 Wibberenz, G., H. Kunow, G. Green, R. Müller-Mellin, H. Hempe, M. Witte, R.A. Mewaldt, E.C. Stone, R.E. Vogt, R. Reinhard, Radial development of a solar cosmic ray event between 0.4 and 1 AU on March 3, 1975 as observed from HELIOS-1 and IMP, in *Proc. 15th Int. Cosmic Ray Conf.*, Vol. 5, Plovdiv, Bulgaria, 188, 1977.

11.326 Wild, J.P., S.F. Smerd, A.A. Weiss, Solar bursts, Ann. Rev. Astr. and Ap., **1**, 291, 1963.

11.327 Witte, M., G. Wibberenz, H. Kunow, R. Müller-Mellin, On the rigidity dependence of the mean free path He/p for solar flare particles derived from their proton/helium time variation, in *Proc. 16th Int. Cosmic Ray Conf.*, Vol. 5, Kyoto, Japan, 79, 1979.

11.328 Wong, K.Y., Interplanetary transport of solar particles in the presence of focusing and pitch-angle diffusion, Master's thesis, Dept. of Mathematics, University of Malaya, 1982.

11.329 Zaitsev, V.V., A.V. Stepanov, On the relation between solar flare gama-ray emission and proton escape into interplanetary space, Solar Phys., **99**, 313, 1985.

11.330 Zöllich, F., Interplanetare Beschleunigung am Beispiel korotierender Ereignisse – Analyse der Messungen auf HELIOS-1 und -2 in den Jahren 1975 und 1976, PhD thesis, University of Kiel, 1981.

11.331 Zöllich, F., G. Wibberenz, H. Kunow, G. Green, Corotating events in the energy range 4–13 MeV as observed on board HELIOS 1 and 2 in 1975 and 1976, Adv. Space Res., **1**, 89, 1981.

11.332 Zwickl, R.D., E.C. Roelof, R.E. Gold, S.M. Krimigis, T.P. Armstrong, Z-rich solar particle event characteristics, 1972–1976, Astrophys. J., **225**, 281, 1978.

11.333 Zwickl, R.D., W.R. Webber, Solar particles from 1 to ∼5 AU, Solar Phys., **54**, 457, 1977.

Subject Index

Solar wind
- acceleration   107
- expansion   87, 100
- fast streams   316–318, 325–328
- flow speed   205
- fluctuations   169
- parameters   51
- slow streams   325
- stream structure   243
- streams   197
- structures   110
Soliton   142
- collapse   143
Sound speed   29, 35, 171
Sound wave   229
Spatial inhomogeneity   163
Specific angular momentum   112
Specific energy   115
Specific entropy   92
Spectra   202, 205, 210, 211, 217, 225, 300
- galactic protons   248
- helium   300, 307
- nonrelativistic electrons   300
- protons   300, 307, 324
- relativistic electrons   300
- stochastic acceleration   297
Spectral
- analysis   24
- code   217, 219
- density   201, 224
- flux function   105
- index   180, 201, 202, 205, 206, 209, 214, 215, 226
- slope   200, 202, 227, 228
- transfer function   226
Spherical expansion   113
Spiral angle   263
Spiral pattern   111
Spitzer's law   120
Stability analysis   68–70, 76
Stability of magnetic clouds   11, 12
Standard model   264, 269, 271
Stationarity   160, 178
Steepening   24, 25, 31, 42, 202, 206, 212
Stellar wind acceleration   160
Stirring turbulence   195
Stochastic acceleration   297, 300
Strahl   55, 56, 58, 59, 64, 66, 67
- electrons   67, 71, 76, 120
- strength   59, 60
Stream interaction regions   167
Stream structure   29, 46, 60, 88, 97, 99, 161, 164, 170, 187, 189, 194, 207, 208, 214
Streaming   260

Streamline constant   113, 117
Strength of a shock   32
Structure of magnetic clouds   5, 8
Structureless solar wind   111
Sub-Alfvénic solar wind region   26
Sunspot number   247
Supersonic regime   116
Suprathermal electrons   13, 41, 64, 65, 76

Tangential discontinuity   24, 28, 35, 36, 41, 192
Temperature   54, 59, 88
- anisotropy   46, 49, 53, 55, 63, 68, 74, 79, 97
- differences   88
- equilibrium   107
- gradients   55, 98
- maximum   100
- plateau   102
- profiles   94, 97, 101
- ratio   61, 63, 72, 74, 91
- velocity relationship   100
Thermal
- diffusion   108
- equilibrium   61, 135, 154
- potential   115, 119–121
- spread   90
Thermally driven wind   115
Topology of a magnetic cloud   13
Torus   9
Total
- cross-helicity   217
- energy   177
- energy flux   113
- ion heat flux   94, 95
- plasma pressure   172, 229
- pressure   36
- solar wind heat flux   94
Trace of the correlation matrix   201
Transfer equation   220, 224, 225
Transition zone   108
Transport
- coefficients   67, 87
- equation   259, 262, 273
- models   258
- theory   66, 88, 120, 121, 233
Triple-correlations   224, 226
Turbulence   16, 47, 104
- generation   216, 218
- spectra   178
Turbulent
- cascade   105, 215
- coronal envelope   100
- events   109